Organelles, Genomes and Eukaryote Phylogeny

An Evolutionary Synthesis in the Age of Genomics

The Systematics Association Special Volume Series

Series Editor

Alan Warren

Department of Zoology, The Natural History Museum,
Cromwell Road, London SW7 5BD, UK.

The Systematics Association promotes all aspects of systematic biology by organizing conferences and workshops on key themes in systematics, publishing books and awarding modest grants in support of systematics research. Membership of the Association is open to internationally based professionals and amateurs with an interest in any branch of biology including palaeobiology. Members are entitled to attend conferences at discounted rates, to apply for grants and to receive the newsletters and mailed information; they also receive a generous discount on the purchase of all volumes produced by the Association.

The first of the Systematics Association's publications *The New Systematics* (1940) was a classic work edited by its then-president Sir Julian Huxley, that set out the problems facing general biologists in deciding which kinds of data would most effectively progress systematics. Since then, more than 70 volumes have been published, often in rapidly expanding areas of science where a modern synthesis is required.

The *modus operandi* of the Association is to encourage leading researchers to organize symposia that result in a multi-authored volume. In 1997 the Association organized the first of its international Biennial Conferences. This and subsequent Biennial Conferences, which are designed to provide for systematists of all kinds, included themed symposia that resulted in further publications. The Association also publishes volumes that are not specifically linked to meetings and encourages new publications in a broad range of systematics topics.

Anyone wishing to learn more about the Systematics Association and its publications should refer to our website at www.systass.org.

Other Systematics Association publications are listed after the index for this volume.

The Systematics Association Special Volume Series 68

Organelles, Genomes and Eukaryote Phylogeny

An Evolutionary Synthesis in the Age of Genomics

Edited by

Robert P. Hirt

Department of Zoology
The Natural History Museum
London, U.K.

David S. Horner

Dipartimento di Scienze Biomolecolari e Biotecnologie
University of Milan
Milan, Italy

CRC Press
Taylor & Francis Group
Boca Raton London New York

CRC Press is an imprint of the
Taylor & Francis Group, an **informa** business

CRC Press
Taylor & Francis Group
6000 Broken Sound Parkway NW, Suite 300
Boca Raton, FL 33487-2742

First issued in paperback 2019

ISBN-13: 978-0-415-29904-6 (hbk)
ISBN-13: 978-0-367-39412-7 (pbk)

Library of Congress Card Number 2003070023

Library of Congress Cataloging-in-Publication Data

Organelles, genomes, and eukaryote phylogeny: an evolutionary synthesis in the age of genomics / [edited by] Robert P. Hirt, David S. Horner.
p. cm. — (Sytematics Association special volume ; no. 68)
Includes bibliographical references and index.
ISBN 0-415-29904-7 (alk. paper)
1. Phylogeny. 2. Cells—Evolution. 3. Genomics. 4. Eukaryotic cells. I. Hirt, Robert P. II. Horner, David S. III Series
[DNLM: 1. Hepatitis B virus. QW 710 G289h]
QH367.5.074 2004
576. 8′8—dc22 2003070023

Visit the Taylor & Francis Web site at
http://www.taylorandfrancis.com

and the CRC Press Web site at
http://www.crcpress.com

Table of Contents

Contributors

John M. Archibald, University of British Columbia, Vancouver

Sandra L. Baldauf, University of York, U.K.

Robert G. Beiko, University of Ottawa, Ontario

Thomas Cavalier-Smith, University of Oxford, U.K.

Robert L. Charlebois, University of Ottawa, Ontario, and Canadian Institute for Advanced Research Canada and GenomeAtlantic Halifax, Nova Scotia, and NeuroGadgets, Inc.

Deborah Charlesworth, University of Edinburgh, U.K.

Joel B. Dacks, Dalhousie University, Halifax, Nova Scotia

Mark C. Field, Imperial College, London, U.K.

Hrvoje Fulgosi, Rudjer Boskovic Institute, Zagreb, Croatia

Simonetta Gribaldo, Université Pierre et Marie Curie, Paris, France

Robert P. Hirt, The Natural History Museum, London, U.K.

David S. Horner, University of Milan, Italy

Laurence D. Hurst, University of Bath, U.K.

Masami Inaba-Sulpice, Universität München, Germany

Patrick J. Keeling, University of British Columbia, Vancouver

Dario Leister, Max-Planck-Institut für Züchtungsforschung, Köln, Germany

Guy Méténier, Université Blaise Pascal, Aubière, France

Csaba Pál, University of Bath, U.K.

Hervé Philippe, Université de Montréal, Québec

Mark A. Ragan, University of Queensland, Brisbane, Australia and Canadian Institute for Advanced Research

Andrew J. Roger, Canadian Institute of Advanced Research Program in Evolutionary Biology and Dalhousie University, Halifax, Nova Scotia

Anja Schneider, Botanisches Institut der Universität zu Köln, Germany

Alastair G.B. Simpson, Canadian Institute of Advanced Research Program in Evolutionary Biology and Dalhousie University, Halifax, Nova Scotia

Jürgen Soll, Universität München, Germany

Emma T. Steenkamp, University of York, U.K.

John W. Stiller, East Carolina University, Greenville, North Carolina

Jorge Tovar, Royal Holloway University of London, Egham, U.K.

Mark van der Giezen, Royal Holloway University of London, Egham, U.K.

Christian P. Vivares, Université Blaise Pascal, Aubière, France

Chapter 1

An Overview of Eukaryote Origins and Evolution: The Beauty of the Cell and the Fabulous Gene Phylogenies

*David S. Horner and Robert P. Hirt**

CONTENTS

Abstract

Eukaryotic organismal diversity is underpinned by an extraordinary organizational diversity at the cellular and molecular levels. Comparative cell and molecular biological studies on a broad range of eukaryotes are yielding a large number of surprising results. This chapter presents an overview of some of these results and illustrates how some of this diversity is now appreciated within an evolutionary framework. It discusses some of the most interesting

*DSH: david.horner@unimi.it; RPH: rch@nhm.ac.uk

0-415-29904-7/04/$0.00+$1.50

new hypotheses on eukaryote evolution and the methodological advances needed to rigorously test them, and in so doing highlights the significance of the other chapters making up this book. It focuses on the impact of studies of microbial eukaryotes and organelles of endosymbiotic origin on our current understanding of both the origin and the evolution of eukaryotes. We speculate that comparative biology (not restricted to comparative genomics) of a broad diversity of microbial eukaryotes will be critical to the development of biological science in the 21st century.

1.1 Introduction

The past 10 years have seen dramatic advances in the understanding of eukaryote diversity, evolution and cellular and genomic organization. Much of this progress has come through the unprecedented rate of production of molecular sequence data from an increasingly wide range of organisms, presenting biologists with new opportunities and challenges in the investigation of eukaryote evolution and origin. This, in turn, has clearly demonstrated the importance and value of comparative and evolutionary biology in investigating and understanding genome data. Recent years have also seen an increase in the cross-talk among disparate areas of biological expertise, with multidisciplinary approaches to investigating the bewildering and beautiful complexity of cellular life becoming the norm. The objective of this book is to provide a synthesis of current views on (1) eukaryote diversity and phylogeny, (2) molecular and organellar diversity among eukaryotes, (3) the role of endosymbiosis in the evolution of eukaryotes, (4) the importance of evolutionary and molecular systematic perspectives in analysis and exploitation of genome sequence data and (5) the role and importance of epigenetics, including membrane heredity, in the evolution of the cell. The latter is an often-neglected topic of great interest and importance for the biology of the 21st century, since genomes do not contain all the information required to bring about cells and their phenotypes (e.g., Kirschner et al., 2000; Jablonka and Lamb, 1995).

By way of introducing the various contributions that have been selected for this book, we hope to provide a succinct history of molecular phylogenetic investigations into eukaryote origins and evolution and show how a sound molecular systematic framework for eukaryotes and prokaryotes is key for the comparative analysis of eukaryote genomes. By doing so, we hope to emphasize the plastic and chimeric nature of eukaryote genomes and membranes and the complex interplay between these two entities during eukaryote evolution. Finally, we hope to illustrate how little is known about several major eukaryotes lineages, for some of which our knowledge is limited to a few gene sequences.

1.2 The Ribosomal RNA Paradigm and the Archezoa Hypothesis

More than a quarter of a century ago, Carl Woese and his co-workers published a seminal series of papers that indicated, using a universally distributed molecular phylogenetic marker (the small subunit ribosomal RNA gene, SSUrDNA) and molecular cell biological features, that extant life on earth is divided into three primary lines of descent (Woese, 1987) rather than two as in the traditional prokaryote–eukaryote divide (Doolittle, 1996), with the eubacteria (or Bacteria), archaebacteria (Archaebacteria or Archaea) and eukaryotes (spelled either Eucaria or Eukarya) forming the so-called three Domains of life (Woese et al., 1990). By the mid-1980s, incorporation of SSUrDNA sequences from mitochondria and chloroplasts into the tree of life substantiated the long-suspected (Mereschkowsky, 1905; Margulis, 1970, 1993; Martin et al., 2001) endosymbiotic origin of these organelles from alphaproteobacteria and cyanobacteria, respectively (Woese, 1987). The number of organisms and taxonomic groups represented in the SSUrDNA tree has increased at an impressive rate; there are currently in excess of 5000 sequences for eukaryotes alone (Cole et al., 2003).

The SSUrDNA tree also represents an important reference point for investigating the phylogeny of uncultured eukaryotes. Sequences isolated directly from environmental samples (rather than from specific organisms) are beginning to accumulate and these could potentially have an important impact on the perception of eukaryote diversity and evolution by identifying significant new lineages (Moreira and López-García, 2002; Stoeck and Epstein, 2003) as has been the case for environmentally derived SSUrDNA sequences from Bacteria and Archaea (DeLong and Pace, 2001). These new sequences can then be linked with specific organisms to allow further investigations of their biology, possibly leading to surprising discoveries about their cellular and genomic features. Such findings might even increase our understanding of eukaryote origins.

SSUrDNA trees have also provided a stimulating framework for the comprehension of early eukaryote evolution by recovering a group of sequences from a disparate collection of amitochondriate microbial eukaryotes at the base of the tree, rather distant (in terms of branch length) from the best-known and most intensively investigated animals, fungi and plants (Woese, 1987; Sogin, 1991; Hedges, 2002). The latter were defined as the crown taxa and the former as early branching eukaryotes (e.g., Sogin et al., 1996; Figure 1.1A).

The perceived simplicity of early branching eukaryotes (organisms such as diplomonads and microsporidia apparently lacked mitochondria, peroxisomes, Golgi dictyosomes and, in the case of microsporidia, flagella), combined with their early branching position in the eukaryotic SSUrDNA tree, was seen as consistent with the existing notion that these lineages were primitively amitochondriate and that simpler cells precede more complex ones (e.g., Patterson, 1994; Sogin et al., 1996; Cavalier-Smith, 1993). This engaging parallel between cellular simplicity and basal phylogenetic position was formalized as the Archezoa hypothesis (e.g., Cavalier-Smith, 1993; also see Chapter 2).

Several laboratories were tempted to test the clear premises and implications of the Archezoa hypothesis and initiated research programs on the phylogenetics, comparative genomics and cell biology of several Archezoa (e.g., Patterson and Sogin, 1993; Embley and Hirt, 1998; Roger, 1999; Philippe et al., 2000b; also see Chapters 2, 10 and 13). During the 1990s, a series of experiments demonstrated that the apparently amitochondriate entamoebids (Clark and Roger, 1995), parabasalids (Bui et al., 1996; Germot et al., 1996; Horner et al., 1996; Roger et al., 1996), diplomonads (Roger et al., 1998; Horner and Embley, 2001) and microsporidia (Hirt et al., 1997; Germot et al., 1997; Peyretaillade et al ., 1998) harbored genes encoding heat shock proteins that were apparently of mitochondrial origin. Furthermore, microsporidia had been shown not to be early branching eukaryotes but to be related to, or even to branch within, Fungi through phylogenies derived from various protein-coding genes and large subunit rDNA (e.g., Edlind et al., 1996; Hirt et al., 1999; van de Peer et al., 2000a; Keeling, 2003). Cell biological studies also revealed that the hydrogenosomes of parabasalids (e.g., Dyall and Johnson, 2000) and hitherto-unrecognized double-membrane-bound organelles in entamoebids (Tovar et al., 1999; Mai et al., 1999) and microsporidians (Williams et al., 2002) likely descended from the same endosymbiotic event that gave rise to mitochondria. More recent evidence suggests that diplomonads also contain relict mitochondria (Tovar et al., 2003). Together, these data clearly falsify the Archezoa hypothesis, push back the probable timing of the mitochondrial endosymbiosis to before the divergence of any known extant eukaryotic lineage and make it difficult to establish whether the nucleus or mitochondria arose first.

1.3 From SSUrDNA Trees to Protein-Coding Gene Trees

The testing of the Archezoa hypothesis illustrated both the significance (in providing key primary phylogenetic hypotheses) and the weaknesses of the stimulating and intuitively appealing eukaryotic SSUrDNA tree topologies. The implication that SSUrDNA trees might

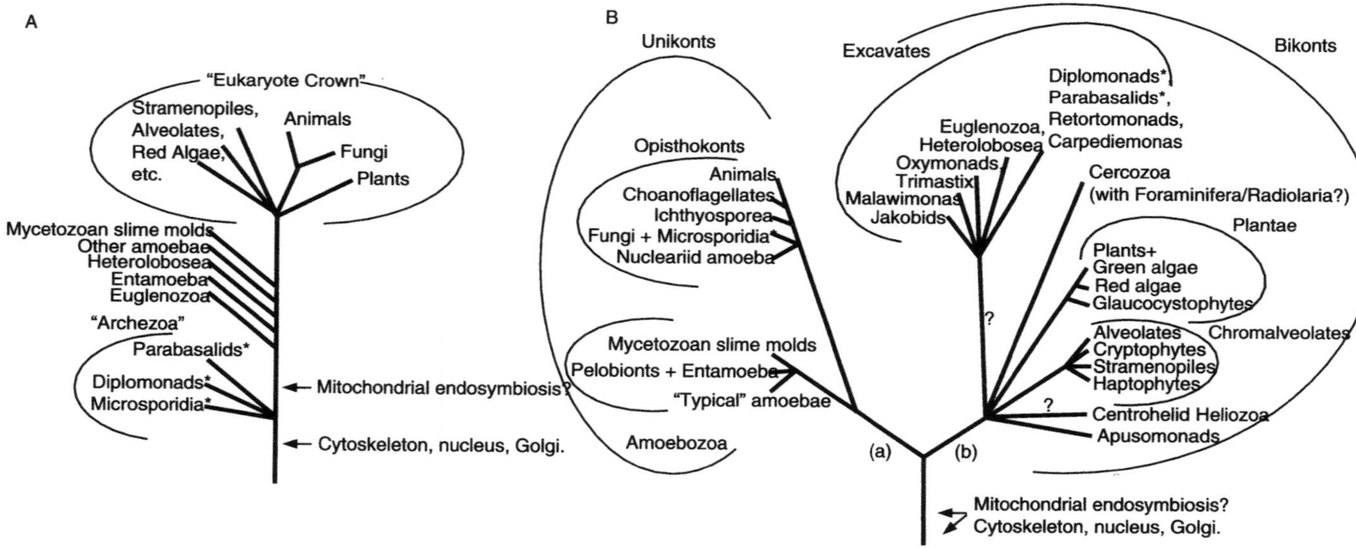

FIGURE 1.1 The early SSUrDNA-based eukaryotic phylogeny contrasted with the new-generation trees recovered from new phylogenetic analyses and other considerations. (Adapted from Figure 1 in Simpson and Roger (2002) R691–R693 and Figure 1 in Stechmann and Cavalier-Smith (2002a) 89–91. (A) Schematic SSUrDNA tree from the early 1990s [e.g., Sogin (1991) 457–463; Sogin et al. (1996) 167–184]. Such trees were often obtained by using distance methods and little consideration was given to the fit of the substitution model to properties of the data under consideration (e.g., base composition and rate heterogeneity between sites — see text and Chapter 6). Distant prokaryotic sequences were typically used as outgroups. These analyses are now thought to have been affected greatly by long-branch artifacts (see Chapter 6). The three major Archezoa clades at the base of the tree are highlighted with the symbol *. (B) More recent views on eukaryote phylogeny based on various analyses of protein alignments [e.g., Baldauf et al. (2000) 972–977; Lang et al. (2002) 1773–1778; Bapteste et al. (2002) 1414–1419], reanalyses of SSUrDNA datasets [e.g., Van de Peer et al. (2000b) 565–576; Cavalier-Smith (2002a) 297–354) and protein gene fusion events [Stechmann and Cavalier-Smith (2002) 89–91; (2003) 665–R664]. Two sets of gene fusion events suggest that the root of the eukaryotic tree might be positioned between two major clades, defined as the unikonts (uniciliates) and bikonts (bicilia) [Cavalier-Smith 2002a; Stechmann and Cavalier-Smith (2003) R665–R664]. A fusion of the genes encoding carbamoyl phosphate synthase, dihydroorotase (DHO) and aspartate carbamoyltransferase (ACT) supports the monophyly of the unikonts, branch labeled (a). A fusion of the genes encoding dihydrofolate reductase (DHFR) and thymidylate synthase (TS), branch labeled (b), suggests monophyly of the bikonts [Stechmann and Cavalier-Smith (2003) R665–R664]. The new positions of three former Archezoa are highlighted by the symbol *.

not be infallible prompted a serious reappraisal of the choice of phylogenetic markers and a more critical use of phylogenetic reconstruction methodologies (e.g., Embley and Hirt, 1998; Philippe and Adoutte, 1998). Protein-coding genes were increasingly used to investigate the global phylogeny of eukaryotes with both nuclear (e.g., Baldauf et al., 2000; Bapteste et al., 2002) and mitochondrial coding genes (Lang et al., 2002) contributing to new hypotheses for the phylogeny of eukaryotes (see later and Figure 1.1B).

1.3.1 Availability of Molecular Phylogenetic Markers and the Advent of Genomics

Although the primary gene cloning efforts were on a gene-by-gene basis, genome sequencing projects along with partial genome surveys [genome sequence surveys (GSSs) and expressed sequence tags (ESTs)] increasingly represent the principal sources of molecular sequence data. The first eukaryotic nuclear genomes to be completed were those of the classical model organisms *Saccharomyces cerevisiae* (1997), *Caenorhabditis elegans* (1998), *Drosophila melanogaster* (2000), *Arabidopsis thaliana* (2000) and *Homo sapiens* (2001), for which a large amount of functional and physiological data had already accumulated. More recently, molecular phylogenetic considerations and the fact that most of the early branching taxa are parasites have stimulated the selection of several of these organisms for genome sequencing. The phylogenetic questions about these organisms make them interesting and important genomic models (Sogin, 1993; Hedges, 2002). These include the microsporidia, represented by the genome of *Encephalitozoon cuniculi* (Katinka et al., 2001; Chapter 10), the diplomonad *Giardia lamblia* (McArthur et al., 2000 — The Marine Biology Laboratory: http://jbpc.mbl.edu/Giardia-HTML/index2.html), the amoeba *Entamoeba histolytica* and relatives (TIGR: http://www.tigr.org/tdb/e2k1/eha1/ and the Sanger Institute: http://www.sanger.ac.uk/Projects/E_histolytica/) and more recently the parabasalid *Trichomonas vaginalis* (TIGR: http://www.tigr.org/tdb/e2k1/tvg/). In addition to these genome projects, other microbial eukaryote genomes (including additional fungi and algae) have been, or are currently being, fully sequenced or investigated through EST and GSS projects (e.g., Bapteste et al., 2002). A recently launched Canadian EST program covering several major eukaryotic lineages (see list at http://megasun.bch.umontreal.ca/pepdb/pep.html) is particularly notable in this respect. For a listing of such projects, readers are referred to the informative genome online database (GOLD) at http://wit.integratedgenomics.com/GOLD/ (Bernal et al., 2001) and to the GOBASE database of organellar genome projects (O'Brien et al., 2003). The databases housing these and future genome sequences represent a rich source of data for investigating eukaryote phylogeny and cell and genome evolution.

One important conclusion from genome surveys is that the long-suspected, but poorly characterized, phenomenon of horizontal (or lateral) gene transfer (HGT) plays an important role in microbial genome evolution and could, at least for prokaryotes, represent a genuine threat to the established phylogenetic world view (Doolittle et al., 2003). The idea of a complex net of life, with numerous genes being exchanged between genomes, is now emerging as a plausible model for prokaryotes (Doolittle et al., 2003). The impact of HGT on eukaryote genome evolution is clearly real, as illustrated by genes of endosymbiotic origin (e.g., Martin et al., 2002), but what is less clear is the impact of HGT from nonendosymbiotic sources (Doolittle et al., 2003). Evidence from a small collection of organisms, mainly parasitic protozoa, suggests that HGT also plays a role in shaping eukaryotic genomes (Andersson et al., 2003; Doolittle et al., 2003; Katz, 2002; Richards et al., 2003). It will be necessary to assess a broad collection of microbial eukaryote genomes for a real understanding of the generality and importance of this process for eukaryotes. Although the most extreme claims (for prokaryotes) have been somewhat tamed and genome phylogenies recover trees consistent with single-gene trees (see Chapter 9), the data and analytical tools that will allow us to address these questions fully are still being developed in sequencing centers and minds of bioinformaticians and evolutionary biologists, respectively (see Chapter 9).

Equipped with a putative phylogenetic framework for all forms of life (e.g., SSUrDNA) and an expanding list of complete genomes from representatives of all major eukaryotic lineages (and numerous prokaryotes), evolutionary biologists now have extensive resources to test and develop theories for both the origin and diversification of eukaryotes. Systematists and evolutionary biologists have seized the opportunity to pursue their research in new ways, especially as evolutionary biology has moved to center stage as a key component of the postgenomic age (Koonin and Galperin, 2003; Eisen and Fraser, 2003; also see Chapters 7 and 9). Comparisons between a broad range of eukaryotes and prokaryotes can also guide and complement experimental work in functional and structural studies (Hedges, 2002). These developments also serve to remind us that experimental molecular cell biology focusing on a few model systems can miss important aspects of the formidable richness and diversity of life forms on earth (Baldauf, 2003; Hedges, 2002) and present a highly biased view of the nature of the eukaryotic cell.

1.3.2 Phylogenetic Methodologies

The advent of molecular systematics was expected to mark a fundamental breakthrough in the phylogenetic classification of eukaryotes. It was anticipated that comparative analysis of relatively few genes from a diverse range of organisms would allow the development of a definitive picture of eukaryote evolution; clarifying relationships within and between previously identified major taxa. To some extent, these expectations have been fulfilled. Early analyses of SSUrDNA sequences provided independent confirmation of several groups of taxa whose phylogenetic coherence had previously been postulated on the basis of morphological considerations (e.g., Patterson and Sogin, 1993). In addition, reanalyses of the SSUrDNA dataset with rate-calibrated distances also recovered trees congruent with those obtained from protein-coding genes (e.g., van de Peer et al., 2000b).

However, analyses associated with the testing of the Archezoa hypothesis also highlighted important problems and limitations of both SSUrDNA and protein-coding genes. For example, it was shown that tree topologies or support values for particular nodes, or both, could change radically with variation in taxa sampling and sequence alignments as well as with inference methods used. Some dramatic effects were also observed with the removal of constant or fast-evolving sites in both protein-coding genes and SSUrDNA (e.g., Hirt et al., 1999; Philippe et al., 2000b). Other workers began to ask how single-gene trees compared with trees obtained from concatenated protein-coding genes (Baldauf et al., 2000; Lang et al., 2002) or trees shown to be most compatible between numerous genes considered in parallel (Bapteste et al., 2002).

The first trees derived from analysis of protein-coding genes, such as translation elongation factors (Hashimoto and Hasegawa, 1996), actin (Bricheux and Brugerolle, 1997; Horner et al., 1998) and tubulin (Keeling and Doolittle, 1996; Horner et al., 1998), concurred with several parts of the SSUrDNA trees. However, some important incongruences began to emerge; for example, tubulin trees consistently indicated that microsporidia rather than being basal eukaryotes might be related to fungi (Edlind et al., 1996; Keeling and Doolittle, 1996). In fact, few genes gave strong support for particular sets of relationships between major taxa; different branching orders emerged from different genes or from different analyses of the same genes (e.g., Philippe and Adoutte, 1998; Embley and Hirt, 1998).

Estimating molecular phylogenetic trees is a complex exercise. It encompasses the careful choice of genes and aligned positions (which characters are homologous?) and the selection of appropriate methods to yield the best results with the data at hand (Holder and Lewis,

2003; see Chapter 6). Molecular datasets (as well as others) are often noisy and can have complex biases, including long branches, base or amino acid composition biases, site-rate heterogeneity and mutational saturation. Furthermore, these problems are typically found in combination, making inference of tree topologies very difficult, if not meaningless, when data become essentially randomized during geological time (Yang, 1996; Chapter 6). In all cases, one should be aware of such potential limitations (Chapters 6 and 8) and use molecular phylogenetic trees as working hypotheses to be carefully tested with more data and additional analyses. For distance, likelihood and Bayesian methods, the substitution model used should as adequately as possible describe the characteristics of the dataset (e.g., Goldman, 1993). The quality and reliability of the trees obtained should also be assessed. Congruence from multiple datasets, with as broad as possible properties and origins (different genes from various pathways or subcellular components, characters other than molecular sequence data, etc.), is the best source of confidence in support for a given tree topology (Darwin, 1859).

Molecular phylogenetic methodologies have essentially undergone four major developments in recent years, all of which have benefited from the rapid increase in computational capacity and speed:

1. Models of sequence evolution have become more sophisticated and parameter rich (Whelan et al., 2001), and it has become routine to choose DNA models carefully in relation to datasets under analysis (Posada and Crandall, 2001; also see Bollback, 2002, for more recent developments in this field). The effects of compositional variation and interlineage rate variation are also regularly tested for both DNA and protein sequences. Furthermore, variation of rates of substitution among sites is commonly modeled with the gamma distribution or discrete approximations to it (Buckley et al., 2001). Another development is the implementation of heterogeneous substitution models for different parts of the sequence data or for different parts of the tree (Yang and Roberts, 1995; Galtier and Gouy, 1998; Foster, 2003). This is an important and promising development, because most methods implemented until recently use exclusively homogenous models, i.e., use the same model for all sites in the alignment and branches in the tree. Clearly, homogenous models do not fit the observed heterogeneity in the structural organization and mode of evolution of biomolecules between organisms.
2. There is a capacity to perform more efficient evaluations and searches of parameter space, including tree space. This has benefited from approaches based on the implementation of Bayesian methodologies to phylogenetics (e.g., Huelsenbeck and Ronquist, 2001; Huelsenbeck et al., 2001; Holder and Lewis, 2003; Foster, 2003, 2004).
3. Measures of tree topology robustness and significance of difference between alternative tree topologies have also become more sophisticated, with statistically based tests of alternative topologies becoming the norm (Shimodaira and Hasegawa, 2001; Shimodaira, 2002; Foster, 2003, 2004).
4. New approaches to the construction of trees from genome data other than primary gene and protein sequences (see Chapters 7 to 9) are also in development to complement sequence-alignment-derived approaches. These can be viewed as analogous to efforts to examine molecular phylogenies in the context of morphological characters and those derived from the paleoarcheological record where appropriate. Some workers have stressed the potential significance of discrete molecular characters, rare events such as insertions, deletions and gene fusions, to help resolve branching orders and relationships (e.g., Rivera and Lake, 1992; Baldauf, 1999; Gupta, 1998; Stechmann and Cavalier-Smith, 2002, 2003; see Chapter 6 for a critical assessment of their use). Another fascinating possibility, now becoming plausible because of the increase in the number of whole genome sequences available, is to consider the distribution of entire genetic systems, such as developmental or gene regulatory pathways, as individual cladisitic

markers. The rationale here is that such characters, once developed and integrated into the fabric of an organism's biology, might not be commonly secondarily lost (see Chapter 8). The value of all these complementary approaches is their relative independence from analysis of primary sequence data. Thus, hypotheses derived from sequence-alignments-based phylogenies can be tested by these methods and vice versa.

The rapidly increasing amount of data now available has allowed the analysis of concatenated and multigene datasets, in some cases thousands of amino acids long (Baldauf et al., 2000; Bapteste et al., 2002; Lang et al., 2002). The rationale behind these approaches is that the phylogenetic signal (if it exists) should be additive whereas noise should not be and is likely to cancel itself out if enough characters are analyzed. However, although the increased availability of data and sophistication of phylogenetic reconstruction methods are likely to contribute to the resolution of deep-level phylogenies, they constitute only some of the methodological progress made in the last 10 years.

Perhaps as important as the considerations outlined previously has been a new open-mindedness and willingness to consider the weaknesses of available methods (and datasets) and to acknowledge that no single gene or method of analysis is likely to provide the magic bullet that will resolve all the questions at hand. Many workers who have previously concentrated on resolving phylogenies are now devoting a significant amount of their efforts to underlining and illustrating the potential weaknesses of those very conclusions and the methodologies used to infer them (e.g., Philippe and Adoutte, 1998; Embley and Hirt, 1998). Thus, for example, there is now a growing awareness of the problems caused by both mutational saturation (the phenomenon of multiple unseen substitutions) and composition bias (of bases or amino acids) in the reconstruction of phylogenies. Furthermore, the possibility that existing methods of accommodating site-by-site variation in substitution rates might not be adequate is increasingly the object of meticulous quantitative analysis (e.g., Buckley et al., 2001; Horner and Pesole, 2003). Several groups are now examining the phenomena of covarions and heterotachy, whereby individual sites undergo substitution at different rates and under different processes in different lineages (Penny et al., 2001; see Chapter 6).

1.4 Eukaryote Diversity and Phylogeny

In contrast to the situation confronting early bacterial systematists (Woese, 1987), the study of metazoans, multicellular fungi and higher plants provided a huge number of morphological characters that could be used for phylogenetics and classification. The development of cladistic methodologies and the early perception of the importance of developmental characters led to the formulation of coherent phylogenetic hypotheses before the advent of molecular systematics. Improvements in light microscopic techniques and the invention of the electron microscope allowed extensive anatomical characterization of many groups of protists; systematic analysis of data derived from such studies resulted in grouping of many, if not most, known protists into phylogenetically coherent taxa (Patterson, 1994; Patterson and Sogin, 1993). However, both the interrelationships between many major groups of taxa and the phylogenetic placement of many unusual organisms, especially those that apparently lacked key organellar structures, continued to pose problems to systematic protistologists for several years. Furthermore, objective criteria for rooting the eukaryotic tree were largely lacking before the emergence of molecular phylogenetics, which provided the methods to address global eukaryote phylogeny and the position of the eukaryote root (see Chapter 2).

With the aforementioned considerations in mind, we can briefly examine the evolution of ideas regarding eukaryote phylogeny and how they stand in 2003. What is broadly accepted? What is hotly contested? What remains in a state of flux? For simplicity, we begin with well-accepted conclusions drawn from molecular and other considerations and follow with more contentious hypotheses and questions regarding the position of the root of the eukaryotic tree.

1.4.1 Eukaryote Molecular Phylogeny

A relationship among animals, choanoflagellates and fungi, to the exclusion of most other eukaryote groups, is now well accepted and has been formalized as the clade named opisthokonts (see Chapter 5). Several molecular datasets suggest this relationship, and increase of taxa and number of characters used tend to result in increased statistical support for this conclusion (Baldauf, 2003; Chapter 5). Several morphological and molecular analyses have also concluded that some choanoflagellates are likely to be the sister group to animals (Lang et al. 2002; Chapter 5).

Interestingly the combination of data from recent analyses that use large protein datasets (in either combined evaluations of trees or concatenated) and protein gene fusion events have provided a compelling hypothesis that many amoeboid lineages, grouped together in a clade often called the Amoebozoa (see Chapter 5), such as *Entamoeba*, *Mastigamaoeba* and *Acanthamoeba*, are related to the slime molds and together are a sister taxa to the opisthokonts (Lang et al., 2002; Bapteste et al., 2002; Stechmann and Cavalier-Smith, 2003; Chapter 5).

The monophyly of chromists (including heterokonts also called stramenopiles, e.g., Patterson and Sogin, 1993) and alveolates to form the chromalveolate clade, a formidably diverse group of eukaryotes, is becoming more widely accepted (see Chapter 4). This group probably shared an ancestral secondary endosymbiotic plastid, perhaps derived from a red alga (see Chapters 3 and 4). However, the deeper-level relationships of this group are more poorly resolved (see Chapter 4). Whereas little controversy is associated with the relationship between green algae and higher plants, the subject of red algal relationships remains contentious (see Chapters 3 and 8) although much of the most recent evidences argues for a red-green clade (see Chapter 3). For example, several recent molecular analyses suggest a sister group relationship between red and green algae whereas others do not, and several other factors beyond the distribution of chlorophyll types underline the distinctness of these groups (see Chapter 8).

As discussed in more detail later in this chapter and in other chapters, the placement of many groups of protists originally thought to be representatives of primitive lines of eukaryotic descent (members at one time or another of the Archezoa) is currently uncertain. Several lines of evidence suggest that many of these parasitic protists might in fact be highly derived members of large, relatively late-branching lineages, named the Excavates (see Chapter 2 and Figure 1.1B).

One of the great challenge of future investigation of eukaryote diversity and phylogeny is to obtain stable and well-supported clades based on both molecular and morphological considerations (e.g., Baldauf, 2003 ; Patterson and Sogin, 1993), which implies the identification of the root of the eukaryotic tree.

1.4.2 Rooting of the Eukaryotic Tree

To be able to map evolutionary changes or polarize them (give them directionality) and to define monophyletic groupings (or clades) on a phylogeny, the position of the root on the tree has to be determined. This is one of the major difficulties in molecular phylogenetics (Smith, 1994), and eukaryote phylogeny is no exception (Embley and Hirt, 1998; Baldauf, 2003; Simpson and Roger, 2002; Chapters 2 and 6). Rooting any of the three Domains of life by using outgroups is a major problem because gene sequences between Domains are typically highly divergent and can vary in base or amino acid composition, adding an additional level of complication for phylogenetic inference (Chapter 6). One of the most dramatic and best examples illustrating this problem comes from analysis of the phylogenetic position of the spectacular misplacement of microsporidia in the SSUrDNA phylogeny, as discussed previously. This case, among others, clearly illustrates the need to be critical of the conclusions derived from single datasets and that congruent trees from diverse data and analyses are the best source of confidence in phylogenetic conclusions.

With these potential difficulties in mind, two interesting developments in the eukaryotic tree rooting were recently published. One is concerned with the identification of gene fusion events of protein-encoding genes in a small collection of taxa. The distribution of two gene fusion events splits the eukaryotic tree into two major clades, suggesting a position of the root between them (Stechman and Cavalier-Smith, 2002, 2003). The data suggest on the one hand an opisthokont–Amoebozoa clade and on the other side of the tree plants with numerous protists lineages representing the great majority of eukaryote diversity (e.g., rhodophytes, jakobids, chromists, alveolates). The authors formalize these two major clades with the names of unikonts (opisthokonts plus Amoebozoa) and bikonts (plants plus the rest of eukaryotes) (Stechman and Cavalier-Smith, 2003; Figure 1.1B). The distribution of these fusion events will have to be tested among broader taxonomic samples. Very similar trees were previously suggested from a composite analysis of various protein-coding genes and morphological features (Dacks and Roger, 1999; Embley and Hirt, 1998).

Lang and coworkers, using a very different source of gene data (concatenation of proteins encoded by mitochondrial genomes), recovered a well-supported tree (Lang et al., 2002), which is consistent with the unrooted tree obtained from concatenated nuclear genes (Baldauf et al., 2000; Baldauf, 2003) and the rooting suggested by gene fusion events (Stechman and Cavalier-Smith, 2002, 2003). An interesting feature of the concatenated mitochondrial dataset is that the sequences used to root the eukaryotic tree are the relatively similar homologues of the alpha-proteobacteria, potentially minimizing the effects of the long-branch attraction artifact.

Where does the former Archezoa fit in these rooted trees? At present, there is no definitive answer (Baldauf, 2003; Chapter 2). There is no data on the relevant gene fusion events, and diplomonads and parabasala lack mitochondrial genomes despite the presence of mitochondria-derived organelles (Embley et al., 2003a,b; Chapters 2 and 13). However, inference from other phylogenetic analyses that have several taxa in common with the rooted analyses suggests that diplomonads and parabasala are part of the group known as excavates, which are related to, or included within, the bikonts (Baldauf, 2003; Stechman and Cavalier-Smith, 2003; Chapter 2).

To us the position of the root suggested by the aforementioned analyses and the implied bikont and unikont clades represent excellent working hypotheses that will have be tested in the future by analyses of the data (and methods) becoming available through genome surveys and whole genome-sequencing projects.

1.5 Organelles of Endosymbiotic Origins, Their Genomes and Proteomes

With the availability of complete nuclear and organellar genome sequences and proteomics has come a unique opportunity to estimate the complete proteomic complement of endosymbiotic organelles and further understand the extent and mechanism of endosymbiotic contributions to eukaryote genomes evolution. The first complete genomes to be sequenced from eukaryotic cells were those of the organelles whose small size allows easier manipulation and sequencing (Lang et al., 1999). These sequence data have amply confirmed that these organelles have descended from bacterial ancestors and that the mitochondria arose from a single source (Lang et al., 1999).

1.5.1 Mitochondria and Their Alpha-Protebacterial Origins

A large diversity of mitochondrial functions beyond the usual paradigm of the Krebs cycle and oxidative phosphorylation is now evident (Tielens et al., 2002; Embley et al., 2003a,b; Chapters 10 and 13). Mitochondria or organelles derived from them are probably present in all extant eukaryotes. However, anaerobic eukaryotes exhibit strikingly diverse mitochondrial biochemistry. The conversion of conventional mitochondria into hydrogenosomes, anaerobic genomeless energy generating organelles that excrete molecular hydrogen, is observed in many independent lineages (Embley et al., 1997, 2003a,b; Chapters 2 and 13). The origin of the distinctive hydrogenosomal biochemistry and its distribution among aerobic taxa have been the subjects of intense research and debate (Rotte et al., 2001; Horner et al., 1999, 2000, 2002; Müller, 1998). Even more striking is the presence in other anaerobic organisms of apparently energetically inert mitochondria-derived organelles. Such structures have been described in *Entamoeba* (Mai et al., 1999; Tovar et al., 1999) wherein energy metabolic electron transport is believed to be cytosolic, and in microsporidia wherein the nature of energy metabolism remains somewhat obscure (Williams et al., 2002; Chapters 10 and 13). Whether other functions, perhaps shared with aerobic mitochondria, such as the iron–sulfur cluster assembly, provide a selective pressure for organelle retention in these systems is the subject of current research (Mülenhoff and Lill, 2000; see Chapters 10 and 13).

Although the distribution and diversity of mitochondrial biochemistry have been studied for many years, the advent of mitochondrial and whole genome-sequencing efforts and proteomic studies with improved *in-silico* methodologies have raised other questions. These include variation in mitochondrial proteome between organisms, the extent and mechanism of organelle-to-nucleus gene transfer and the origin of genes whose products are localized to and function in the mitochondrion. Early studies suggest that an important fraction of the mitochondrial proteome might not be encoded by genes that originate with the mitochondrial symbiont, but might be the products of eukaryote-specific genes that have evolved since the incorporation of mitochondria (Kuan and Saier, 1993; Andersson et al., 2003). Although many questions remain unanswered, the endosymbiotic origin of the mitochondrial compartment and genome from within the proteobacteria, and in particular the alpha-proteobacteria, is uncontroversial. This conclusion, originally based on biochemical considerations (John and Whatley, 1975), has been consistently supported by molecular phylogenetic analysis of mitochondrial SSUrDNA (Woese, 1987) and other genes such as those of the heat-shock proteins Hsp60 (cpn60) and Hsp70 (Boorstein et al., 1994; Falah and Gupta, 1994; Viale and Arakaki, 1994; Lang et al., 1999).

1.5.2 Plastids

Although it is generally accepted that the first plastids appeared later in eukaryote evolution than mitochondria, there has been much debate over many aspects of plastid evolution, including their possible early acquisition and more widespread distribution in early stages of eukaryote evolution (e.g., Andersson and Roger, 2002). One of the most significant questions has always been about the number of primary plastid endosymbiotic events. Three major lines of eukaryotes bear primary plastids: green algae and higher plants (Chlorophytes), glaucocystophytes and red algae (Rhodophytes). Recent changes in ideas of relationships among major eukaryotic groups, as well as the recovery of genes that appear to display cyanobacterial ancestry from nonplastid-bearing organisms (Andersson and Roger, 2002), have further complicated this issue. Furthermore, it is clear that violations of assumptions made in phylogenetic methodologies might specifically be expected to affect results from analysis of plastid gene sequences intended to address this question (Lockhart et al., 1992). It remains to be seen whether the loss or reduction of primary plastid function is as common as seems to be the case for mitochondria.

In addition to a better understanding of the primary plastid endosymbiosis and symbioses, it is also critical that a more complete picture of the distribution and mechanisms behind secondary (and tertiary) plastid symbiosis be developed. It is clear that secondary plastid acquisition (a secondary plastid is one derived from the endosymbiotic acquisition of a plastid-containing algae by another eukaryote) has occurred several times. Secondary plastids are often surrounded by three or four membranes that correspond to plastid membranes and membranes derived from the original algal host and potentially from the ultimate host. Elaborate protein import systems have evolved to allow import of nuclear-encoded gene products (see Chapters 3 and 4). Furthermore, in some cases, a vestigial nucleus (corresponding to that of the original algal host nucleus) is retained and might encode genes crucial to plastid function (see Chapter 4). The investigation of mechanisms of secondary plastid endosymbiosis promises great insight not only into plastid and host function and evolution but also into organelle-to-host and eukaryote-to-eukaryote lateral gene transfer (see Chapters 3 and 4).

1.6 Eukaryote Origins

Perhaps the most obvious distinction between eukaryotes and prokaryotes is the one implied by the terms themselves — the presence or absence of a nucleus. Other commonly cited distinctions include the presence in eukaryotes of mitochondria, plastids, endoplasmic reticulum, Golgi, cytoskeleton and introns, as well as fundamental differences in the types of flagella found in prokaryotes and eukaryotes and the absence of true operons in eukaryotes. A recurring theme in this list is the high level of endomembrane complexity of eukaryotes with its underlying functional compartmentalization. Other distinctions include genome structure and size, metabolic diversity and the capacity to generate differentiated multicellular forms.

Of these factors, the presence of the nucleus and its associated complex endomembrane system, making up the exocytic and endocytic membranes, as well as the cytoskeleton have classically been considered the most significant. The eukaryotic flagellar apparatus and the ability to perform phagocytosis (and thus engulf the bacteria that became mitochondria and plastids) were considered to be likely consequences of the presence of a cytoskeleton. Whereas some workers favored an exogenous (endosymbiotic) origin for the nucleus, there is no support for such hypotheses as the organization and topology of the nuclear membrane are not easily explained by endosymbiotic mechanisms; the best-available hypothesis to date is an endogenous origin of the nuclear membrane and its linked membrane of the exocytic and endocytic system (Martin, 1999).

Advances in phylogenetics and comparative genomics of both prokaryotes and eukaryotes have led to a series of reappraisals of eukaryote phylogeny and have ultimately prompted the formulation of radical new hypotheses on the origin of eukaryotes. Much of the upheaval in our understanding of the biology and relationships among eukaryotes has been the result of closer study of endosymbiotically derived organelles. Many workers now believe that establishment of the mitochondrial endosymbiosis rather than the evolution of the nucleus might have been the key event in the origin of eukaryotes (e.g., Martin and Müller, 1998; Vellai and Vida, 1999; Lang et al., 1999) that initiated and eventually brought about the development of the complex eukaryotic cell as it is known now.

All available molecular data are inevitably from extant organisms. One can only work with data at hand. However, speculations have often been based on the basis of hypothetical evidence or data, such as assumed absence of organelles or other features. Along the same lines, there has also been a tendency to construct hypothetical organisms for which there is no direct evidence. These sometimes become "real" as if there is proof that they actually existed at some point in the past. We have tried to be as aware as possible of the distinction between facts and hypotheses and when discussing the latter try to clearly label them as such with an attempt to judge and review the evidence supporting them. With these considerations in mind, we would like to quote Tom Fenchel, who wrote in the preface of his recent book (Fenchel, 2003):

> I cannot believe that anyone would be so naïve as to think that a title containing the phrase 'origin of life' means that the book actually explains how life originated — because this is not known. But preoccupation with the problem gives insight into what life is.

The same can be said of the origin of eukaryotes; it is certainly difficult to explain, but the investigation can help learn about what eukaryotes are.

A very interesting piece in the jigsaw of eukaryote origins came with the realization that the root of the universal tree of life could theoretically be found through the use of anciently duplicated gene trees to root one another (Iwabe et al., 1989; Gogarten et al., 1989; Brown and Doolittle, 1997). The few genes that allow such analyses have led to the conclusion that Archaea and Eukarya are more closely related to one another than either are to Bacteria, indicating that the ancestral eukaryote (or the prokaryotic cell) was derived from a last common ancestor of eukaryotes and Archaea (Iwabe et al., 1989; Gogarten et al., 1989; Brown and Doolittle, 1997). However, recent evaluations of these analyses suggest that this rooting is not well supported, if at all reliable (Philippe and Forterre, 1999). This inference, also considering prokaryote metabolic diversity, which includes numerous chemolitotrophes, is consistent with the hypothesis that bacteria were the first forms of cellular life (e.g., Martin et al., 2003). An important issue to consider in discussions of eukaryote origins is that there is no need to invoke a phagocytic cell (assumed by some to be a eukaryotic cell, e.g., Cavalier-Smith, 2002a) to allow a prokaryote to be found inside the cytosol of another cell (e.g., Margulis, 1993; Martin, 2002).

In the following paragraphs we provide an overview of some of the most influential and sometimes controversial models for the origin of eukaryotes. The comprehensive examination of the questions involved and the interpretations of phylogenetic and other data deserve inclusion in another book, and it is beyond the scope of this book to discuss the huge and heterogeneous literature on the subject. We therefore treat the subject in a cursory manner and do not provide a detailed treatment of individual hypotheses. Some might disagree with the way we have grouped models together — we concede that this is not an objective assessment but rather a way of introducing some of the diversity of thought in this exciting area of theoretical and evolutionary biology (Figure 1.2).

1.6.1 The Serial Endosymbiosis Theory and Related Hypotheses

The serial endosymbiosis theory (SET) of Lynn Margulis (Sagan, 1967; Margulis, 1970) has been, since its inception in the 1960s, the most widely accepted model of eukaryogenesis. This hypothesis has undergone several updates, but has remained essentially constant in its major components. In its present incarnation (Margulis et al., 2000), it is suggested that a motile consortium of a wall-less *Thermosplasma*-like archaeon and a *Spirochaete*-like eubacterium became a chimeric archezoan organism after a mixing of genomic material and the evolution of a protonucleus associated with the karyomastigont (flagellar and microtubule organizing apparatus). According to the SET hypothesis, this primitive eukaryote subsequently acquired mitochondria from the alpha-proteobacterial lineage. The energetic advan-

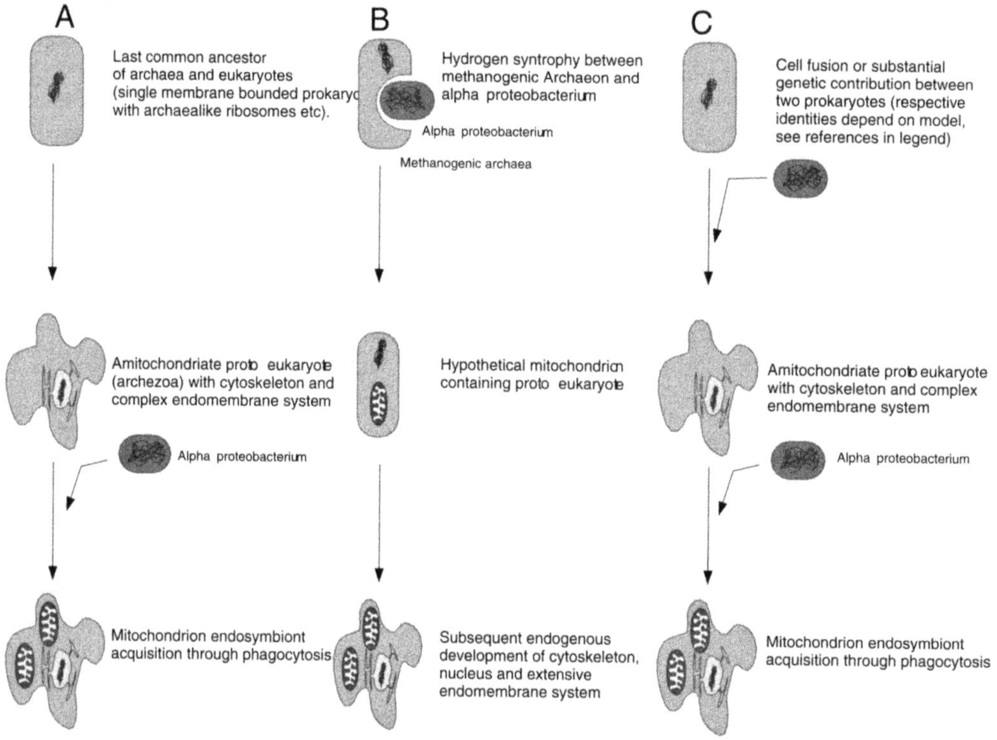

FIGURE 1.2 Three major classes of model for eukaryogenesis. (A) Models in which endogenous origin of cytoskeleton and phagotrophy are followed by endosymbiotic origin of mitochondria [e.g., Cavalier-Smith, (2002a) 52: 297–354]. (B) The hydrogen hypothesis. Metabolic syntrophy between a methanogenic archaeon and the protomitochondrion leads to the formation of a heterotrophic organism with eukaryote-like metabolism. Subsequent endogenous origins of nucleus, other endomembranes and cytoskeleton are invoked in this model [Martin and Müller (1998) 37–41]. (C) Models that indicate two major bacterial contributions to all extant eukaryotes. The identity of the first contribution varies between models (e.g., Sogin (1991) 457-463; Doolittle (1995) 235–240; Gupta (1998) 1435–1491; Hartman Fedorov (2002) 1420–1425; Canback et al. (2002) 6097–6102; Sagan (1967) 255–274; Margulis et al. (2000) 6954–6959; Moreira and López-García (1998) 517–530). For a broad and detailed review on endosymbiosis in eukaryogenesis, see Martin et al. (2001) 1521–1539 and Margulis (1993).

tage of oxidative phosphorylation is stated as the selective advantage favoring the establishment of the organelle in an environment with increasing atmospheric oxygen levels. However, many molecular and cell biologists take issue with several of her assumptions (e.g., Martin et al., 2001).

However, in spite of — or indeed because of — its lack of mechanistic specificity, the SET theory provides a good framework against which to contrast other current ideas about eukaryogenesis.

1.6.2 The Hydrogen Hypothesis

Perhaps the most interesting and radical response to the rejection of the Archezoa hypothesis was the formulation of the hydrogen hypothesis for the origin of eukaryotes (Martin and Müller, 1998). Martin and Müller were also concerned that ATP could not have been the currency of the early relationship between mitochondrion and host. They were informed particularly by studies on hydrogenosomes, modified hydrogen-producing mitochondria found in some anaerobic protists (see Chapter 13). They noted the coexistence of methanogenic archeabacteria with hydrogenosomes in some ciliates and proposed a plausible metabolic syntrophy based on the exchange of metabolites to explain the origin of eukaryotes. In their scenario, an autotrophic, obligatory anaerobic methanogenic archaebacterium (the host) developed a mutually beneficial association with a facultative anaerobic, heterotrophic alpha-proteobacterium (the protomitochondrial symbiont). The eubacterium would have provided its waste products (hydrogen, CO_2 and possibly acetate) to the methanogen and thereby increased its capacity to exist in other than completely anaerobic environments. In more aerobic environments, the bacterial oxidative respiration might have been able to protect the archaebacterium's anaerobic metabolism from oxidative damage. The hydrogen hypothesis suggests that rather than phagocytosis, selection for hosts with large surface area (and hence the capacity to develop a high surface area of contact with the symbiont) ultimately led to the inclusion of the symbiont within the host. During this stage of the association, Martin and Müller point out that loss of methanogenesis by the host, coupled with the transfer of genes for glycolysis and several transporters of reduced organic molecules from the symbiont to the host, might serve to generate a protoeukaryote that (metabolically) had the properties of both mitochondriate and hydrogenosomal eukaryotes (and from which both types of eukaryote metabolism might be derived through differential loss). Thus, the hydrogen hypothesis posits that development of the nucleus and cytoskeleton occurred subsequently to the establishment of the eukaryotic metabolic type. The hypothesis neatly and parsimoniously explains the eubacterial nature of eukaryotic energy metabolism and (through its assumption of a single major eubacterial contribution to the early eukaryote) also makes the testable prediction that eukaryotic energy metabolic pathways, and other eubacteria-like proteins ubiquitous among eukaryotes should be derived from the protomitochondrial symbiont.

As perhaps the most innovative view of eukaryogenesis for many years, the hydrogen hypothesis has been the subject of fierce debate. For example, its critics argue that it does not adequately explain the origin of the nucleus and cytoskeleton. To some extent this is justified, although Martin and Müller point out that it does not attempt to do this. Indeed, other hypotheses do not provide such plausible accounts for the origin of eukaryotic energy metabolism either. Thus, the hydrogen hypothesis was the first of a new generation of models of eukaryogenesis invoking true metabolic syntrophies as the precursors of stable endosymbiotic relationships.

Others have invoked alternative metabolic syntrophies at the origin of eukaryotes. The syntrophic hypothesis of Moreira and López-García suggests that hydrogen syntrophy between a methanogen and a delta-proteobacteria gave rise to an endosymbiotically derived nucleus within an archezoa-like organism (Moreira and López-García, 1998, 1999). In this scenario, the mitochondrion was the result of a subsequent endosymbiosis. Although it has been argued that the model of nuclear evolution proposed by these workers is not sound (e.g., Cavalier-Smith, 2002a; see also arguments in Martin, 1999) others might sympathize with such two-symbiont models, pointing to the failure of molecular phylogenetics to identify an alpha-proteobacterial origin for glycolysis (Canback et al., 2002), the enzymes of anaerobic energy metabolism found in hydrogenosomes (Horner et al., 1999, 2000, 2002) and cytosolic and endoplasmic reticular heat-shock proteins that are of bacterial but not unambiguously alpha-proteobacterial ancestry (Gupta, 1998).

1.6.3 Other Theories of Eukaryogenesis

Other hypotheses for eukaryogenesis have been developed from the perspective of explaining the evolution of the cytoskeleton. This is not in itself straightforward, because although prokaryotic homologues of both actin and tubulin have now been identified (reviewed in van den Ent et al., 2001, and Doolittle and York, 2002), the degree of primary sequence similarity between prokaryotic and eukaryotic proteins is extremely low and these relationships have been established mostly through structural and mechanistic similarities. In fact, there are a large number of proteins that are widespread among eukaryotes (and in some cases, these genes are evolving extremely slowly), but for which homologues in Bacteria or Archaea are absent or extremely hard to recognize (Hartman and Fedorov, 2002; Doolittle, 1995). This situation is exemplified by actin and tubulin. These are among the most slowly evolving proteins in eukaryotes but share very little sequence similarity with mreB and ftsZ — the slowly evolving prokaryotic proteins with which they appear to share common ancestry. The prokaryotic gene products with which they appear to share common ancestry (mreB and ftsZ) also evolve at a relatively slow rate among prokaryotes. Why then is it so hard to recognize the homology between prokaryotic and eukaryotic forms of these proteins, when, for example, the similarity between prokaryotic and eukaryotic metabolic enzymes — which evolve at faster rates than cytoskeletal proteins — is much more pronounced (Doolittle, 1995)? One possible explanation for this observation is that genes for eukaryotic cytoskeletal proteins have been evolving independently of their prokaryotic counterparts for much longer than have other genes, such as metabolic enzymes and the transcription and translation apparatus. Thus, several models suggest that the cytoskeleton [and potentially many other eukaryote-specific systems, such as gene splicing, and some intracellular signaling pathways (Hartman and Fedorov, 2002)] are not derived from either a eubacteria-like or archaebacteria-like ancestor, but rather from a third lineage that must have diverged from the prokaryotes before the diversification of extant eubacterial and archaeabacterial groups (Sogin, 1991; Hartman and Fedorov, 2002; Doolittle, 1995; Doolittle and York, 2002). It has been suggested that such an organism might have possessed an RNA-based genetic system (Sogin, 1991), gene splicing having evolved as a replication error correction system (Doolittle, 1995). Such hypotheses posit various combinations of additional endosymbiotic contributions to the eukaryote lineage from archaebacteria and eubacteria. Such events would have been facilitated, argue proponents of these models, by the capacity to perform phagocytosis, a consequence of the presence of the cytoskeleton. Although these models are theoretically appealing, the main criticism posted against them is that they invoke a major evolutionary lineage, fundamentally different from other prokaryotes, for which no direct evidence has been uncovered.

The degree of divergence between actin and tubulin with their prokaryotic homologues has also been justified as being the result of a phase of extremely rapid evolution in a protoeukaryotic lineage that was a eubacteria-like sister group of archaebacteria (rather than an archaebacterium as envisaged by the hydrogen and serial endosymbiotic hypotheses; Cavalier-Smith, 2002a). This scheme, argued elegantly on the basis of distribution of various gene fusions and membrane characteristics requires reassessment of the position of the root of the universal tree of life (Cavalier-Smith, 2002b).

Of the hypotheses for the origin of eukaryotes that have been formulated in the last quarter of a century or so, many have aspects in common with others while possessing striking individual features (Figure 1.2). Inherent in the field of phylogenetics is the difficulty of inferring events that have potentially occurred billions of years ago, and although criticisms of some mechanistic details of some hypotheses might be valid, these do not necessarily invalidate whole perspectives (e.g., the hydrogen hypothesis that does not directly address the origin of the nucleus and cytoskeleton). It therefore seems reasonable at this stage to have an open mind on many questions relating to eukaryogenesis. A pick-and-mix approach of preferred ideas might not always be popular with the proponents of some of the hypotheses outlined here, but in many cases the basic tenets of the published models do not preclude incorporation of some aspects of competing hypotheses.

1.7 The Cell – A Final Comment

Although most of the data and ideas that have been discussed so far (and the questions addressed in subsequent chapters) concern molecular sequence data, the fundamental unit of life remains the cell (or multicellular organisms made of cells) — the entity that integrates genomes, membranes, cytoskeleton (Cavalier-Smith 2002a; Chapter 15) and the effects of environmental stimuli (Kirschner et al., 2000) into the numerous and amazing phenotypes of life. Although the bulk of hereditable characteristics seemed to be mediated by the various genomes resident within eukaryotic cells, there is a growing awareness of the importance of non-DNA-coded information or epigenetic information (Jablonka and Lamb, 1995; Kirschner et al., 2000). At the population genetic level, epigenetic inheritance of DNA methylation patterns has been shown to be an important mechanism (Chapter 16). However, the significance of membrane inheritance (see Chapter 15) has not often been considered at the level of eukaryote origins and evolution. This long-neglected area will provide interesting characters and fascinating insight into the development and evolution of the eukaryotic condition and on the nature of life itself. (Kirschner et al., 2000, gives some interesting examples.) John Keats in his poem "Ode on a Grecian Urn" proclaimed: "Beauty is truth, truth beauty, that is all ye know on earth and all ye need to know." In the case of evolutionary history, the truth can be hard to uncover. However, the ongoing synthesis of comparative biological data into new and refined theories of eukaryote origins and evolution will also serve to highlight the beauty and diversity of the eukaryote condition.

Acknowledgments

We appreciate critical comments from Martin Embley, Peter Foster and Alina Goldberg on an earlier version of this chapter. We also thank Martin Embley for his encouragement and for the stimulating research environment he provided at the Natural History Museum, where from 1995 to 2001 we developed our interests in microbial eukaryote diversity and evolution. DSH was supported by a Natural History Museum (London, U.K.) Research Fellowship, and is currently funded by a European Union Marie Curie Category 30 fellowship (MCFI-2001-00634). RPH was supported by a Wellcome Trust Career Development Fellowship and is currently supported by a Wellcome Trust University Award.

References

Andersson, J.O. and Roger, A.J. (2002) A cyanobacterial gene in nonphotosynthetic protists–an early chloroplast acquisition in eukaryotes? *Curr. Biol.* 12: 115–119.

Andersson, S.G., Karlberg, O., Canback, B. and Kurland, C.G. (2003) On the origin of mitochondria: A genomics perspective. *Philos. Trans. R. Soc. Lond. B Biol. Sci.* 358: 165–177.

Andersson, J.O., Sjogren, A.M., Davis, L.A., Embley, T.M. and Roger, A.J. (2003) Phylogenetic analyses of diplomonad genes reveal frequent lateral gene transfers affecting eukaryotes. *Curr. Biol.* 13: 94–104.

Baldauf, S.L. (1999) A search for the origin of animals and fungi: Comparing and combining molecular data. *Am. Nat.* 154(suppl.): S179–S188.

Baldauf, S.L. (2003) The deep roots of eukaryotes. *Science* 300: 1703–1706.

Baldauf, S.L., Roger, A.J., Wenk-Siefert, I. and Doolittle, W.F. (2000) A kingdom-level phylogeny of eukaryotes based on combined protein data. *Science* 290: 972–977.

Bapteste, E., Brinkmann, H., Lee, J.A., Moore, D.V., Sensen, C.W., Gordon, P., Durufle, L., Gaasterland, T., Lopez, P., Muller, M. and Philippe, H. (2002) The analysis of 100 genes supports the grouping of three highly divergent amoebae: *Dictyostelium*, *Entamoeba*, and *Mastigamoeba*. *Proc. Natl. Acad. Sci. USA* 99: 1414–1419.

Bernal, A., Ear, U. and Kyrpides, N. (2001) Genomes OnLine Database (GOLD): A monitor of genome projects worldwide. *Nucl. Acids Res.* 29: 126–127.

Bollback, J.P. (2002) Bayesian model adequacy and choice in phylogenetics. *Mol. Biol. Evol.* 19: 1171–1180.

Boorstein, W.R., Ziegelhoffer, T. and Craig, E.A. (1994) Molecular evolution of the HSP70 multigene family. *J. Mol. Evol.* 38: 1–17.

Bricheux, G. and Brugerolle, G. (1997) Molecular cloning of actin genes in *Trichomonas vaginalis* and phylogeny inferred from actin sequences. *FEMS Microbiol. Lett.* 153: 205–213.

Brown, J.R. and Doolittle, W.F. (1997) Archaea and the prokaryote-to-eukaryote transition. *Microb. Mol. Biol. Rev.* 61: 456–502.

Buckley, T.R., Simon, C. and Chambers, G.K. (2001) Exploring among-site rate variation models in a maximum likelihood framework using empirical data: effects of model assumptions on estimates of topology, branch lengths, and bootstrap support. *Syst. Biol.* 50: 67–86.

Bui, E.T.N., Bradley, P.J. and Johnson, P.J. (1996) A common evolutionary origin for mitochondria and hydrogenosomes. *Proc. Natl. Acad. Sci. USA* 93: 9651–9656.

Canback, B., Andersson, S.G. and Kurland, C.G. (2002) The global phylogeny of glycolytic enzymes. *Proc. Natl. Acad. Sci. USA* 99: 6097–6102.

Cavalier-Smith, T. (1993) Kingdom protozoa and its 18 phyla. *Microbiol. Rev.* 57: 953–994.

Cavalier-Smith, T. (2002a) The phagotrophic origin of eukaryotes and phylogenetic classification of Protozoa. *Int. J. Syst. Evol. Microbiol.* 52: 297–354.

Cavalier-Smith, T. (2002b) The neomuran origin of archaebacteria, the negibacterial root of the universal tree and bacterial megaclassification. *Int. J. Syst. Evol. Microbiol.* 52: 7–76.

Cavalier-Smith, T. and Chao, E.E. (2003) Phylogeny of choanozoa, apusozoa, and other protozoa and early eukaryote megaevolution. *J. Mol. Evol.* 56: 540–563.

Clark, C.G. and Roger, A.J. (1995) Direct evidence for secondary loss of mitochondria in *Entamoeba histolytica. Proc. Natl. Acad. Sci. USA* 92: 6518–6521.

Cole, J.R., Chai, B., Marsh, T.L., Farris, R.J., Wang, Q., Kulam, S.A., Chandra, S., McGarrell, D.M., Schmidt, T.M., Garrity, G.M., Tiedje, J.M. (2003) Ribosomal Database Project. The Ribosomal Database Project (RDP-II): previewing a new autoaligner that allows regular updates and the new prokaryotic taxonomy. *Nucl. Acids Res.* 31: 442–443.

Dacks, J. and Roger, A. J. (1999) The first sexual lineage and the relevance of facultative sex. *J. Mol. Evol.* 48: 779–783.

Darwin, C. (1859) *On the Origin of Species by Means of Natural Selection, or the Preservation of Favoured Races in the Struggle for Life*, Murray, London.

DeLong, E.F. and Pace, N.R. (2001) Environmental diversity of Bacteria and Archaea. *Syst. Biol.* 50: 470–478.

Doolittle, R.F. (1995) The origins and evolution of eukaryotic proteins. *Philos. Trans. R. Soc. Lond. B Biol. Sci.* 349: 235–240.

Doolittle, R.F. and York, A.L. (2002) Bacterial actins? An evolutionary perspective. *Bioessays* 24: 293–296.

Doolittle, W.F. (1996) Some aspects of the biology of cells and their possible evolutionary significance. In *Evolution of Microbial Life* (Roberts, D.McL., Sharp, P., Alderson, G. and Collins, M., Eds.), Cambridge University Press, Cambridge, pp.1–21.

Doolittle, W.F., Boucher, Y., Nesbo, C.L., Douady, C.J., Andersson, J.O. and Roger, A.J. (2003) How big is the iceberg of which organellar genes in nuclear genomes are but the tip? *Phil. Trans. R. Soc. Lond. B* 358: 39–58.

Dyall, S.D. and Johnson, P.J. (2000) Origins of hydrogenosomes and mitochondria: evolution and organelle biogenesis. *Curr. Opin. Microbiol.* 3: 404–411.

Edlind, T.D., Li, J., Visvesvara, G.S., Vodkin, M.H., McLaughlin, G.L. and Katiyar, S.K. (1996) Phylogenetic analysis of beta-tubulin sequences from amitochondrial protozoa. *Mol. Phylogenet. Evol.* 5: 359–367.

Eisen, J.A. and Fraser, C.M. (2003). Phylogenomics: intersection of evolution and genomics. *Science* 300: 1706–1707.

Embley, T.M. and Hirt, R.P. (1998) Early branching eukaryotes? *Curr. Opin. Genet Dev.* 8: 624–629.

Embley, T.M., Horner, D.S. and Hirt, R.P. (1997) Anaerobic eukaryote evolution: hydrogenosomes as biochemically modified mitochondria? *Tr. Ecol. Evol.* 12: 437–440.

Embley, T.M., van der Giezen, M., Horner, D.S., Dyal, P.L. and Foster, P. (2003a) Mitochondria and hydrogenosomes are two forms of the same fundamental organelle. *Philos. Trans. R. Soc. Lond. B Biol. Sci.* 358: 191–201.

Embley, T.M., van der Giezen, M., Horner, D.S., Dyal, P.L. Samantha, B. and Foster, P. (2003b) Hydrogenosomes, mitochondria and early eukaryote evolution. *IUBMB Life* 55: 387–395.

Falah, M. and Gupta R.S. (1994) Cloning of the hsp70 (dnaK) genes from *Rhizobium meliloti* and *Pseudomonas cepacia*: phylogenetic analyses of mitochondrial origin based on a highly conserved protein sequence. *J. Bacteriol.* 176: 7748–7753.

Fenchel, T. (2003) *The Origin and Early Evolution of Life,* Oxford University Press, Oxford.

Foster, P.G. (2003) P4, http://www.nhm.ac.uk/zoology/external/foster/p4man/index.html.

Foster, P.G. (2004). Modeling compositional heterogeneity. *Syst. Biol.* In press.

Galtier, N. and Gouy, M. (1998) Inferring pattern and process: maximum-likelihood implementation of a nonhomogeneous model of DNA sequence evolution for phylogenetic analysis. *Mol. Biol. Evol.* 15: 871–879.

Germot, A., Philippe, H. and Le Guyader, H. (1996) Presence of a mitochondrial-type 70-kDa heat shock protein in *Trichomonas vaginalis* suggests a very early mitochondrial endosymbiosis in eukaryotes. *Proc. Natl. Acad. Sci. USA* 93: 14614–14617.

Germot, A., Philippe, H. and Le Guyader, H. (1997) Evidence for loss of mitochondria in Microsporidia from a mitochondrial-type HSP70 in *Nosema locustae. Mol. Biochem. Parasitol.* 87: 159–168.

Gogarten, J.P., Kibak, H., Dittrich, P., Taiz, L., Bowman, E.J., Bowman, B.J., Manolson, M.F., Poole, R.J., Date, T. Oshima, T. et al. (1989) Evolution of the vacuolar H+-ATPase: Implications for the origin of eukaryotes. *Proc. Natl. Acad. Sci. USA* 86: 6661–6665.

Goldman, N. (1993) Statistical tests of models of DNA substitution. *J. Mol. Evol.* 36: 182–198.

Gupta, R.S. (1998) Protein phylogenies and signature sequences: a reappraisal of evolutionary relationships among archaebacteria, eubacteria, and eukaryotes. *Microbiol. Mol. Biol. Rev.* 62: 1435–1491.

Hartman, H. and Fedorov, A. (2002) The origin of the eukaryotic cell: a genomic investigation. *Proc. Natl. Acad. Sci. USA* 99: 1420–1425.

Hashimoto, T. and Hasegawa, M. (1996) Origin and early evolution of eukaryotes inferred from the amino acid sequences of translation elongation factors 1alpha/Tu and 2/G. *Adv. Biophys.* 32: 73–120.

Hedges, S.B. (2002) The origin and evolution of model organisms. *Nat. Rev. Genet.* 3: 838-849.

Hirt, R.P., Healy, B., Vossbrinck, C.R., Canning, E.U. and Embley T.M. (1997) A mitochondrial Hsp70 orthologue in *Vairimorpha necatrix*: molecular evidence that microsporidia once contained mitochondria. *Curr. Biol.* 7: 995–958.

Hirt, R.P., Logsdon, J.M. Jr., Healy, B., Dorey, M.W., Doolittle, W.F. and Embley, T.M. (1999) Microsporidia are related to Fungi: evidence from the largest subunit of RNA polymerase II and other proteins. *Proc. Natl. Acad. Sci. USA* 96: 580–585.

Holder, M. and Lewis, P.O. (2003) Phylogeny estimation: Traditional and Bayesian approaches. *Nat. Rev. Genet.* 4: 275–284.

Horner, D.S. and Embley, T.M. (2001) Chaperonin 60 phylogeny provides further evidence for secondary loss of mitochondria among putative early-branching eukaryotes. *Mol. Biol. Evol.* 18: 1970–1975.

Horner, D.S., Foster, P.G. and Embley, T.M. (2000) Iron hydrogenases and the evolution of anaerobic eukaryotes. *Mol. Biol. Evol.* 17: 1695–1709.

Horner, D.S., Heil, B., Happe, T. and Embley, T.M. (2002) Iron hydrogenases: Ancient enzymes in modern eukaryotes.*Tr. Biochem. Sci.* 27: 148–153.

Horner, D.S., Hirt, R.P. and Embley, T.M. (1998) Phylogeny of *Trichomonas* and *Naegleria*: A comparative analysis of molecular sequence data. In *Evolutionary Relationships among Protozoa* (Coombs, G.H., Vickerman, K., Sleigh, M.A. and Warren, A.) Kluwer Academic Publishers, Dordrecht, pp.133–148.

Horner, D.S., Hirt, R.P. and Embley, T.M. (1999) A single eubacterial origin of eukaryotic pyruvate:ferredoxin oxidoreductase genes: Implications for the evolution of anaerobic eukaryotes. *Mol. Biol. Evol.* 16: 1280–1291.

Horner, D.S., Hirt, R.P., Kilvington, S., Lloyd, D. and Embley, T.M. (1996) Molecular data suggest an early acquisition of the mitochondrion endosymbiont. *Proc. R. Soc. Lond. B Biol. Sci.* 263: 1053–1059.

Horner, D.S. and Pesole, G. (2003) The estimation of relative site variability among aligned homologous protein sequences. *Bioinformatics* 19: 600–606.

Huelsenbeck, J.P. and Ronquist, F. (2001) MRBAYES: Bayesian inference of phylogenetic trees. *Bioinformatics* 17: 754–755.

Huelsenbeck, J.P., Ronquist, F., Nielsen, R. and Bollback, J.P. (2001) Bayesian inference of phylogeny and its impact on evolutionary biology. *Science* 294: 2310–2314.

Iwabe, N., Kuma, K., Hasegawa, M., Osawa, S. and Miyata, T. (1989) Evolutionary relationship of archaebacteria, eubacteria, and eukaryotes inferred from phylogenetic trees of duplicated genes. *Proc. Natl. Acad. Sci. USA* 86: 9355–9359.

Jablonka, E. and Lamb, M. (1995) *Epigenetic Inheritance and Evolution*, Oxford University Press, Oxford.

John, P. and Whatley, F.R. (1975) *Paracoccus denitrificans* and the evolutionary origin of the mitochondrion. *Nature* 254: 495–498.

Katinka, M. D., Duprat, S., Cornillot, E., Méténier, G., Thomarat, F., Prensier, G., Barbe, V., Peyretaillade, E., Brottier, P., Wincker, P. et al. (2001) Genome sequence and gene compaction of the eukaryote parasite *Encephalitozoon cuniculi*. *Nature* 414: 450–453.

Katz, L.A. (2002) Lateral gene transfers and the evolution of eukaryotes: Theories and data. *Int. J. Syst. Evol. Microbiol.* 52: 1893–1900.

Keeling, P.J. (2003) Congruent evidence from alpha-tubulin and beta-tubulin gene phylogenies for a zygomycete origin of microsporidia. *Fung. Genet. Biol.* 38: 298–309.

Keeling, P.J. and Doolittle, W.F. (1996) Alpha-tubulin from early-diverging eukaryotic lineages and the evolution of the tubulin family. *Mol. Biol. Evol.* 13: 1297–1305.

Kirschner, M., Gerhart, J. and Mitchison, T. (2000) Molecular "vitalism." *Cell* 100: 79-88.

Koonin, E.V. and Galperin, M.Y. (2003) *Sequence-Evolution-Function: Computational Approaches in Comparative Genomics*, Kluwer Academic, Boston

Kuan, J. and Saier, M.H. Jr. (1993) The mitochondrial carrier family of transport proteins: structural, functional, and evolutionary relationships. *Crit. Rev. Biochem. Mol. Biol.* 28: 209–233.

Lang, B.F., Gray, M.W. and Burger, G. (1999) Mitochondrial genome evolution and the origin of eukaryotes. *Annu. Rev. Genet.* 33: 351–397.

Lang, B.F., O'Kelly, C., Nerad, T., Gray, M.W. and Burger, G. (2002) The closest unicellular relatives of animals. *Curr. Biol.* 12: 1773–1778.

Lockhart, P.J., Penny, D., Hendy, M.D., Howe, C.J., Beanland, T.J. and Larkum, A.W. (1992) Controversy on chloroplast origins. *FEBS Lett.* 301: 127–131.

López-García, P. and Moreira, D. (1999) Metabolic symbiosis at the origin of eukaryotes. *Tr. Biochem. Sci.* 24: 88–93.

Mai, Z., Ghosh, S., Frisardi, M., Rosenthal, B., Rogers, R. and Samuelson, J. (1999) Hsp60 is targeted to a cryptic mitochondrion-derived organelle ("crypton") in the microaerophilic protozoan parasite *Entamoeba histolytica*. *Mol. Cell Biol.* 19: 2198-2205.

Margulis, L. (1970) *Origin of Eukaryotic Cells*, Yale University Press, New Haven, CT.

Margulis, L. (1993) *Symbiosis in Cell Evolution*, 2nd ed., Freeman, San Francisco.

Margulis, L., Dolan, M.F. and Guerrero, R. (2000) The chimeric eukaryote: origin of the nucleus from the karyomastigont in amitochondriate protists. *Proc. Natl. Acad. Sci. USA* 97: 6954–6959.

Martin, M.O. (2002) Predatory prokaryotes: an emerging research opportunity. *J. Mol. Microbiol. Biotechnol.* 4: 467–477.

Martin, W. (1999). A briefly argued case that mitochondria and plastids are descendants of endosymbionts, but that the nuclear compartment is not. *Proc. R. Soc. Lond. B Biol. Sci.* 266: 1387–1395.

Martin, W. and Müller, M. (1998) The hydrogen hypothesis for the first eukaryote. *Nature* 392: 37–41.

Martin, W. Hoffmeister, M. Rotte, C. and Henze, K. (2001) An overview of endosymbiotic models for the origins of eukaryotes, their ATP-producing organelles (mitochondria and hydrogenosomes), and their heterotrophic lifestyle. *Biol. Chem.* 382: 1521–1539.

Martin, W., Rujan T., Richly E., Hansen A., Cornelsen S., Lins T., Leister D., Stoebe B., Hasegawa M. and Penny D. (2002) Evolutionary analysis of *Arabidopsis*, cyanobacterial, and chloroplast genomes reveals plastid phylogeny and thousands of cyanobacterial genes in the nucleus. *Proc. Natl. Acad. Sci. USA* 99:12246–12251.

Martin, W., Rotte, C., Hoffmeister, M., Theissen, U., Gelius-Dietrich, G., Ahr, S. and Henze, K. (2003) Early cell evolution, eukaryotes, anoxia, sulfide, oxygen, fungi first (?), and a tree of genomes revisited. *IUBMB Life* 55: 193–204.

McArthur, A.G., Morrison, H.G., Nixon, J.E., Passamaneck, N.Q., Kim, U., Hinkle, G., Crocker, M.K., Holder, M.E., Farr, R., Reich, C.I., Olsen, G.E., Aley, S.B., Adam, R.D., Gillin, F.D. and Sogin, M.L. (2000) The *Giardia* genome project database. *FEMS Microbiol. Lett.* 189: 271–273.

Mereschkowsky, C. (1905). Über Natur und Ursprung der Chromatophoren. im Pflanzenreiche. *Biol. Centralbl.* 25: 593-604. [English translation in Martin, W. and Kowallik, K.V. (1999). Annotated English translation of Mereschkowsky's 1905 paper 'Über Natur und Ursprung der Chromatophoren im Pflanzenreiche.' *Eur. J. Phycol.* 34: 287–295].

Moreira, D. and López-García, P. (1998) Symbiosis between methanogenic archaea and delta-proteobacteria as the origin of eukaryotes: The syntrophic hypothesis. *J. Mol. Evol.* 47: 517–530.

Moreira, D. and López-García, P. (2002) The molecular ecology of microbial eukaryotes unveils a hidden world. *Tr. Microbiol.* 10: 31-38.

Mühlenhoff, U. and Lill, R. (2000) Biogenesis of iron-sulfur proteins in eukaryotes: A novel task of mitochondria that is inherited from bacteria. *Biochem. Biophys. Acta* 1459: 370–382.

Müller, M. (1998) Enzymes and compartments of core energy metabolism of anaerobic protests — a special case of in eukaryote evolution? In *Evolutionary Relationships among Protozoa* (Coombs, G.H., Vickerman, K., Sleigh, M.A. and Warren, A., Eds.) Kluwer Academic, Dordrecht, pp.109–132.

O'Brien, E.A., Badidi, E., Barbasiewicz, A., deSousa, C., Lang, B.F. and Burger, G. (2003) GOBASE: a database of mitochondrial and chloroplast information. *Nucl. Acids Res.* 31: 176-178.

Patterson, D.J. (1994) Protozoa: evolution and systematics. In *Progress in Protozoology* (Hausmann, K. and Hülsmann, N., Eds.), Gustav Fisher Verlag, Stuttgart, pp. 1–14.

Patterson, D.J. and Sogin, M.L. (1993) Eukaryote origins and protistan diversity. In *The Origin and Evolution of the Cell* (Hartman, H. and Matsuno, K., Eds.), World Scientific, Singapore, pp. 13–46.

Penny, D., McComish, B.J., Charleston, M.A. and Hendy, M.D. (2001) Mathematical elegance with biochemical realism: the covarion model of molecular evolution. *J. Mol. Evol.* 53: 711–723.

Peyretaillade, E., Broussolle, V., Peyret, P., Méténier, G., Gouy, M. and Vivarès, C.P. (1998) Microsporidia, amitochondrial protists, possess a 70-kDa heat shock protein gene of mitochondrial evolutionary origin. *Mol. Biol. Evol.* 15: 683–689.

Philippe, H. and Adoutte, A. (1998) The molecular phylogeny of eukaryota: solid facts and uncertainties. In *Evolutionary Relationships among Protozoa* (Coombs, G.H., Vickerman, K., Sleigh, M.A. and Warren, A., Eds.), Kluwer Academic Publishers, Dordrecht, pp. 133–148.

Philippe, H. and Forterre, P. (1999) The rooting of the universal tree of life is not reliable. *J. Mol. Evol.* 49: 509–523

Philippe, H., Germot, A. and Moreira, D. (2000a) The new phylogeny of eukaryotes. *Curr. Opin. Genet. Dev.* 10: 596–601.

Philippe, H., Lopez, P., Brinkmann, H., Budin, K., Germot, A., Laurent, J., Moreira, D., Müller, M. and Le Guyader, H. (2000b) Early-branching of fast evolving eukaryotes? An answer based on slowly evolving positions. *Proc. R. Soc. Lond. B* 267: 1213–1221.

Posada, D. and Crandall, K.A. (2001) Selecting the best-fit model of nucleotide substitution. *Syst. Biol.* 50: 580–601.

Richards, T.A., Hirt, R.P., Williams, B.A. and Embley, T.M. (2003) Horizontal gene transfer and the evolution of parasitic protozoa. *Protist* 154: 17–32.

Rivera, M.C. and Lake, J.A. (1992) Evidence that eukaryotes and eocyte prokaryotes are immediate relatives. *Science* 257: 74–76.

Roger, A.J. (1999) Reconstructing early events in eukaryotic evolution. *Am. Nat.* 154(suppl.): S146-S163.

Roger, A.J., Clark, C.G. and Doolittle, W.F. (1996) A possible mitochondrial gene in the early-branching amitochondriate protist *Trichomonas vaginalis*. *Proc. Natl. Acad. Sci. USA* 93: 14618–14622.

Roger, A.J., Svärd, S.G., Tovar, J., Clark, C.G., Smith, M.W., Gillin, F.D. and Sogin, M.L. (1998) A mitochondrial-like chaperonin 60 gene in *Giardia lamblia*: evidence that diplomonads once harbored an endosymbiont related to the progenitor of mitochondria. *Proc. Natl. Acad. Sci. USA* 95: 229–234.

Rotte, C., Stejskal, F., Zhu, G., Keithly, J.S. and Martin, W. (2001) Pyruvate : NADP+ oxidoreductase from the mitochondrion of *Euglena gracilis* and from the apicomplexan *Cryptosporidium parvum*: a biochemical relic linking pyruvate metabolism in mitochondriate and amitochondriate protists. *Mol. Biol. Evol.* 18: 710–720

Sagan, L. (1967) On the origin of mitosing cells. *J. Theor. Biol.* 14: 255–274.

Shimodaira, H. (2002) An approximately unbiased test of phylogenetic tree selection. *Syst. Biol.* 51: 492–508.

Shimodaira, H. and Hasegawa, M. (2001) CONSEL: for assessing the confidence of phylogenetic tree selection. *Bioinformatics* 17: 1246–1247.

Simpson, A.G. and Roger, A.J. (2002) Eukaryotic evolution: getting to the root of the problem. *Curr Biol.* 12: R691–R693.

Smith, A.B. (2003) Rooting molecular trees. *Biol. J. Linn. Soc.* 24: 153–188.

Sogin, M.L. (1991) Early evolution and the origin of eukaryotes. *Curr. Opin. Genet. Dev.* 1: 457–463.

Sogin, M.L. (1993) Comment on genome sequencing. In *The Origin and Evolution of the Cell* (Hartman, H. and Matsuno, K., Eds.), World Scientific, Singapore, pp. 387–389.

Sogin, M.L., Silberman, J.D., Hinkle, G. and Morrison, H.G. (1996) Problems with molecular diversity in the eukarya. In *Evolution of Microbial Life* (Roberts, D. McL., Sharp, P., Alderson, G. and Collins, M., Eds.), Cambridge University Press, Cambridge, pp.167–184.

Stechmann, A. and Cavalier-Smith, T. (2002) Rooting the eukaryote tree by using a derived gene fusion. *Science* 297: 89–91.

Stechmann, A. and Cavalier-Smith, T. (2003) The root of the eukaryote tree pinpointed. *Curr. Biol.* 13: R665–R664.

Stoeck, T. and Epstein, S. (2003) Novel eukaryotic lineages inferred from small-subunit rRNA analyses of oxygen-depleted marine environments. *Appl. Environ. Microbiol.* 69: 26572663.

Tielens, A.G., Rotte, C., van Hellemond, J.J. and Martin, W. (2002) Mitochondria as we don't know them. *Tr. Biochem. Sci.* 27: 564–572.

Tovar, J., Fischerm A. and Clark, C.G. (1999) The mitosome, a novel organelle related to mitochondria in the amitochondrial parasite *Entamoeba histolytica*. *Mol. Microbiol.* 32: 1013–1021.

Tovar, J., León-Avila, G., Sánchez, L., Sutak, R., Tachezy, J., van der Giezen, M., Hernández, M., Müller, M. and Lucocq, J.M. (2003) Mitochondrial remnant organelles of *Giardia* function in iron-sulfur protein maturation. *Nature* 426: 172–176.

van den Ent, F., Amos, L. and Lowe, J. (2001) Bacterial ancestry of actin and tubulin. *Curr. Opin. Microbiol.* 4: 634–638.

Van de Peer, Y., Ben Ali, A. and Meyer, A. (2000a) Microsporidia: accumulating molecular evidence that a group of amitochondriate and suspectedly primitive eukaryotes are just curious fungi. *Gene* 246: 1–8.

Van de Peer, Y., Baldauf, S.L., Doolittle, W.F. and Meyer, A. (2000b) An updated and comprehensive rRNA phylogeny of (crown) eukaryotes based on rate-calibrated evolutionary distances. *J. Mol. Evol.* 51: 565-576.

Vellai, T. and Vida, G. (1999) The origin of eukaryotes: the difference between prokaryotic and eukaryotic cells. *Proc. R. Soc. Lond. B Biol. Sci.* 266: 1571–1577.

Viale, A.M. and Arakaki, A.K. (1994) The chaperone connection to the origins of the eukaryotic organelles. *FEBS Lett.* 341: 146–151.

Whelan, S., Lio, P. and Goldman, N. (2001) Molecular phylogenetics: state-of-the-art methods for looking into the past. *Tr. Genet.* 17: 262–272.

Williams, B.A., Hirt, R.P., Lucocq, J.M. and Embley, T.M. (2002) A mitochondrial remnant in the microsporidian *Trachipleistophora hominis*. *Nature* 418: 865-869.

Woese, C.R. (1987) Bacterial evolution. *Microbiol. Rev.* 51: 221–271.

Woese, C.R., Kandler, O. and Wheelis, M.L. (1990) Towards a natural system of organisms: proposal for the Domains Archaea, Bacteria, and Eucarya. *Proc. Natl. Acad. Sci. USA* 87: 4576–4579.

Yang, Z. (1996) Phylogenetic analysis using parsimony and likelihood methods. *J. Mol. Evol.* 42: 294–307.

Yang, Z. and Roberts, D.McL. (1995) On the use of nucleic acid sequences to infer early branchings in the tree of life. *Mol. Biol. Evol.* 12: 451–458.

Section I

Eukaryote Diversity and Phylogeny

Chapter 2

Excavata and the Origin of Amitochondriate Eukaryotes

Alastair G.B. Simpson and Andrew J. Roger

CONTENTS

Abstract

Ever since the distinction between prokaryotic and eukaryotic cells became clear (Stanier, 1970) there has been great interest in understanding the early history of the eukaryotic cell and the nature of the first eukaryotes. For much of the past three decades, attention has been focused on mitochondria as a key (and comparatively late) acquisition by the eukaryotic

0-415-29904-7/04/$0.00+$1.50

lineage, and on mitochondrion-lacking groups as potentially primitive or deep-branching eukaryotes. This paradigm reached its zenith in the early 1990s with the small subunit ribosomal rRNA (SSUrRNA) supported version of the archezoa hypothesis. Over the past decade, better estimates of the relationships among eukaryotes and independent elucidation of the probable mitochondrial ancestry of many mitochondrion-lacking groups have forced a reevaluation of the history of mitochondrial mode in eukaryotes. The classic mitochondrion-lacking lineages originally considered to be archezoa probably fall within three major clades: opisthokonts, Amoebozoa and Excavata. Excavata includes diplomonads, retortamonads, oxymonads and parabasalids from the original archezoa, as well as two new amitochondriate taxa *Carpediemonas* and *Trimastix*, and several mitochondrion-bearing groups, namely jakobids, *Malawimonas*, Heterolobosea and Euglenozoa. Most excavates share a distinctive feeding groove, which identifies the group. Five groups — retortamonads, *Carpediemonas*, *Trimastix*, jakobids and *Malawimonas* — share several discrete cytoskeletal characters unique to excavates, strongly suggesting that they have descended from a similar common ancestor. Molecular phylogenies (SSUrRNA or multiple protein analyses, or both) indicate specific relationships between each of diplomonads, oxymonads, parabasalids, Heterolobosea and Euglenozoa, and at least one of the first five taxa. However, the same molecular phylogenies generally do not support the monophyly of Excavata as a whole. There are three possible scenarios: (1) Excavata is a clade and available molecular phylogenies are incorrect, (2) the excavate morphology has arisen convergently several times or (3) the excavates are actually the plesiomorphic stem group for most or all living eukaryotes. We favor the first of these scenarios, noting that many excavates have unusually divergent gene sequences that might be expected to foster artifacts in molecular phylogenetic analyses. Regardless, the complete resolution of the relationships among various Excavata should provide the phylogenetic basis for resolving many outstanding questions over the evolution of the amitochondriate state and perhaps the nature of early eukaryotic cells.

2.1 The Search for Primitive Eukaryotes

Before the widespread acceptance of the endosymbiotic origin of plastids, the distribution of oxygenic photosynthesis in both prokaryotes and eukaryotes seemed to require eukaryotes to have directly descended from cyanobacteria, with the first eukaryotes also being phototrophs (Dougherty and Allen, 1960). Under this paradigm, rhodophytes (red algae) were usually posited as the earliest diverging living eukaryotes, because their photosynthetic machinery appeared most similar to that of cyanobacteria, with single thylakoid membranes and phycobilisomes (Allsopp, 1969; Cavalier-Smith, 1975; Dougherty and Allen, 1960; Klein and Cronquist, 1967). Rhodophytes also lack eukaryotic flagella and centrioles, easily rationalized as primitive absences. Alternatively, dinoflagellates were regarded as more primitive by some workers. Most dinoflagellates have nuclei that apparently lack histones, and this was sometimes interpreted as an organization intermediate between prokaryotes and true eukaryotes (Dodge, 1965; Herzog et al., 1984; Loeblich, 1976).

Acceptance of the endosymbiotic origins of plastids had profound implications for understanding the early history of eukaryotes. As plastids had been acquired through endocytobiosis rather than by direct descent, there was no need to assume that the earliest eukaryotes were photosynthetic (Margulis, 1970; Sagan, 1967). From the late 1960s, morphologically simple heterotrophic organisms began to be seriously proposed as the earliest eukaryotes. Ascomycete fungi, which lack flagella, Golgi dictyosomes and endocytosis, were championed by some (Cavalier-Smith, 1980, 1981). Others, however, focused on the giant amoeboid cell *Pelomyxa*. In addition to apparently lacking eukaryotic flagella (since disproved) and Golgi dictyosomes, *Pelomyxa* seemed to lack mitochondria (Griffin, 1988), whereas some accounts suggested that the nucleus might divide by budding rather than mitosis (Daniels and Breyer, 1967). All these traits (with the frequent exception of the last, which was widely disbelieved) were viewed as possible primitive states (Bovee and Jahn, 1973; Margulis, 1970; Whatley, 1976; Whatley et al., 1979).

2.2 The Archezoa Hypothesis

Pelomyxa is only one of several well-known groups of eukaryotes that uniformly lack readily distinguishable mitochondria. Others include the other pelobionts (e.g., *Mastigamoeba*), the amoeba *Entamoeba* and the spore-forming microsporidian parasites. Also, four well-known mitochondrion-lacking groups that are predominantly flagellates: diplomonads (e.g., *Giardia, Spironucleus, Hexamita, Trepomonas*), retortamonads (e.g., *Retortamonas, Chilomastix*), oxymonads (e.g., *Pyrsonympha, Monocercomonoides*) and parabasalids (e.g., *Trichomonas*). Most members of the four groups are animal parasites or commensals, e.g., the human parasites *Giardia lamblia* and *Trichomonas vaginalis* (Kulda and Nohynková, 1978), but some are free living (Bernard et al., 2000; Brugerolle and Müller, 2000). Cavalier-Smith (1983, 1987) formulated the explicit hypothesis that these seven or so groups diverged before the acquisition of the mitochondrial symbiont. He united these groups as a taxon Archezoa, and we refer to Cavalier-Smith's proposal as the *archezoa hypothesis* (see Roger, 1999). To avoid confusion, it should be noted that Cavalier-Smith has lately used the term *Archezoa* to refer to particular amitochondriate groups that are explicitly proposed to derive from mitochondriate ancestors (Cavalier-Smith, 1999, 2002). The terms *archezoa* and *archezoa hypothesis* referred to hereafter do *not* refer to this latter conception of a taxon Archezoa.

2.3 Candidate Archezoa as Deep Branches

The archezoa hypothesis coincided with the emergence of universal molecular phylogenies based on SSUrRNA sequences. During the 1980s, the database of full-length SSUrRNA sequences expanded to include an ever-wider diversity of protists. Phylogenies based on simple models of sequence evolution or parsimony suggested a ladder-like evolutionary tree of eukaryotes. Some unusual but mitochondrion-bearing protists, such as Euglenozoa, Heterolobosea and *Dictyostelium*, formed a succession of branches diverging before a crown-like radiation of animals, plants, fungi and some other protists (Sogin, 1989; Sogin et al., 1986).

In the late 1980s, SSUrRNA sequences for some of the groups argued to be archezoa were analyzed for the first time. Strikingly, first microsporidia (Vossbrinck et al., 1987), then diplomonads and parabasalids (Sogin, 1989; Sogin et al., 1989), formed the earliest branches in the ladder-like tree, diverging before any mitochondrion-bearing taxa. The basal positions for diplomonads, parabasalids and microsporidia were supported by subsequent analyses incorporating other SSUrRNA sequences from these groups (e.g., Gunderson et al., 1995; Leipe et al., 1993). Further support was lent by phylogenetic analyses of some other genes. In particular, trees of elongation factor proteins (EF-1α and EF-2) also placed diplomonads, and, eventually, parabasalids and microsporidia at the base of eukaryotes (Hashimoto et al., 1994, 1995; Kamaishi et al., 1996; Yamamoto et al., 1997).

Not all archezoa survived these early tests. Early SSUrRNA and elongation factor analyses gave contradictory estimates of whether *Entamoeba* was more deeply diverging than Euglenozoa (Hasegawa and Hashimoto, 1993; Hasegawa et al., 1993; Sogin, 1989). A single relatively late-branching SSUrRNA sequence seemed to disqualify pelobionts from being archezoa (Cavalier-Smith, 1995, 1997; Hinkle et al., 1994). For a long time, no molecular data were available for oxymonads and retortamonads. Nonetheless, these molecular phylogenies, especially those of SSUrRNA, appeared to confirm the general archezoa hypothesis, crystallized around diplomonads, parabasalids and microsporidia as the living representatives from the earliest stage of eukaryotic evolution (Patterson and Sogin, 1992). Accessible organisms from within these three groups, in particular the diplomonad *Giardia*, were (and still are) looked to as model systems for the molecular and cellular biology of early eukaryotes (Gillin et al., 1996; Hehl et al., 2003; Upcroft and Upcroft, 1998).

2.4 Challenges to the Archezoa Hypothesis

2.4.1 Genes of Mitochondrial Origin

Although the eukaryote mitochondrion is derived from a prokaryote and retains a reduced genome, many of the genes originating from the ancestral endosymbiont have relocated to the nucleus (Gray et al., 1999). If a eukaryote has secondarily lost or modified its mitochondria, the prior presence of the organelles can be inferred by detecting genes of mitochondrial origin retained by the nucleus (Clark and Roger, 1995). One or more of several genes of probable mitochondrial origin, including mitochondrial isoform chaperonin 60 (cpn60), mitochondrial isoform heat-shock protein 70 (hsp70), and pyridoxal-5′-phosphate-dependent cysteine desulfurase (IscS), have now been found in *Entamoeba*, parabasalids, diplomonads and microsporidia (Bui et al., 1996; Germot et al., 1996, 1997; Hirt et al., 1997; Horner and Embley, 2001; Horner et al., 1996; Peyretaillade et al., 1998; Roger et al., 1996, 1998; Tachezy et al., 2001). This suggests the prior presence of a mitochondrion in all these taxa, independently and irrespective of evidence from organismal phylogeny. One or more of these proteins has now been localized to double-membrane-bounded organelles (the presumed physical relics of the mitochondrion) in all these taxa (see Bui et al., 1996; Mai et al., 1999; Tovar et al., 1999, 2003; Williams et al., 2002). In parabasalids, these organelles turn out to be the hydrogenosomes, hydrogen-evolving, energy-generating organelles whose origins were previously unclear (Bui et al., 1996; Cavalier-Smith, 1993; Müller, 1988).

2.4.2 Are the Deep Branches Truly Deep?

The first major challenge to the SSUrRNA-based tree of eukaryotes involved the position of microsporidia. The first α- and β-tubulin phylogenies including microsporidia placed them with fungi (Edlind et al., 1996; Keeling and Doolittle, 1996; Li et al., 1996), a position that was highly inconsistent with classical SSUrRNA trees, in which fungi were a late-branching group (Sogin, 1991). Gradually, other gene trees emerged that also supported a fungal affinity for microsporidia (Fast et al., 1999; Germot et al., 1997; Hirt et al., 1999; Keeling et al., 2000; van de Peer et al., 2000; Weiss et al., 1999). At present, it is widely accepted that the basal placements of microsporidia in typical rRNA and elongation factor trees are analysis artifacts caused by extremely aberrant sequence evolution by these genes in this lineage (Hirt et al., 1999; van de Peer et al., 2000).

Recently, the phylogenetic positions of *Entamoeba* and pelobionts has also been substantially clarified. SSUrRNA analyses with improved taxon sampling and evolutionary models and fructose-1,6-bisphosphate aldolase phylogenies indicate that entamoebae and pelobionts are closely related to each other (Edgcomb et al., 2002; Sánchez et al., 2002; Silberman et al., 1999). Concatenated protein analyses strongly recover this relationship, and, furthermore, demonstrate that these two amitochondriate groups form a clade with the mitochondrion-bearing mycetozoan slime mold *Dictyostelium* (Arisue et al., 2002; Bapteste et al., 2002). Pelobionts and entamoebae almost certainly fall within a large, otherwise mitochondrion-bearing clade of amoeboid organisms referred to as Amoebozoa and are not closely related to either microsporidia or other amitochondriate taxa.

The discovery that microsporidia were incorrectly placed hinted that the deep positions of diplomonads and parabasalids might also be artifactual. Philippe and coworkers (Philippe and Adoutte, 1998; Philippe and Laurent, 1998) argued convincingly that many genes used for phylogenetic analysis of eukaryotes, including SSUrRNA, are mutationally saturated. It could be difficult to correctly recover deep-level phylogenies with single genes, because artifacts caused by aberrant sequence evolution (e.g. long branch attraction) would swamp the little-remaining historical signal. Detailed analyses suggest that the deepest-branching eukaryotic sequences in many gene trees, including those from diplomonads and parabasalids, exhibit very high rates of evolution, and thus that their basal placement probably reflects analysis artifact (Dacks et al., 2002; Germot and Philippe, 1999; Hirt et al., 1999; Philippe et al., 2000; Roger et al., 1999b; Stiller and Hall, 1999; Stiller et al., 1998). However, unlike the case of microsporidia, this possible artifact has not been convincingly overcome by a consistently supported and plausible alternative position for diplomonads and parabasalids.

2.5 Excavata — A Home for Many Reformed Archezoa?

Molecular phylogenetic examinations have been unable to provide convincing close relatives for diplomonads and parabasalids from among the well-studied eukaryotes. Until recently, the positions of the retortamonads and oxymonads were also unclear (Brugerolle and Müller, 2000). The search for possible relatives of these organisms, both mitochondriate and amitochondriate, has involved the characterization of several novel protists and more detailed investigations of morphological data. Considering both molecular phylogenies and cytoskeletal morphology, a reasonable case can now be made that diplomonads, retortamonads, parabasalids and oxymonads (Figure 2.1a–d and Figure 2.2 a,b) belong within a novel eukaryotic supergroup called Excavata.

During the reign of the archezoa hypothesis, some workers (Cavalier-Smith, 1992a; Sleigh, 1989) had noted a broad commonality in the cytoskeletal organizations of retortamonads and some unusual diplomonads (Figure 2.1a,b and Figure 2.2a,b), and flagellate forms of the mitochondrion-bearing taxon Heterolobosea (Figure 2.1e). O'Kelly (1993) subsequently outlined more detailed similarities between these organisms and what turned out to be two distinct groups of poorly known mitochondriate flagellates, jakobids (= core jakobids *sensu* Simpson and Patterson, 1999) and *Malawimonas* (Figure 2.1g,h and Figure 2.2d). In particular, many of these organisms possess a conspicuous feeding groove on the cell surface that collects suspended particles (usually bacteria) for consumption. This groove is usually supported by a distinctive set of microtubular and nonmicrotubular structures (see later). Recently other new flagellates *Trimastix* and *Carpediemonas* have been characterized with this morphology (Figure 2.1i,j and Figure 2.2e,f). Both of these lack classical mitochondria (Brugerolle and Patterson, 1997; O'Kelly et al., 1999; Simpson and Patterson, 1999; Simpson et al., 2000).

Under the archezoa hypothesis, unusual morphological homologies between diplomonads and retortamonads and mitochondriate taxa, such as the feeding groove, were interpreted as ancestral features of the original host of the mitochondrial symbiont. This host was thought to be a diplomonad-like or retortamonad-like amitochondriate organism, with one or more of Heterolobosea, jakobids and *Malawimonas* being the most plesiomorphic of mitochondriate lineages (Cavalier-Smith, 1991, 1992a,b; O'Kelly, 1993). However, the eclipse of the archezoa hypothesis has allowed these same homologies to be considered as possible shared-derived characters of a more restricted clade that would include both the diplomonads and the retortamonads, and the mitochondriate taxa to which they are morphologically similar. The putative clade stemming from a common ancestor with the suspension feeding groove is now named Excavata (Cavalier-Smith, 2002; Simpson, 2003).

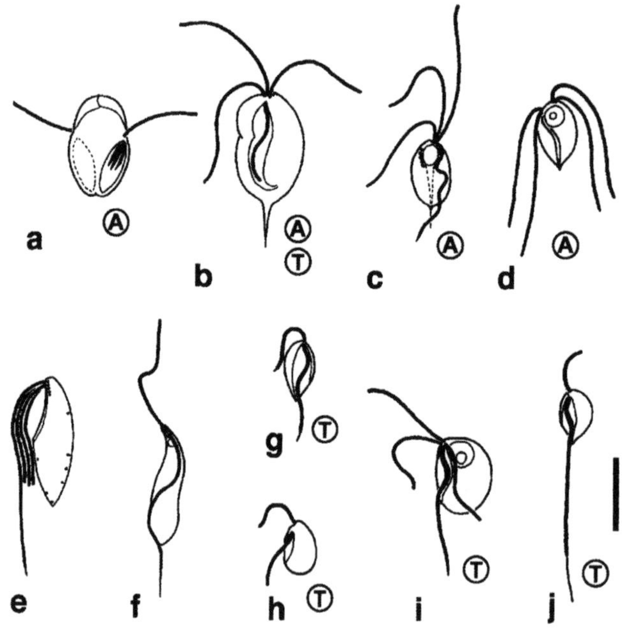

FIGURE 2.1 Light microscopical appearance of various Excavata. (a) Diplomonad; *Trepomonas agilis.* (b) Retortamonad; *Chilomastix cuspidata.* (c) Parabasalid: *Pseudotrichomonas keilini.* (d) Oxymonad; *Monocercomonoides hausmanni.* (e) Heteroloboseid; *Percolomonas descissus.* (f) Euglenozoon; *Dimastigella trypaniformis.* (g) Jakobid; *Jakoba incarcerata.* (h) *Malawimonas*; *Malawimonas jakobiformis.* (i) *Trimastix*; *Trimastix pyriformis.* (j) *Carpediemonas*; *Carpediemonas membranifera.* "A" signifies groups originally considered to be archezoa [Cavalier-Smith (1983), 1027–1034)]. "T" signifies typical excavate groups. Scale bar represents 10 μm. [(a)–(c), (e)–(g), (i),(j) adapted from Bernard et al. (2000), 113–142; (d) and (h) adapted from Simpson et al. (2002b).]

Assuming that this form of feeding groove evolved once, Excavata would include all groove-bearing taxa, plus any organisms specifically related to subsets of these by other evidence (e.g., molecular phylogenies — see later). The groups that lack a feeding groove but are suspected nonetheless to fall within Excavata are the mitochondrion-lacking oxymonads and parabasalids (Figure 2.1c,d) and the mitochondrion-bearing Euglenozoa (Figure 2.1f).

2.6 Morphological Evidence for Excavata

Five excavates (retortamonads, jakobids, *Malawimonas*, *Trimastix* and *Carpediemonas*) have closely comparable cytoskeletons, and for convenience are herein referred to as typical excavates (see Figure 2.1 and Simpson, 2003). They share at least seven distinct morphological features in addition to the feeding groove itself, none of which have been identified in nonexcavate eukaryotes (Figure 2.3 and Table 2.1). These features include several different nonmicrotubular fibers associated with the groove margins (I, B, C and composite fibers), a splitting of the R1 microtubular root into discrete inner and outer portions, an extra singlet microtubular root originating close to R1 and a distinctive type of flagellar vane (O'Kelly and Nerad, 1999; Simpson, 2003; Simpson and Patterson, 1999, 2001). These similarities strongly suggest that these five groups at least are descended from a common typical excavate ancestor. The morphological evidence is weaker for a shared ancestry with the other five groups. Diplomonads, heteroloboseids and oxymonads each include organisms

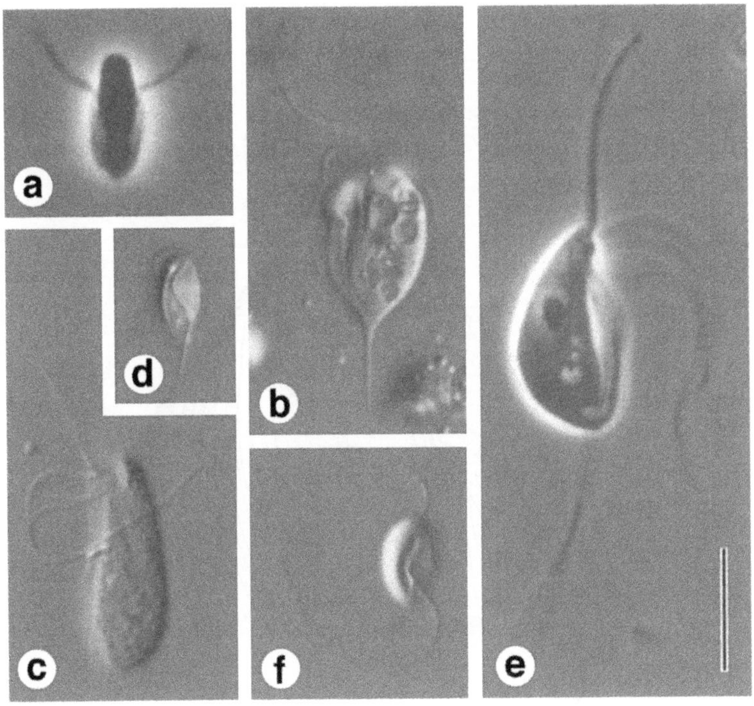

FIGURE 2.2 Light micrographs of some more poorly known excavates. (a) *Trepomonas agilis*, a free-living, groove-bearing diplomonad, shown "edge-on," with grooves appearing as lucent regions on either side of the posterior half of the cell (phase contrast). (b) *Chilomastix cuspidata*, a free-living retortamonad (differential interference contrast, DIC). (c) *Percolomonas descissus*, a groove-bearing flagellated heteroloboseid with no known amoeboid phase (DIC). (d) *Jakoba incarcerata*, a jakobid, shown in ventral view, looking directly into the groove (DIC). (e) *Trimastix marina*, lateral view (phase contrast). (f) *Carpediemonas membranifera*, ventral view (DIC). Scale bar represents 10 μm. [Adapted from Bernard, et al. (2000) 113–142.]

that share some (three or four) but not all the features (Table 2.1), even though oxymonads lack the feeding groove itself (Simpson, 2003; Simpson et al., 2002b). It is arguable whether any of the distinctive excavate features are present in parabasalids or Euglenozoa. The cases for inclusion of these latter five taxa in Excavata stem primarily from molecular data (see later).

2.7 Molecular Phylogenies and Excavata

A molecular understanding of excavate relationships has been hampered by the limited data available for many of the constituent groups, especially the typical excavates. Nonetheless, some comprehensive accounts of excavate relationships are emerging. One data set, SSUrRNA, now includes sequences from all 10 excavate groups (Simpson et al., 2002c). Good coverage of excavates is available for several nuclear-encoded protein markers (α- and β-tubulin, cytosolic heat-shock proteins 70 and 90, CCTα and EF-1α), although species sampling is often limited (e.g., Archibald et al., 2002; Keeling and Leander, 2003; Simpson et al., 2002c). There has been an emphasis on excavates in mitochondrial genome sequencing efforts (Gray et al., 1998, 2001; Lang et al., 1997); however, six of the ten excavate groups lack classical mitochondria, and therefore probably lack mitochondrial genomes. Of great

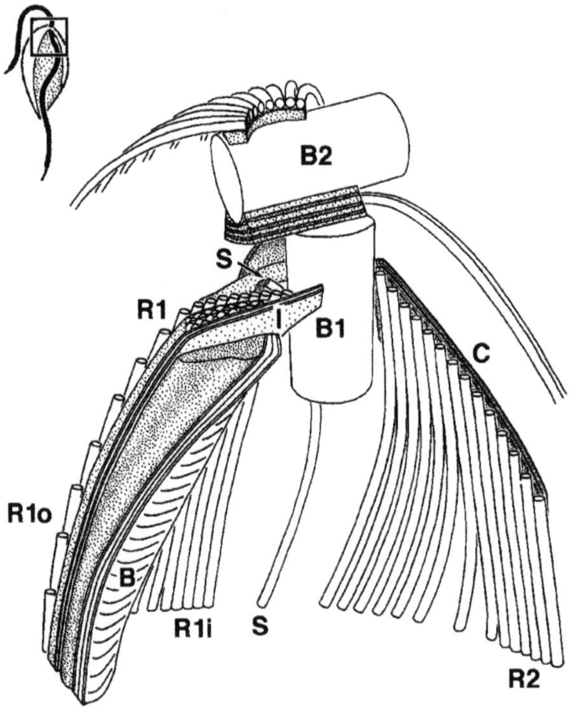

FIGURE 2.3 Diagrammatic view of the anterior cytoskeleton of the jakobid *Jakoba incarcerata.* Diagram includes structures from the region of the cell boxed in the drawing at the top left (groove shown in stippling), with the emergent portions of the flagella omitted. The basal bodies of the two flagella are labeled (B1, B2), as are the two major microtubular roots associated with B1 (R1, R2). Several of the features distinctive for excavates are also labeled (see Table 2.1): the nonmicrotubular "I" and "B" fibers associated with R1, the nonmicrotubular "C" fiber associated with R2, the singlet microtubular root (S) originating dorsal to R1 and the division of R1 into two distinct portions ($R1_o$ and $R1_i$). The flagellar vanes, composite fiber and the feeding groove are not shown. [Adapted and relabeled from Simpson and Patterson (2001), 480-492.]

value is the increasing number of excavates subjected to EST or genome sequencing, allowing their inclusion in larger-scale concatenated gene phylogenies.

In almost all molecular phylogenetic analyses performed to date, excavates do not form a single monophyletic group (but see Cavalier-Smith, 2003). In SSUrRNA phylogenies, diplomonads, parabasalids, Heterolobosea, Euglenozoa, *Carpediemonas* and retortamonads exhibit long terminal branches and fall in the basal portions of the canonical SSUrRNA tree (Figure 2.4). Other newly examined excavates (*Trimastix*, oxymonads, jakobids and *Malawimonas*) are shorter branches and generally form several (up to four) separate clades in the crown of the tree (note that unrooted eukaryotic trees from analyses that account for among-site rate variation, such as in Figure 2.4, might not divide into clear crown and base portions). Analyses of α- and β- tubulins including six to eight excavate groups place them as four to eight clades (Edgcomb et al., 2001; Keeling and Leander, 2003; Simpson et al., 2002c). CCTα phylogenies including six excavate groups distribute them into at least three weakly supported clades (Archibald et al., 2002). We also present here a preliminary analysis of four concatenated proteins (α-tubulin, β-tubulin, hsp70, hsp90) including eight excavate groups. The maximum likelihood tree for this data set again places excavates in three

TABLE 2.1 The Distribution of Eight Distinctive Morphological Features in the 10 Established Groups of Excavata

Group	Feeding Groove	I Fiber	B Fiber	C Fiber	Split Root I	Singlet Root	Flagellar vanes	Composite Fiber
Jakobids	X	X	X	X	X	X	X	X
Malawimonas	X	X	X	X	X	X	X	N.D.
Trimastix	X	X	X	X	X	X	X	X
Carpediemonas	X	X	X	X	X	X	X	X
Retortamonads	X	X	X	X	X	X	X	X
Diplomonads	X	X			X	?		
Heterolobosea	X	X			X			
Oxymonads		X	X	X		X		
Parabasalids				?				
Euglenozoa								

Note: X — presence of a feature; ? — an arguable homology; N.D. — no data available. Typical excavates, which seem to possess all eight distinctive features, are set in bold.

Source: Adapted from Simpson (2003), 1759–1777.

separated clades (Figure 2.5). Importantly, each one of these clades contains one or more of the typical excavates that are conspicuously similar in their cytoskeletal morphology.

How can morphological and molecular evidence be reconciled? There are three simple alternatives:

1. The morphological evidence indicating that excavates (especially typical excavates) descend from a recent common ancestor is reliable, and the molecular phylogenies that split them up are misleading, i.e., Excavata is a clade.
2. The molecular phylogenies are correct in not recovering an excavate clade, and the typical excavate cytoskeletal organization evolved convergently more than once in eukaryotic history, i.e., Excavata is polyphyletic.
3. The molecular phylogenies are essentially correct, but the typical excavate cytoskeleton evolved only once. This scenario implies that Excavata, as envisaged here, is paraphyletic, and is the stem group for most or all extant eukaryotes.

On current evidence we favor the first scenario and consider Excavata as a clade. Although it is difficult to place objective values on the strength of morphological evidence, we know of no case in which a cytoskeletal organization as distinct as that of typical excavates is known to have evolved convergently in multiple lineages. By contrast, the evidence from molecular phylogenies indicating that excavates are not monophyletic is problematic without even considering its incongruity with morphology. In the case of SSUrRNA data, analyses that use evolutionary models incorporating among-site rate heterogeneity generally recover little robust structure in the backbone of the tree (see Figure 2.4; Kumar and Rzhetsky, 1996; Silberman et al., 1999; Simpson et al., 2002c). The branches separating excavates are poorly supported by bootstrap analyses and sensitive to taxonomic sampling (Dacks et al., 2001; Keeling and Leander, 2003; Silberman et al., 2002; Simpson et al., 2002c). In the SSUrRNA analysis reported here, excavate monophyly is not rejected by statistical tests (see Figure 2.4 legend). In our concatenated protein analysis, the clade grouping diplomonads, *Carpediemonas* and

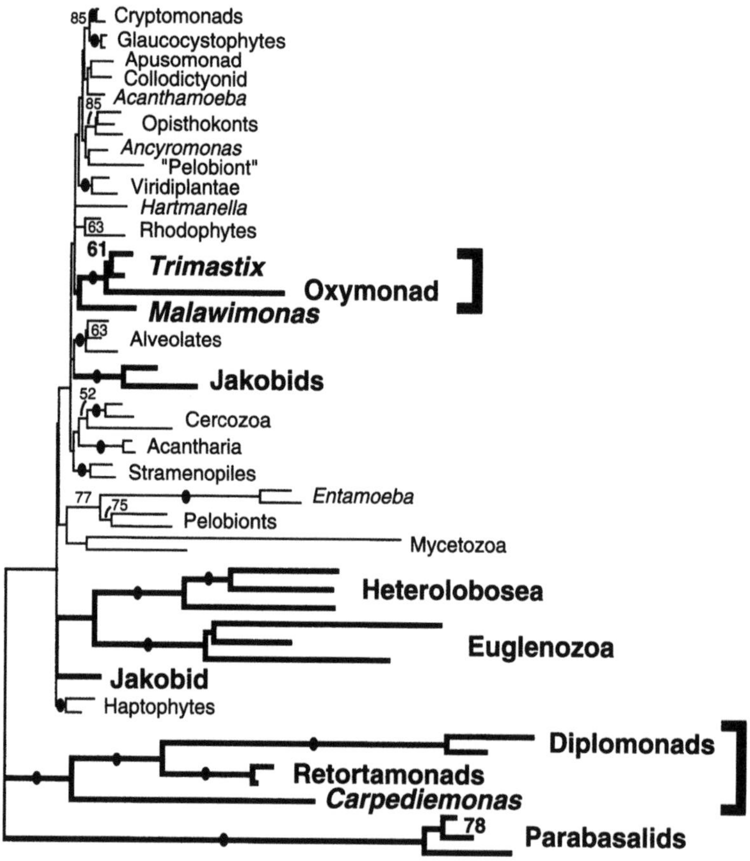

FIGURE 2.4 Maximum likelihood distances phylogeny of SSUrRNA sequences from diverse eukaryotes, including all 10 excavate groups (large bold type). The analysis used 997 sites, with the Tamura–Nei substitution model and a gamma distribution plus invariable sites model of among-site rate heterogeneity. Bootstrap support values for internal edges more than 50% are shown (2000 replicates). Filled circles denote edges with >90% support. Excavates fall in several places across the tree separated by weakly supported internal nodes. The brackets on the right delineate the two clades uniting different excavate groups that are consistently recovered. Maximum likelihood analysis of a restricted taxon set (29 taxa) gives a broadly similar tree. With this smaller data set, trees in which excavates are constrained to form a clade are not rejected by Shimodaira's approximately unbiased test ($p = 0.44$) when compared to 50 other good trees (the ML tree and the best trees from 49 bootstrap replicates).

parabasalids with animals and fungi receives strong bootstrap support (Figure 2.5). However, this support is largely contributed by only one of the four included genes, α-tubulin (AGBS, unpublished), and is inconsistent with an emerging consensus from other molecular trees that animals and fungi branch with an Amoebozoa clade including *Dictyostelium* (see Chapter 5). We also note other inconsistencies between individual molecular trees; for example, our SSUrRNA analysis provides no support for a jakobid–Heterolobosea–Euglenozoa clade (Figure 2.4), whereas this grouping of three excavates is fairly well supported by our concatenated protein data (Figure 2.5; see later).

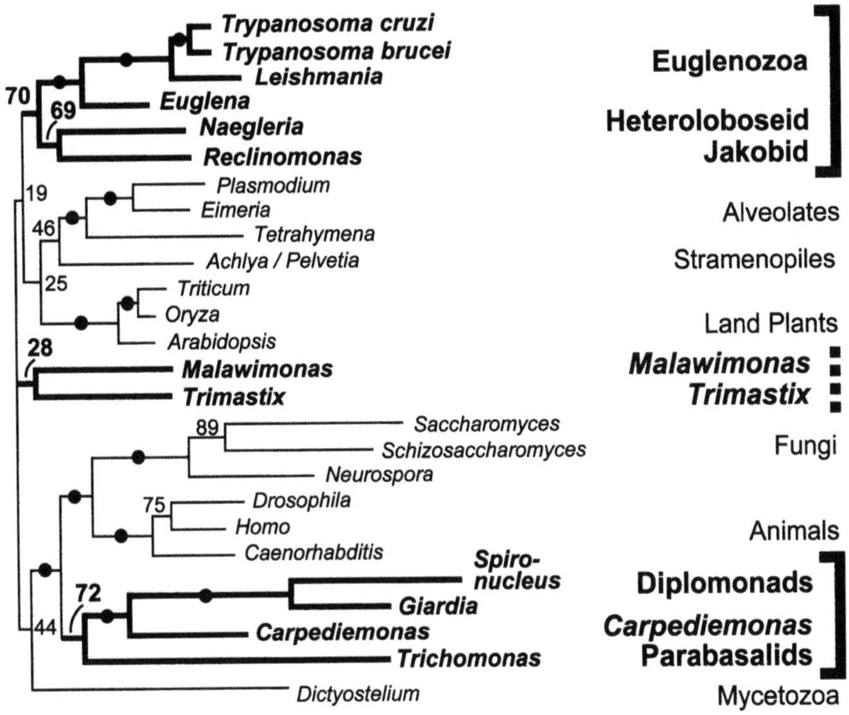

FIGURE 2.5 Maximum likelihood analysis of concatenated proteins from diverse eukaryotes, including most excavate groups. Four proteins (α-tubulin, β-tubulin, cytosolic hsp90 and cytosolic hsp70) were included, totaling 1668 alignable sites. The Jones–Taylor–Thornton amino acid substitution matrix was used, with among-site rate variation modeled by a gamma distribution. Maximum likelihood bootstrap support values are shown (200 replicates). Filled circles represent >90% bootstrap support. Excavates fall as three separated clades (bracketed), one of which, the *Malawimonas–Trimastix* clade, is poorly supported.

2.8 Relationships among Excavata

As with examining excavate monophyly, uncovering the possible relationships among excavate taxa is complicated by the paucity of data from many excavate groups. Here we present the molecular evidence supporting various groupings within Excavata (morphological analyses do not recover robust clades uniting different excavate groups — see Simpson, 2003). We draw a distinction between evidence from excavate-poor data sets (with four or fewer of the ten excavate groups included) and recent excavate-rich analyses, such as those shown in Figure 2.4 and Figure 2.5.

2.8.1 Eopharyngia — Diplomonads and Retortamonads

Molecular phylogenies have recently confirmed a close relationship between diplomonads and retortamonads, two groups that have often been linked (although not necessarily considered sisters) on the basis of morphology (Brugerolle, 1973, 1991; Cavalier-Smith, 1987). These groups form a strongly supported clade in both SSUrRNA analyses (Figure 2.4) and hsp90 protein phylogenies (Simpson et al., unpublished). SSUrRNA analyses often place retortamonads inside diplomonads when *Giardia* is included (Silberman et al., 2002), whereas hsp90 trees place them as sister to diplomonads. The diplomonad–retortamonad clade is recognized formally as the taxon Eopharyngia (Cavalier-Smith, 1993).

2.8.2 Carpediemonas

The diplomonad–retortamonad clade is most closely related to *Carpediemonas*. This clade is formally named Fornicata, after a potential apomorphy, a left-to-right arched proximal portion of the B fiber [fornix (Latin) means *arch*, see Simpson, 2003). Support is strong in most SSUrRNA analyses (Figure 2.4), even when prokaryotic outgroups are included (Simpson et al., 2002c). The clade is also strongly supported in α-tubulin trees (Simpson et al., 2002c) and moderately supported in β-tubulin, hsp70 and hsp90 analyses (Simpson et al., 2002a,c). Our concatenated protein analysis provides very strong support for a clade of *Carpediemonas* with Eopharyngia, represented by diplomonads (Figure 2.5).

2.8.3 *Parabasalids*

A close relationship between diplomonads and parabasalids has been proposed for some time on the basis of several excavate-poor gene phylogenies (Baldauf et al., 2000; Embley and Hirt, 1998). Most convincing are data sets in which diplomonads and parabasalids form a clade even when eukaryotes are rooted by outgroups, e.g., valyl tRNA synthetase, malate dehydrogenase, mitochondrial cpn60, β-tubulin and often EF-1α and EF-2 (Dacks and Roger, 1999; Embley and Hirt, 1998; Hashimoto et al., 1998; Hirt et al., 1999; Horner and Embley, 2001; Keeling et al., 2000; Roger et al., 1998, 1999a). Parabasalids and diplomonads also share an unusual glucose-6-phosphate isomerase gene, probably acquired by common lateral transfer from a prokaryote, although the only excavate group positively known to have the normal eukaryotic form is Euglenozoa (Henze et al., 2001).

Excavate-rich SSUrRNA analyses often recover a parabasalids–Fornicata grouping, at least in unrooted trees (Figure 2.4; Cavalier-Smith, 2000, 2002, 2003; Dacks et al., 2001; Keeling and Leander, 2003). A weakly supported clade is also recovered in excavate-rich β-tubulin and combined α + β-tubulin analyses (Edgcomb et al., 2001; Simpson et al., 2002c). Our concatenated protein analysis also recovers a Parabasalia–Fornicata grouping with moderately strong bootstrap support (Figure 2.5). No analysis to date indicates a position for parabasalids within Fornicata (i.e., closer to diplomonads than are retortamonads or *Carpediemonas*).

2.8.4 *Preaxostyla — Oxymonads and* Trimastix

The position of oxymonads had been uncertain for some time, even after the first gene sequences (EF1α and α-tubulin) were analyzed (Dacks and Roger, 1999; Moriya et al., 1998, 2001). However, more recent SSUrRNA phylogenies show strong support for a close relationship between oxymonads and *Trimastix* (Dacks et al., 2001; Keeling and Leander, 2003; Moriya et al., 2003; Stingl and Brune, 2003). When all excavates are considered, these groups remain as a clade (Figure 2.4). Oxymonads have yet to be included in excavate-rich concatenated protein analyses, although hsp90 and tubulin trees confirm the close relationship with *Trimastix* (AGBS, AJR, unpublished; Keeling, personal communication). The *Trimastix*–oxymonad clade is formally designated Preaxostyla (Simpson, 2003).

2.8.5 Malawimonas

SSUrRNA trees often show a specific relationship between *Malawimonas* and Preaxostyla (Figure 2.4), although the clade receives only very weak bootstrap support (Simpson et al., 2002c). A similar topology is seen is some β-tubulin phylogenies, but again with only weak bootstrap support unless taxon sampling is heavily restricted (AJR, unpublished). Our concatenated protein trees also show a weakly supported clustering of *Malawimonas* with Preaxostyla, represented by *Trimastix* (Figure 2.5).

2.8.6 Heterolobosea, Euglenozoa and Jakobids

A close relationship between Euglenozoa and Heterolobosea has been posited for some time, based on their unusual discoid mitochondrial cristae (Patterson, 1988; Patterson and Sogin, 1992) and on SSUrRNA trees wherein these two long-branching lineages diverge at similar points, or actually form a clade (Figure 2.3; and Hinkle and Sogin, 1993; Patterson and Sogin, 1992; Sogin, 1989). The name Discicristata is now sometimes applied to this potential group (Baldauf et al., 2000; Cavalier-Smith, 1998). An Euglenozoa–Heterolobosea clade is recovered in some excavate-poor tubulin phylogenies and combined protein trees (Baldauf, 1999; Baldauf et al., 2000; Keeling and Doolittle, 1996).

More recently, this result has been complicated by the position of jakobids. Sequences from several jakobids are available for SSUrRNA, α-tubulin and β-tubulin. Analyses of these data sets fail to recover jakobid monophyly, minimally with the divergent sequences from *Jakoba incarcerata* falling separately to other jakobids (Edgcomb et al., 2001; Keeling and Leander, 2003; Simpson et al., 2002c). In analyses of combined α + β-tubulins, the jakobids *Reclinomonas americana* and *Jakoba libera* form a sister group to a clade of Heterolobosea and Euglenozoa (Edgcomb et al., 2001; Simpson et al., 2002c). However, unrooted CCTα trees place jakobids specifically with Heterolobosea, with Euglenozoa falling separately (Archibald et al., 2002). As with combined tubulin analyses, our concatenated protein trees recover a clade of Euglenozoa, Heterolobosea and *Reclinomonas*, now with moderately strong bootstrap support (Figure 2.5). However, as with some CCTα trees, Heterolobosea and jakobids form a moderately strong clade to the exclusion of Euglenozoa. This topology is inconsistent with a parsimonious explanation of mitochondrial cristae evolution (i.e., a common origin for the discoid cristae of Heterolobosea and Euglenozoa and no secondary reversion in jakobids).

2.8.7 Relationships in Summary

Current molecular phylogenetic evidence suggests three basic groups of Excavata: (1) diplomonads, retortamonads, *Carpediemonas* and parabasalids; (2) oxymonads, *Trimastix* and *Malawimonas*; and (3) jakobids, Heterolobosea and Euglenozoa. The second is poorly supported, although the *Trimastix*–oxymonads clade it contains seems robust. Available evidence suggests that parabasalids are basal within the first group. We await further data to confirm the branching order within the third group. Irrespective, although molecular phylogenies generally do not support the monophyly of excavates in whole, each of the basic three groups includes one or more of the typical excavates. In other words, strong molecular data group all nontypical excavates into clades with typical excavates (Figure 2.6). Thus, if it is accepted that typical excavates share a recent common ancestor by virtue of their similar cytoskeletal features, it is reasonable to assume that all excavates have descended from this same ancestor.

2.9 Excavates as Early Eukaryotes?

If unrooted trees were the only information available, it would, in principle, be possible to claim that *any* apparent eukaryotic clade was actually the paraphyletic stem group for all other eukaryotes. Such a claim implies that the group represents the ancestral condition for all eukaryotes (see previously), and also that some of the group might be more early diverged than others. Despite the widespread rejection of the archezoa hypothesis, this stem-group

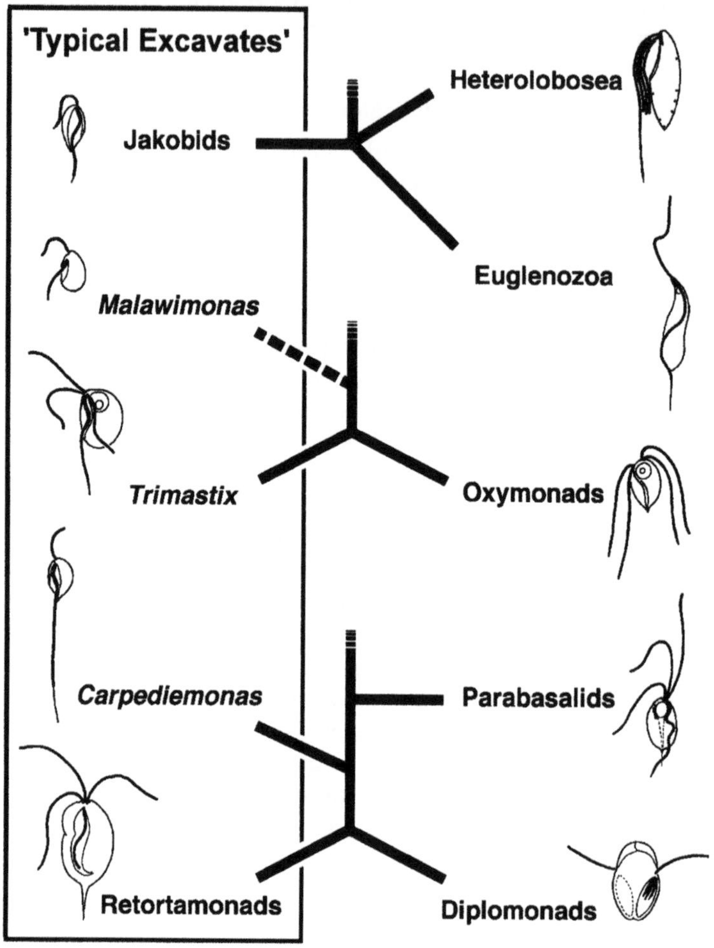

FIGURE 2.6 Diagram summarizing the morphological and molecular evidence for a common origin for all Excavata. Typical excavates (boxed, left) have similar cytoskeletal architectures, strongly suggesting descent from a common excavate ancestor (see Table 2.1). The other five groups of excavates (right) are each specifically related to one or more of the typical excavates in SSUrRNA trees or protein phylogenies, or both, as indicated by the trees figures in the center of the diagram (see also Figures 2.4 and 2.5). The root of each tree is attached vertically. The dashed branch leading to *Malawimonas* indicates the weak support for the position of this taxon.

inference is still made for Excavata (or subgroups of Excavata) much more often than for other major groups of eukaryotes. Here we review recent proposals for a especially early diverged status for particular excavate groups.

2.9.1 *Diplomonads and Parabasalids Early?*

Post-archezoa, at least three lines of evidence have led to suggestions that diplomonads or parabasalids, or both, are early diverged eukaryotes; however, each is problematic.

The potential for long branches to be artifactually attracted to the outgroup branch in deep molecular phylogenies is now widely recognized (Philippe and Laurent, 1998). Nevertheless, it is possible that a long branch is also a bona fide basal branch. The fact remains that diplomonads or parabasalids, or both, are among the deepest branches in many rooted phylogenetic trees of eukaryotes, including most of the most commonly used marker genes, e.g., rRNA, actin, RNA polymerase II, EF-1α, enolase and hsp70 (Dacks et al., 2002; Germot and Philippe, 1999; Keeling, 1998; Keeling and Palmer, 2000; Philippe and Adoutte, 1998; Roger et al., 1999b; Simpson et al., 2002b). This placement is usually recovered even when complex models of sequence evolution are used or many genes are analyzed, or both (Bapteste et al., 2002). In the absence of strong evidence placing other groups at the base of the tree, diplomonads and parabasalids are often considered as the eukaryotes most likely to be basal. Although the appeal of this argument is understandable, we believe that it takes an overly optimistic view of the practical power of phylogenetic analysis.

Second, two indels are shared by the enolase genes of prokaryotes and several parabasalids to the exclusion of those from all other studied eukaryotes (Keeling and Palmer, 2000). These indels could be synapomorphies for all eukaryotes except parabasalids, which would therefore be basal (gene phylogenies argue strongly against the possibility that the whole enolase gene of parabasalids is of recent prokaryotic origin). However, the phylogenetic reliability of such data might be questionable, as evidenced by cases in which different indels suggest irreconcilable clades (Bapteste and Philippe, 2002). Subgene-level lateral transfer and gene conversion are processes that result in different segments of the same gene having different histories. These mechanisms might explain the nonvertical inheritance of such indels, and are invoked to account for otherwise implausible phylogenetic distributions of other indels in eukaryotic enolase genes (Keeling and Palmer, 2001). The apparent shared lateral transfer of the glucose-6-phosphate isomerase genes of parabasalids and diplomonads also argues against scenarios for parabasalids being deeper than diplomonads, or vice versa (Henze et al., 2001; Stechmann and Cavalier-Smith, 2002).

Third, in phylogenies of two tRNA-synthetases (alanyl and prolyl), sequences from diplomonads form the sister group to Archaea, whereas those from all other eukaryotes except *Entamoeba* fall as a clade within Eubacteria (Andersson et al., 2003; Bunjun et al., 2000; Chihade et al., 2000). One proposed explanation is that diplomonads have retained the ancestral eukaryotic forms of these proteins, whereas these have been replaced by eubacterial forms in a common ancestor of all other eukaryotes (Bunjun et al., 2000; Chihade et al., 2000). This would place diplomonads basal to most or all other well-studied eukaryotes, including Euglenozoa. However, comparatively recent lateral transfers of these genes to diplomonads from Archaea are plausible and less extraordinary explanations, providing one accepts that analysis artifact could explain the basal placement of diplomonad proteins in the diplomonad–Archaea clade (Andersson et al., 2003).

2.9.2 *Jakobids Early?*

Other evidence points to jakobids as potentially very early diverged eukaryotes. The mitochondrial genomes of jakobids, especially *Reclinomonas americana,* contain more protein-encoding genes than for other studied eukaryotes (Gray et al., 1999, 2001; Lang et al., 1997). Most importantly, jakobid mitochondria include some or all of the genes for a multisubunit bacterial-type RNA polymerase, presumably inherited directly from the mitochondrial symbiont. All other studied eukaryotes, including Euglenozoa and Heterolobosea, seem to use a nuclear-encoded single-subunit RNA polymerase related to T7 phage forms (Cermakian et al., 1996; Clement and Koslowsky, 2001). On the basis of the mitochondrial genome data alone, the simplest explanation is that jakobids diverged before all other (mitochondriate) eukaryotes, with a common ancestor of other mitochondriate eukaryotes later replacing the bacterial RNA polymerase with the phage type. Clearly, this scenario cannot be reconciled with a monophyletic Excavata, but also cannot be reconciled with the moderately strong, and otherwise plausible, jakobid–Heterolobosea–Euglenozoa clade (Figure 2.5).

Alternative scenarios to explain the distribution of RNA polymerases can be conceived. Perhaps both phage type and bacterial type were present in the last common ancestor of mitochondriate eukaryotes, with the latter then being lost (several times, independently) in ancestors of eukaryotes other than jakobids (Stechmann and Cavalier-Smith, 2002). This possibility is not unreasonable, given that parallel gene loss is commonplace in mitochondria (Gray et al., 1999). Alternatively, but more improbably, the phage-type polymerase alone could be ancestral in mitochondriate eukaryotes, with the bacterial-type RNA polymerase genes in jakobids representing an almost unprecedented, recent lateral transfer into the mitochondrial genome of this lineage. Finally, the bacterial-type RNA polymerase might have been ancestral to all Excavata, but then secondarily lost in an ancestor or ancestors of Euglenozoa and Heterolobosea after they acquired the phage-type polymerase by lateral gene transfer from other eukaryotes. In any case, a position near the base of eukaryotes remains a reasonable possibility for jakobids, and the phylogenetic distribution of mitochondrial RNA polymerases is perhaps the most important evidence contradicting an excavate clade.

2.10 Alternatives to Mitochondria in Excavates

Even leaving aside the considerable biochemical diversity in mitochondria of different eukaryotic lineages (Tielens et al., 2002), excavates display several alternatives to true mitochondria. In parabasalids, the mitochondrial homologues are clearly present in the form of hydrogenosomes (Figure 2.7a). These large but cristae- and genome-lacking organelles with closely adpressed bounding membranes have similar biogenesis characteristics and protein import machinery to that of mitochondria, but their energy metabolism is exclusively anaerobic (Dyall and Johnson, 2000; Müller, 1988, 1998; Rotte et al., 2000). In parabasalids (and other unrelated taxa such as certain ciliates and chytrid fungi), hydrogenosomes convert pyruvate to acetate in an energy-yielding pathway that evolves molecular hydrogen. Two organelle-localized enzymes are key to this pathway: pyruvate:ferredoxin oxidoreductase (PFO) and hydrogenase. Although most Heterolobosea have mitochondria, some such as *Psalteriomonas* and *Lyromonas* have instead hydrogenosomes of presumed mitochondrial origin (Broers et al., 1990, 1993).

Until recently, there was no conclusive evidence that diplomonads possess a physical remnant of the mitochondrion. There were indications of mitochondrial-like membrane compartment activity in *Giardia* (Lloyd et al., 2002b), whereas a short N-terminal extension on mitochondrial isoform cpn60 from *Spironucleus* is found to be weakly similar to mitochondrial and hydrogenosomal targeting signals (Horner and Embley, 2001). However, tiny mitochondrion homologues (mitosomes — see below) have now been found in *Giardia* (Tovar et al., 2003). These organelles are not hydrogenosomes, for although *Giardia* produces some molecular hydrogen, diplomonad PFO and hydrogenases do not appear to be targeted to a mitochondrial-like organelle (Horner et al., 1999, 2000; Lloyd et al., 2002a; Müller, 1998). Very poorly studied by comparison, there is still no strong evidence for mitochondrial remnants in either retortamonads or oxymonads. Isolated accounts of electron-dense bodies in both groups are more likely to represent endobiotic bacteria (Bernard et al., 1997; Bloodgood et al., 1974).

Carpediemonas and *Trimastix* both contain organelles with two closely adpressed bounding membranes but no cristae (Figure 2.7b–d). Their functions are as yet unknown, and molecular and biochemical support for homology with mitochondria is not available. Whereas the organelles of *Carpediemonas* are comparable in conspicuousness to parabasalid hydrogenosomes, those in *Trimastix marina* are small (Figure 2.7d) and few in number. Cryptic organelles of mitochondrial origin that lack hydrogenosomal function are known

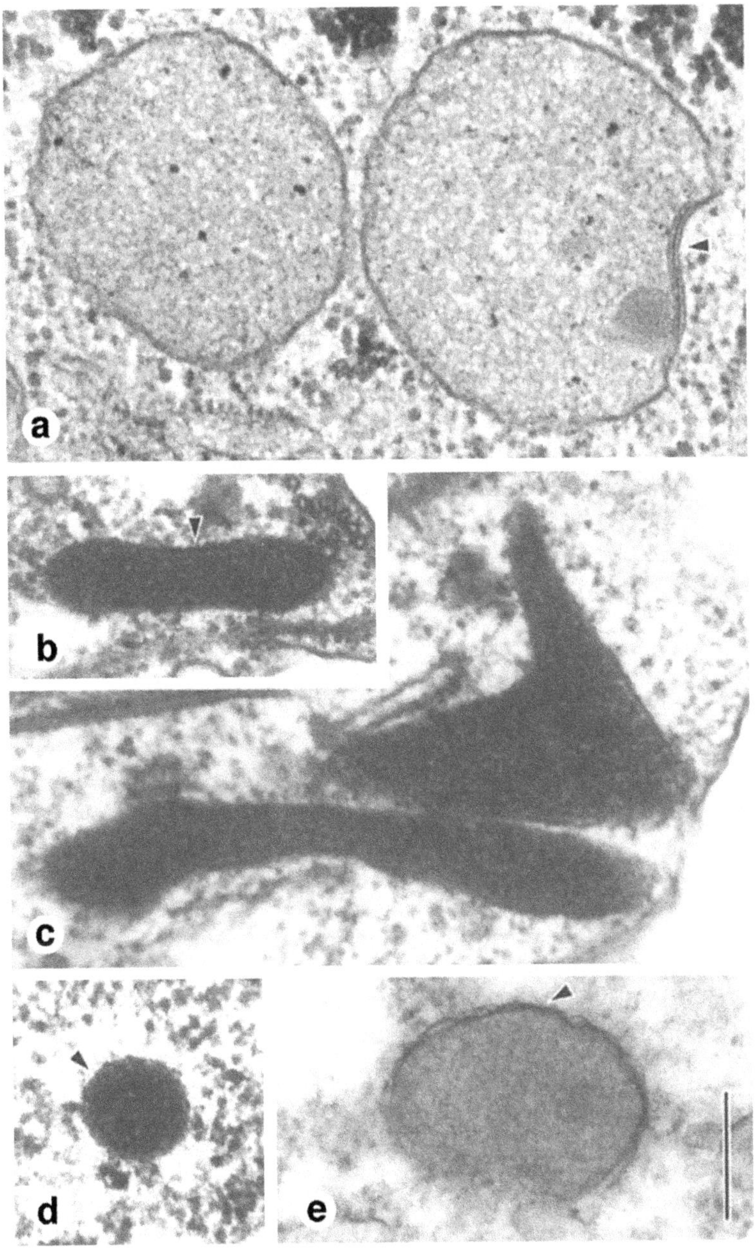

FIGURE 2.7 Transmission electron micrographs showing the mitochondrial homologues (and suspected mitochondrial homologues) from various amitochondriate eukaryotes. Note that all micrographs are at the same scale, but differences in matrix electron-density can result from different fixation conditions. (a) Parabasalid hydrogenosomes (*Tritrichomonas foetus*), approximately the same size as mitochondria, but lacking cristae and with closely adpressed bounding membranes (arrowhead). (b) *Carpediemonas* dense organelle profile showing the bounding membranes. (c) *Carpediemonas* dense organelle showing the extensive, but often narrow, network (*Carpediemonas membranifera*). (d) *Trimastix* dense organelle in cross-section (*Trimastix marina*). e. Pelobiont dense organelle (*Mastigamoeba balamuthi*). Scale bar represents 200 nm. [(a) adapted from Müller (1980); (b) and (c) adapted from Simpson and Patterson (1999), 353–370; (d) adapted from Simpson et al. (2000), 229–252.]

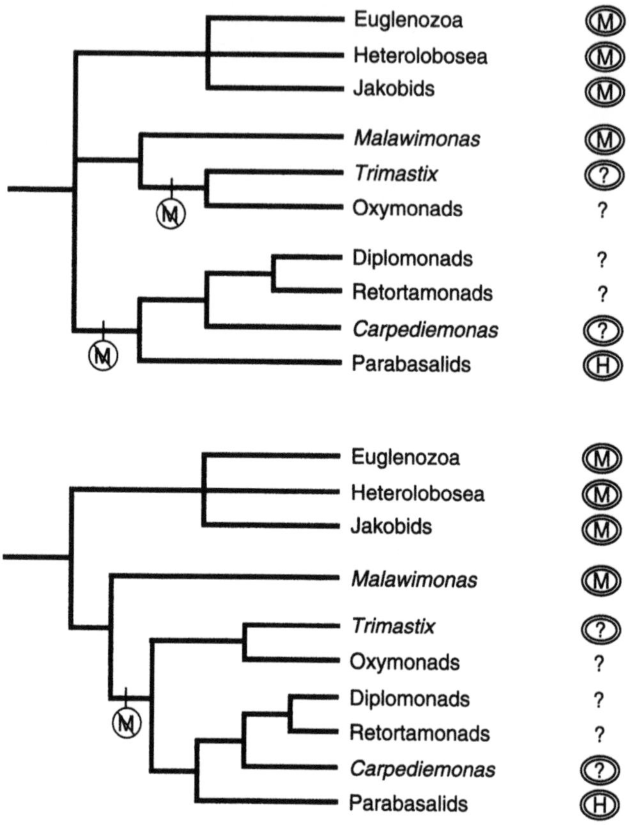

FIGURE 2.8 Two different scenarios for the relationships among excavates, and implications for mitochondrial evolution. Cartoons alongside taxon names illustrate their ancestral mitochondrial mode. M: classical mitochondria, H: hydrogenosomes*: mitosomes (circled): organelles of unknown function and origin, ? (not circled): no organelle known. (a) Default scenario, wherein each of the three groupings suggested by molecular phylogenies is a clade. This scenario requires a minimum of two loss events to explain the phylogenetic distribution of amitochondriate conditions. (b) An alternative scenario, the neoarchezoa hypothesis, wherein all major amitochondriate excavate groups form a clade, suggesting a single loss of mitochondrial function in their common ancestor.

from some nonexcavate eukaryotes, namely *Entamoeba* (Mai et al., 1999; Tovar et al., 1999), microsporidia (Williams et al., 2002) and now, from *Giardia* (Tovar et al., 2003). The name *mitosome* is most commonly used for these organelles (Katinka et al., 2001; Tovar et al., 1999). One mitochondrial function that has been suggested to be retained by such organelles is a role in iron–sulfur cluster assembly, as a number of genes from this pathway are encoded in the genome of the microsporidian *Encephalitozoon* (Katinka et al., 2001; Roger and Silberman, 2002) and localize to the mitosome of *Giardia* (Tovar et al., 2003). The double-membrane-bound organelles in *Trimastix* and *Carpediemonas* could turn out to be functionally more similar to mitosomes than to the hydrogenosomes of parabasalids.

2.11 Excavate Phylogeny and Mitochondrial Loss — Neoarchezoa?

If excavates are in fact a monophyletic group, what are the implications for understanding the evolutionary loss of mitochondria in eukaryotes? This depends on the deepest-level relationships among excavates, and, as discussed previously, these are not well understood at present. From available data, the simplest scenario would be that each of the three excavate groupings suggested by concatenated protein phylogenies is a true clade (Figure 2.8a). If we make two reasonable assumptions — (1) there was a single mitochondrial symbiosis in eukaryotes (Gray et al., 1999) and (2) the last common ancestor of all true mitochondria was mitochondrial in nature (rather than hydrogenosomal or mitosomal) — this scenario implies that classical mitochondrial functions were lost on as few as two occasions early in the evolutionary history of excavates. One of these events would have involved a common ancestor of diplomonads, retortamonads, *Carpediemonas* and parabasalids. The other would have involved common ancestors of *Trimastix* and oxymonads after their divergence from the *Malawimonas* lineage. Additional recent loss events are required to explain the absence of mitochondria in a few Heterolobosea, e.g., *Psalteriomonas* and *Lyromonas* (Broers et al., 1990, 1993), and a few Euglenozoa, e.g., *Postgaardi* (Simpson et al., 1997).

It remains possible, however, that one of the three basic groups of Excavates suggested by molecular phylogenies is actually the ancestral stem group for one or both of the other two groups. The most plausible and intriguing scenario is that the *Malawimonas–Trimastix*–oxymonads group. Interestingly, two SSUrRNA trees have been published that are consistent with this topology, although statistical support is negligible in each case (Simpson et al., 2002c; Cavalier-Smith, 2003). which is only weakly supported, actually contains the diplomonad–retortamonad–*Carpediemonas*–parabasalid group. Under this scenario, all six ancestrally mitochondrion-lacking excavate groups can form a single clade (Figure 2.8b). If so, a single loss of classical mitochondrial function could have occurred in a common ancestor of all excavate former archezoa, as well as *Carpediemonas* and *Trimastix* (Figure 2.8b). We refer to this scenario as the *neoarchezoa hypothesis* [seeCavalier-Smith (2003) for an independent formulation of this proposal].

In both these scenarios, organisms with different alternatives to classical mitochondria share a recent history. First, the possibly organelle-lacking diplomonads and retortamonads group with the organelle-possessing parabasalids and diplomonads (which themselves have organelles with different functions) and *Carpediemonas*. Second, the possibly organelle-lacking oxymonads are sisters to the organelle-possessing *Trimastix*. We anticipate that further advances in our understanding of organismal phylogeny and organelle function in excavates will resolve much of the current uncertainty of mitochondrial evolution in eukaryotes.

Acknowledgments

We thank Miklos Müller and Erin E. Gill for Figure 2.7a and Figure 2.7e, respectively; Robert P. Hirt, David S. Horner and Joel B. Dacks for critical comments; and the Canadian Institute for Advanced Research, Program in Evolutionary Biology for fellowship support to AGBS and AJR. This research was also supported by Grant 227085-00 awarded to AJR by the Natural Sciences and Engineering Research Council of Canada.

References

Allsopp, A. (1969) Phylogenetic relationships of the Prokaryota and the origin of the eukaryotic cell. *New Phytol.* 68: 591–612.

Andersson, J.O., Sjögren, A.M., Davis, L.A. M., Embley, T.M. and Roger, A.J. (2003) Phylogenetic analyses of diplomonad genes reveal frequent lateral gene transfers affecting eukaryotes. *Curr. Biol.* 13: 1–20.

Archibald, J.M., O'Kelly, C.J. and Doolittle, W.F. (2002) The chaperonin genes of jakobid and jakobid-like flagellates: Implications for eukaryotic evolution. *Mol. Biol. Evol.* 19: 422–431.

Arisue, N., Hashimoto, T., Lee, J.A., Moore, D.V., Gordon, P., Sensen, C.W., Gaasterland, T., Hasegawa, M. and Müller, M. (2002) The phylogenetic position of the pelobiont *Mastigamoeba balamuthi* based on sequences of rDNA and translation elongation factors EF-1alpha and EF-2. *J. Euk. Microbiol.* 49: 1–10.

Baldauf, S.L. (1999) A search for the origins of animals and fungi: Comparing and combining molecular data. *Am. Nat.* 154: S178–S188.

Baldauf, S.L., Roger, A.J., Wenk-Siefert, I. and Doolittle, W.F. (2000) A kingdom-level phylogeny of eukaryotes based on combined protein data. *Science* 290: 972–977.

Bapteste, E., Brinkmann, H., Lee, J.A., Moore, D.V., Sensen, C.W., Gordon, P., Durufle, L., Gaasterland, T., Lopez, P., Müller, M. and Philippe, H. (2002) The analysis of 100 genes supports the grouping of three highly divergent amoebae: *Dictyostelium*, *Entamoeba*, and *Mastigamoeba*. *Proc. Natl. Acad. Sci. USA* 99: 1414–1419.

Bapteste, E. and Philippe, H. (2002) The potential value of indels as phylogenetic markers: Position of trichomonads as a case study. *Mol. Biol. Evol.* 19: 972–972.

Bernard, C., Simpson, A.G.B. and Patterson, D.J. (1997) An ultrastructural study of a free-living retortamonad, *Chilomastix cuspidata* (Larsen & Patterson, 1990) n. comb. (Retortamonadida, Protista). *Eur. J. Protistol.* 33: 254–265.

Bernard, C., Simpson, A.G.B. and Patterson, D.J. (2000) Some free-living flagellates (Protista) from anoxic habitats. *Ophelia* 52: 113–142.

Bloodgood, R.A., Miller, K.R., Fitzharris, T.P. and McIntosh, J.R. (1974) The ultrastructure of *Pyrsonympha* and its associated microorganisms. *J. Morphol.* 143: 77–106.

Bovee, E.C. and Jahn, T.L. (1973) Taxonomy and phylogeny. In *The Biology of Amoeba* (Jeon, K.W., Ed.), Academic Press, New York, pp. 38–82.

Broers, C.A.M., Meijers, H.H.M., Symens, J.C., Stumm, C.K., Vogels, G.D. and Brugerolle, G. (1993) Symbiotic association of *Psalteriomonas vulgaris*, n. spec. with *Methanobacterium formicicum*. *Eur. J. Protistol.* 29: 98–105.

Broers, C.A.M., Stumm, C.K., Vogels, G.D. and Brugerolle, G. (1990) *Psalteriomonas lanterna* gen. nov., sp. nov., a free-living amoeboflagellate isolated from freshwater anaerobic sediments. *Eur. J. Protistol.* 25: 369–380.

Brugerolle, G. (1973) Etude ultrastructurale du trophozoite et du kyste chez le genre *Chilomastix* Aléxéieff, 1910 (Zoomastigophorea, Retortamonadida Grassé, 1952). *J. Protozool.* 20: 574–585.

Brugerolle, G. (1991) Flagellar and cytoskeletal systems in amitochondrial flagellates: Archamoeba, Metamonada and Parabasala. *Protoplasma* 164: 70–90.

Brugerolle, G. and Müller, M. (2000) Amitochondriate flagellates. In *The Flagellates: Unity, Diversity and Evolution* (Leadbeater, B. S. C. and Green, J. C., Eds.), Taylor & Francis, London, pp. 166–189.

Brugerolle, G. and Patterson, D. J. (1997) Ultrastructure of *Trimastix convexa* Hollande, an amitochondriate anaerobic flagellate with a previously undescribed organization. *Eur. J. Protistol.* 33: 121–130.

Bui, E.T.N., Bradley, P.J. and Johnson, P.J. (1996) A common evolutionary origin for mitochondria and hydrogenosomes. *Proc. Natl. Acad. Sci. USA* 93: 9651–9656.

Bunjun, S., Stathopoulos, C., Graham, D., Min, B., Kitabatake, M., Wang, A.L., Wang, C.C., Vivarès, C.P., Weiss, L.M. and Söll, D. (2000) A dual-specificity aminoacyl-tRNA synthetase in the deep-rooted eukaryote *Giardia lamblia*. *Proc. Natl. Acad. Sci. USA* 97: 12997–13002.

Cavalier-Smith, T. (1975) The origin of nuclei and of eukaryotic cells. *Nature* 256: 463–468.

Cavalier-Smith, T. (1980) Cell compartmentation and the origin of eukaryote membranous organelles. In *Endocytobiology: Endosymbiosis and Cell Biology* (Schwemmler, W. and Schenk, H.E.A., Eds.), Walter deGruyter, Berlin, pp. 893–916.

Cavalier-Smith, T. (1981) Eukaryote kingdoms: Seven or nine? *BioSystems* 14: 461–481.

Cavalier-Smith, T. (1983) A 6-kingdom classification and a unified phylogeny. In *Endocytobiology II* (Schwemmler, W. and Schenk, H.E.A., Eds.), Walter deGruyter, Berlin, pp. 1027–1034.

Cavalier-Smith, T. (1987) Eukaryotes with no mitochondria. *Nature* 326: 332–333.

Cavalier-Smith, T. (1991) Cell diversification in heterotrophic flagellates. In *The Biology of Free-Living Heterotrophic Flagellates* (Patterson, D.J. and Larsen, J., Eds.), Clarendon Press, Oxford, pp. 113–131.
Cavalier-Smith, T. (1992a) Percolozoa and the symbiotic origin of the metakaryote cell. In *Endocytobiology V* (Sato, S., Ishida, M. and Ishikawa, H., Eds.), Tübingen University Press, Tübingen, pp. 399–406.
Cavalier-Smith, T. (1992b) Origin of the cytoskeleton. In *The Origin and Evolution of the Cell* (Hartman, H. and Matsumo, K., Eds.), World Scientific, Singapore, pp. 79–106.
Cavalier-Smith, T. (1993) Kingdom Protozoa and its 18 phyla. *Microbiol. Rev.* 57: 953–994.
Cavalier-Smith, T. (1995) Zooflagellate phylogeny and classification. *Tsitologia* 37: 1010–1029.
Cavalier-Smith, T. (1997) Amoeboflagellates and mitochondrial cristae in eukaryote evolution: Megasystematics of the new protozoan subkingdoms Eozoa and Neozoa. *Arch. Protistenkd.* 147: 237–258.
Cavalier-Smith, T. (1998) A revised six-kingdom system of life. *Biol. Rev.* 73: 203–266.
Cavalier-Smith, T. (1999) Principles of protein and lipid targeting in secondary symbiogenesis: Euglenoid, dinoflagellate, and sporozoan plastid origins and the eukaryotic family tree. *J. Euk. Microbiol.* 46: 347–366.
Cavalier-Smith, T. (2000) Flagellate megaevolution: The basis for eukaryote diversification. In *The Flagellates: Unity, Diversity and Evolution* (Leadbeater, B. S. C. and Green, J. C., Eds.), Taylor & Francis, London, pp. 361–390.
Cavalier-Smith, T. (2002) The phagotrophic origin of eukaryotes and phylogenetic classification of Protozoa. *Int. J. Syst. Evol. Microbiol.* 52: 297–354.
Cavalier-Smith, T. (2003) The excavate protozoan phyla Metamonada Grassé emend. (Anaeromonadea, Parabasalia, Carpediemonas, Eopharyngia) and Loukozoa emend. (Jakobea, Malawimonas): their evolutionary affinities and new higher taxa. *Int. J. Syst. Evol. Microbiol.* 53: 1741-1758.
Cermakian, N., Ikeda, T. M., Cedergren, R. and Gray, M. W. (1996) Sequences homologous to yeast mitochondrial and bacteriophage T3 and T7 RNA polymerases are widespread throughout the eukaryotic lineage. *Nucleic Acids Res.* 24: 648–654.
Chihade, J.W., Brown, J.R., Schimmel, P.R. and Ribas de Pouplana, L. (2000) Origin of mitochondria in relation to evolutionary history of eukaryotic alanyl-tRNA synthetase. *Proc. Natl. Acad. Sci. USA* 97: 12153–12157.
Clark, C.G. and Roger, A.J. (1995) Direct evidence for secondary loss of mitochondria in *Entamoeba histolytica. Proc. Natl. Acad. Sci. USA* 92: 6518–6521.
Clement, S.L. and Koslowsky, D.J. (2001) Unusual organization of a developmentally regulated mitochondrial RNA polymerase (TBMTRNAP) gene in *Trypanosoma brucei. Gene* 272: 209–218.
Dacks, J.B., Marinets, A., Doolittle, W.F., Cavalier-Smith, T. and Logsdon, J.M. (2002) Analysis of RNA polymerase II genes from free-living protists: Phylogeny, long branch attraction and the eukaryotic big bang. *Mol. Biol. Evol.* 19: 830–840.
Dacks, J.B., Silberman, J.D., Simpson, A.G.B., Moruya, S., Kudo, T., Ohkuma, M. and Redfield, R. (2001) Oxymonads are closely related to the excavate taxon Trimastix. *Mol. Biol. Evol.* 18: 1034–1044.
Dacks, J. and Roger, A.J. (1999) The first sexual lineage and the relevance of facultative sex. *J. Mol. Evol.* 48: 779–783.
Daniels, E.W. and Breyer, E.P. (1967) Ultrastructure of the giant amoeba *Pelomyxa palustris. J. Protozool.* 14: 167–179.
Dodge, J.D. (1965) Chromosome structure in the dinoflagellates and the problem of the mesokaryotic cell. *Excerpta Med. Int. Congr. Ser.* 91: 339–341.
Dougherty, E.C. and Allen, M.B. (1960) Is pigmentation a clue to protistan phylogeny? In *Comparative Biochemistry of Photoreactive Systems* (Allen, M.B., Ed.), Academic Press, New York, pp. 129–144.

Dyall, S.D. and Johnson, P.J. (2000) Origins of hydrogenosomes and mitochondria: Evolution and organelle biogenesis. *Curr. Opin. Microbiol.* 3: 404–411.

Edgcomb, V.P., Roger, A.J., Simpson, A.G.B., Kysela, D. and Sogin, M.L. (2001) Evolutionary relationships among "jakobid" flagellates as indicated by alpha- and beta-tubulin phylogenies. *Mol. Biol. Evol.* 18: 514–522.

Edgcomb, V.P., Simpson, A.G.B., Zettler, L.A., Nerad, T.A., Patterson, D.J., Holder, M.E. and Sogin, M.L. (2002) Pelobionts are degenerate protists: Insights from molecules and morphology. *Mol. Biol. Evol.* 19: 978–982.

Edlind, T.D., Li, J., Visvesvara, G.S., Vodkin, M.H., McLaughlin, G.L. and Katiyar, S.K. (1996) Phylogenetic analysis of β-tubulin sequences from amitochondrial protozoa. *Mol. Phylogenet. Evol.* 5: 359–367.

Embley, T.M. and Hirt, R.P. (1998) Early branching eukaryotes? *Curr. Opin. Genet. Dev.* 8: 624–629.

Fast, N.M., Logsdon, J.M. and Doolittle, W.F. (1999) Phylogenetic analysis of the TATA box binding protein (TBP) gene from *Nosema locustae*: Evidence for a microsporidia-fungi relationship and spliceosomal intron loss. *Mol. Biol. Evol.* 16: 1415–1419.

Germot, A. and Philippe, H. (1999) Critical analysis of eukaryotic phylogeny: A case study based on the HSP70 family. *J. Euk. Microbiol.* 46: 116–124.

Germot, A., Philippe, H. and Le Guyader, H. (1996) Presence of a mitochondrial-type 70 kDa heat shock protein in *Trichomonas vaginalis* suggests a very early mitochondrial endosymbiosis in eukaryotes. *Proc. Natl. Acad. Sci. USA* 93: 14614–14617.

Germot, A., Philippe, H. and Le Guyader, H. (1997) Evidence for loss of mitochondria in Microsporidia from a mitochondrial-type HSP70 in *Nosema locustae*. *Mol. Biochem. Parasitol.* 87: 159–168.

Gillin, F.D., Reiner, D.S. and McCaffery, J.M. (1996) Cell biology of the primitive eukaryote *Giardia lamblia*. *Annu. Rev. Microbiol.* 50: 679–705.

Gray, M. W., Burger, G. and Lang, B.F. (1999) Mitochondrial evolution. *Science* 283: 1476–1481.

Gray, M. W., Burger, G. and Lang, B.F. (2001) The origin and early evolution of mitochondria. *Genome Biol.* 2: 1018.1–1018.6.

Gray, M.W., Lang, B.F., Cedergren, R., Golding, G.B., Lemieux, C., Sankoff, D., Turmel, M., Brossard, N., Delage, E., Littlejohn, T.G., Plante, I., Rioux, P., Saint-Louis, D., Zhu, Y. and Burger, G. (1998) Genome structure and gene content in protist mitochondrial DNAs. *Nucl. Acids Res.* 26: 865–878.

Griffin, J.L. (1988) Fine structure and taxonomic position of the giant amoeboid flagellate *Pelomyxa palustris*. *J. Protozool.* 35: 300–315.

Gunderson, J., Hinkle, G., Leipe, D., Morrison, H.G., Stickel, S.K., Odelson, D.A., Breznak, J.A., Nerad, T.A., Müller, M. and Sogin, M.L. (1995) Phylogeny of trichomonads inferred from small-subunit rRNA sequences. *J. Euk. Microbiol.* 42: 411–415.

Hasegawa, M. and Hashimoto, T. (1993) Ribosomal RNA trees misleading? *Nature* 361: 23.

Hasegawa, M., Hashimoto, T., Adachi, J., Iwabe, N. and Miyata, T. (1993) Early branchings in the evolution of eukaryotes: Ancient divergence of *Entamoeba* that lacks mitochondria revealed by protein sequence data. *J. Mol. Evol.* 36: 380–388.

Hashimoto, T., Nakamura, Y., Kamaishi, T., Nakamura, F., Adachi, J., Okamoto, K. and Hasegawa, M. (1995) Phylogenetic place of mitochondrion-lacking protozoan, *Giardia lamblia*, inferred from amino acid sequences of elongation factor 2. *Mol. Biol. Evol.* 12: 782–793.

Hashimoto, T., Nakamura, Y., Nakamura, F., Shirakura, T., Adachi, J., Goto, N., Okamoto, K. and Hasegawa, M. (1994) Protein phylogeny gives a robust estimation for early divergences of eukaryotes: Phylogenetic place of a mitochondria-lacking protozoan, *Giardia lamblia*. *Mol. Biol. Evol.* 11: 65–71.

Hashimoto, T., Sanchez, L.B., Shirakura, T., Müller, M. and Hasegawa, M. (1998) Secondary absence of mitochondria in *Giardia lamblia* and *Trichomonas vaginalis* revealed by valyl-tRNA synthetase phylogeny. *Proc. Natl. Acad. Sci. USA* 95: 6860–6865.

Hehl, A.B., Marti, M., Li, Y., Schraner, E.M., Wild, P. and Kohler, P. (2003) The secretory apparatus of an ancient eukaryote: Protein sorting to separate export pathways occurs prior to formation of transient Golgi-like compartments. *Mol. Biol. Cell.* 14:1433–1447.

Henze, K., Horner, D.S., Suguri, S., Moore, D.V., Sanchez, L.B., Müller, M. and Embley, T.M. (2001) Unique phylogenetic relationships of glucokinase and glucosephosphate isomerase of the amitochondriate eukaryotes *Giardia intestinalis*, *Spironucleus barkhanus* and *Trichomonas vaginalis*. *Gene* 281: 123–131.

Herzog, M., von Boletzky, S. and Soyer, M.-O. (1984) Ultrastructural and biochemical nuclear aspects of eukaryote classification: Independent evolution of the dinoflagellates as a sister group of the actual eukaryotes? *Orig. Life* 13: 205–215.

Hinkle, G., Leipe, D.D., Nerad, T.A. and Sogin, M.L. (1994) The unusually long small subunit ribosomal RNA of *Phreatamoeba balamuthi*. *Nucl. Acids Res.* 22: 465–469.

Hinkle, G. and Sogin, M.L. (1993) The evolution of the Vahlkampfiidae as deduced from 16S-like ribosomal RNA analysis. *J. Euk. Microbiol.* 40: 599–603.

Hirt, R.P., Healy, B., Vossbrinck, C.R., Canning, E.U. and Embley, T.M. (1997) A mitochondrial Hsp70 orthologue in *Vairimorpha necatrix*: Molecular evidence that microsporidia once contained mitochondria. *Curr. Biol.* 7: 995–998.

Hirt, R.P., Logsdon, J.M., Healy, B., Dorey, M. W., Doolittle, W.F. and Embley, T.M. (1999) Microsporidia are related to fungi: Evidence from the largest subunit of RNA polymerase II and other proteins. *Proc. Natl. Acad. Sci. USA* 96: 580–585.

Horner, D.S. and Embley, T.M. (2001) Chaperonin 60 phylogeny provides further evidence for secondary loss of mitochondria among putative early-branching eukaryotes. *Mol. Biol. Evol.* 18: 1970–1975.

Horner, D.S., Foster, P.G. and Embley, T.M. (2000) Iron hydrogenases and the evolution of anaerobic eukaryotes. *Mol. Biol. Evol.* 17: 1695–1709.

Horner, D.S., Hirt, R.P. and Embley, T.M. (1999) A single eubacterial origin of eukaryotic pyruvate:ferredoxin oxidoreductase genes: Implications for the evolution of anaerobic eukaryotes. *Mol. Biol. Evol.* 16: 1280–1291.

Horner, D.S., Hirt, R.P., Kilvington, S., Lloyd, D. and Embley, T.M. (1996) Molecular data suggest an early acquisition of the mitochondrion endosymbiont. *Proc. R. Soc. Lond. B* 263: 1053–1059.

Kamaishi, T., Hashimoto, T., Nakamura, Y., Nakamura, F., Murata, S., Okada, N., Okamoto, K., Shimizu, M. and Hasegawa, M. (1996) Protein phylogeny of translation elongation factor ef-1-alpha suggests microsporidians are extremely ancient eukaryotes. *J. Mol. Evol.* 42: 257–263.

Katinka, M.D., Duprat, S., Cornillot, E., Méténier, G., Thomarat, F., Prensier, G., Barbe, V., Peyretaillade, E., Brottier, P., Wincker, P., Delbac, F., El Alaoui, H., Peyret, P., Saurin, W., Gouy, M., Weissenbach, J. and Vivarès, C.P. (2001) Genome sequence and gene compaction of the eukaryote parasite *Encephalitozoon cuniculi*. *Nature* 414: 450–453.

Keeling, P.J. (1998) A kingdoms progress: Archezoa and the origin of eukaryotes. *Bioessays* 20: 87–95.

Keeling, P.J. and Doolittle, W.F. (1996) Alpha-tubulin from early-diverging eukaryotic lineages and the evolution of the tubulin family. *Mol. Biol. Evol.* 13: 1297–1305.

Keeling, P.J. and Leander, B.S. (2003) Characterisation of a non-canonical genetic code in the oxymonad *Streblomastix strix*. *J. Mol. Biol.* 326: 1337–1349.

Keeling, P.J., Luker, M.A. and Palmer, J.D. (2000) Evidence from beta-tubulin phylogeny that microsporidia evolved from within the fungi. *Mol. Biol. Evol.* 17: 23–31.

Keeling, P.J. and Palmer, J.D. (2000) Parabasalian flagellates are ancient eukaryotes. *Nature* 405: 635–637.

Keeling, P.J. and Palmer, J.D. (2001) Lateral transfer at the gene and subgenic levels in the evolution of eukaryotic enolase. *Proc. Natl. Acad. Sci. USA* 98: 10745–10750.

Klein, R.M. and Cronquist, A. (1967) A consideration of the evolutionary and taxonomic significance of some biochemical, micromorphological and physiological characters in the thallophytes. *Q. Rev. Biol.* 42: 105–296.

Kulda, J. and Nohynková, E. (1978) Flagellates of the human intestine and of intestines of other species. In *Parasitic Protozoa*, Vol. II (Kreier, J. P., Ed.), Academic Press, San Diego, pp. 1–138.

Kumar, S. and Rzhetsky, A. (1996) Evolutionary relationships of eukaryotic kingdoms. *J. Mol. Evol.* 42: 183–193.

Lang, B.F., Burger, G., O'Kelly, C.J., Cedergren, R., Golding, G.B., Lemieux, C., Sankoff, D., Turmel, M. and Gray, M.W. (1997) An ancestral mitochondrial DNA resembling a eubacterial genome in miniature. *Nature* 387: 493–497.

Leipe, D.D., Gunderson, J.H., Nerad, T.A. and Sogin, M.L. (1993) Small subunit ribosomal RNA of *Hexamita inflata* and the quest for the first branch in the eukaryotic tree. *Mol. Biochem. Parasitol.* 59: 41–48.

Li, J., Katiyar, S.K., Hamelin, A., Visvesvara, G.S. and Edlind, T.D. (1996) Tubulin genes from AIDS-associated microsporidia and implications for phylogeny and benzimidazole sensitivity. *Mol. Biochem. Parasitol.* 78: 289–295.

Lloyd, D., Harris, J.C., Maroulis, S., Wadley, R.B., Ralphs, J.R., Hann, A.C., Turner, M.P. and Edwards, M.R. (2002b) The "primitive" microaerophile *Giardia intestinalis* (syn. *lamblia, duodenalis*) has specialized membranes with electron transport and membrane-potential-generating functions. *Microbiology* 148: 1349–1354.

Lloyd, D., Ralphs, J.R. and Harris, J.C. (2002a) *Giardia intestinalis*, a eukaryote without hydrogenosomes, produces hydrogen. *Microbiology* 148: 727–733.

Loeblich, A.R. (1976) Dinoflagellate evolution: Speculation and evidence. *J. Protozool.* 23: 13–28.

Mai, Z.M., Ghosh, S., Frisardi, M., Rosenthal, B., Rogers, R. and Samuelson, J. (1999) Hsp60 is targeted to a cryptic mitochondrion-derived organelle ("crypton") in the microaerophilic protozoan parasite *Entamoeba histolytica. Mol. Cell Biol.* 19: 2198–2205.

Margulis, L. (1970) *Origin of Eukaryotic Cells.* Yale University Press, New Haven, CT.

Moriya, S., Dacks, J.B., Tagaki, A., Noda, S., Ohkuma, M., Doolittle, W.F. and Kudo, T. (2003) Molecular phylogeny of three oxymonad genera: Pyrsonympha, Dinenympha and Oxymonas. *J. Euk. Microbiol.* 50: 190–197.

Moriya, S., Ohkuma, M. and Kudo, T. (1998) Phylogenetic position of symbiotic protist *Dinemympha exilis* in the hindgut of the termite *Reticulitermes speratus* inferred from the protein phylogeny of elongation factor 1-alpha. *Gene* 210: 221–227.

Moriya, S., Tanaka, K., Ohkuma, M., Sugano, S. and Kudo, T. (2001) Diversification of the microtubule system in the early stage of eukaryote evolution: Elongation factor 1α and α-tubulin protein phylogeny of termite symbiotic oxymonad and hypermastigote protists. *J. Mol. Evol.* 52: 6–16.

Müller, M. (1988) Energy metabolism of protozoa without mitochondria. *Annu. Rev. Microbiol.* 42: 465–488.

Müller, M. (1998) Enzymes and compartmentation of core energy metabolism of anaerobic protists: A special case in eukaryotic evolution. In *Evolutionary Relationships among Protozoa* (Coombs, G.H., Vickerman, K., Sleigh, M.A. and Warren, A., Eds.), Kluwer Academic, Dordrecht, pp. 109–127.

O'Kelly, C.J. (1993) The jakobid flagellates: Structural features of *Jakoba*, *Reclinomonas* and *Histiona* and implications for the early diversification of eukaryotes. *J. Euk. Microbiol.* 40: 627–636.

O'Kelly, C.J., Farmer, M.A. and Nerad, T.A. (1999) Ultrastructure of *Trimastix pyriformis* (Klebs) Bernard et al.: Similarities of *Trimastix* species with retortamonad and jakobid flagellates. *Protist* 150: 149–162.

O'Kelly, C.J. and Nerad, T.A. (1999) *Malawimonas jakobiformis* n. gen., n. sp. (Malawimonadidae fam. nov.): A Jakoba-like heterotrophic nanoflagellate with discoidal mitochondrial cristae. *J. Euk. Microbiol.* 46: 522–531.

Patterson, D.J. (1988) The evolution of protozoa. *Mem. Inst. Oswaldo Cruz.* 83: 580–600.

Patterson, D.J. and Sogin, M.L. (1992) Eukaryote origins and protistan diversity. In *The Origin and Evolution of the Cell* (Hartman, H. and Matsumo, K.. Eds.), World Scientific, Singapore, pp. 14–46.

Peyretaillade, E., Broussolle, V., Peyret, P., Méténier, G., Gouy, M. and Vivarès, C.P. (1998) Microsporidia, amitochondrial protists, possess a 70-kDa heat shock protein gene of mitochondrial evolutionary origin. *Mol. Biol. Evol.* 15: 683–689.

Philippe, H. and Adoutte, A. (1998) The molecular phylogeny of Eukaryota: Solid facts and uncertainties. In *Evolutionary Relationships among Protozoa* (Coombs, G.H., Vickerman, K., Sleigh, M.A. and Warren, A., Eds.), Kluwer Academic, Dordrecht, pp. 25–56.

Philippe, H. and Laurent, J. (1998) How good are deep phylogenetic trees? *Curr. Opin. Genet. Dev.* 8: 616–623.

Philippe, H., Lopez, P., Brinkmann, H., Budin, K., Germot, A., Laurent, J., Moreira, D., Müller, M. and Le Guyader, H. (2000) Early-branching or fast-evolving eukaryotes? An answer based on slowly evolving positions. *Proc. R. Soc. Lond. B* 267: 1213–1221.

Roger, A.J. (1999) Reconstructing early events in eukaryotic evolution. *Am. Nat.* 154: S146–S163.

Roger, A.J., Clark, C.G. and Doolittle, W.F. (1996) A possible mitochondrial gene in the early-branching amitochondriate protist *Trichomonas vaginalis*. *Proc. Natl. Acad. Sci. USA* 93: 14618–14622.

Roger, A.J., Morrison, H.G. and Sogin, M.L. (1999a) Primary structure and phylogenetic relationships of a malate dehydrogenase gene from *Giardia lamblia*. *J. Mol. Evol.* 48: 750–755.

Roger, A.J., Sandblom, O., Doolittle, W.F. and Philippe, H. (1999b) An evaluation of elongation factor 1 alpha as a phylogenetic marker for eukaryotes. *Mol. Biol. Evol.* 16: 218–233.

Roger, A.J. and Silberman, J.D. (2002) Mitochondria in hiding. *Nature* 418: 827–829.

Roger, A.J., Svard, S.G., Tovar, J., Clark, C.G., Smith, M.W., Gillin, F.D. and Sogin, M.L. (1998) A mitochondrial-like chaperonin 60 gene in *Giardia lamblia*: Evidence that diplomonads once harbored an endosymbiont related to the progenitor of mitochondria. *Proc. Natl. Acad. Sci. USA* 95: 229–234.

Rotte, C., Henze, K., Müller, M. and Martin, W. (2000) Origins of hydrogenosomes and mitochondria. *Curr. Opin. Microbiol.* 3: 481–486.

Sagan, L. (1967) On the origin of mitosing cells. *J. Theor. Biol.* 14: 225–274.

Sánchez, L.B., Horner, D.S., Moore, D.V., Henze, K., Embley, T.M. and Müller, M. (2002) Fructose-1,6-bisphosphate aldolases in amitochondriate protists constitute a single protein subfamily with eubacterial relationships. *Gene* 295: 51–59.

Silberman, J D., Clark, C.G., Diamond, L.S. and Sogin, M.S. (1999) Phylogeny of the genera *Entamoeba* and *Endolimax* as deduced from small-subunit ribosomal RNA sequences. *Mol. Biol. Evol.* 16: 1740–1751.

Silberman, J.D., Simpson, A.G.B., Kulda, J., Cepicka, I., Hampl, V., Johnson, P.J. and Roger, A.J. (2002) Retortamonad flagellates are closely related to diplomonads: Implications for the history of mitochondrial function in eukaryote evolution. *Mol. Biol. Evol.* 19: 777–786.

Simpson, A.G.B. (2003) The composition, cytoskeletal organisation, and phylogenetic relationships of Excavata (Eukaryota). *Int. J. Syst. Evol. Microbiol.* 53: 1759–1777.

Simpson, A.G.B., Bernard, C. and Patterson, D.J. (2000) The ultrastructure of *Trimastix marina* Kent, 1880 (Eukaryota), an excavate flagellate. *Eur. J. Protistol.* 36: 229–252.

Simpson, A.G.B., MacQuarrie, E.K. and Roger, A.J. (2002a) Early origin of canonical introns. *Nature* 419: 270.

Simpson, A.G.B. and Patterson, D.J. (1999) The ultrastructure of *Carpediemonas membranifera* (Eukaryota) with reference to the excavate hypothesis. *Eur. J. Protistol.* 35: 353–370.

Simpson, A.G.B. and Patterson, D.J. (2001) On core jakobids and excavate taxa: The ultrastructure of *Jakoba incarcerata*. *J. Euk.Microbiol.* 48: 480–492.

Simpson, A.G.B., Radek, R., Dacks, J.B. and O'Kelly, C.J. (2002b) How oxymonads lost their groove: An ultrastructural comparison of Monocercomonoides and excavate taxa. *J. Euk. Microbiol.* 49: 239–248.

Simpson, A.G.B., Roger, A.J., Silberman, J.D., Leipe, D., Edgcomb, V.P., Jermiin, L.S., Patterson, D.J. and Sogin, M.L. (2002c) Evolutionary history of "early diverging" eukaryotes: The excavate taxon *Carpediemonas* is closely related to *Giardia*. *Mol. Biol. Evol.* 19: 1782–1791.

Simpson, A.G.B., van den Hoff, J., Bernard, C., Burton, H.R. and Patterson, D.J. (1997) The ultrastructure and systematic position of the euglenozoon *Postgaardi mariagerensis*, Fenchel et al. *Arch. Protistenkd.* 147: 213–225.

Sleigh, M. (1989) *Protozoa and Other Protists*, Edward Arnold, London.

Sogin, M.L. (1989) Evolution of eukaryotic microorganisms and their small subunit ribosomal RNAs. *Am. Zool.* 29: 487–499.

Sogin, M.L. (1991) Early evolution and the origin of eukaryotes. *Curr. Opin. Genet. Dev.* 1: 457–463.

Sogin, M.L., Elwood, H.J. and Gunderson, J.H. (1986) Evolutionary diversity of eukaryotic small-subunit rRNA genes. *Proc. Natl. Acad. Sci. USA* 83: 1383–1387.

Sogin, M.L., Gunderson, J.H., Elwood, H.J., Alonso, R.A. and Peattie, D.A. (1989) Phylogenetic significance of the kingdom concept: An unusual eukaryotic 16S-like ribosomal RNA from *Giardia lamblia*. *Science* 243: 75–77.

Stanier, R.Y. (1970) Some aspects of the biology of cells and their possible evolutionary significance. In *Organization and Control in Prokaryotic and Eukaryotic Cells* (Charles, H. P. and Knight, B. D., Eds.), Cambridge University Press, London, pp. 1–38.

Stechmann, A. and Cavalier-Smith, T. (2002) Rooting the eukaryote tree by using a derived gene fusion. *Science* 297: 89–91.

Stiller, J. W., Duffield, E.C.S. and Hall, B.D. (1998) Amitochondriate amoebae and the evolution of DNA-dependent RNA polymerase II. *Proc. Natl. Acad. Sci. USA* 95: 11769–11774.

Stiller, J.W. and Hall, B.D. (1999) Long-branch attraction and the rDNA model of early eukaryotic evolution. *Mol. Biol. Evol.* 16: 1270–1279.

Stingl, U. and Brune, A. (2003) Phylogenetic diversity and whole-cell hybridization of oxymonad flagellates from the hindgut of the wood-feeding lower termite *Reticulitermes flavipes*. *Protist* 154: 147–155.

Tachezy, J., Sánchez, L.B. and Müller, M. (2001) Mitochondrial type iron-sulfur cluster assembly in the amitochondriate eukaryotes *Trichomonas vaginalis* and *Giardia intestinalis*, as indicated by the phylogeny of IscS. *Mol. Biol. Evol.* 18: 1919–1928.

Tielens, A.G., Rotte, C., van Hellemond, J.J. and Martin, W. (2002) Mitochondria as we don't know them. *Tr. Biochem. Sci.* 27: 564–572.

Tovar, J., Fischer, A. and Clark, C.G. (1999) The mitosome, a novel organelle related to mitochondria in the amitochondrial parasite *Entamoeba histolytica*. *Mol. Microbiol.* 32: 1013–1021.

Tovar, J., Leon-Avila, G., Sánchez, L.B., Sutak. R., Tachezy, J., van der Giezen, M., Hernandez, M., Müller, M. and Lucocq, J.M. (2003) Mitochondrial remnant organelles of *Giardia* function in iron-sulphur protein maturation. *Nature* 426: 172-176.

Upcroft, J. and Upcroft, P. (1998) My favourite cell: *Giardia*. *Bioessays* 20: 256–263.

van de Peer, Y., Ben Ali, A. and Meyer, A. (2000) Microsporidia: Accumulating molecular evidence that a group of amitochondriate and suspectedly primitive eukaryotes are just curious fungi. *Gene* 246: 1–8.

Vossbrinck, C.R., Maddox, J.V., Friedman, S., Debrunner-Vossbrinck, B.A. and Woese, C.R. (1987) Ribosomal RNA sequence suggests microsporidia are extremely ancient eukaryotes. *Nature* 326: 411–414.

Weiss, L.M., Edlind, T.D., Vossbrinck, C.R. and Hashimoto, T. (1999) Microsporidian molecular phylogeny: The fungal connection. *J. Euk.Microbiol.* 46: 17S–18S.

Whatley (1976) Bacteria and nuclei in *Pelomyxa palustris*: Comments on the theory of serial endosymbiosis. *New Phytol.* 76: 111–120.

Whatley, J.M., John, P. and Whatley, F.R. (1979) From extracellular to intracellular: The establishment of mitochondria and chloroplasts. *Proc. R. Soc. Lond. B* 204: 165–187.

Williams, B.A., Hirt, R.P., Lucocq, J.M. and Embley, T.M. (2002) A mitochondrial remnant in the microsporidian *Trachipleistophora hominis*. *Nature* 418: 865–869.
Yamamoto, A., Hashimoto, T., Asaga, E., Hasegawa, M. and Goto, N. (1997) Phylogenetic position of the mitochondrion-lacking protozoan *Trichomonas tenax*, based on amino acid sequences of elongation factors 1-alpha and 2. *J. Mol. Evol.* 44: 98–105.

Chapter 3

The Evolutionary History of Plastids: A Molecular Phylogenetic Perspective

John M. Archibald and Patrick J. Keeling

CONTENTS

Abstract

Plastids, the light-harvesting organelles of photosynthetic eukaryotes, are derived from an ancient symbiosis between a eukaryote and a cyanobacterium. This process is called primary endosymbiosis, and accounts for plastids in glaucocystophytes, red algae, green algae and land plants. All other plastid-containing eukaryotes acquired their plastids from either a red or a green alga by secondary endosymbiosis, in which a eukaryotic cell swallows a second phototrophic eukaryote and retains its photosynthetic machinery. Secondary endosymbiosis accounts for the plastids found in most of the diversity of eukaryotic algae, including the chlorarachniophytes, euglenids, cryptomonads, haptophytes, heterokonts, dinoflagellates and apicomplexan parasites. This chapter discusses the changing views with respect to the origin and evolution of plastid-containing organisms, with an emphasis on the molecular phylogenetic evidence bearing on the number of primary and secondary endosymbioses that have occurred during eukaryotic evolution.

0-415-29904-7/04/$0.00+$1.50

3.1 Introduction

The origin of photosynthesis was one of the major turning points in eukaryotic evolution. Beyond forming the foundation of most marine, freshwater and terrestrial ecosystems, the diversification of plants and algae has impacted most aspects of eukaryotic diversity and evolution in some way. Both the origin and the spread of photosynthesis in eukaryotes have complicated histories: the history of our knowledge on these subjects, and also, as it turns out, the evolutionary history of photosynthetic organelles themselves are unexpectedly complex. It is now widely accepted that the photosynthetic organelles of eukaryotes (plastids or chloroplasts) are derived from once free-living cyanobacteria. The similarities between plastids and cyanobacteria had been noted as early as the end of the 19th century (Schimper, 1883), and in 1905, Mereschkowsky formally proposed that the organelle was derived from a prokaryotic cell (Mereschkowsky, 1905). Nearly a century later, a wealth of ultrastructural, biochemical and molecular data now support this conclusion.

Although the prokaryotic nature of plastids is now patently obvious, many questions remain as to how the two cells integrated, what kinds of cells were involved in the endosymbiotic partnership and when and how many times these events took place. Part of the difficulty in resolving these questions lies in the extraordinary diversity of extant photosynthetic eukaryotes. Plastids are found in organisms ranging from benthic and planktonic unicells to soil microorganisms, 30-m kelps on the ocean floor and the trees and other plants that inhabit dry land. Plastid-bearing organisms are not only morphologically diverse but also heterogeneous with respect to their mode of nutrition: many are strict phototrophs, others maintain both phototrophic and heterotrophic lifestyles and still others are nonphotosynthetic parasites or predators. Because of this diversity, inferring the evolutionary origins of plastids and tracing their history among modern-day photosynthetic organisms have proven difficult. The situation has been further confounded by plastids having spread laterally among unrelated eukaryotic lineages. This process, called secondary endosymbiosis, has generated the bulk of algal biodiversity.

This chapter discusses the origin and evolution of primary and secondary plastids. In particular, it focuses on molecular phylogenetic evidence bearing on the question of the number of times each of these processes has occurred during eukaryotic evolution and the nature of the cells involved.

3.2 Primary Plastids

Oxidative photosynthesis first evolved in the ancestor of cyanobacteria, a ubiquitous and diverse group of photosynthetic prokaryotes. Several prokaryotic lineages use some form of photosynthesis, but the system found in cyanobacteria is distinctive as it uses two consecutive light-harnessing photosystems (Photosystems I and II) and oxygen as the terminal electron acceptor (Blankenship, 1994). The origin of this very powerful system was a critical event in evolution: it not only resulted in the drastic alteration of the earth's atmosphere by forming substantial quantities of free oxygen gas but also set the stage for photosynthesis in eukaryotes.

As diverse and abundant as the cyanobacteria are, they still account for only a small fraction of primary production at present. The vast majority of primary production is carried out by eukaryotic phototrophs (plants and algae) that acquired their photosynthetic apparatus by engulfing and retaining a cyanobacterium. This process, illustrated in Figure 3.1, began with a phagotrophic eukaryote swallowing a cyanobacterium, likely intended to be a meal for the predator (Figure 3.1a). Rather than being digested, however, the cyanobacterium was retained in the host cell, where it continued to photosynthesize (Figure 3.1b). Initially, the host probably took up endosymbionts transiently for a short-term gain of some

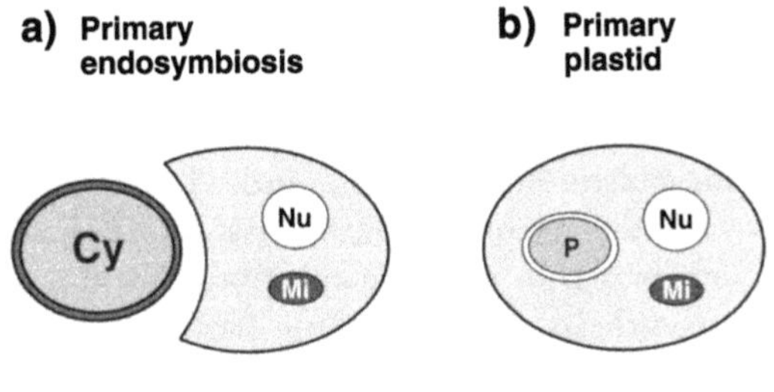

FIGURE 3.1 The process of primary endosymbiosis. (a) Primary endosymbiosis involves the uptake of a cyanobacterium by a phagotrophic eukaryote. (b) Primary-plastid-containing organism with two membranes surrounding its photosynthetic apparatus. The plastid resides in the cytosol and its two membranes are derived from the inner and outer membranes of the Gram-negative bacterium (see text). Abbreviations: Cy, cyanobacterium; Nu, nucleus; Mi, mitochondrion; P, plastid.

free carbohydrate, but at some point this partnership was fixed. The process of phagocytosis would result in a three-membrane-bound inclusion: two membranes from the Gram-negative cyanobacterium and the phagosomal membrane of the host. However, primary plastids are bound by two membranes only and the protein composition of these membranes (Jarvis and Soll, 2001) suggests that they correspond to the inner and outer membranes of the cyanobacterium. It has been suggested therefore that the endosymbiont was not digested because the phagosomal membrane ruptured, leaving the endosymbiont relatively safe in the host cytosol. This might be the case, but it is also possible that a prolonged series of transient endosymbioses allowed the host to adapt to the maintenance of an endosymbiont, and that the fixation of the partnership was finally ensured by some other event.

Whatever the initial forces promoting the fixation of the endosymbiosis, the cyanobacterium and host ultimately became mutually dependent by genetic integration. While an endosymbiont lives within its host there is constant potential for endosymbiont genes to move to the host genome. If these genes are integrated into the host chromosomes, transcribed and translated, and their protein products somehow transported back to the endosymbiont, then the original endosymbiont-encoded copies of the gene are redundant and can be lost without deleterious effect. Once a mechanism is in place to ensure that a class of proteins is targeted to the endosymbiont, it opens the door to further gene migrations, and the endosymbiont becomes increasingly dependent on the host for its full complement of proteins. Although this process seems staggeringly unlikely, it must have happened thousands of times. All plastids retain a genome, but in all cases it is a mere shadow of the original cyanobacterial genome and only encodes a small fraction of the proteins needed to maintain the organelle. The majority of plastid proteins are encoded in the nuclear genomes of plants and algae. In most cases, these genes are clearly of cyanobacterial origin, but have been transferred to the nuclear genome. [A recent analysis of the genome of the flowering plant *Arabidopsis* estimates that ca. 4500 genes are derived from the plastid endosymbiont, and about half of these encode proteins that appear to be targeted to the plastid (Martin et al., 2002)]. These proteins are expressed using the host transcription and translational apparatus and are posttranslationally targeted to the plastid. This targeting relies on an N-terminal extension on each of these proteins, called a transit peptide. The transit peptide is the flag that tells the cell that the protein belongs in the plastid, and it is transported across the two plastid membranes by a complex protein translocation system embedded on the outer and inner membranes.

On the other side of the equation, the host might also become dependent on the endosymbiont. Photosynthesis is clearly an enviable trait to maintain, but it can be lost, and this has occurred in several plants and a large number of algal lineages. However, the endosymbiont was not merely a photosynthetic machine; initially it was a biochemically complex cell. As it integrated with the host, a few pathways not related to photosynthesis were also retained by the plastid, and it is possible that the corresponding pathways were lost in the host, making the plastid indispensable. Indeed, plastids of plants and algae have been documented to carry out diverse biochemical roles, such as biosynthesis of amino acids, fatty acids, isoprenoids and heme, in addition to photosynthesis and various biosynthetic pathways related to photosynthesis. Whether these pathways have been partially or fully lost in various hosts is unclear, but it is clear that each makes absolute loss of the plastid (as opposed to loss of photosynthesis, which is relatively common) a very difficult undertaking. In no case has plastid loss been unambiguously documented, although nonphotosynthetic plastids can be very difficult to detect, and it is hard to say what kind of evidence would be necessary to prove that a plastid was absent.

3.3 One Origin

In spite of the huge variety in morphology, biochemistry and pigment content of modern plastids, it has become clear that they all evolved from a single endosymbiotic event. This was not always thought to be the case. The diversity of plastid types has always left some suspicion that plastids arose multiple times independently by several endosymbioses involving different cyanobacteria. The recognition that plastids were being passed around among otherwise-unrelated eukaryotes by secondary endosymbiosis (see later) simplified this problem to a great extent as it revealed that several algal lineages derived plastids from red or green algae. With a full understanding of the extent of secondary endosymbiosis, it was clear that only three lineages harbor plastids derived from primary endosymbiotic events with a cyanobacterium: red algae, green algae and glaucocystophytes. Molecular data immediately supported the notion that the plastids harbored in all three lineages were closely related to one another to the exclusion of all cyanobacteria. Phylogenies based on genes encoding plastid small subunit ribosomal RNA (SSUrRNA) and elongation factor Tu (EF-Tu), and an intron found in plastid $tRNA^{Leu}$, supported the monophyly of these primary plastids (e.g., Besendahl et al., 2000; Delwiche et al., 1995; Helmchen et al., 1995). In addition to these phylogenetic studies, there are reports that plastids possess a novel class of three-helix light-harvesting antennae proteins (Durnford et al., 1999) and plastid genomes share a number of unique characteristics, most importantly the presence of an inverted repeat structure consisting of the rRNA operon and tRNAs and the organization of ribosomal protein operons (McFadden and Waller, 1997; Stoebe and Kowallik, 1999). This evidence notwithstanding, phylogenies based on genes from the host nuclear genomes appeared to challenge the monophyly of plastids. The first gene trees (based on SSUrRNA) that included sequences from red, green and glaucocystophyte algae did not show them to be a monophyletic group (Bhattacharya et al., 1995a), leading to various hypotheses regarding independent plastid origins. This was also the case with a few protein-coding genes, including β-tubulin and actin (in some analyses; Bhattacharya and Weber, 1997; Keeling et al., 1999). Although these trees did not support the monophyly of primary algae, they did not really argue against it either, because there was virtually no support for the position of these lineages. Several weak gene trees failing to support a group do not equate to strong evidence against the group. What is needed is a strong gene tree supporting some other relationship. One gene phylogeny apparently did support an alternative story: analysis of the largest subunit of RNA polymerase II (RPB1) showed strong statistical support for a separation between red and green algae (Stiller and Hall, 1997). This was used to support a model of plastid evolution in which only one of red, green or glaucocystophyte algae contained a primary plastid, and the other two contained secondary plastids, which had simply reduced their membrane compliment to two. This is very difficult to imagine in light of the protein-trafficking system in secondary plastids (see later).

Subsequently, other genes have been analyzed individually and in combination, and altogether the case for multiple independent plastid origins has evaporated. Several nuclear genes have shown a weak relationship between red and green algae, including enolase; actin (in some analyses); α-tubulin; hsp70; hsp90; α and β subunits of vacuolar ATPase; and a combined analysis of α- and β-tubulins, EF-1α and actin (Archibald et al., 2001; Baldauf et al., 2000; Keeling, 2001; Keeling et al., 1999; Keeling and Palmer, 2001; Moreira et al., 2000; Stibitz et al., 2000). Moreover, the phylogeny of translation elongation factor 2 (EF-2) was shown to support the monophyly of red and green algae with very high statistical support, as was a combined analysis of 13 nuclear-encoded cytosolic genes (Moreira et al., 2000). Conversely, a reanalysis of the RPB1 gene failed to reject the possibility of a monophyletic red and green algae (Moreira et al., 2000). Recent analyses of RPB1 have also shown the red and green algae to branch together, albeit weakly (D. Longet, JMA, PJK and J. Pawlowski, unpublished data). For glaucocystophytes, there is little molecular data from nuclear genes, but some genes show them branching with the red or green algae, or both (e.g., Stibitz et al., 2000), and two multigene analyses also concluded that the red and green algae were closely related to the glaucocystophytes (Baldauf et al., 2000; Moreira et al., 2000). Altogether, the consistent picture coming from plastid genes, mitochondrial genes and nuclear genes is the same: red, green and glaucocystophyte algae share a common ancestor, and the origin of primary plastids can reasonably be inferred to trace back to a single endosymbiotic event.

3.4 Three Lineages

At present, primary plastids are found in only three lineages of eukaryotes: the glaucocystophytes, the red algae and the green algae (and their land plant relatives). Land plants and green algae are a pervasive and dominant group of eukaryotes. Green algae are found in almost every aquatic and marine environment known and are an extremely diverse and specious group. The green algal lineage is deeply divided into two major subgroups, the chlorophytes and streptophytes, and land plants evolved from within the streptophyte clade. Green algae and land plants are characterized by the presence of chlorophylls *a* and *b* in their plastid. Red algae are also a very diverse and specious group. They are predominantly marine but are also present in freshwater environments, and some red algae inhabit relatively hostile environments such as high heat or high salinity. Certain red algae have evolved a complex form of multicellularity not unlike plants in many respects. Red algal plastids are also distinctive as they contain chlorophyll *a* and phycobilins, the latter being accessory pigments associated with structures called phycobilisomes that are visible by electron microscopy on the outside of the inner plastid membrane. Glaucocystophytes are a relatively small group of algae, with only three genera and a handful of species. The glaucocystophytes also contain chlorophyll *a* and phycobilins, but their plastid is dramatically distinguished from other primary plastids in that it has retained the peptidoglycan wall between the inner and outer membranes.

The peptidoglycan wall of the cyanobacterial endosymbiont has been lost in red and green algae, and this has led to speculation that the glaucocystophytes might be the descendents of an early stage of the primary endosymbiosis (Herdman and Stanier, 1977). However, every other possible relationship between the three primary algal lineages has been proposed based on different lines of evidence (Cavalier-Smith, 1982; Delwiche et al., 1995; Kowallik, 1997; Martin et al., 1998; Valentin and Zetsche, 1990). Initially, molecular phylogenies with data from all three groups shed little light on this problem, as many phylogenies failed to even unite the three lineages (see previously), whereas others showed little consistency in the relative order of the three clades. In phylogenies based on plastid EF-Tu, green algae are sister to a red algal–glaucocystophyte clade (Delwiche et al., 1995). In contrast, rubisco (ribulose-1,5-bisphosphate carboxylase/oxygenase) phylogenies show that red algae contain one type of rubisco, whereas glaucocystophytes and green algae share another distantly related type (Delwiche and Palmer, 1996; Valentin and Zetsche, 1990). The molecule most extensively applied to this question is plastid SSUrRNA, and phylogenies based on this gene generally show glaucocystophytes branching before green and red algae (Helmchen et al., 1995; Turner et al., 1999). However, different analyses result in different topologies and varying levels of support (Nelissen et al., 1995), leaving the matter open to speculation.

One approach that has been used to resolve these relationships with greater certainty is to analyze very large data sets of concatenated genes. Recent analyses of plastid-encoded genes identified glaucocystophytes as the earliest primary-plastid-containing lineage, and in most phylogenies this position is well supported (Martin et al., 1998, 2002). Even with this amount of data, however, maximum likelihood analysis fails to unambiguously resolve the relative position of glaucocystophytes to green and red algae. This region of the tree is clearly recalcitrant to phylogenetic reconstruction, but for the most part, existing data seem to support the notion that glaucocystophytes are the deepest lineage of primary-plastid-containing eukaryotes.

3.5 Secondary Plastids

Although all plastids appear to be the product of a single primary endosymbiosis, primary-plastid-containing organisms account for only a fraction of eukaryotic photosynthetic diversity. A vast array of algae have acquired plastids through a process called secondary endosymbiosis, in which a phagotrophic eukaryote engulfs a eukaryotic cell with a primary plastid and retains its photosynthetic machinery (Figure 3.2a,b). Secondary (or complex) plastids differ from primary plastids in several important respects. First, and as a natural consequence of eukaryotic phagocytosis, secondary plastids are characterized by the presence of additional membranes. Whereas primary plastids have a double-membrane envelope, secondary plastids have three or four membranes. The outermost membrane corresponds to the phagosomal membrane of the secondary host cell and the third membrane of four-membrane plastids is thought to be derived from the plasma membrane of the algal endosymbiont (McFadden, 1999). Secondary plastids therefore reside within the lumen of the host cell's endomembrane system, unlike primary plastids, which are in the cytosol. Second, the differences in membrane structure between primary and secondary plastids result in different mechanisms for targeting proteins to the organelle. The vast majority of plastid proteins in plants and algae are encoded in the nucleus. These proteins are translated in the cytosol and imported into the plastid via a transit peptide. In algae with secondary plastids, the proteins must cross one or more additional membranes, for which they make use of the signal-peptide secretion system. Plastid proteins are first targeted to the endomembrane system by using a signal peptide, then localized to the plastid (by an unknown mechanism) and finally sent across the remaining two membranes by a standard transit peptide, as in primary plastids (Cavalier-Smith, 1999; Ishida et al., 2000; Kishore et al., 1993; McFadden 1999).

The space between the second and third membranes of four-membrane plastids corresponds to the remnant cytosol of the engulfed algal cell. In two different algal groups, this cellular compartment harbors the "smoking gun" of the process of secondary endosymbiosis: the relict nucleus of the eukaryotic endosymbiont. First identified as a double-membrane-bound body in early microscopic studies of cryptomonads (Greenwood et al., 1977), the nucleomorph was later discovered in the same compartment of chlorarachniophyte algae (Hibberd and Norris, 1984) and was found to contain DNA (Ludwig and Gibbs, 1989).

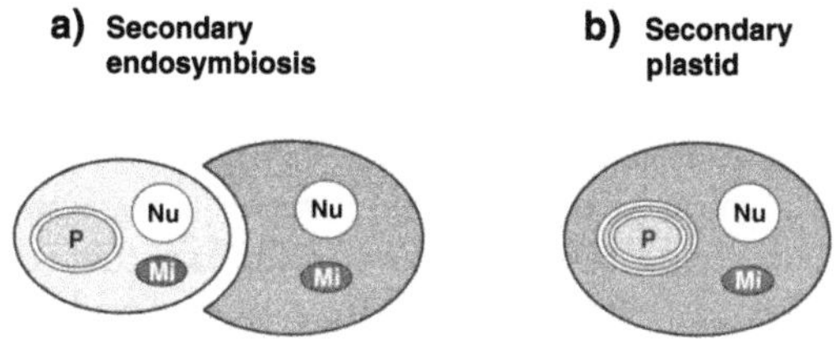

FIGURE 3.2 The general features and end product of secondary endosymbiosis. (a) Secondary endosymbiosis involves a phagotrophic, nonphotosynthetic eukaryote engulfing a primary-plastid-containing eukaryote. (b) A secondary-plastid-containing organism. The plastid resides within the endomembrane system of the host cell and is bound by additional membranes, typically four. The outermost (fourth) membrane corresponds to the phagosomal membrane of the secondary host, and the third membrane is derived from the plasma membrane of the engulfed algal cell (see text). Abbreviations: Cy, cyanobacterium; Nu, nucleus; Mi, mitochondrion; P, plastid.

Nucleomorphs have since been shown to be genuine eukaryotic nuclei, although with highly reduced, AT-rich and generally divergent genomes (Douglas et al., 2001; Gilson et al., 1997; Gilson and McFadden, 1996). Nucleomorph genome sequencing projects (Douglas et al., 2001; Gilson and McFadden, 2002) have revealed that large numbers of genes encoding plastid proteins have been transferred from the nucleomorph to the host nucleus in cryptomonads and chlorarachniophytes. This process has presumably gone to completion in all other algae that possess secondary plastids but lack nucleomorphs.

3.6 Red and Green Endosymbionts

The most obvious division among secondary plastids is between those derived from red or green algal endosymbionts. Two distinct groups of algae have green algal secondary plastids: euglenids and chlorarachniophytes. Euglenids are a common group of flagellates in both freshwater and marine environments, whereas the chlorarachniophytes are a more rare class of amoeboflagellate algae characterized by long and sometimes reticulating pseudopodia. A green algal origin for their respective plastids seemed likely given their photosynthetic pigment composition: both contain the signature pigmentation of green algae and land plants — chlorophylls *a* and *b*. This has since been confirmed by molecular data. Phylogenies of plastid-, and, in the case of the chlorarachniophytes, nucleomorph-encoded genes demonstrate that both plastids are derived from green algae (e.g., Durnford et al., 1999; Ishida et al., 1997, 1999; McFadden et al., 1995; Turmel et al., 1999). In neither case, however, has a particular green algal lineage been identified as the unambiguous source of the organelle. Although phylogenies of chloroplast EF-Tu have suggested that the chlorarachniophyte endosymbiont was an ulvophycean green alga (Ishida et al., 1997), nucleomorph SSUrRNA trees have been used to argue that the endosymbiont was an ulvophyte (Ishida et al., 1999) or, alternatively, a trebouxiophyte green alga (Van de Peer et al., 1996). The use of nucleomorph-encoded genes has been hampered by the fact that they are typically fast evolving (Gilson and McFadden, 2002) and therefore prone to misplacement in molecular phylogenies. Phylogenetic trees inferred from plastid SSUrRNA and the large subunit of rubisco have placed the chlorarachniophytes basal to the chlorophytes (to which the ulvophytes and trebouxiophytes belong), the streptophytes and the euglenids (McFadden et al., 1995). To complicate matters further, the chlorarachniophytes have recently been shown to possess a gene encoding the metabolic enzyme enolase that shows affinity to streptophyte enolases, a result that is supported by several homologous amino acid insertions (Keeling and Palmer,

2001). The origin of the chlorarachniophyte endosymbiont within green algae is thus at present an open question.

The exact origin of the euglenid plastid is similarly ambiguous. The plastid genome of *Euglena gracilis* has been completely sequenced (Hallick et al., 1993) and shares many features with other green algal plastid genomes. However, no specific group has been singled out as a close relative of the endosymbiont, beyond the fact that it was likely a member of the Chlorophyta (Turmel et al., 1999). The plastids in euglenids and chlorarachniophytes are surrounded by three and four membranes, respectively, and the two groups are generally considered to have acquired their plastids independently (see later).

By comparison, a much larger number of algal groups possess secondary plastids derived from red algae. The cryptomonads are an abundant group of flagellated unicells that are significant in possessing, together with the chlorarachniophytes, a nucleomorph. The nucleomorph genome of the cryptomonad *Guillardia theta* has been completely sequenced and is a mere 551 kb in size, partitioned among three similarly sized chromosomes (Douglas et al., 2001). The heterokonts (or stramenopiles) and haptophytes are two very important algal groups that together make up a large fraction of the earth's aquatic photosynthesizers. From a cell biological perspective, the cryptomonad, haptophyte and heterokont plastids are unique in that the outermost membrane surrounding the plastid is continuous with the endoplasmic reticulum (ER) and the outer membrane of the nuclear envelope (Gibbs, 1981). Molecular evidence for a red algal origin for their secondary plastids is extensive and involves consideration of conserved features of plastid genome organization (Douglas and Penny, 1999) as well as phylogenetic analyses of plastid and nucleomorph rRNAs and several proteins (e.g., Archibald et al., 2001; Daugbjerg and Andersen, 1997; Van de Peer and De Wachter, 1997; Van der Auwera et al., 1998). The phylogenies of the large and small subunits of rubisco have been particularly informative. In contrast to green algae, chlorarachniophytes and euglenids, which contain a rubisco enzyme derived from cyanobacteria, red algae contain a distantly related proteobacterial type of rubisco that appears to have been acquired via lateral gene transfer (Delwiche, 1999; Delwiche and Palmer, 1996). Significantly, the cryptomonads, haptophytes and heterokonts contain the proteobacterial or red algal form of rubisco.

The dinoflagellates and apicomplexan parasites, which, together with ciliates, make up a higher-order group of eukaryotes referred to as the alveolates (Wolters, 1991), also appear to harbor plastids derived from red algae. In the case of the apicomplexans, the presence of a plastid was most unexpected (McFadden et al., 1996; Wilson et al., 1996), because these organisms are nonphotosynthetic, intracellular parasites of animals. Determining the source of the relict organelle has been difficult, in large part because all of the genes encoding proteins directly involved in photosynthesis have been lost, and the ones that remain are extremely divergent. Although early phylogenies based on EF-Tu suggested that the apicomplexan plastid was derived from a green alga (Köhler et al., 1997), a number of observations on the gene content of the genome and phylogenetic analyses based on other genes suggested that the plastid was red algal (Williamson et al., 1994; McFadden and Waller, 1997; Blanchard et al ., 1999; Gardner et al., 2002). More recently, a green algal origin for the apicomplexan plastid has been suggested based on characteristics of the mitochondrial protein COXII. In nearly all eukaryotes, COXII is found as a single protein encoded in the mitochondrial genome, but apicomplexa and certain green algal COXIIs are split into two pieces encoded by the nuclear genes *cox2a* and *cox2b* (Funes et al., 2002). Interestingly, Funes et al. show that the split genes in apicomplexa are closely related to their split homologues in green algae, leading to the conclusion that apicomplexa inherited these split genes from a green algal endosymbiont that gave rise to the plastid (Funes et al., 2002). However, these authors did not analyze ciliate COXIIs, which are critical because ciliates are close relatives of apicomplexa. When these genes are included in the phylogeny, the ciliates and apicomplexa form a clade to the exclusion of green algae (unpublished observations). Moreover, the ciliate proteins contain a very large (up to 300 amino acids) insertion in exactly the same position as the split in the apicomplexan and green algal homologues, which seems unlikely to be a coincidence. It will be interesting to investigate the nature of dinoflagellate COXIIs (none are known as yet), but considering all the information available at present, there is no evolutionary link between the COXIIs of apicomplexa and green algae.

In general, the overall characteristics of the apicomplexan plastid genome seem more consistent with a red algal origin. Analyses of plastid gene sequences from dinoflagellates support a red ancestry for their plastid (Takishita and Uchida, 1999; Zhang et al., 1999) and also suggest that the dinoflagellate and apicomplexan plastids are specifically related (Zhang et al., 2000; see later), consistent with the known relationship between the two host cells. With the exception of the dinoflagellate plastid, which is bound by three membranes, all other red algal secondary plastids are surrounded by four membranes. The cryptomonads, haptophytes, heterokonts and dinoflagellates are distinctive in that their plastids contain the secondary photosynthetic pigment chlorophyll *c*. (The Apicomplexa are nonphotosynthetic and thus lack pigmentation.) Table 3.1 summarizes the full spectrum of primary- and secondary-plastid-containing organisms and the general features of their plastids.

3.7 How Many Secondary Endosymbioses?

The fact that modern-day algae harbor secondary plastids derived from both red and green algal endosymbionts necessitates that secondary endosymbiosis has occurred at least twice during eukaryotic evolution. However, the morphological and biochemical diversity of secondary-plastid-containing algae, particularly among those with red algal plastids, is such that it has been difficult to determine the exact number of events. Depending on the type of data considered, estimates have ranged from two to as many as seven separate secondary endosymbioses (Cavalier-Smith, 1999; Delwiche and Palmer, 1997). Fortunately, recent molecular data have begun to clarify this controversial issue.

TABLE 3.1 General Characteristics of Primary and Secondary Plastids

Lineage	*Plastid Type*	*Plastid Membranes*	*Pigments*
Green algae	1°	2	Chlorophyll a + b
Red algae	1°	2	Chlorophyll a, phycobilins
Glaucocystophytes	1°	2[a]	Chlorophyll a, phycobilins
Chlorarachniophytes[b]	2° (green)	4	Chlorophyll a + b
Euglenids	2° (green)	3	Chlorophyll a + b
Cryptomonads[b]	2° (red)	4	Chlorophyll a + c, phycobilins
Heterokonts	2° (red)	4	Chlorophyll a + c
Haptophytes	2° (red)	4	Chlorophyll a + c
Dinoflagellates	2° (red)	3	Chlorophyll a + c, peridinin
Apicomplexa	2° (red)	4	N/A[c]

[a] Glaucocystophyte plastids are also bound by a peptidoglycan layer between the two membranes.

[b] Groups that have retained the nucleus of the algal endosymbiont (the nucleomorph).

[c] Apicomplexan parasites are nonphotosynthetic; their plastids contain no light-harvesting pigments.

With respect to the green plastids of chlorarachniophytes and euglenids, although it has been suggested that their photosynthetic organelles trace back to a single secondary endosymbiotic event (Cavalier-Smith, 1999, 2000), the most widely held view is that they were acquired separately from different green algae. This is based on a general lack of morphological similarity shared between the host cells (Hibberd and Norris, 1984) and on both their hosts and plastids appearing unrelated in molecular phylogenies (Bhattacharya et al., 1995b; Ishida et al., 1997; Keeling, 2001; McFadden et al., 1995). The host component of the chlorarachniophyte algae is a member of the Cercozoa, a large assemblage of protists that includes the euglyphid amoebae, cercomonad flagellates and plasmodiophorid plant pathogens. The Cercozoa are an extremely diverse lineage, and they have only recently been recognized as a monophyletic group based on molecular phylogenies (Bhattacharya et al., 1995b; Cavalier-Smith, 1998; Cavalier-Smith and Chao, 1997; Keeling et al., 1998). Cercozoa have more recently been suggested to be related to the Foraminifera, a group of nonphotosynthetic, pseudopod-forming, predominantly testate protists, based on phylogenies of actin (Keeling, 2001) and a shared insertion in their polyubiquitin genes (Archibald et al., 2002). The euglenid host, on the other hand, belongs to the Euglenozoa, a large eukaryotic group that also includes kinetoplastids and diplonemids. Within the Euglenozoa, the photosynthetic euglenids (such as *Euglena*) are clearly nested within heterotrophic lineages (Preisfeld et al., 2001) and the inferred series of cytoskeletal transformations in the group indicates that phototrophy is a relatively recent adaptation within euglenozoan evolution (Leander et al., 2001a,b). At present, the data is most consistent with the hypothesis that the euglenid and chlorarachniophyte host cells are not specifically related and that their plastids are of independent origin.

The origin and evolution of red secondary plastids have been more difficult to discern. A large number of biochemical and ultrastructural features are suggestive of a specific relationship between some or all chlorophyll *a* + *c*-containing algae and the possibility that their plastids share a common origin. In particular, the heterokonts, haptophytes and cryptomonads possess four-membrane plastids that reside in the lumen of the ER. Cavalier-Smith (1981, 1982) placed these organisms together in the kingdom Chromista on the basis of this characteristic, arguing that fusion of the outermost plastid membrane with the outer nuclear envelope is improbable and therefore likely occurred only once in their common ancestor (Cavalier-Smith, 1986). The tubular mastigonemes (hairs) on the flagella of both heterokonts and cryptomonads have been argued to be homologous (Cavalier-Smith, 1986), and with respect to the heterokonts and haptophytes, each contain fucoxanthin and chrysolaminaran. Starch is present in both cryptomonads and dinoflagellates (stored in the cytosol of dinoflagellates but in the periplastid space in cryptomonads), and, finally, heterokonts, haptophytes and dinoflagellates each have three thylakoids per stack in their respective plastids as well as tubular mitochondrial cristae (cryptomonads have flat cristae). Of the four lineages with chlorophyll *a* + *c*-pigmented plastids, the heterokonts and haptophytes are most similar from a morphological and biochemical perspective, and have historically been grouped together as chromophyte algae (Andersen, 1991).

Despite these similarities, early molecular studies painted a different picture. Phylogenetic analyses of rubisco and plastid SSUrRNA showed that cryptomonads, heterokonts and haptophytes did not form a monophyletic group within the red algae (Daugbjerg and Andersen, 1997; Medlin et al., 1995; Müller et al., 2001; Oliveira and Bhattacharya, 2000). Nuclear SSUrRNA trees also failed to unite the host components of the three groups (Bhattacharya et al., 1995a), and together these results supported the idea that the haptophyte, heterokont and cryptomonad plastids were derived from separate endosymbiotic events. However, more recent molecular data are more in line with morphological and biochemical considerations and a single origin for their plastids. A comprehensive analysis of a concatenated dataset containing five plastid-encoded genes (SSUrRNA, *psa*A, *psb*A, *rbc*L and *tuf*A) showed the cryptomonads, heterokonts and haptophytes to be a highly supported monophyletic group (Yoon et al., 2002b).

The dinoflagellates and apicomplexans have been particularly difficult to fit into the puzzle of plastid origins. Until recently, no gene sequences were available from dinoflagellate plastid genomes and there has been considerable debate on whether the relict apicomplexan plastid is of green or red algal origin (Blanchard and Hicks, 1999; Köhler et al., 1997; McFadden and Waller, 1997). Several dinoflagellate plastid genes have now been characterized (Takishita and Uchida, 1999; Zhang et al., 1999). Curiously, these genes are encoded on single-gene minicircles, unlike all other known plastid genomes (Zhang et al., 1999). In molecular phylogenies of rRNA, dinoflagellate and apicomplexan sequences branch together, suggesting that their plastids share a common ancestry (Zhang et al., 2000). However, this result has been interpreted with caution: both the dinoflagellate and the apicomplexan genes are divergent and AT rich, leaving open the possibility that their affinity in phylogenetic trees is due to methodological artifact.

More recently, analysis of the metabolic enzyme glyceraldehyde-3-phosphate dehydrogenase (GAPDH) has provided additional support for the hypothesis that the dinoflagellate and apicomplexan plastids are related and, significantly, suggests that the red algal plastids of heterokonts, cryptomonads and alveolates share a single origin. All algae and plants have two nuclear-encoded forms of GAPDH: one that functions in the plastid and another that functions in the cytosol. As expected, the GAPDH isoform targeted to the primary plastids of red and green algae is related to the GAPDH found in cyanobacteria. In stark contrast, the dinoflagellate, apicomplexan, heterokont and cryptomonad plastid GAPDH is not cyanobacterial, but is related to the eukaryotic cytosolic isoform (Fast et al., 2001). Apparently, the gene encoding the cytosolic GAPDH was duplicated, acquired the information necessary to target its protein product to the plastid and took over the role of the cyanobacterial homologue. This process is known as endosymbiotic gene replacement (Martin and Schnarrenberger, 1997), and in the case of GAPDH the data suggest that the replacement took place only once, in a common ancestor shared by apicomplexa, dinoflagellates, heterokonts and probably cryptomonads (Fast et al., 2001). Curiously, the dinoflagellate *Pyrocystis* possesses a cyanobacterial-like GAPDH that is closely related to that of *Euglena* (Fagan and Hastings, 2002); the origin of this gene is unclear. Nevertheless, the overall picture of red secondary plastid evolution painted by GAPDH is increasingly supported by analyses of host molecules. Phylogenetic analyses of nuclear-encoded rRNA suggest that the alveolates and heterokonts are a monophyletic group (Ben Ali et al., 2001; Van de Peer and De Wachter, 1997), a result that is also found in analyses of protein-coding genes (Baldauf et al., 2000).

Figure 3.3 summarizes the current consensus for the evolutionary history of primary and secondary plastids. Following a single endosymbiotic event between a heterotrophic eukaryote and a cyanobacterium, three primary-plastid-containing groups diverged from one another: the glaucocystophytes, the red algae and the green algae. Two secondary endosymbioses involving different hosts and different green algae gave rise to the euglenids and chlorarachniophytes, and a single secondary endosymbiosis occurred between a red alga and an ancestor of the cryptomonads, heterokonts, haptophytes and alveolates. All algae containing secondary plastids of red algal origin thus appear to comprise a eukaryotic supergroup dubbed as the chromalveolates (Cavalier-Smith, 1999).

3.8 Loss of Photosynthesis: How Common Is It?

The chromalveolate theory of plastid evolution has important ramifications for the evolution and cell biology of a large number of important eukaryotic lineages. Perhaps most significant is the fact that it invokes extensive loss of photosynthesis. Within alveolates, only approximately half of known dinoflagellate species are photosynthetic (Taylor, 1987), and the

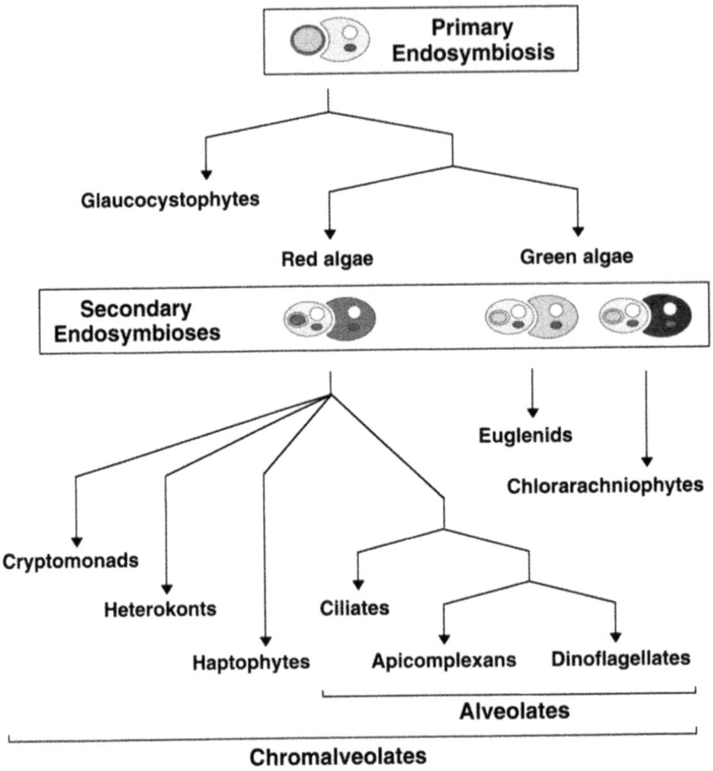

FIGURE 3.3 One scheme for the evolution of photosynthetic eukaryotes by primary and secondary endosymbiosis. A single primary endosymbiosis between a phagotrophic eukaryote and a cyanobacterium (top) gave rise to the three primary-plastid-containing lineages: glaucocystophytes, red algae and green algae. Subsequent to the diversification of these three groups, three secondary endosymbiotic events took place. Two separate endosymbioses involving different hosts and different green algae led to the chlorarachniophytes and the photosynthetic members of the euglenids. A single, ancient secondary endosymbiotic event between a phagotrophic eukaryote and a red alga spawned a lineage from which the cryptomonads, heterokonts, haptophytes, ciliates, apicomplexans and dinoflagellates evolved. The ciliates and apicomplexans are entirely nonphotosynthetic and loss of photosynthesis is rampant in dinoflagellates and heterokonts (see text).

ciliates, a major group of protist predators, are an entirely nonphotosynthetic lineage. If the heterokont and alveolate common ancestor had a plastid, then ciliates must have evolved from a plastid-containing organism, and might still have a plastid or plastid-derived genes (Fast et al., 2001). The heterokonts themselves also contain a large number of nonphotosynthetic members, many of which branch basal to photosynthetic lineages in phylogenetic trees. This pattern has previously been interpreted as evidence for the idea that the group as a whole was ancestrally nonphotosynthetic, and that a plastid (and photosynthesis) was acquired at a later point in the evolution of the heterokonts (Leipe et al., 1994). However, loss of photosynthesis has been invoked for some heterokonts (e.g., Cavalier-Smith et al., 1995), and a plastid-derived 6-phosphogluconate dehydrogenase gene has recently been identified in the nonphotosynthetic heterokont plant pathogen *Phytophthera infestans* (Andersson and Roger 2002), suggesting that the group as a whole had a photosynthetic ancestor. Within dinoflagellates, a recent analysis of nuclear SSUrRNA genes allowed inference of eight independent instances of loss of photosynthesis (Saldarriaga et al., 2001), a number that is sure to grow as more nonphotosynthetic lineages are sampled. Loss of photosynthesis — which should be distinguished from outright plastid loss — now appears to be far more common than previously appreciated (Archibald and Keeling, 2002; Cavalier-Smith, 1999).

3.9 Tertiary Endosymbiosis, Serial Secondary Endosymbiosis

The story of plastid evolution does not end with secondary endosymbiosis. When one considers the history of plastids within the dinoflagellates, an extraordinarily diverse and ecologically significant lineage, it is clear that the situation is even more complex. Most photosynthetic dinoflagellates contain three-membrane plastids with peridinin and chlorophylls *a* and *c*. Some dinoflagellates, however, have replaced their peridinin-containing plastids with an unrelated one through a process known as tertiary endosymbiosis. Remarkably, three of the five red secondary-plastid-containing lineages have had their plastids "stolen" by at least one dinoflagellate group. The endosymbiont of *Peridinium balticum* and *P. foliacium* is a diatom (heterokont) (Chesnick et al., 1997), that of *Gymnodinium acidotum* a cryptomonad (Farmer and Roberts, 1990; Wilcox and Wedemayer, 1985) and *Gymnodinium* and *Gyrodinium* harbor plastids of haptophyte origin (Tengs et al., 2000). It has recently been suggested that the standard peridinin plastid in dinoflagellates is itself derived from a haptophyte via tertiary endosymbiosis, on the basis of *psa*A, *psb*A and *rbc*L (large subunit of rubisco) gene phylogenies (Yoon et al., 2002a). Similarly, the apicomplexan plastid has also recently been suggested to be a product of a plastid replacement (Palmer, 2003), based on the apparent conflict between data suggesting that it is derived from a red alga (Fast et al., 2001; McFadden and Waller, 1997) and data suggesting it is derived from a green alga (Funes et al., 2002; Köhler et al., 1997). Although neither case can be proven or disproven definitively, the strongest evidence at present suggests that the plastid is red in nature and no such replacement took place. Comparing the full sequence of the *Plasmodium* genome (Gardner et al., 2002) with other chromalveolate genomes to come will doubtless settle this debate.

Still other dinoflagellates appear to have replaced their ancestral secondary plastid with a primary plastid in what is, strictly speaking, a serial secondary endosymbiosis. *Lepidodinium viride* has a chlorophyll *a* + *b*-containing plastid that appears to be derived from a green alga (Watanabe et al., 1990). The degree to which the various tertiary and serial secondary endosymbionts and plastids have integrated with their dinoflagellate hosts varies from transient endosymbioses to fully integrated and reduced organelles (Delwiche, 1999).

It is not clear why dinoflagellates have on the one hand repeatedly lost photosynthesis and on the other substituted their peridinin-containing plastids with plastids from other algae. However, the answer might lie, at least in part, in the fact that many dinoflagellates are mixotrophs: they practice both heterotrophy and phototrophy. The maintenance of a dual mode of nutrition might allows plastids (and photosynthesis) to come and go. Phagocytosis provides not only a constant supply of potential replacement plastids from prey algae but also an alternative energy source in instances where photosynthesis is insufficient. Regardless of the reasons for such rampant plastid shuffling, tertiary endosymbiosis adds another layer of complexity to an already complex array of nuclear- and plastid-encoded proteins that service the dinoflagellate plastid. In many instances, the nuclear-encoded, plastid-targeted proteins that functioned in the peridinin plastid can simply be recycled and imported into the newly acquired organelle. However, it is also possible that the genes encoding such proteins will be transferred from the endosymbiont nucleus to that of the host before it is lost, replacing the host copy. This has recently been demonstrated. In the dinoflagellate *Karenia brevis* (= *Gymnodinium breve*), the original nuclear-encoded oxygen-evolving enhancer 1 (PsbO) plastid protein from the peridinin-type plastid has been replaced by a haptophyte version that came in with the tertiary endosymbiont (Ishida and Green, 2002). Tertiary plastids are therefore evolutionary chimaeras in terms of their complement of plastid proteins.

3.10 A Second Primary Endosymbiosis?

Endosymbiosis has played a very important role in several of the critical steps in eukaryotic evolution. However, the process is often difficult to study because it is by nature a relatively rare and catastrophic event that often leads to sweeping changes in the host and endosymbiont. These changes erase many of the traces of how the event took place, or even who was involved. This is one reason why the process of secondary endosymbiosis is attractive to study: because secondary plastids have evolved multiple times independently, one can compare the results of different events and determine what some of the general principles of the process might be. In the case of the original primary plastid (and the origin of mitochondria), this powerful comparative approach is lost because the organelle arose from a unique endosymbiotic partnership, and hence there is nothing with which to compare the process — or is there?

One very interesting but very poorly understood organism that can provide a wealth of comparative information on the origin of plastids is the filose amoeba *Paulinella chromatophora*. Filose amoebae are a group of testate, or shelled, amoebae that are part of the Cercozoa, the same large and diverse lineage to which the chlorarachniophytes belong (although, within Cercozoa, *Paulinella* and other filose amoebae are not closely related to chlorarachniophytes). *P. chromatophora* is of interest because it might represent an independent origin of plastids. Each *P. chromatophora* cell harbors two kidney-shaped endosymbionts that have been shown by light microscopy and ultrastructural studies to be cyanobacteria (Kies, 1974). Interestingly, the endosymbionts could not be cultivated outside the host, and are known to divide synchronously with host cell division. (Each daughter of host division gets an endosymbiont, which then divides once to reconstitute the natural compliment of two endosymbionts.) The cyanobacteria have been suggested to retain their peptidoglycan wall and reside within a host vacuole (Kies, 1974); however, it is not exactly clear how many membranes actually surround the endosymbionts. *P. chromatophora* has not been observed to feed, and digestive vacuoles have never been observed (Kies, 1974), suggesting that the endosymbionts satisfy the energy requirements of the cell. Indeed, early biochemical observations indicated that the cyanobacteria are actively photosynthetic, and that photosynthate is transferred to the host where it is converted into macromolecules (Kies and Kremer, 1979). The cyanobacteria do not resemble extant plastids, suggesting that they are not derived from a secondary endosymbiosis, but instead appear to resemble *Synechococcus* (Kies, 1974), implying that they have been derived from an independent endosymbiosis. Interestingly, another species of *Paulinella*, *P. ovalis*, is nearly identical to *P. chromatophora* except that it is nonphotosynthetic and lacks any evidence of the endosymbionts. *P. ovalis* is phagotrophic, contains a clear digestive vacuole and feeds on *Synechococcus* (Johnson et al., 1988). One can see a possible evolutionary trajectory wherein a heterotrophic amoeba feeding on unicellular cyanobacteria captured a prey cell, and for some reason failed to digest it, resulting in something like *P. chromatophora* (Johnson et al., 1988). Judging from the similarity between *P. ovalis* and *P. chromatophora*, this event must have taken place very recently, so *P. chromatophora* would appear to be an ideal organism in which to study the early stages of a possible primary endosymbiosis and compare the features of this partnership with modern and fully developed plastids.

Acknowledgments

We thank J. Saldarriaga and G. I. McFadden for translations from German, and B. Leander and J. T. Harper for helpful comments on the manuscript. This work was supported by a grant from the Canadian Institutes of Health Research (CIHR) to PJK. JMA was supported by postdoctoral fellowships from CIHR and the Killam Foundation (University of British Columbia). PJK is a scholar of the Canadian Institute for Advanced Research, CIHR, and the Michael Smith Foundation for Health Research.

References

Andersen, R. A. (1991) The cytoskeleton of chromophyte algae. *Protoplasma* 164: 143–159.

Andersson, J. O. and Roger, A. J. (2002) A cyanobacterial gene in nonphotosynthetic protists: An early chloroplast acquisition in eukaryotes? *Curr. Biol.* 12: 115–119.

Archibald, J. M., Cavalier-Smith, T., Maier, U. and Douglas, S. (2001) Molecular chaperones encoded by a reduced nucleus- the cryptomonad nucleomorph. *J. Mol. Evol.* 52: 490–501.

Archibald, J. M. and Keeling, P. J. (2002) Recycled plastids: A green movement in eukaryotic evolution. *Trends Genet.* 18: 577–584.

Archibald, J. M., Longet, D., Pawlowski, J. and Keeling, P. J. (2003) A novel polyubiquitin structure in Cercozoa and Foraminifera: Evidence for a new eukaryotic supergroup. *Mol. Biol. Evol.* 20: 62–66.

Baldauf, S. L., Roger, A. J., Wenk-Siefert, I. and Doolittle, W. F. (2000) A kingdom-level phylogeny of eukaryotes based on combined protein data. *Science* 290: 972–977.

Ben Ali A., De Baere, R., Van der Auwera,G., De Wachter, R and Van de Peer, Y. (2001) Phylogenetic relationships among algae based on complete large-subunit rRNA sequences. *Int. J. Syst. Evol. Microbiol.* 51: 737–749.

Besendahl, A., Qiu, Y. L., Lee, J., Palmer, J.D and Bhattacharya, D. (2000) The cyanobacterial origin and vertical transmission of the plastid tRNA(Leu) group-I intron. *Curr. Genet.* 37: 12–23.

Bhattacharya, D., Helmchen, T., Bibeau, C. and Melkonian, M. (1995a) Comparison of nuclear-encoded small-subunit ribosomal RNAs reveal the evolutionary position of the Glaucocystophyta. *Mol. Biol. Evol.* 12: 415–420.

Bhattacharya, D., Helmchen, T. and Melkonian, M. (1995b) Molecular evolutionary analyses of nuclear-encoded small subunit ribosomal RNA identify an independent rhizopod lineage containing the Euglyphidae and the Chlorarachniophyta. *J. Euk. Microbiol.* 42:64–68.

Bhattacharya, D. and Weber, K. (1997) The actin gene of the glaucocystophyte *Cyanophora paradoxa*: Analysis of the coding region and introns, and an actin phylogeny of eukaryotes. *Curr. Genet.* 31: 439–446.

Blanchard J. L. and Hicks J. S. (1999) The non-photosynthetic plastid in malarial parasites and other apicomplexans is derived from outside the green plastid lineage. *J. Euk. Microbiol.* 46: 367–375.

Blankenship, R. E. (1994) Protein structure, electron transfer and evolution of prokaryotic photosynthetic reaction centers. *Ant. Van Leeuwen.* 65: 311–329.

Cavalier-Smith, T. (1981) Eukaryote kingdoms: Seven or nine? *Biosystems* 14: 461–481.

Cavalier-Smith, T. (1982) The origins of plastids. *Biol. J. Linn. Soc.* 17: 289–306.

Cavalier-Smith, T. (1986) The kingdom Chromista: Origin and systematics. *Progr. Phycol. Res.* 4: 309–347.

Cavalier-Smith, T. (1998) A revised six-kingdom system of life. *Biol. Rev. Camb. Philos. Soc.* 73: 203–266.

Cavalier-Smith, T. (1999) Principles of protein and lipid targeting in secondary symbiogenesis: Euglenoid, dinoflagellate, and sporozoan plastid origins and the eukaryote family tree. *J. Euk. Microbiol.* 46: 347–366.

Cavalier-Smith, T. (2000) Membrane heredity and early chloroplast evolution. *Trends Plant Sci.* 5: 174–182.

Cavalier-Smith, T. and Chao, E. E. (1997) Sarcomonad ribosomal RNA sequences, rhizopod phylogeny, and the origin of euglyphid amoebae. *Arch. Protistenkd.* 147: 227–236.

Cavalier-Smith, T., Chao, E. E. and Allsopp, M. T. E. P. (1995) Ribosomal RNA evidence for chloroplast loss within heterokonta: Pedinellid relationships and a revised classification of ochristan algae. *Arch. Protistenkd.* 145: 209–220.

Chesnick, J. M., Hooistra, W. H., Wellbrock, U. and Medlin, L. K. (1997) Ribosomal RNA analysis indicates a benthic pennate diatom ancestry for the endosymbionts of the dinoflagellates *Peridinium foliaceum* and *Peridinium balticum* (Pyrrhophyta). *J. Euk. Microbiol.* 44: 314–320.

Daugbjerg, N and Andersen, R. A. (1997) Phylogenetic analyses of the *rbcL* sequences from haptophytes and heterokont algae suggest their chloroplasts are unrelated. *Mol. Biol. Evol.* 14: 1242–1251.

Delwiche, C. F. (1999) Tracing the thread of plastid diversity through the tapestry of life. *Am. Nat.* 154 (suppl.): S164–S177.

Delwiche, C. F., Kuhsel, M. and Palmer, J. D. (1995) Phylogenetic analysis of *tuf*A sequences indicates a cyanobacterial origin of all plastids. *Mol. Phylogenet. Evol.* 4: 110–128.

Delwiche, C. F. and Palmer, J. D. (1996) Rampant horizontal transfer and duplication of rubisco genes in eubacteria and plastids. *Mol. Biol. Evol.* 13: 873–882.

Delwiche, C. F. and Palmer, J. D. (1997) The origin of plastids and their spread via secondary endosymbiosis. In *Origins of Algae and Their Plastids* (Bhattacharya, D., Ed.). Springer-Verlag, Wien, pp. 53–86.

Douglas, S. E. and Penny, S. L. (1999) The plastid genome from the cryptomonad alga, *Guillardia theta*: Complete sequence and conserved synteny groups confirm its common ancestry with red algae. *J. Mol. Evol.* 48: 236–244.

Douglas, S. E., Zauner S., Fraunholz, M., Beaton, M., Penny S., Deng, L., Wu X., Reith, M., Cavalier-Smith, T. and Maier, U. -G. (2001) The highly reduced genome of an enslaved algal nucleus. *Nature* 410: 1091–1096.

Durnford, D. G., Deane, J. A., Tan, S., McFadden, G. I., Gantt, E. and Green, B. R. (1999) A phylogenetic assessment of the eukaryotic light-harvesting antenna proteins, with implications for plastid evolution. *J. Mol. Evol.* 48: 59–68.

Fagan, T. M. and Hastings, J. W. (2002) Phylogenetic analysis indicates multiple origins of chloroplast glyceraldehyde-3-phosphate dehydrogenase genes in dinoflagellates. *Mol. Biol. Evol.* 19: 1203–1207.

Farmer, M. A. and Roberts, K. R. (1990) Organelle loss in the endosymbiont of *Gymnodinium acidotum* (Dinophyceae). *Protoplasma* 153: 178–185.

Fast, N. M., Kissinger, J. C., Roos, D. S. and Keeling, P. J. (2001) Nuclear-encoded, plastid-targeted genes suggest a single common origin for apicomplexan and dinoflagellate plastids. *Mol. Biol. Evol.* 18: 418–426.

Funes, S., Davidson, E., Reyes-Prieto, A., Magallón, S., Herion, P., King, M. P. and Gonzalez-Halphen, D. (2002) Apicomplexan split cox2 genes indicate a green algal apicoplast ancestor. *Science* 298: 2155.

Gardner, M. J., Goldman, N., Barnett, P., Moore, P. W., Rangachari, K., Strath, M., Whyte, A. Williamson, D. H. and Wilson, R. J. (1994) Phylogenetic analysis of the *rpo*B gene from the plastid-like DNA of *Plasmodium falciparum*. *Mol. Biochem. Parasitol.* 66: 221–231.

Gardner, M. J., Hall, N., Fung, E., White, O., Berriman, M., Hyman, R. W., Carlton, J. M., Pain, A., Nelson, K. E., Bowman, S., Paulsen, I. T., James, K., Eisen, J. A., Rutherford, K., Salzberg, S. L., Craig, A., Kyes, S., Chan, M. S., Nene, V., Shallom, S. J., Suh, B., Peterson, J, Angiuoli, S., Pertea, M., Allen J., Selengut, J, Haft, D., Mather, M. W., Vaidya, A. B., Martin, D. M., Fairlamb, A. H., Fraunholz, M. J., Roos, D. S., Ralph, S. A., McFadden, G. I., Cummings, L. M., Subramanian, G.M., Mungall C., Venter, J.C., Carucci, D.J., Hoffman, S.L., Newbold, C., Davis, R. W., Fraser, C.M. and Barrell, B. (2002) Genome sequence of the human malaria parasite *Plasmodium falciparum*. *Nature* 419: 498–511.

Gibbs, S. P. (1981) The chloroplast endoplasmic reticulum: structure, function, and evolutionary significance. *Int. Rev. Cytol.* 72: 49–99.

Gilson, P. R., Maier, U. G. and McFadden, G. I. (1997) Size isn't everything: Lessons in genetic miniaturisation from nucleomorphs. *Curr. Opin. Genet. Dev.* 7: 800–806.

Gilson, P. R. and McFadden, G. I. (1996) The miniaturized nuclear genome of a eukaryotic endosymbiont contains genes that overlap, genes that are cotranscribed, and the smallest known spliceosomal introns. *Proc. Natl. Acad. Sci. USA* 93: 7737–7742.

Gilson, P. R. and McFadden, G. I. (2002) Jam packed genomes: A preliminary, comparative analysis of nucleomorphs. *Genetica* 115: 13–28.

Greenwood, A. D., Griffiths, H. B. and Santore, U. J. (1977) Chloroplasts and cell compartments in Cryptophyceae. *Br. Phycol. J.* 12: 119.

Hallick, R. B., Hong, L., Drager, R. G., Favreau, M. R., Monfort A., Orsat, B., Spielmann, A. and Stutz, E. (1993) Complete sequence of *Euglena gracilis* chloroplast DNA. *Nucleic Acids Res.* 21: 3537–3544.

Helmchen, T. A., Bhattacharya, D. and Melkonian, M (1995) Analyses of ribosomal RNA sequences from glaucocystophyte cyanelles provide new insights into the evolutionary relationships of plastids. *J. Mol. Evol.* 41: 203–210.

Herdman, M. and Stanier, R. (1977) The cyanelle: Chloroplast or endosymbiotic procaryote? *FEMS Microbiol. Lett.* 1: 7–12.

Hibberd, D. J. and Norris, R. E. (1984) Cytology and ultrastructure of *Chlorarachnion reptans* (Chlorarachniophyta divisio nova, Chlorarachniophyceae classis nova). *J. Phycol.* 20: 310–330.

Ishida, K., Cao Y., Hasegawa, M., Okada, N. and Hara, Y. (1997) The origin of chlorarachniophyte plastids, as inferred from phylogenetic comparisons of amino acid sequences of EF-Tu. *J. Mol. Evol.* 45: 682–687.

Ishida, K. and Green, B. R. (2002) Second- and third-hand chloroplasts in dinoflagellates: Phylogeny of oxygen-evolving enhancer 1 (PsbO) protein reveals replacement of a nuclear-encoded plastid gene by that of a haptophyte tertiary endosymbiont. *Proc. Natl. Acad. Sci. USA* 99: 9294–9299.

Ishida, K., Green, B. R. and Cavalier-Smith, T. (1999) Diversification of a chimaeric algal group, the chlorarachniophytes: Phylogeny of nuclear and nucleomorph small-subunit rRNA genes. *Mol. Biol. Evol.* 16: 321–331.

Ishida, K., Green, B. R. and Cavalier-Smith, T. (2000) Endomembrane structure and the protein targeting pathway in *Heterosigma akashiwo* (Raphidophyceae, Chromista). *J. Phycol.* 36: 1135–1144.

Jarvis, P. and Soll, J. (2001) Toc, Tic, and chloroplast protein import. *Biochim. Biophys. Acta* 1541: 64–79.

Johnson, P. W., Hargraves, P. E. and Sieburth, J. M. (1988) Ultrastructure and ecology of *Calycomonas ovalis* Wulff, 1919 (Chrysophyceae) and its redescription as a testate rhizopod, *Paulinella ovalis* n. comb. (Filosea: Euglyphina). *J. Protozool.* 35: 618–626.

Keeling, P. J. (2001) Foraminifera and Cercozoa are related in actin phylogeny: Two orphans find a home? *Mol. Biol. Evol.* 18: 1551–1557.

Keeling, P. J., Deane, J. A., Hink-Schauer, C., Douglas, S. E., Maier. U. G. and McFadden, G. I. (1999) The secondary endosymbiont of the cryptomonad *Guillardia theta* contains alpha, beta-, and gamma-tubulin genes. *Mol. Biol. Evol.* 16: 1308–1313.

Keeling, P. J., Deane, J. A. and McFadden, G. I. (1998) The phylogenetic position of alpha- and beta-tubulins from the *Chlorarachnion* host and *Cercomonas* (Cercozoa). *J. Euk. Microbiol.* 45: 561–570.

Keeling, P. J. and Palmer, J. D. (2001) Lateral transfer at the gene and subgenic levels in the evolution of eukaryotic enolase. *Proc. Natl. Acad. Sci. USA* 98: 10745–10750.

Kies, L. (1974) Elektronenmikroskopische Untersuchungen an *Paulinella chromatophora* Lauterborn, einer Thekamöbe mit blau-grünen Endosymbionten (Cyanellen). *Protoplasma* 80: 69–89.

Kies, L. and Kremer, B. P. (1979) Function of cyanelles in the Tecamoeba *Paulinella chromatophora*. *Naturewissenschaften* 66: 578–579.

Kishore, R., Muchhal, U. S. and Schwartzbach, S. D. (1993) The presequence of *Euglena* LHCPII, a cytoplasmically synthesized chloroplast protein, contains a functional endoplasmic reticulum-targeting domain. *Proc. Natl. Acad. Sci. USA* 90: 11845–11849.

Köhler, S., Delwiche, C. F., Denny, P. W., Tilney, L. G., Webster, P., Wilson, R. J. M., Palmer, J. D. and Roos, D. S. (1997) A plastid of probable green algal origin in apicomplexan parasites. *Science* 275: 1485–1489.

Kowallik, K. V. (1997) Origin and evolution of chloroplasts: Current status and future perspectives. In *Eukaryotism and Symbiosis: Intertaxonic Combination versus Symbiotic Adaptation* (Schenk, H. E., Herrmann, R. G., Jeon, K. W., Müller, N. E. and Schwemmler, W., Eds.), Springer, Berlin, pp. 3–23.

Leander, B. S., Triemer, R. E. and Farmer, M. A. (2001a) Character evolution in heterotrophic euglenids. *Eur. J. Protistol.* 37: 337-356.

Leander, B. S., Witek, R. P. and Farmer, M. A. (2001b) Trends in the evolution of the euglenid pellicle. *Evolution* 55: 2215–2235.

Leipe, D. D. and Wainright, P. O., Gunderson, J. H., Porter, D., Patterson, D. J., Valois, F., Himmerich. S. and Sogin, M. L. (1994) The stramenopiles from a molecular perspective: 16S-like rRNA sequences from *Labyrinthuloides minuta* and *Cafeteria roenbergensis. Phycologia* 33: 369–377.

Ludwig, M. and Gibbs, S. P. (1989) Evidence that nucleomorphs of *Chlorarachnion reptans* (Chlorarachniophyceae) are vestigial nuclei: Morphology, division and DNA-DAPI fluorescence. *J. Phycol.* 25: 385–394.

Martin, W., Rujan, T., Richly, E., Hansen, A., Cornelsen, S., Lins, T., Leister, D., Stoebe, B., Hasegawa, M. and Penny, D. (2002) Evolutionary analysis of *Arabidopsis*, cyanobacterial, and chloroplast genomes reveals plastid phylogeny and thousands of cyanobacterial genes in the nucleus. *Proc. Natl. Acad. Sci. USA* 99: 12246–12251.

Martin, W. and Schnarrenberger, C. (1997) The evolution of the Calvin cycle from prokaryotic to eukaryotic chromosomes: A case study of functional redundancy in ancient pathways through endosymbiosis. *Curr. Genet.* 32: 1–18.

Martin, W., Stoebe, B., Goremykin, V., Hansmann, S., Hasegawa, M., and Kowallik, K. V. (1998) Gene transfer to the nucleus and the evolution of chloroplasts. *Nature* 393: 162–165.

McFadden, G. I. (1999) Plastids and protein targeting. *J. Euk. Microbiol.* 46: 339–346.

McFadden, G. I., Gilson, P. R. and Waller, R. F. (1995) Molecular phylogeny of chlorarachniophytes based on plastid rRNA and *rbcL* sequences. *Arch. Protistenkd.* 145: 231–239.

McFadden, G. I , Reith, M., Munholland, J. and Lang-Unnasch, N. (1996) Plastid in human parasites. *Nature* 381: 482.

McFadden, G. I. And Waller, R. F. (1997) Plastids in parasites of humans. *Bioessays* 19: 1033–1040.

Medlin, L. K., Cooper, A., Hill, C., Wrieden, S. and Wellbrock, U. (1995) Phylogenetic position of the Chromista plastids based on small subunit rRNA coding regions. *Curr. Genet.* 28: 560–565.

Mereschkowsky, C. (1905) Über Natur und Ursprung der Chromatophoren im Pflanzenreiche. *Biol. Centralbl.* 25:593–604. English translation in Martin,W. and Kowallik, K. V. (1999) Annotated English translation of Mereschkowsky's 1905 paper "Über Natur und Ursprung der Chromatophoren im Pflanzenreiche." *Eur. J. Phycol.* 34: 287–295.

Moreira, D., Le Guyader, H. and Phillippe, H. (2000) The origin of red algae and the evolution of chloroplasts. *Nature* 405: 69–72.

Müller, K. M., Oliveira, M. C., Sheath, R. G. and Bhattacharya, D, (2001) Ribosomal DNA phylogeny of the Bangiophycidae (Rhodophyta) and the origin of secondary plastids. *Am. J. Bot.* 88: 1390–1400.

Nelissen, B., Van de Peer, Y., Wilmotte, A. and De Wachter, R. (1995) An early origin of plastids within the cyanobacterial divergence is suggested by evolutionary trees based on complete 16S rRNA sequences. *Mol. Biol. Evol.* 12: 1166–1173.

Oliveira, M. C. and Bhattacharya D. (2000) Phylogeny of the Bangiophycidae (Rhodophyta) and the secondary endosymbiotic origin of algal plastids. *Am. J. Bot.* 87: 482–492.

Palmer, J. D. (2003) The symbiotic birth and spread of plastids: How many times and whodunnit? *J. Phycol.* 39, 4–11.

Preisfeld, A., Busse, I., Klingberg, M., Talke S. and Ruppel, H. G. (2001) Phylogenetic position and inter-relationships of the osmotrophic euglenids based on SSU rDNA data, with emphasis on the Rhabdomonadales (Euglenozoa). *Int. J. Syst. Evol. Microbiol.* 51: 751–758.

Saldarriaga, J. F., Taylor, F. J. R., Keeling, P. J. and Cavalier-Smith, T. (2001) Dinoflagellate nuclear SSUrRNA phylogeny suggests multiple plastid losses and replacements. *J. Mol. Evol.* 53: 204–213.

Schimper, A. F. W. (1883) Ueber die Entwickelung der Chlorophyllkörner und Farbkörper. *Bot. Zeit.* 41: 105–114, 121–131, 137–146, 153–162.

Stibitz, T. B., Keeling, P. J. and Bhattacharya, D. (2000) Symbiotic origin of a novel actin gene in the cryptophyte *Pyrenomonas helgolandii*. *Mol. Biol. Evol.* 17: 1731–1738.

Stiller, J. W. and Hall, B. D. (1997) The origin of red algae: implications for plastid evolution. *Proc. Natl. Acad. Sci. USA* 94: 4520–4525.

Stoebe, B. and Kowallik, K. V. (1999) Gene-cluster analysis in chloroplast genomics. *Trends Genet.* 15: 344–347.

Takishita, K. and Uchida, A. (1999) Molecular cloning and nucleotide sequence analysis of *psb*A from the dinoflagellates: Origin of the dinoflagellate plastid. *Phycol. Res.* 47: 207–216.

Taylor, F. J. R. (Ed.) (1987) *The Biology of Dinoflagellates*, Blackwell Scientific Publications, Oxford.

Tengs, T., Dahlberg, O. J., Shalchian-Tabrizi, K., Klaveness, D., Rudi, K., Delwiche, C. F. and Jakobsen, K.S. (2000) Phylogenetic analyses indicate that the 19'hexanoyloxy-fucoxanthin-containing dinoflagellates have tertiary plastids of haptophyte origin. *Mol. Biol. Evol.* 17: 718–729.

Turmel, M., Otis, C. and Lemieux, C. (1999) The complete chloroplast DNA sequence of the green alga *Nephroselmis olivacea*: insights into the architecture of ancestral chloroplast genomes. *Proc. Natl. Acad. Sci. USA* 96: 10248–10253.

Turner, S., Pryer, K. M., Miao, V. P. and Palmer, J. D. (1999) Investigating deep phylogenetic relationships among cyanobacteria and plastids by small subunit rRNA sequence analysis. *J. Euk. Microbiol.* 46: 327–338.

Valentin, K. and Zetsche, K. (1990) Nucleotide sequence of the gene for the large subunit of Rubisco from *Cyanophora paradoxa:* Phylogenetic implications. *Curr. Genet.* 18: 199–202.

Van de Peer, Y. and De Wachter, R. (1997) Evolutionary relationships among eukaryotic crown taxa taking into account site-to-site variation in 18S rRNA. *J. Mol. Evol.* 45: 619–630.

Van de Peer, Y., Rensing, S. A. and Maier, U.-G. (1996) Substitution rate calibration of small subunit ribosomal RNA identifies chlorarachniophyte endosymbionts as remnants of green algae. *Proc. Natl. Acad. Sci. USA* 93: 7732–7736.

Van der Auwera, G., Hofmann, C. J. B., De Rijk, P., De Wachter, R. (1998) The origin of red algae and cryptomonad nucleomorphs: A comparative phylogeny based on small and large subunit rRNA sequences of *Palmaria palmata*, *Gracilaria verrucosa*, and the *Guillardia theta* nucleomorph. *Mol. Phylogenet. Evol.* 10: 333–342.

Watanabe, M. M., Suda, S., Inouye, I., Sawaguchi, I. and Chihara, M. (1990) *Lepidodinium viride* gen et sp. nov. (Gymnodiniales, Dinophyta), a green dinoflagellate with a chlorophyll *a*- and *b*-containing endosymbiont. *J. Phycol.* 26: 741–751.

Wilcox L. W. and Wedemayer G. J. (1985) Dinoflagellate with blue-green chloroplasts derived from an endosymbiotic eukaryote. *Science* 227: 192–194.

Williamson, DH., Gardner, M.J., Preiser, P., Moore, D.J., Rangachari, K. and Wilson, R.J. (1994) The evolutionary origin of the 35 kb circular DNA of *Plasmodium falciparum*: new evidence supports a possible rhodophyte ancestry. *Mol. Gen. Genet.* 243: 249–252.

Wilson, R. J. M. I., Denny P. W., Preiser, D.J., Rangachari, K., Roberts, K., Roy, A., Whyte, A., Strath, M., Moore, D.J., Moore, P.W. and Williamson, D.H. (1996) Complete gene map of the plastid-like DNA of the malaria parasite *Plasmodium falciparum*. *J. Mol. Biol.* 261: 155–172.

Wolters, J. (1991) The troublesome parasites: Molecular and morphological evidence that Apicomplexa belong to the dinoflagellate-ciliate clade. *Biosystems* 25: 75–84.

Yoon, H. S., Hackett, J.D. and Bhattacharya, D. (2002a) A single origin of the peridinin- and fucoxanthin-containing plastids in dinoflagellates through tertiary endosymbiosis. *Proc. Natl. Acad. Sci. USA* 99: 11724–11729.

Yoon, H.S., Hackett, J.D., Pinto, G. and Bhattacharya, D. (2002b) A single, ancient origin of the plastid in the Chromista. *Proc. Natl. Acad. Sci. USA* 99: 15507–15512.

Zhang, Z., Green, B.R. and Cavalier-Smith, T. (1999) Single gene circles in dinoflagellate chloroplast genomes. *Nature* 400: 155–159.

Zhang, Z., Green, B.R. and Cavalier-Smith, T. (2000) Phylogeny of ultra-rapidly evolving dinoflagellate chloroplast genes: A possible common origin for sporozoan and dinoflagellate plastids. *J. Mol. Evol.* 51: 26–40.

Chapter 4

Chromalveolate Diversity and Cell Megaevolution: Interplay of Membranes, Genomes and Cytoskeleton

Thomas Cavalier-Smith

CONTENTS

Abstract

Eukaryote cell architecture depends on specific topogenic proteins that control protein targeting to different genetic membranes and cell compartments or mediate mechanical attachments between them, the cytoskeleton and chromosomes. Chromalveolates are a major branch of the eukaryote tree whose cells are often marvellously more complex — and immensely more disparate — than those of animals and plants, e.g., the four-genomed cryptophytes, diatoms or dinoflagellates, but sometimes highly simplified such as the degen-

0-415-29904-7/04/$0.00+$1.50

erate *Blastocystis* parasite of human guts. Their origin and diversification are discussed in relation to molecular and ultrastructural evidence. Chromalveolates evolved when a phagotrophic biciliate protozoan cell enslaved and merged with a unicellular red alga more than 530 million years ago. The algal plasma membrane was converted into the periplastid membrane by inserting duplicated chloroplast outer membrane translocons for importing nuclear-coded plastid proteins through it and the chloroplast envelope. Chromalveolates comprise the protozoan infrakingdom Alveolata (Myzozoa, Ciliophora) and the kingdom Chromista (Cryptista, Chromobiota, the latter comprising heterokonts and haptophytes), in which fusion of the former food vacuole membrane and nuclear envelope placed the periplastid membrane and enclosed plastid inside the rough endoplasmic reticulum. Both groups lost photosynthesis and probably plastids several times, and therefore comprise phototrophs (chromophyte algae), phagotrophs (e.g., ciliate protozoa), parasites (e.g., *Plasmodium*, the malaria agent) and saprotrophs. The ancestral chromalveolate probably had simple ciliary hairs, still common in the haploid biciliate Myzozoa (Dinozoa, Apicomplexa) but lost — like chloroplasts — by their diploid or multiploid multiciliate sisters the Ciliophora (ciliates, suctorians); these evolved into rigid bipartite tubular glycoprotein hairs in chromists (lost by haptophytes when predation using the haptonema evolved and by a few other chromists). Chromalveolates might be sisters of kingdom Plantae, cortical alveoli probably having evolved to create their corticate ancestor. The astonishing trophic diversity and cellular virtuosity of chromalveolates give them tremendous ecological and medical importance and unsurpassed evolutionary fascination.

4.1 Introduction: What Are Chromalveolates?

Chromalveolates embrace a major fraction of eukaryotic biodiversity, ranging from minute intracellular parasites of bacterial dimensions to brown seaweeds (giant kelps) longer than a blue whale. Their seven phyla include phototrophs (i.e., chromophyte algae, the major eukaryotic primary producers in the oceans, which cover 70% of the globe), phagotrophs, photophagotrophic mixotrophs, parasites such as the world's leading nonviral killer of humans (*Plasmodium*) and saprotrophs. Their unity lies in their common evolutionary history and not in their superficial phenotype or trophic habits, so it has taken much longer to be recognized (Cavalier-Smith, 1999) than that of the better-known and trophically more homogeneous animals, fungi and plants. They comprise two major groups: the kingdom Chromista (Cavalier-Smith, 1981, 1986, 1989) and the protozoan infrakingdom Alveolata (Cavalier-Smith, 1991a, 1993b). Each has very distinctive cellular properties that differ from those of the three more familiar higher kingdoms and offer difficult but exciting challenges to evolutionary cell biology. Phylogenetically, chromalveolates can be defined as the chromophyte algae (those ancestrally having chloroplasts with chlorophyll *c*, Christensen, 1989) and all their disparate nonphotosynthetic descendants. As Fungi comprise only four phyla, even including Microsporidia (Cavalier-Smith, 1998, 2000) and plants but five (Cavalier-Smith, 1998), the systematic importance of the seven chromalveolate phyla is considerable. There are 123,000 or more described species (Corliss, 2000), more than half of all protists, and perhaps as many undescribed ones. Table 4.1 summarizes their classification into 47 classes and Figure 4.1 their mutual relationships. Together with the plant kingdom and the protozoan infrakingdoms Excavata and Rhizaria, chromalveolates constitute the most megadiverse eukaryotic clade, the bikonts — the ancestrally biciliate eukaryotes (Figure 4.2). It is beyond the scope of the chapter to illustrate their remarkable structural diversity; intrigued readers should consult Graham and Wilcox (2000) and Lee et al. (2002).

Chromists were originally defined (Cavalier-Smith, 1981) as organisms having one or both of two key characters: (1) rigid tubular hairs on at least one of their typically two cilia, and (2) plastids having an additional smooth membrane (now called the periplastid

TABLE 4.1 Revised Classification of the 47 Classes of Chromalveolates

Kingdom CHROMISTA Cavalier-Smith 1981

Subkingdom 1. Cryptista Cavalier-Smith 1989

Phylum Cryptista Cavalier-Smith 1986 emend.

Subphylum 1. Cryptomonada subphyl. nov. (periplast; ejectisomes with secondary scroll; mitochondrial cristae flat tubules) [periplasta praesens; ejectisomae bipartitae; cristae mitochondrialis complanatae; cilia villosa]

Class 1. Cryptophyceae em. (= Cryptomonadea sensu Cavalier-Smith 1989) (e.g., *Guillardia*)

Class 2. Goniomonadea (*Goniomonas*)

Subphylum 2. Leucocrypta subphyl. nov. Diagnosis as for sole class:

Class Leucocryptea cl. nov. Diagnosis: biciliates lacking inner and outer periplast but having a rigid two-layered extracellular sheath that extends in more flexible form over the ciliary membrane; ejectisomes without secondary scroll; tubular mitochondrial cristae; cilia without tubular hairs or spurs (*Kathablepharis*, *Leucocryptos*) [Cellulae biciliatae sine periplasto; ejectisomae unipartitae; cellulae et cilia lamina externa tecta; cristae mitochondrialis tubulatae; cilia tereta]

Subkingdom 2. Chromobiota Cavalier-Smith 1989

Infrakingdom 1. Heterokonta Cavalier-Smith 1986

Phylum 1. Ochrophyta Cavalier-Smith 1986

Subphylum 1. Phaeista Cavalier-Smith 1995

Infraphylum 1. Hypogyrista Cavalier-Smith 1995

Class 1. Pelagophyceae (e.g., *Pelagomonas*, *Sarcinochrysis*)

Class 2. Actinochrysea (= Dictyochophyceae) silicoflagellates, pedinellids (e.g., *Pteridomonas*, *Ciliophrys*); possibly Actinophryales

Class 3. Pinguiophyceae (e.g. *Pinguiochrysis*, *Glossomastix*)

Infraphylum 2. Chrysista Cavalier-Smith 1995

Classes 1-8: Raphidophyceae (e.g., *Heterosigma*), Eustigmatophyceae (e.g., *Vischeria*), Chrysophyceae (chrysomonads, e.g., *Ochromonas*, *Synura*, *Spumella*, *Oikomonas*), Chrysomerophyceae (e.g., *Giraudyopsis*), Phaeothamniophyceae (e.g., *Pleurochloridella*), Xanthophyceae (e.g., *Vaucheria*), Phaeophyceae (brown algae, e.g., *Fucus*, *Schizocladia*)

Subphylum 2. Khakista Cavalier-Smith 2000

Class 1. Bolidophyceae (*Bolidomonas*)

Class 2. Diatomeae (e.g., *Coscinodiscus*, *Bacillaria*, *Nitzschia*)

Phylum 2. Bigyra Cavalier-Smith 1998

Subphylum 1. Bigyromonada Cavalier-Smith 1998

Class Bigyromonadea (*Developayella*)

Subphylum 2. Pseudofungi Cavalier-Smith 1986

Class 1. Oomycetes (e.g., *Phytophthora*, *Achlya*)

Class 2. Hyphochytrea (e.g., *Rhizidiomyces*)

Subphylum 3. Opalinata Wenyon 1926 stat. nov. Cavalier-Smith 1997

Class 1. Proteromonadea (*Proteromonas*)

Class 2. Blastocystea (*Blastocystis*)

Class 3. Opalinea (e.g., *Cepedea*, *Opalina*)

Phylum 3. Sagenista Cavalier-Smith 1995

Class 1. Labyrinthulea, (e.g., *Thraustochytrium*, *Labyrinthula*)

Class 2. Bicoecea (e.g., *Bicosoeca Cafeteria*, *Caecitellus*)

Class 3. Placididea (*Placidia*, *Wobblia*, e.g., *Pendulomonas*)

Infrakingdom 2. and **phylum Haptophyta** Cavalier-Smith 1986

Class 1. Pavlovophyceae (e.g., *Pavlova*)

Class 2. Prymnesiophyceae (e.g., *Emiliania*, *Isochrysis*, *Prymnesium*)

Kingdom **PROTOZOA** Owen 1858

Infrakingdom Alveolata Cavalier-Smith 1991

Phylum 1. Myzozoa Cavalier-Smith 2004 = Miozoa Cavalier-Smith 1987 stat. nov. 1999

(Table 4.1 continued next page)

TABLE 4.1 Revised Classification of the 47 Classes of Chromalveolates (continued)

Subphylum 1. Dinozoa Cavalier-Smith 1981 em. (ancestrally with numerous discrete cortical alveoli; often with toxicysts or tripartite trichocysts)

Infraphylum 1. Protalveolata[a] Cavalier-Smith 1991 em.

Class 1. Colponemea (*Colponema, Algovora*)

Class 2. Myzomonadea (*Voromonas, Alphamonas, Chilovora*)

Class 3. Perkinsea (*Perkinsus, Rastrimonas, Parvilucifera, Phagodinium*)

Class 4. Ellobiopsea (e.g., *Ellobiopsis, Thalassomyces*)

Infraphylum 2. Dinoflagellata Bütschli 1885 stat. nov. Cavalier-Smith 1999

Superclass 1. Syndina Cavalier-Smith 1993

Class Syndinea (e.g., *Amoebophrya, ?Hematodinium*)

Superclass 2. Dinokaryota Fensome et al. 1993

Class 1. Noctilucea (e.g., *Noctiluca*)

Class 2. Peridinea 4 subclasses: 1. Peridinoidia[a] (e.g., *Peridinium, Heterocapsa, Prorocentrum, Haplozoon, Amylodinium*); 2. Dinophysoidia (e.g., *Dinophysis*); 3. Gonyaulacoidia Cavalier-Smith subcl. n. (Diagnosis: with gonyaulacoid tabulation of thecal plates: see Fensome et al. (1993) pp. 11–27) order Gonyaulacida Taylor 1980 (e.g., *Ceratium, Crypthecodinium*) 4. Suessioidia subcl. n. (suessioid tabulation of thecal plates) order Suessiida Fensome et al. 1993 (e.g., *Symbiodinium, Polarella*); Oxyrrhia subcl. n. (Diagnosis as order Oxyrrhida Sournia 1984) (*Oxyrrhis*)

Subphylum 2. Apicomplexa Levine 1970 stat. nov. emend. Cavalier-Smith 2003 (with flattenened inner membrane complex trichocysts absent)

Infraphylum 1. Apicomonadea Parasites (*Acrocoelus*) or myzocytotic predators (colpodellids: *Colpodella*)

Class **Apicomonada** Cavalier-Smith stat nov. 2003

Infraphylum 2. Sporozoa Leuckart 1879 stat. nov. Cavalier-Smith 1999 em. auct.

Class 1. Coccidea[a] (e.g., *Hepatozoon, Cryptosporidium, Toxoplasma*)

Class 2. Gregarinea (e.g., *Monocystis, Ophriocystis*)

Class 3. Hematozoa (e.g., *Plasmodium, Babesia, Theileria*)

Phylum 2. Ciliophora Doflein 1901 (ciliates and suctorians)

Subphylum 1. Postciliodesmatophora Gerassimova & Seravin 1976 (desmates)

Class 1. Karyorelictea (e.g., *Kentrophoros, Tracheloraphis, Loxodes, Geleia*)

Class 2. Heterotrichea (e.g., *Stentor, Folliculina, Blepharisma, Condylostoma*)

Subphylum 2. Intramacronucleata Lynn 1996 (amphiesmates)

Infraphylum 1. Spirotrichia infraphylum nov. Diagnosis as for the class: (Lynn and Small, 2002 p. 420)

Class Spirotrichea (e.g., *Protocruzia, Oxytricha, Euplotes, Tintinnus, Stichotricha, Halteria, Metopus*)

Infraphylum 2. Rhabdophora infraphyl. nov. Diagnosis as for class Litostomatea: (Lynn and Small, 2002 p. 477)

Class Litostomatea (e.g., *Didinium, Lacrymaria, Entodinium*)

Infraphylum 3. Ventrata infraphyl. nov. Diagnosis: amphiesmate ciliates ancestrally with ventral cytopharynx

Classes 1–6 Phyllopharyngea (e.g., *Dysteria, Podophrya*), Colpodea[b] (e.g., *Colpoda*), Nassophorea[b] (e.g., *Nassula*), Prostomatea (e.g., *Coleps*), Plagiopylea, Oligohymenophorea (e.g., *Tetrahymena, Paramecium, Vorticella*)

[a] Probably paraphyletic.

[b] If they prove to be sisters, they should be combined into one class, as their morphology is not dramatically different.

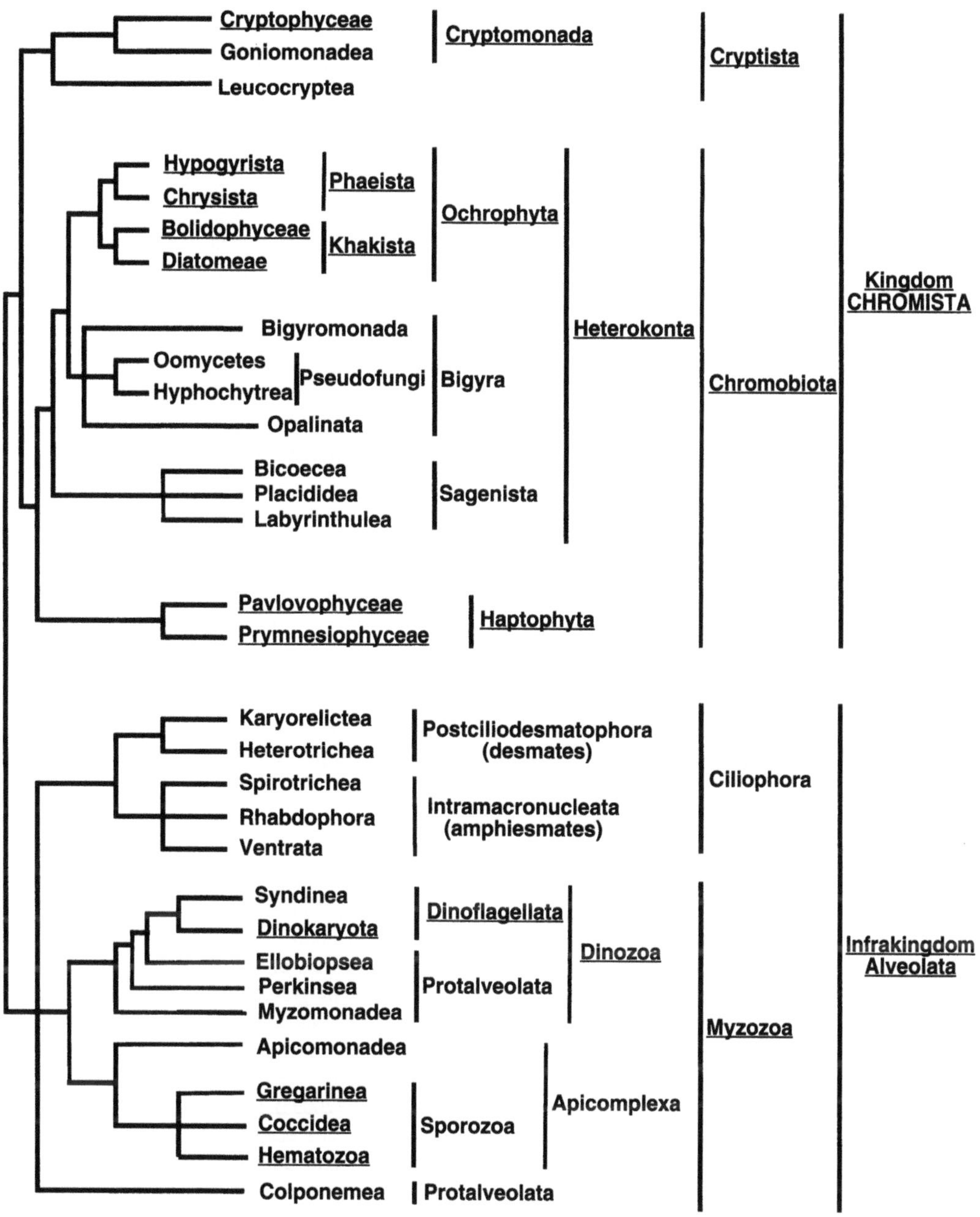

FIGURE 4.1 Probable phylogenetic relationships of the major chromalveolate clades and their grouping into higher taxa. Groups with at least some photosynthetic members or leucoplasts are underlined. The tally of chromalveolate groups might not be complete, e.g., centrohelid heliozoa are possibly sisters to Haptophyta [Cavalier-Smith and Chao (2003a) 387–396].

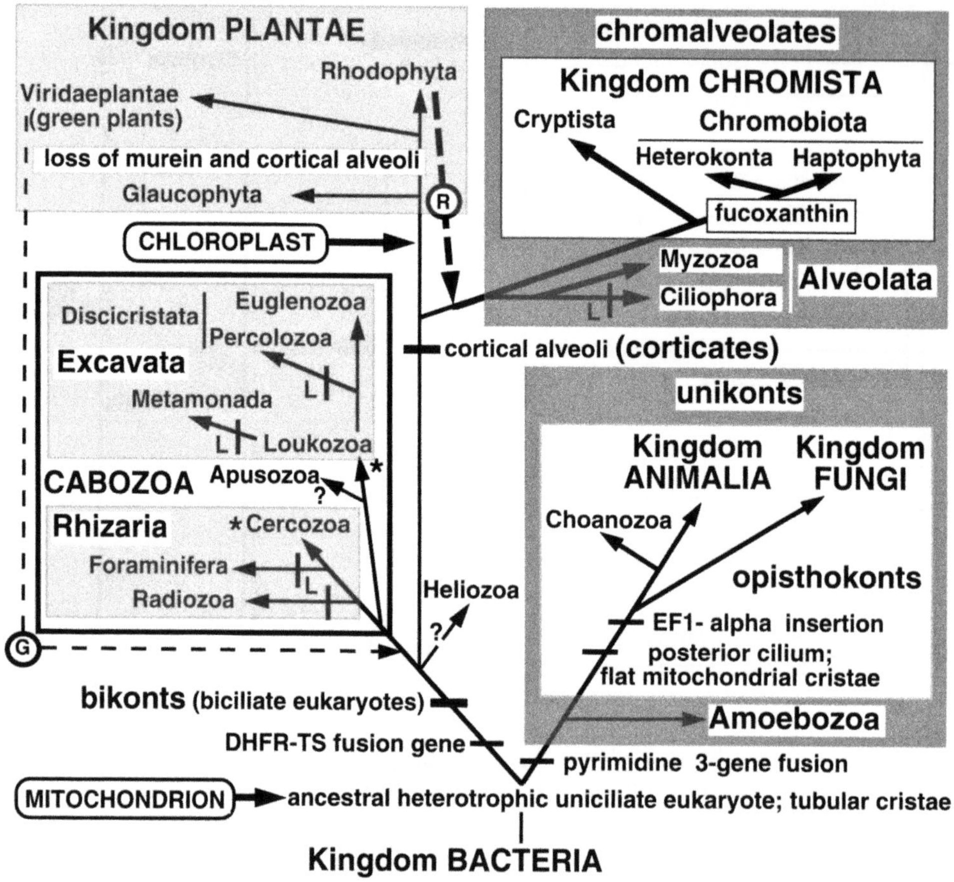

FIGURE 4.2 Position of chromalveolates within the six-kingdom system of life. Chromalveolates arose by the secondary symbiogenetic enslavement of a red alga (R) by a corticate bikont host to form the ancestral chromophyte alga, one of the four most important symbiogenetic events in the history of life. Also shown are the three others: the origins of mitochondria, chloroplasts and cabozoan algae. Independently of chromalveolate origins, green algal plastids were implanted into the ancestors of discicristate protozoa and cercozoan chlorarachnean algae, either independently where shown by asterisks or more likely in their common ancestor (G) in a single secondary symbiogenetic event to generate the clade cabozoa, comparable in importance to the chromalveolates [Cavalier-Smith (1999) 347–366; (2003a) 1741–1758; (2003b) 109–134]; however, the evidence that excavates and cabozoa are clades is much weaker than for chromalveolates [Cavalier-Smith and Chao (2003c) 540–563; Simpson and Roger (2002) R691–R693]. Subsequent plastid losses (L) in chromalveolates and cabozoa generated purely nonphotosynthetic taxa. The basal eukaryotic kingdom Protozoa comprises alveolates, excavates, Rhizaria (Cercozoa, Foraminifera, Radiolaria), the ancestrally nonphotosynthetic phyla Amoebozoa and Choanozoa, and Apusozoa and Heliozoa; the last two might be related as shown, but Heliozoa might be secondarily nonphotosynthetic chromists related to haptophytes [Cavalier-Smith and Chao (2003a) 387–396]. The root of the eukaryote tree is between bikonts and unikonts, as the gene fusion between dihydrofolate reductase (DHFR) and thymidylate synthetase (TS) [Stechmann and Cavalier-Smith (2002) 89–9 and (2003a) R665–R666], the protein synthesis elongation factor (EF1-α) insertion and the fusion between the first three pyrimidine biosynthesis enzymes shared by all unikonts (Nara et al., 2002 209–222) are all derived; this topology is also supported by Hsp90 and concatenated mitochondrial protein trees [Lang et al. (2002) 1773–1778; Stechmann and Cavalier-Smith (2003) 408–419.

membrane, PPM, Cavalier-Smith, 1989) external to their double envelope, this entire periplastid complex being uniquely located inside the lumen of the rough endoplasmic reticulum (RER). I argued that both characters evolved simultaneously to create the ancestral chromist, but tubular hairs were secondarily lost in the ancestral haptophyte (Cavalier-Smith, 1986) when the haptonema evolved (Cavalier-Smith, 1994). Photosynthesis was lost several times to generate phagotrophs (e.g., bicoecids), formerly classified as protozoa, and saprotrophs (e.g., oomycetes and hyphochytrids, collectively Pseudofungi), formerly misclassified as Fungi. Chromists comprise two subkingdoms: Cryptista and Chromobiota. Cryptists comprise cryptophytes (Cryptophyceae), which always have plastids, a PPM, a relict nucleus (the nucleomorph), 80S ribosomes and starch, all in the periplastid space between the PPM and plastid envelope, plus the zooflagellate goniomonads and katablepharids that lack all these structures. (One katablepharid has cytoplasmic starch.)

Naïve interpretations of rRNA trees caused scepticism over chromist monophyly, as they seldom group together on single-gene trees. However, sequence trees based on five different chloroplast genes very robustly show the monophyly of chromists and chromobiotes (Yoon et al., 2002b), vindicating the soundness of earlier deductions from ultrastructural and biochemical evidence, the unique chromistan membrane topology and the principle of parsimony in evolution of complex protein-targeting machinery. These trees indicate that the ancestor of the chromist chloroplast diverged very early in red algal evolution, just after the primary divergence within unicellular red algae — between the thermophilic Cyanidiophyceae and nonthermophilic Rhodellophyceae. This early divergence, coupled with that of the three plant groups (red algae, their green plant sisters, glaucophytes) and of the three chromist groups [Cryptista, Heterokonta (= stramenopiles), Haptophyta], explains why single-gene trees cannot robustly resolve the correct branching order of these six corticate taxa.

Alveolates comprise the phyla Ciliophora (ciliates, suctorians) and Myzozoa (dinoflagellates, protalveolates, apicomonads, sporozoa). Like chromists, they were originally defined by a combination of two characters: (1) membrane-bound cortical alveoli and (2) tubular mitochondrial cristae. Contrary to widespread misconceptions, cortical alveoli are insufficient basis for their definition, as they are also present in glaucophytes, which by contrast have irregularly flattened mitochondrial cristae as in the rest of kingdom Plantae (Cavalier-Smith, 1987). As rRNA trees quickly confirmed the postulated relationship between sporozoa (= Apicomplexa), dinoflagellates and ciliates, Alveolata was much more rapidly accepted than Chromista; its monophyly is confirmed by trees based on numerous proteins (Bapteste et al., 2002; Baldauf et al., 2000; Fast et al., 2002) and by rRNA trees confirming that protalveolates (including ellobiopsids) and apicomonads properly belong there (Cavalier-Smith and Chao, 2004b; Silberman et al., personal communication). Contrary to prevailing views, I long thought that the chromalveolate ancestor was photosynthetic and related to the ancestral plant by the presence of both cortical alveoli and multilayered structure ciliary roots (Cavalier-Smith, 1982). However, discovery of relict plastids in many sporozoa and the remarkable minicircular single-gene chromosomes of dinoflagellate chloroplasts (Zhang et al., 1999) led to radical reevaluation of the origins of alveolate plastids. Sequence evidence, albeit weak because of the dramatically accelerated evolutionary rates of dinoflagellate minicircle genes, suggested that sporozoan and dinoflagellate plastids are sisters, implying photosynthetic ancestry for Myzozoa, and that they might be closer to chromists than to any plant groups (Zhang et al., 2000).

The dominant idea that dinoflagellates got their plastids in a separate symbiogenesis (Whatley et al., 1979; Gibbs, 1981) was always highly unparsimonious with respect to the onerous origin of their chloroplast-specific protein import (Cavalier-Smith, 1982). Later, I argued that chromists and alveolates were sisters and that only one symbiogenetic enslavement of a red alga ever occurred in the history of life, in their common chromalveolate ancestor; Figure 4.2 and Figure 4.3). This was strongly confirmed by the discovery that the

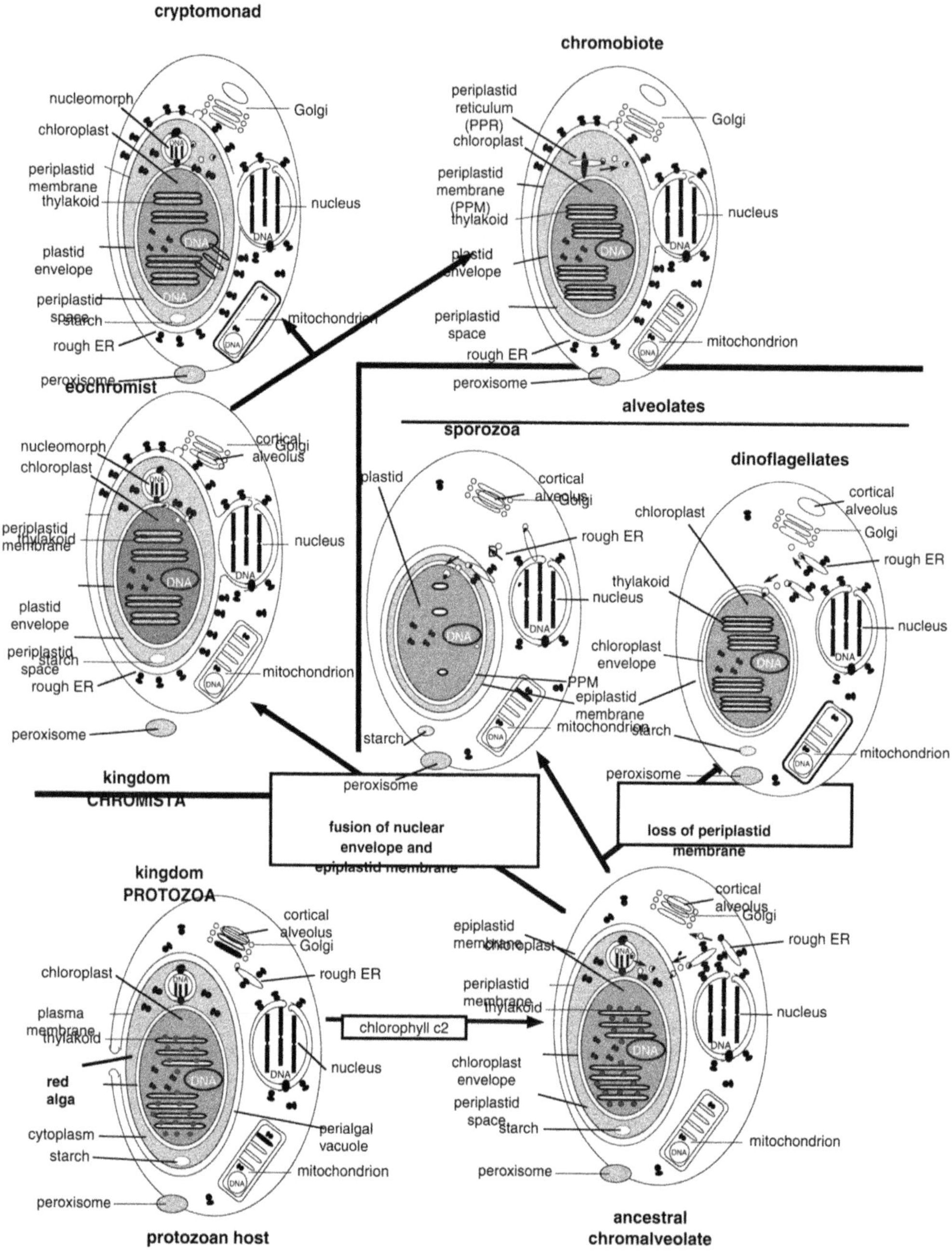

original red algal chloroplast version of glyceraldehyde phosphate dehydrogenase (GAPDH) of all five major chromalveolate groups was replaced by one encoded by a duplicate of the nuclear gene for the host cytosolic version (Fast et al., 2001; Harper and Keeling, 2003). As the host had two nuclear GAPDH genes (for the cytosolic and mitochondrial enzymes) and the symbiont three (for cytosolic, mitochondrial and plastid enzymes), this would have been a remarkable, frankly incredible, coincidence if the five groups (alveolate sporozoa and dinoflagellates; chromist cryptophytes, haptophytes and heterokonts) acquired their plastids

FIGURE 4.3 The origin and diversification of chromalveolate cells emphasizing membrane and genomic changes. The host enslaved the red alga after its phagocytosis into a food vacuole (phagosome) by inserting translocators for extracting useful molecules into the surrounding membranes by fusion of novel ER-derived coated vesicles (probably copII) with the epiplastid membrane. The symbiont nucleus lost essential genes transferred to the nucleus that acquired bipartite plastid targeting sequences (with N-terminal signal sequence for traversing the RER and subterminal transit-like sequence for crossing the PPM and plastid envelope), allowing their proteins to reenter the plastid. The mitochondrion and Golgi (not shown) were lost early from the enslaved alga, but its nucleus remains as the nucleomorph in cryptophytes, where gene transfer to the host nucleus was fortuitously never completed. Chromobiotes and alveolates independently managed to transfer all essential red algal nuclear genes to the host, losing the nucleomorph genome, pore complexes and periplastid ribosomes, but retained its membrane as the PPR perhaps to make membrane lipids for the PPM. The ancestral chromalveolate evolved chlorophyll c_2, and lost phycobilisomes, allowing thylakoid stacking: one phycobiliprotein remained in cryptophytes, retargeted into the thylakoid lumen. In the ancestral chromist, the nuclear envelope fused with the epiplastid membrane adding ribosome receptors, converting it into ordinary RER, thereby dispensing with vesicle transport of chloroplast preproteins. Dinoflagellates instead simplified membrane topology by losing the PPM and PPR. Most of the numerous losses of plastids are not shown.

in five independent symbiogenetic events — by chance all from red algae and, in chromists at least, even the same red algal lineage (Yoon et al., 2002b). One replacement in a single photosynthetic chimaeric ancestor, as predicted by the chromalveolate theory, simply explains how all five groups lost the three symbiont genes and replaced the plastid one by a novel duplicate (specifically of the host cytosolic not the mitochondrial one) that acquired a novel bipartite targeting sequence for import across four (in dinoflagellates three) membranes. Classical evidence for chromist and chromobiote monophyly and the more recent chloroplast gene evidence that haptophyte and heterokont plastids are sisters (mutually closer than to cryptophyte plastids, Yoon et al., 2002b) complement the compelling GAPDH evidence that all chromalveolates arose by one red algal enslavement, not five separate ones as often unparsimoniously assumed (Delwiche, 1999). Sequence trees for multiple nuclear-coded proteins support chromalveolate monophyly (Baldauf et al., 2000; Bapteste et al., 2002), but are insufficiently sampled taxonomically to be definitive. Alveolate plastid and mitochondrial genomes are too divergent in gene composition and evolutionary rate to help conclusively (Zhang et al., 2000).

4.2 Diversity and Unity of Chloroplast Protein Targeting in Chromalveolates

As these topics are discussed fully by Cavalier-Smith (2003b), which should be consulted for detailed arguments and references, I give only a simplified outline.

Under 10% of plastid proteins (probably <1% in dinoflagellates) are encoded by the plastid genome. The rest are nuclear coded and imported into plastids by proteinaceous machinery in the surrounding membranes that recognizes specific topogenic peptide sequences encoded by their messenger RNA. In the plant kingdom, wherein plastids have only a double envelope, import depends on an N-terminal transit peptide recognized by a translocon (protein translocator) made of two parts: Toc in the outer membrane (OM) and Tic in the inner membrane (IM). All five chromalveolate groups with plastids have bipartite import sequences, with a terminal signal sequence and a subterminal transit-peptide-like sequence (Figure 4.4a); the signal sequence directs proteins across the RER membrane and is removed as it passes into the lumen.

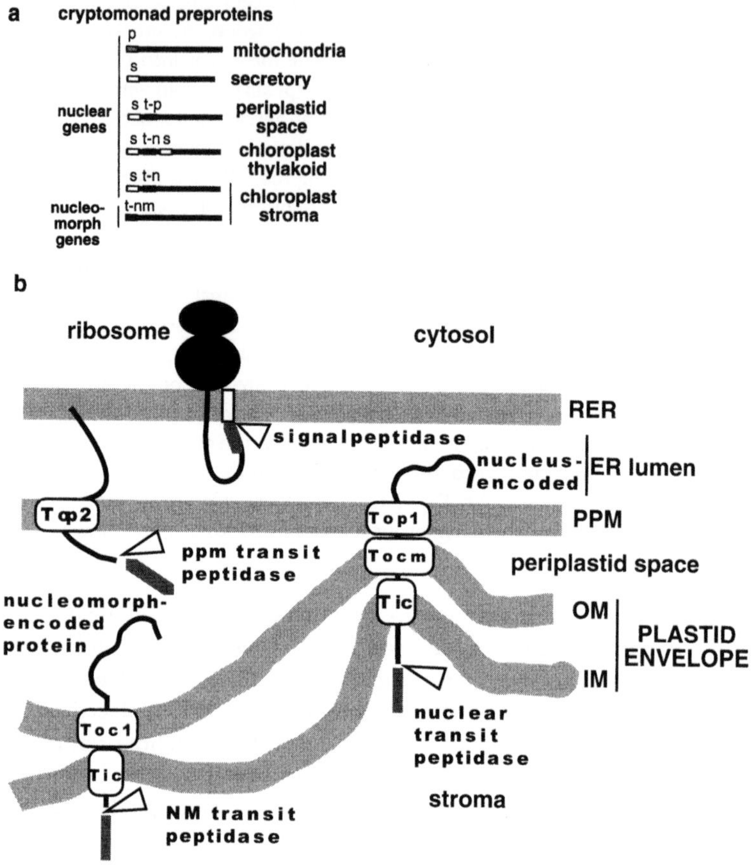

FIGURE 4.4 Protein targeting in cryptophytes. (a) Preproteins targeted to different cryptophyte cell compartments by N-terminal topogenic sequences. T-nm: typical nucleomorph-encoded transit sequences recognized by Toc1 (see Figure 4.4b); T-n: transit-like sequences putatively recognized by Top1; T-p: hypothetical transit-like sequences recognized by Top2. Although ciliary hair proteins and secretory proteins such as those of ejectisomes must share a common signal sequence for crossing the RER membrane, there must be additional sorting sequences to ensure that they subsequently reach appropriate destinations: there are two different kinds of ejectisomes (large ones around the gullet and small ones on the general cell surface) and three sorts of tubular ciliary hair (two on the anterior and one on the ventral cilia). (b) A model for protein targeting into and across the cryptophyte periplastid space. See text for explanation.

The subsequent fate of the preprotein now bearing only the N-terminal transit-like sequence is almost certainly different in alveolates and chromists but not fully elucidated in either. In chromists, in which the plastid and PPM are within the ER lumen, it passes directly across the PPM, the OM and IM of the plastid envelope, and the transit-like peptide is removed by a stromal transit-peptidase. In alveolates, the plastid is apparently not within the ER lumen; preproteins are probably carried to it by vesicle budding and fusion (Cavalier-Smith, 2003b). The host enslaved the red alga by inserting permeases into the epiplastid membrane (EM) and PPM to extract photosynthate. The key molecular innovations enabling this were either specific copII-coated vesicles (plus v- and t-SNAREs) to target them to the EM (Cavalier-Smith, 2003b) or (more simply?) fusion of the former food vacuole membrane with an endomembrane compartment lying between the ER and Golgi through which plastid-directed and secretory proteins might both flow. Subsequently, this protein-translocation machinery also enabled import of proteins coded by former algal genes transferred to the slave-owner's nucleus.

Although chromalveolate targeting mechanisms remain poorly understood, chloroplast preproteins do not enter the Golgi (Joiner and Roos, 2002), whereas they do in euglenoids, which obtained their plastid from green algae by independent enslavement (Cavalier-Smith, 2003b). This shows that eukaryotic algae can be enslaved in two contrasting ways: by novel targeting to the former phagosomal membrane to convert it into an EM either as a pre-Golgi (chromalveolates) or a post-Golgi compartment (euglenoids) (Cavalier-Smith, 2003b). The likelihood that chromalveolates use pre-Golgi and cabozoa post-Golgi vesicles is a further contrasting, but individually unifying, feature for the two lineages of meta-algae (those created by enslaving eukaryotic algae, Cavalier-Smith, 1995b).

The historical accident that chromalveolates hit on the more direct route bypassing the Golgi preadapted them for the origin of the simplified chromist pattern. Using copII vesicles for targeting, unlike in cabozoa, made it possible for such vesicles to fuse with the target EM accidentally without first being fully separated from the donor ER. This single membrane-mutational accident (needing no DNA mutations) would have given rise to the chromist topology in one fell swoop, dispensing with the coated vesicles targeted to the EM. This simplification, not the symbiogenetic enslavement of the red alga, created the first chromist. Targeting directly via ER-derived vesicles provides a simple high-probability mechanism for this long-postulated membrane fusion, making it much easier to understand how the unique intra-RER location of the chromist PPM and plastid was generated by individually adaptive steps. As algal enslavement can in principle occur either via the pre-Golgi or post-Golgi vesicle route, the unique membrane topology of chromists is an even more significant shared derived character than was originally apparent (Cavalier-Smith, 1982, 1986); if the three chromist groups had originated by independent symbiogeneses, as others long assumed, there is no reason why they should always have been via the pre-Golgi route and the same red algal lineage.

Outside their double plastid envelope, chromistan algae and sporozoa have a smooth PPM derived from the red algal plasma membrane; between the PPM and the plastid is the periplastid space, the former cytoplasm of the enslaved red alga. In cryptophytes, this retains a miniaturized nucleus (nucleomorph), 80S ribosomes and starch, but only a smooth periplastid reticulum (PPR) in chromobiotes and (possibly) sporozoa, whose ancestors independently lost the nucleomorph genome and periplastid ribosomes. The mode of PPM biogenesis and its protein translocation machinery are unknown, but could be the same throughout chromalveolates. Possibly the nucleomorph envelope and PPR retained the phospholipid biosynthetic machinery of the red alga and the ability to bud off (putatively copII-coated) vesicles (periplasmic vesicles) able to fuse directly with the PPM for its growth (Cavalier-Smith, 2003b). The PPR is therefore the homologue of the nucleomorph membrane: only its genome and pore complexes were lost by those chromobiotes and sporozoa that retain plastids. Thus, red algal exocytosis is retained, but its Golgi apparatus, copI- and clathrin-coated vesicles and capacity for endocytosis, mitochondria and peroxisomes have been lost. On this view, periplasmic vesicles serve solely for PPM growth and not for protein import from the host cytosol as Gibbs (1979) proposed, and nucleomorph and PPR membranes must contain acyl transferase and other specific proteins. As a gene could not be found for 7S RNA of the signal recognition particle (SRP) in the cryptophyte nucleomorph genome (Douglas et al., 2001), such proteins might be inserted by an RNA-free SRP, as in plastids, which lost the 4.5S SRP-RNA of their cyanobacterial ancestors (Eichacker and Henry, 2001).

The simplest origin for protein translocation across the PPM into the periplasmic space would be to insert a duplicate of the plastid Toc OM translocon into the PPM (Cavalier-Smith, 1999) able to recognize the transit-like sequence of the preprotein (after the signal sequence is removed) and translocate the preprotein serially across the PPM, and the plastid

OM and IM. Figure 4.4b summarizes this model in which three distinct novel duplicates of the Toc machinery arose in the ancestral chromalveolate: two (Top1 and 2) in the PPM able to recognize the transit-like sequence and one (Tocm) remaining in the plastid OM but modified to allow Top1 to attach (for details see Cavalier-Smith, 2003b). Ancestrally, Top1 and 2 recognized the same transit-like sequence, but specialized respectively to import plastid proteins and periplasmic proteins (e.g., acyl transferase, or, in cryptophytes, aminoacyl-tRNA synthetases and nuclear-coded nucleomorph proteins, e.g., DNA polymerase). In the ancestral chromalveolate and cryptophytes, the original Toc1 remained in the plastid OM for import of nucleomorph-coded plastid proteins. *Guillardia theta* has more than 40 such proteins (Douglas et al., 2001). Their transit peptides are mutually similar, relatively conserved and can direct proteins into green plant chloroplasts; by contrast, transit-like peptides of the nuclear-coded cryptophyte proteins differ considerably and cannot do this *in vitro* (Wastl and Maier, 2000), perhaps because of the necessarily substantial modification of Toc into Top1 during secondary symbiogenesis. Probably Toc and Top1 use separate stroma transit-peptidases.

When nucleomorph genomes were lost, chromobiotes and alveolates lost Toc1 but retained Top1, Top2 and Tocm for protein import into the plastid and periplastid space. Dinoflagellates alone were further simplified, losing the PPM and PPR (needed only for PPM biogenesis) and restructuring the Top1–Tocm–Tic complex; PPM loss was probably very early before Tocm changed enough not to recognize transit peptides directly. Some dinoflagellates have an initial transit peptide FXP or FXXP motif, absent in the ancestral red algal transit peptide and the nucleomorph-coded one, which is probably a chromalveolate synapomorphy lost by sporozoa. This shared derived motif is further evidence for the single symbiogenetic origin of chromalveolates and the postulated divergence between Toc and Tops (Cavalier-Smith, 2003b).

GAPDH tells that chromalveolates are monophyletic, but not whether alveolates and chromists are sisters: one might be the paraphyletic ancestor of the other. The probability that ER-vesicle targeting as in alveolates is ancestral to the simplified direct route of chromists does not exclude the possibility that ER–EM fusion occurred more than once. However, independent evidence from the shared autofluorescent posterior cilium, fucoxanthin and intracristal filament of their tubular mitochondrial cristae (Cavalier-Smith, 1994) and concatenated chloroplast gene trees (Yoon et al., 2002b) indicates that chromobiotes are holophyletic; ciliary tubular hair structure and concatenated chloroplast gene trees (Yoon et al., 2002b) agree that chromobiotes are sisters to Cryptista. It is improbable that chromists are paraphyletic ancestors of alveolates; that would require that the periplastid complex escaped from the rough ER into a smooth vesicle and simultaneously reacquired the ancestral chromalveolate protein-targeting mechanisms, both evolutionarily onerous and with no selective advantage. There is also no reason to consider alveolates ancestral to chromists; single and multiple gene sequence trees show alveolates as robustly holophyletic (Baldauf et al., 2000; Bapteste et al., 2002; Cavalier-Smith, 2002; Cavalier-Smith and Chao, 2003b; Fast et al., 2002). Chromalveolates, alveolates, chromists and chromobiotes are almost certainly all holophyletic, as are heterokonts.

4.3 Alveolate Cell and Organellar Diversity

Alveolates comprise two phyla (Table 4.1 and Figure 4.5), the almost exclusively haploid and typically biciliate Myzozoa (Dinozoa, Apicomonadea, Sporozoa) and the dikaryotic Ciliophora with diploid germline nuclei, separate macronuclei and longitudinal rows of cilia (kineties) with centriolar roots embedded in the cell cortex (the infraciliature). Ciliophora comprise ciliate protozoa and suctorians, highly derived tentaculate predators lacking cilia when adult. Ciliophora diverged early into two contrasting subphyla — Postciliodesmatophora and Intramacronucleata — which I colloquially dub desmates and amphiesmates. Amphiesmates have a cell cortex like that of dinoflagellates, comprising the plasma membrane, numerous largely discrete cortical alveoli bonded to it, an associated microtubular skeleton and often an underlying epiplasm. In dinoflagellates, this complex is called the amphiesma; here I apply this term equally to intramacronucleate ciliates and those protalveolates with discrete cortical alveoli (also called amphiesmal vesicles), e.g., *Colponema*, for this type of cortex was almost certainly ancestral for all alveolates.

The idea that glaucophyte and alveolate cortical alveoli are homologues (Cavalier-Smith, 1982) depends on similarity in structure. This is not so complex that convergence can be totally ruled out, but until the molecular biology of cortical alveoli is understood, a common origin in the common ancestor of Plantae and chromalveolates (grouped informally as corticates, Cavalier-Smith and Chao, 2003c; Cavalier-Smith, 2003b) and multiple losses is the simplest hypothesis. An alternative possibility raised by the tendency of cryptophytes to group with glaucophytes on rRNA trees is that Plantae are paraphyletic and chromalveolates sisters only to Glaucophyta; this would involve one less loss of cortical alveoli, but one more for phagotrophy.

The amphiesma gives the alveolate cell cortex an elastic stiffness; its mechanical virtues created a major new adaptive zone and tremendous success for the major two groups bearing it (dinoflagellates and amphiesmate ciliates) as the largest highly mobile single-celled predators (Cavalier-Smith, 1991a). The contrasting requirements of giant and small cell size are a recurring theme throughout alveolate evolution.

4.3.1 Cortical Contractility and the Degenerative Simplification of Desmate Ciliates

Colponema, with toxicyst extrusomes like those of amphiesmate ciliates, might be the sister group to ciliates and the best guide to the nature of their biciliate ancestor. Desmates differ greatly from this archetypal pattern; Karyorelictea, long, vermiform flattened ciliates that favor anaerobic or interstitial marine habitats, lack cortical alveoli, and Heterotrichea have only vestiges. This implies extreme reduction in their common ancestor and total loss in karyorelictids — I suggest as a coevolutionary response to evolving strong cortical contractility. Heterotrichs include giant highly contractile ciliates, such as *Stentor*, *Blepharisma*, *Spirostomum* and the loricate *Folliculina*, having thick cortical contractile bands, here designated as desms. I suggest they are homologous with those of karyorelictids, wherein two of the three orders [Protostomatida, Protoheterotrichida: absurdly flattened interstitial dwellers (Lynn and Small, 2002)] are even more dramatically contractile. Molecular sequence evidence that karyorelictids are sisters to heterotrichs (Riley and Katz, 2001), which have ordinary polyploid dividing macronuclei, not to all other ciliates, implies that karyorelictid nuclei are degenerate evolutionary dead ends, as their nondivision long suggested, not primitive precursors of the polyploid dividing macronuclei of other ciliates (Raikov, 1982).

I suggest that ancestral karyorelictids were highly contractile flattened creatures that replaced the few normal giant ciliate polyploid macronuclei by numerous small diploid ones, because giant ones as in *Stentor* or *Blepharisma* would have been far too thick to allow their mechanical snaking as tenuously thin strips between sand grains. The thesis that diploid macronuclei are derived not primitive is supported by their independent evolution in Protocruziidia (amphiesmate ciliates), according to sequence trees (Hammerschmidt et al., 1996). A corollary is that noncontractility of Loxodida is secondary; interestingly, they never reevolved dividing polyploid macronuclei. The macronucleus of ciliates other than karyorelictids and *Protocruzia* is often misleadingly called polyploid. However, it is not truly polyploid; instead, there is unequal representation of different components (Jahn and Klobutcher, 2002), and this selective representation in high copy numbers is designated as multiploidy (Cavalier-Smith, 1985). Small cell diameter explains why karyorelictid and small size why *Protocruzia* macronuclei became diploid; Section 4.3.2 describes why they do not divide.

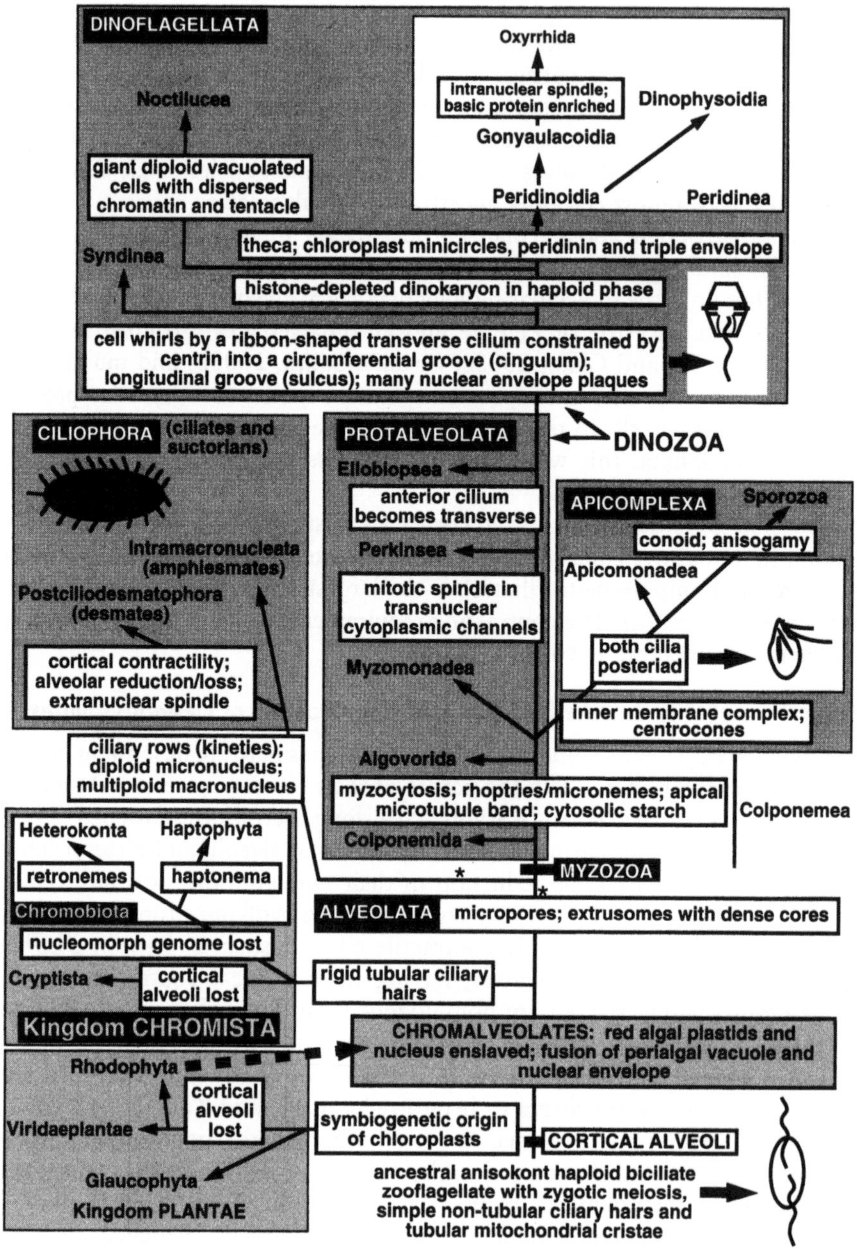

FIGURE 4.5 Major events in chromalveolate diversification. The dashed line shows the pivotal event in the origin of chromalveolates — the unique enslavement of a red algal nucleus (now the cryptophyte nucleomorph), plasma membrane (now the PPM) and chloroplasts. The thumbnail sketches indicate the ancestral ciliary state (anisokont) with one forwardly directed and one posteriad cilium (still found in Colponemea and Perkinsea) and three major derived ciliary states: the multiciliate condition of Ciliophora, the two posterior pointing cilia of Sporozoa plus Apicomonadea and the modification of the anterior cilium into a circumferential one in Dinozoa. A dinokaryon is the nucleus with interphase condensed chromosomes lacking nucleosomal histones that typify Peridinea (except *Oxyrrhis*) and the probably haploid dinoflagellate phase of Noctilucea. The unspecialized alveolate ultrastructural characters of *Colponema* do not predict whether it should branch as shown or in one of the positions marked by asterisks.

Lynn (1996) assumed that amphiesmate intramacronuclear microtubules are derived. On the contrary, intranuclear spindles were almost certainly ancestral for Myzozoa, alveolates and chromalveolates. The unique heterotrich extranuclear spindle is derived, requiring adaptive explanation. I suggest that an extranuclear microtubular investment arose as mechanical protection against nuclear fragmentation by their novel derived condition of extreme cortical contractility, together with extra proteinaceous strengthening inside the macronuclear envelope. Extranuclear spindles always evolved in cells prone to gigantism (dinoflagellates, Parabasalia, heterotrichs), needing extra intranuclear cortical support, carried to the extreme in the immense somatic giant nucleus of the dinoflagellate *Noctiluca*.

Desmates are not the ancestral state but a retreat into more specialized niches, where they compete with contractile animals — nematodes, flatworms and gastrotrichs — all at an advantage because of compartmentation into separate cells, especially muscle, nerve, sensory cells and differentiated gut, and thus fill many more niches. The desmates' loss of the stiff semirigid two-dimensional amphiesmal structure required the postciliary centriolar microtubules to elongate and overlap by cross-bridges to form longitudinal elastic elements spanning their entire length — the postciliodesmata. Any contractile machine requires such elastic elements (in muscle provided intracellularly by titin and extracellularly by collagen or elastin in connective tissue). Postciliodesmata are not the ancestral state but a mechanical adaptation for cytoplasmic contractility.

4.3.2 Why Ciliates Have Separate Germline and Somatic Nuclei: Making Giant Cells with Rapid Growth Rates

Lynn (1996) was puzzled about what synapomorphy made amphiesmates diversify more than desmates; if the desmate condition is derived not ancestral, the answer is obvious. The huge success and diversity of amphiesmate ciliates stem from their unique combination of three things: (1) the amphiesma itself, allowing gigantism to be mechanically successful; (2) invention of kineties and transverse fission, allowing greatly increased somatic and oral ciliary complexity (important for more efficient feeding and more rapid locomotion — about six times faster than in flagellates of the same size, Fenchel, 1987); and (3) invention of somatic multiploid macronuclei — increasing gene dosage enabled their giant cells to grow much faster than can haploids or diploids of comparable size (Cavalier-Smith, 1985). Thus the two ancestral ciliate synapomorphies (kineties, macronuclear multiploidy) explain amphiesmate success, not an undiscovered property of amphiesmate ciliate cellular organization (Lynn and Small, 2002).

It has been suggested that a *sine qua non* for complex organisms is a uninucleate germline stage in the life cycle on which selection can act to purge somatically selfish or degenerate mutations (Crow, 1988); this would be most efficient if the nucleus were haploid (Cavalier-Smith, 1995a) and regularly transcribed. Yet ciliates, unlike animals and multicellular plants, lack a small-celled uninucleate gamete or spore life-cycle stage, having become large by ciliary multiplication as kineties that must be transmitted transgenerationally as preformed cell structure to maintain their complex soma (Frankel, 1989; Chapter 15 and Chapter 16). Their subdivision into germline and soma is intracellular by suppressing the long-term inheritance of macronuclei; division of numerous equal diploid nuclei would be genetically more dangerous, favoring mutational degeneration. Crow's principle might suggest that division is blocked in karyorelictid diploid macronuclei to exclude them from the germline, reducing the accumulation of harmful mutations by restricting the number of germline nuclei per cell. However, such Panglossian population genetic perfectionism involves a very weak long-term selective force. The evolution of gene scrambling (Section 4.3.2) shows that ciliates are far from optimized to minimize harmful mutations.

A more powerful short-term benefit is increasing growth rates, the very selective force that originated macronuclei (Cavalier-Smith, 1985). By never dividing, karyorelictid diploid macronuclei can be continuously transcribed throughout the cell cycle. Even if transcriptional shutdown during mitosis (and gene-by-gene interruption during replication) lasted only 5% of the cell cycle, this could give cells a 5% growth rate advantage every cell cycle, a far stronger immediate selective force than the avoidance in the distant future of a few more harmful mutations. Thus, karyorelictids could recover some of the sacrifice of transcriptional intensity made by reverting from multiploid to diploid macronuclei. Macronuclear nondivision and multiploidy are alternative ways of increasing growth rates of larger cells, multiploidy being quantitatively superior but not feasible if cells have to be very narrow and efficiently contractile. In karyorelictids and *Protocruzia*, micronuclei specialize in replication and macronuclei in transcription; such extreme compartmentation also economizes on the total synthesis per cell cycle of replication, repair and transcriptional enzymes and mitotic proteins.

Physical limitations of the interstitial environment necessarily make clearance rates by filter feeding on bacteria up to 10-fold lower than in open water (Fenchel, 1987); so when the ancestors of karyorelictids first colonized them they were forced to grow more slowly, weakening the selective force for retaining the very high transcriptional output allowed by polyploidy compared with other ciliates. Thus, the ability of their macronuclei to become secondarily diploid to make smaller nuclei for mechanical reasons is comprehensible, as macronuclear diploidy would reduce their actual growth rates relatively much less than for noninterstitial ciliates.

Next, I discuss how the short-term functional adaptation that led to multiploidy increased the power of historical contingency and mutation pressure to create bizarre genomic properties that no all-powerful intelligent designer or creator would contemplate.

4.3.3 Genomic Consequences of Evolving Macronuclei: Multiploidy, Chromatin Diminution and Gene-Fragment Scrambling

Multiploidy comes about by eliminating many noncoding sequences by mechanisms partially related to those silencing higher eukaryote heterochromatin (Taverna et al., 2002). Some ciliates show only slight departure from true polyploidy, but in the largest (stichotrichs and hypotrichs) it is very great, only ca. 7% of the genome remaining. After internal eliminated sequences (IES) are removed, most species rejoin numerous DNA ends, yielding very high-molecular-weight chromosomes. In hypotrichs and stichotrichs, many are not rejoined, yielding tens of thousands of gene-sized or oligogenic microchromosomes; these imply gene numbers ca. 27,000 in the stichotrich *Sterkiella* (formerly *Oxytricha) nova* (Prescott et al., 2002), not many fewer than in mammals, attesting to the tremendous structural complexity of these giant chromalveolate cells. Phyllopharyngean, armophorid and clevelandellid macronuclear chromosomes also often dramatically fragment into small pieces (Riley and Katz, 2001). This chromatin elimination, and the gene scrambling discussed later, are consequences of the separation in function between germline micronuclei, which have to remain capable of meiosis but are not transcribed significantly during vegetative growth, and macronuclei specialized for massive vegetative transcription. By eliminating much noncoding DNA (equivalent to heterochromatin), its replicative cost is reduced during vegetative growth, allowing much greater amplification of the genes themselves and therefore much more rapid growth than possible with a polyploid nucleus of the same size (which would have fewer gene copies; copy number can become rate limiting in massive cells). This is the functional benefit of multiploidy and chromatin diminution (Cavalier-Smith, 1985).

However, separation into germline and somatic nuclei has a genetic cost or load. Vegetative transcriptional silencing of the micronucleus weakens purifying selection against harmful mutations therein. (This reduction is less severe in animal and plant germline cells wherein many genes are expressed every cell cycle.) Potentially lethal mutations can accumulate in the micronucleus during vegetative growth and can be eliminated only during or after meiosis when expressed and the cells bearing them die. Harmful, nonlethal mutations also easily accumulate and are slower to be eliminated. Because of the long time during vegetative growth when harmful micronuclear mutations can persist, suppressor mutations have more chance to occur before purifying selection recommences at the next sexual bout. If another gene mutates into a suppressor during vegetative latency, recombinants bearing both the lethal or harmful mutation and its suppressor might be at no significant disadvantage compared with the wild type lacking both mutations. Therefore, the lethal mutation could persist or spread through the population by drift or hitchhiking on a linked beneficial mutation. If the suppressor mutation could correct not only the phenotype of the original harmful mutation but also the phenotypes of a whole class of mutations, then similarly correctable, otherwise harmful mutations could persist in many genes with relative impunity and low cost (Cavalier-Smith, 1993a). This mildly harmful evolutionary consequence of vegetative silencing of the micronucleus simply explains the widespread gene scrambling in ciliate macronuclei; it is not a beneficial mechanism to "facilitate faster evolution of new genes" as often supposed (Prescott, 1999a,b, 2000).

Originally, DNA excision arose to eliminate intergenic heterochromatin, but if the method of rejoining DNA ends after elimination could be used for any DNA, it could also join intragenic DNA to phenotypically correct lethal mutations caused by accidental IES insertion into genes (Hogan et al., 2001). If the rejoining mechanism did not require joinable ends to be neighbors before excision, it would be relatively easy for gene segments with inserted IES to become scrambled by transposition or other chromosome rearrangements, because they could still be correctly rejoined during macronucleus formation. *Sterkiella* rejoins about 150,000 separated gene fragments (Prescott, 1999b). This useless cost of chromatin diminution is analogous to the useless costs of RNA pan-editing in trypanosomes or intron removal generally (Cavalier-Smith, 1993a). All three mechanisms arose originally as generalized phenotypic corrections of harmful mutations, but making it easier for other harmful mutations to spread; short-term benefit caused long-term harm. Seeing any of them as fundamentally beneficial for evolution or adaptive is a basic mistake; they are simply genetic burdens that arose and spread by mutation pressure and were perpetuated by the virtual impossibility of eliminating them once numerous genes became dependent on them — a genetic analogue of disorder-introducing entropy. Once evolved, any of these three mechanisms might be recruited occasionally in particular genes for a useful function, e.g., differential splicing, but the vast majority of instances of IES or intron insertion are probably mildly harmful, not beneficial.

4.3.4 Mouth Evolution and Ciliate Structural Diversity

The origin of the mouth was of primary importance to the origin of ciliates: it enables them to outcompete flagellates in bacteria-rich habitats (Fenchel, 1987). It was virtually necessitated by the amphiesmal condition, its rigidity impeding ingestion over most of the cell surface — probably the ancestral ciliate was a ventral-mouthed feeder on bacteria. If the ancestor was a biciliate such as *Colponema* with a ventral groove, it could directly have evolved into the ancestral ciliate mouth. Oral evolution was central to ciliate diversification. Ventral mouths are particularly characteristic of mobile filter-feeding bacteria eaters; apical mouths are secondary adaptations to more discriminating raptorial feeding on larger prey or filter feeding by sessile cells, and probably arose several times to exploit the faster feeding that sessility allows (Fenchel, 1987). Even in karyorelictids, with apical mouths sometimes considered archetypal for ciliates, the mouth is assembled ventrally and later moves to the cell apex.

Amphiesmates diversified into infraphyla Spirotrichia, Rhabdophora and Ventrata. Spirotrichs include the degenerate *Protocruzia*; the conical Choreotrichia (e.g., tintinnids, abundant oceanic lorica dwellers) with few body cilia and complex apical oral kineties for swimming and feeding; and oligotrichs, hypotrichs and stichotrichs, all with apicolateral mouths and immensely more complex, highly differentiated somatic cilia or cirri than most ciliates. Rhabdophora have an apical mouth, ancestrally surrounded by a complex radially symmetric pharyngeal basket, the rhabdos; they comprise only Litostomatea, subdivided into the free-living Haptoria, armed with toxicysts, e.g., *Didinium* that swallows whole paramecia and *Lacrymaria* with a marvellously extensible neck, and the delightfully tufted Trichostomatia ("hairy mouths") that mostly dwell in vertebrate guts and unsurprisingly lost toxicysts.

Ventrates comprise six classes of often morphologically duller ciliates (e.g., *Tetrahymena*, *Paramecium*, in the most disparate class, Oligohymenophorea) ancestrally characterized by a ventral mouth, but dramatically modified in suctorians, prostomates and peritrichs. Their pharyngeal cytoskeleton was ancestrally a cyrtos, comprising a distal fibrous and proximal dense annulus encompassing a cylindrical palisade of microtubular rods (nematodesmata) united by a fibrous sheath. This was lost or partially disorganized in Oligohymenophorea/Plagiopylea; some components are absent or ancillary ones present in some other lineages. Oligohymenophorea, Colpodea and Nassophorea have bicentriolar somatic kinetids, but the other classes simplified them to one centriole, probably independently in Plagiopylea/Prostomatea (related to Oligohymenophorea and Colpodea) and Phyllopharyngea (typically attached, e.g., the strikingly tentaculate suctorians and curious chonotrichs or thigmotactic; characterized by pharyngeal radial microtubular ribbons — phyllae — within the nematodesmata, if present). That the mouth of Prostomatea is assembled ventrally but is apical in adults is consistent with its being ancestrally ventral and its apicality being convergent with that of Rhabdophora and Karyorelictea. The delightful peritrichs (e.g., the sessile *Vorticella*, with its contractile centrin-related stalk) through evolving sessility also moved their mouth apically, a state retained when some became secondarily mobile as hovercraft-like gliders on animal surfaces (e.g., *Trichodina*).

I do not agree that mouthlessness was ancestral (Lynn and Small, 2002), considering all mouthless ciliates degenerate. Mouthlessness in the oligohymenophorean subclass Apostomatia of annelid guts presumably arose because they were so well supplied with predigested food that it could be taken up at micropores by clathrin-mediated endocytosis, dispensing with phagocytosis of prey. The loss of cortical alveoli by free-living karyorelictids, together with ciliary simplification creating bare surface areas, allowed many to phagocytose bacteria at bare soft regions and dispense with morphogenetically expensive bulky mouths.

4.3.5 Myzozoa and Myzocytosis: Apical Suctions and Cortical Artistry

Oral traditions also played a major role in evolution of Myzozoa, which comprise Dinozoa, Apicomonada and Sporozoa. Most free-living Myzozoa are dinoflagellates. Purely photosynthetic dinoflagellates need no mouth, but are degenerate as the ancestral dinoflagellate was almost certainly a photophagotroph. For Peridinea (the typical dinoflagellate majority without histones), the ancestral amphiesma was armored by cellulose plates within the cortical alveoli, as in Peridiniida and Gonyaulacida. Naked former Gymnodiniales polyphyletically lost this armor (Saldarriaga et al., 2001, 2004), possibly to allow smaller, more rapidly dividing cells with numerous little cortical alveoli rather than lumbering rhinoceros-like armored giants. The ancestral armored peridinean needed an unarmored region for feeding: in contrast to Ciliophora, not a depression, the cytopharynx, but a projection, the extensible peduncle, which emanates from the sulcal groove sheltering the posterior longitudinal cilium.

Dinoflagellates and Sporozoa are not sisters: each evolved independently from basal protalveolates and apicomonad ancestors, respectively. Molecular trees weakly suggest that the apicomonads are sisters of Sporozoa, whereas the only protalveolate with sequences available (*Perkinsus, Voromonas*) are sisters to dinoflagellates or dinoflagellates plus Sporozoa and apicomonads. Myzocytotic protalveolates (Myzomonadea) have members with discrete cortical alveoli of the amphiesmal type like dinoflagellates (*Voromonas pontica*), whereas apicomonads (*Colpodella, Acrocoelus*) have a compressed and continuous inner-membrane complex indistinguishable from that of Sporozoa (Cavalier-Smith and Chao, 2004). *Perkinsus* has neither type, but simple anterior alveoli.

Apicomonads and myzomonads feed by myzocytosis, using an apical rostrum supported by a microtubular skeleton. Dinoflagellates use the peduncle for myzocytosis; probably it evolved from the myzomonad apical rostrum. Myzocytosis differs from phagocytosis in that the rostrum or peduncle makes a circular break in the prey's plasma membrane, which reseals around it; the apical plasma membrane now in direct contact with the prey's cytosol sucks it out in gobbets into food vacuoles (Elbrächter, 1991). Apicomonad myzocytosis is important for understanding the origin of Sporozoa. Attachment of extracellular gregarine Sporozoa to the host cell surface and the initial attachment of intracellular Sporozoa (coccidians, Hematozoa) before self-insertion into a parasitophorous vacuole resemble the initial stages of myzocytosis. Possibly the parasitic Sporozoa evolved from a free-living myzocytotic predatory biciliate such as *Colpodella* (Cavalier-Smith and Chao, 2004). Apicomonads and protalveolates lack an apical complex with a conoid, which probably evolved to mediate injection into the host cell only when a *Colpodella*-like apicomonad became the first sporozoan. The rostral microtubular skeleton of early myzocytotic myzozoan predators was probably ancestral independently to the apical complex skeleton and the dinoflagellate peduncle.

Conoids are extrusible apical organelles for host penetration, unique to Sporozoa: symmetrical truncated cones made of a peripheral lattice of two criss-cross spirals of trough-like tubulin rods and two central microtubules. Each rod differs profoundly from a microtubule, being a long curved sheet of about nine tubulin protofilaments arranged like a comma or asymmetric trough in cross section (Hu et al., 2002). This novel structure decisively validates earlier cogent criticisms (Vivier, 1982; Siddall et al., 1997) of the notion that the rostral microtubular ribbon of *Perkinsus* is a conoid (Levine, 1978). The *Perkinsus* ribbon resembles the very likely homologous components of the microtubular basket of the dinoflagellate peduncle (Jacobson and Anderson, 1992). The microtubular attachment rings of *Colpodella* resemble neither the *Perkinsus*/dinoflagellate rostral ribbons nor the conoid. The rostral skeleton of *Voromonas* differs from all three. Probably the conoid evolved from an apicomonad-like rostral skeleton of some sort, but the transformation to form the novel trough-like rods and conical lattice was dramatic. Conoids and conoidal rings are not universal in Sporozoa (lost by Hematozoa).

The apical location of the rostral complex and posteriad orientation of the apicomonad cilia both probably evolved to enable these small flagellates to dock into their often larger (e.g., ciliate) prey — preadaptations to the sporozoan mode of parasitism (Cavalier-Smith and Chao, 2004). The contrasting anisokont ciliary arrangement in *Perkinsus*, *Colponema* and myzomonads is probably ancestral for Myzozoa and to the dinoflagellate dinokont arrangement. The resemblances between *Colpodella* and Sporozoa on the one hand and *Voromonas* and dinoflagellates on the other make myzozyotic predators a bridge between Sporozoa and dinoflagellates; both are now in the same phylum, Myzozoa. As *Colpodella* is much more similar to Sporozoa than is *Perkinsus*, grouping *Perkinsus* alone with Sporozoa, as Apicomplexa (Levine, 1978), was artificial; as Simpson and Patterson (1996) stress, there are no obvious synapomorphies for Sporozoa plus perkinsids that exclude colpodellids. One reason for the unnecessary multiple renaming of Sporozoa as Polannulifera and then Apicomplexa was antipathy to the term *spores* for sporozoan cysts (Levine, 1970), but as not all have an apical complex (of which conoids, conoidal rings and rhoptries are the only unique parts), Apicomplexa and Polannulifera were equally inappropriate. Although the name Apicomplexa was therefore originally not needed, it is retained in the latest classification of Myzozoa as a subphylum that includes Apicomonadea and Sporozoa (Cavalier-Smith and Chao, 2004)

Sporozoa have four key structural differences from apicomonads: the conoid, conoidal rings (both secondarily lost in Hematozoa), centrocones and rhoptries. Although like Sporozoa, *Colpodella* have long dense extrusomes known as micronemes, they lack rhoptries. Possibly rhoptries evolved from lysosomes, not otherwise present in Sporozoa. Apicomonads need lysosomes (the ancestral myzozoan state) for intracellular digestion of food vacuoles produced by myzocytosis; when Sporozoa evolved parasitism, they retained the first stage of myzocytosis, penetration of the prey–host membrane, but abandoned the second, phagocytosis of cytoplasmic gobbets, replacing it by extracellular digestion by enzymes secreted at the cell apex by rhoptries (within the parasitophorous vacuole of intracellular species or the anterior vacuole of extracellular ones, e.g., many gregarines). The apical end of the extracellular gregarines is topologically equivalently located to intracellular species; strictly speaking, they are not entirely extracellular, but partly intra- and partly extracellular, so I will call them ambicellular (Latin *ambi-* "both sides"). Ancestrally, Sporozoa were probably small cells like apicomonads and intracellular, like Coccidea, originally infecting other much larger long-lived protozoa such as Foraminifera and Radiolaria. Gregarine ambicellularity was probably a secondary adaptation after colonizing animals to allow much larger cells by protruding their rear into host body cavities (gut or coelom) from which rich food was available by coated-vesicle pinocytosis through micropores (a universal alveolate feature presumably arising with the amphiesma).

Intracellularity led to marked genomic reduction, *Plasmodium* having among the smallest genomes of all eukaryotes (Gardner et al., 2002). Ambicellularity and associated cellular gigantism entailed tremendous genomic expansion, as nuclear genome size necessarily coevolves with cell size to allow balanced growth (Cavalier-Smith, 1978, 1982, 2004). The distinguished geneticist (Goldschmidt, 1955) used the dramatic contrast between the minute coccidian chromosomes and the giant ones of amphicellular gregarines (bigger than those of humans) to argue that genes cannot be made of DNA and that DNA has another, probably structural, role. Most geneticists have assumed that as genes are made of DNA, DNA cannot have a structural function. Both conclusions are mistaken; all nuclear DNA has a skeletal function in nuclear assembly and in determining nuclear volume (Cavalier-Smith, 1985, 1991b, 2004), but an important (typically minor) fraction also has the classical genic function. The universal genic function of DNA began soon after life began, before the first bacteria, but the additional skeletal function of nuclear DNA arose only with the origin of the nuclear envelope and the segregation of transcription and translation into separate compartments.

In contrast to Sporozoa, dinoflagellates probably arose from a biciliate with one anteriorly directed cilium such as *Colponema* and *Perkinsus*. The key step was the modification of the anterior cilium by the contractile centrin paraxonemal rod, wrapping it around the cell transversely into the cingulum, a novel spiral groove within which it beats to spin the cell, driven forward by the posterior one. Spinning would gyroscopically stabilize giant motile biciliate cells and broaden the search path for feeding via the peduncle. Origin of the cingulum involved rearrangements of cortical microtubules to support each rim, and of the typical myzozoan longitudinal ones into two fields in the two halves of the cell (hypocone and epicone). Cortical alveoli were specially shaped to fit into these three cellular regions, perhaps by patterned assembly of amorphous lumenal material that developmentally precedes cellulose addition to the thecal plates of armored dinoflagellates. The discrete alveoli of *Voromonas* have similar material (Cavalier-Smith and Chao, 2004); the plateins constituting ciliate alveolar plates appear fairly fast evolving (Kloetzel, 1991; Kloetzel et al. 1992), which might hinder tests of their homology with myzozoan ones.

Dinoflagellates are the only alveolates that combine photosynthesis and phagotrophy and the most diverse protists in this mixotrophic adaptive zone. The dinoflagellate amphiesma gives exceptional mechanical ability to successfully exploit the large end of the protist size spectrum, where they are in competition with ciliates, also having an amphiesma, and Retaria (Foraminifera, Radiolaria), which evolved a central capsule and internal skeletons to support large size and exploit reticulopodial feeding, and some Amoebozoa that use lobopodia. The most diverse animal group that combines phagotrophy with photosynthesis (corals) does so by cultivating suessiid dinoflagellates intracellularly. The only other ecologically important protist photophagotrophs are also chromalveolates — chrysomonad and actinochrysean heterokonts and haptophytes. Through retaining phagotrophy these chromists also spawned purely heterotrophic predatory subgroups by losing photosynthesis, but had to concentrate on the small end of the size spectrum as they lack an amphiesma.

Peduncular myzocytosis was probably the ancestral mode of phagotrophy for Peridinea. The more specialized pallial feeding of some Peridiniida probably evolved from it; after contacting prey the peduncle tip extends as a very thin veil-like pallium that can completely surround it — even diatoms larger than the dinoflagellate itself. The prey is digested extrasomally within the pallium; only digested food products, not food vacuoles, are transported back into the cell body before the pallium is retracted (Elbrächter, 1991; Jacobson and Anderson, 1992). This is an adaptation for taking giant prey impossible for armored cells to ingest. Peridinea originated in the Triassic, radiating immensely in the Jurassic synchronously with the origin of planktonic centric diatoms (Tappan, 1980); dinoflagellates' great success might partially stem from their ability to predate large diatoms and microzooplankton while being active swimmers. The Mesozoic radiation of both chromalveolate groups — thereafter the dominant larger protist marine plankton — followed the massive Permian extinction that terminated the Palaeozoic, inducing the greatest biotic turnover in the whole Phanerozoic.

Dinoflagellates are the only eukaryotes known to have replaced their chloroplasts by foreign ones (Delwiche, 1999), probably at least twice (Saldarriaga et al., 2001). This exceptional evolutionary capacity is probably because, alone among chromalveolates, they retained both phagotrophy (needed for acquiring replacement plastids) and (putatively) ER-vesicle targeting to the epiplastid membrane (which would preadapt them for enslaving a foreign plastid much more simply than could the ancestral chromalveolate or cabozoan). The suggestion that the peridinin-containing plastid arose by tertiary replacement by a haptophyte plastid (Yoon et al., 2002a) not vertically from the ancestral chromalveolate (Cavalier-Smith, 1999) is ill-founded, based solely on single-gene (*psbA*) trees strongly distorted by long-branch attraction because of the bizarrely fast evolution of peridinean minicircles, and contradicted by a more clock-like nuclear gene (*psbO*) tree that supports the chromalveolate theory (Ishida and Green, 2002). Heterotrophic dinoflagellates can also temporarily cultivate foreign plastids intracellularly (Takishita et al., 2002). Peridinea are unique among algae in having transferred nearly all their plastid genes into the nucleus (Hackett et al., 2004), and consequently having evolved bizarrely unique single-gene minicircular DNA for the approximately 15 genes that remain in the chloroplast (Zhang et al., 1999, 2000, 2001, 2002).

Dinoflagellates have been equally successful parasites, their primary bifurcation being between the purely parasitic Syndinea that retain traditional histones and Dinokaryota that lack them (Peridinea; at least for the bulk of their skeletal DNA) or retain them only when growing (Noctilucea). Like many protist parasites, Syndinea evolved multinucleate nonciliate plasmodial feeding stages, but retained a more ancestral biciliate dispersal stage — the dinokont dinospore. Another parasitic group, ellobiopsids, might be sister to dinoflagellates, as they have motile biciliate spores with a transverse cilium (but no cingulum) and a closed mitosis with mitotic spindle in transnuclear cytoplasmic channels (probably an adaptation for large cell size), widespread in subphylum Dinozoa. If the common ancestor of dinoflagellates and ellobiopsids were a myzocytotic photophagotroph, it would have been preadapted to evolve parasitism and lose photosynthesis multiple times, analogously to Sporozoa. This happened several times within Peridinea, forming parasites such as *Hematodinium*, *Amylodinium* and *Blastodinium*. Microscopy suggests that *Blastodinium* has normal histone and chromatin during part of its life cycle (Fensome et al., 1993), but this has not been verified biochemically, unlike in *Oxyrrhis*, in which the abundant basic proteins are not the classic nucleosomal histones (Kato et al., 1997).

Noctilucea evolved much greater somatic contractility than Peridinea, analogously to desmate ciliates, and giant cells by grotesque inflation by internal vacuoles; unlike other dinoflagellates, this feeding stage has a well-defined cytostome for normal phagocytosis and is diploid, unlike all other Myzozoa. The numerous small swarmers produced by multiple fission (Sato et al., 1998) or haploid gametes lack histones (Zingmark, 1970), possibly adaptive to small cell size, analogous to replacement of histones by protamines in many animal sperm. Each of the three families was dramatically modified in different ways; for example, Noctilucidae lost the cingulum and have only one ribbon-shaped cilium in a groove and a huge contractile tentacle.

Dinoflagellates and ellobiopsids were ancestrally large cells, but a major unidentified alveolate lineage apparently sister to the established dinoflagellates has recently been revealed by environmental sequencing of (putatively heterotrophic) picoplankton (López-García et al., 2001; Moon-van der Staay, et al., 2001, Clade 2 is probably Syndinea). The large amounts of peridinin, the unique dinoflagellate antenna carotenoid, in picoplankton (Latasa and Bidigare, 1998) suggest that there might also be many minute uncharacterized, possibly loosely gymnodinealean, peridinean picoalgae.

4.4 Chromist Cell Diversification

The primary divergence within Chromista, ancestrally biciliate, is between cryptists that have no obvious homologues of cortical alveoli and chromobiotes that sometimes do. Chromobiotes were ancestrally photophagotrophs with fucoxanthin-containing plastids and tubular mitochondrial cristae with intracristal filaments like at least some Myzozoa (e.g., some dinoflagellates and *Voromonas*). They diverged early into heterokonts, ancestrally with tripartite rigid thrust-reversing tubular hairs on the anterior cilium (retronemes), and haptophytes, ancestrally with two smooth cilia and a raptorial haptonema emanating from near their centrioles. Fucoxanthin was secondarily lost by a few groups favoring fresh water (xanthophytes, eustigmatophytes, freshwater raphidophytes), suggesting that ancestral chromobiotes were marine. The xanthophylls fucoxanthin and peridinin enable equally efficient light absorption over a much wider spectrum than green plants achieve, especially important for pelagic marine phytoplankton prone to sink into dimmer deep waters, unlike soil and shallow freshwater algae (predominantly green) for which excessive light absorption is hazardous.

4.4.1 Ejecting Ribbons and Cryptist Cell Diversification

Cryptists are biciliates characterized by a rigid nonalveolate cell surface, open mitosis, two parallel subapical centrioles and at least two kinds of ejectisome: scroll-like extrusomes that unfurl on secretion to form a long, tapered strip that inrolls laterally to form a tube (Schuster, 1968); each has small ejectisomes over the body generally and large ones posterior to the cilia associated with a groove in cryptophytes (Kugrens and Lee, 1991) or ridge in katablepharids — one katablepharid has three kinds. Cryptomonads, here treated as a subphylum, comprise cryptophytes with tubular bipartite ciliary hairs (typically with a single terminal filament and two rows on the anterior dorsal cilium and two filaments but a single row on the ventral cilium), a nucleomorph, plastid and PPR, and their goniomonad sisters with solid recurved spines (anterior cilium) or delicate simple hairs (ventral cilium) and no nucleomorph, plastid or PPR. Both have a kinetid with parallel centrioles and flattened-tubular mitochondrial cristae, unlike the anisokont state and tubular mitochondrial cristae with a circular cross-section ancestral for chromists and chromalveolates. Their relationship is solidly supported by molecular evidence, still wanting for their putative sisters, katablepharids (Clay and Kugrens, 1999), here included in Chromista for the first time.

Katablepharid ejectisomes are a single scroll, whereas cryptophyte ejectisomes have a second minor scroll, making the exploded organelle resemble a one-headed pick. Unlike cryptophytes, katablepharids retain the ancestral tubular mitochondrial cristae but lack tubular hairs, I suggest secondarily. Most have an apical microtubule basket suggestive to some of affinities with sporozoa; however, they lack conoid, micronemes, rhoptries or inner-membrane complex, and their ejectisomes favor a relationship with cryptomonads instead. If they are sisters to cryptomonads, their apical microtubular basket is evidence that the common ancestor of cryptists and chromobiotes (the ancestral chromist) had a rostrum with a microtubular skeleton able to generate both it (by multiplying curved apical ribbons and arranging them in a ring) and the haptonemal skeleton. The absence from katablepharids of the cryptomonad periplast is no obstacle to their sisterhood; as it is unique to cryptomonads, like their double ciliary transition region plates, secondarily flattened cristae and second minor ejectisomal scroll, such characters provide no information about their likely sisters. Katablepharids instead have a rigid extracellular wall or sheath that also extends over the cilium. Like the cryptomonad periplast, the origin of this sheath, by providing an alternative stiff layer to the cortical alveoli would have allowed them to be lost.

Goniomonads feed actively on bacteria, using their cytostome at the opposite end of the apical groove from the cilia. Presumably, goniomonads and katablepharids diverged early from cryptophytes before their common ancestor became obligately dependent on the plastid or periplastid space for lipid biosynthesis or other vital functions.

Cryptophytes are mostly phototrophs that flourish in dimmer light, rarely saprotrophs (*Chilomonas*, which retains the nucleomorph, PPM and plastid presumably for starch metabolism and perhaps also lipid biosynthesis); very few appear to be photophagotrophs, but many might utilize dissolved organic material. Ancestral cryptophytes were probably red because of intrathylakoid phycoerythrin 545, and diverged into Pyrenomonadales, with the nucleomorph embedded in a groove in the pyrenoid, and Cryptomonadales, in which it is free in the periplastid space (Deane et al., 2002). Cryptomonadales are more diverse, having evolved novel phycobilins, including a blue clade (e.g., *Chroomonas*) and the large-celled *Cryptomonas–Chilomonas* clade with phycoerythrin 566 that evolved two periplastid complexes per cell. The gullet near the ciliary bases, sometimes opening into or replaced by a furrow, need not be primarily for phagotrophy. It might help orient the extruded large ejectisomes to push the cell backward away from enemies, and be useful for pinocytosis and emptying the contractile vacuole. Its geometry might be partially a consequence of the origin of the relatively rigid periplast coupled with the parallel centriolar orientation, analogous with the reservoir or gullet of euglenoids (also not their "mouth"), which independently evolved a rigid pellicle and parallelized formerly divergent centrioles (Cavalier-Smith, 2003a). Recently, *Chilomonas* has been merged into a broader *Cryptomonas* genus, several species of which show a remarkable dimorphism in periplast structure (Hoef-Emden and Melkonian, 2003).

The most decisive evidence for a positive function for the massive amounts of noncoding nuclear DNA in most nuclei comes from the dramatic difference in scaling with cell size of nuclear and nucleomorph genome sizes in cryptophytes (Beaton and Cavalier-Smith, 1999; Cavalier-Smith and Beaton, 1999; Cavalier-Smith, 2003b, 2004).

4.4.2 Chromobiote Predation: Cell Organellar Novelty through Diverging Modes of Prey Entrapment

The haptonemal axoneme is usually assumed to have evolved *de novo*. However, the horseshoe arrangement of its usually nine microtubules is reminiscent of the curved microtubular bands forming the microtubular cage of the dinoflagellate peduncle (Calado and Moestrup, 1997) and the curved rostral microtubular sheet of perkinsids. Might all be homologues, present in the ancestral chromalveolate? If so, it would have been lost independently by Ciliophora, cryptists and heterokonts.

Like the peduncle, the haptonema was probably ancestrally for feeding. When well developed, it extends ahead of the swimming cell, catches prey by adhesion, then bends back to allow it to be phagocytosed. It is reduced to a small vestige in secondarily purely phototrophic haptophytes that abandoned phagotrophy (the majority). When previously discussing its origin (Cavalier-Smith, 1994), I saw it as an alternative feeding adaptation to that of their sister heterokonts, in which retronemes propel bacteria in powerful water currents to the ciliary bases for ingestion by a mobile microtubular trap. I assumed that the heterokont pattern with this feeding basket (Moestrup and Andersen, 1991) was ancestral and the origin of the haptonema caused retroneme loss, because the presence of bipartite tubular ciliary hairs in cryptophytes suggested that they were ancestral for chromists. However, although the latter is probably true, it does not follow that the heterokont mode of feeding also was ancestral for all chromists.

There is no confirmation or rejection from high-resolution video of my hypothesis (Cavalier-Smith, 1986) that cryptophyte anterior ciliary hairs reverse thrust, like retronemes; my impression from watching cryptophytes swim is that they perhaps do not, but might simply increase thrust. If retronemy is a derived condition for heterokonts alone, the ancestral chromobiote was an anisokont biciliate with nonthrust reversing tubular hairs on both cilia and a myzozoan-like rostrum with horseshoe-shaped apical ribbon. One lineage could have modified this into the heterokont feeding pattern by retronemal water currents, losing the apical ribbon, whereas its sister lineage developed it into the haptonemal axoneme for raptorial feeding, losing the tubular hairs to allow isokont swimming with anteriorly projecting haptonema. As the ventral cilium of katablepharids and *Goniomonas amphinema* points backward, the ancestral cryptist probably had oppositely pointing (anisokont) cilia like ancestral chromobiotes and alveolates, despite secondary parallelism of their centrioles.

As a proper cytopharynx is absent from all chromists but *Goniomonas*, and present only in ciliates and Noctilucea among alveolates, the ancestral chromalveolate probably lacked a cytopharynx. It might have had a rostrum for feeding, but fed phagocytically, like katablepharids which have an apical cytostome but no cytopharynx (Lee et al., 1991), not myzocytotically, and could ingest eukaryote algae to become the first chromalveolate. *Cytopharyngeal cylinder* (Vørs, 1992) is an inappropriate term for the inner microtubular basket of the katablepharid ingestion apparatus, which lacks any cytopharynx-like invagination. The slanted angle with subterminal cilia of cryptophyte anteriors might be a rostral relic frozen in shape when their periplast evolved. The cytostome or cytopharynx of goniomonads probably evolved independently of that of ciliates, perhaps from a rostral ingestion area by reshaping the cell anterior to its present truncated shape, quite different from cryptophytes; the cytostome opens into a pharynx (vestibulum) on the left side, separate from the central groove. The cryptophyte microtubule cytoskeleton is varied, possibly through simplification in smaller cells. The large-celled *Chilomonas*, unlike most cryptophytes, has a 12-microtubule, slightly curved anterior band, resembling the myzozoan rostral microtubular sheet more closely than the parabasalid pelta with which Roberts et al. (1981) compared it; is it homologous with the haptonemal skeleton? Thus, the ancestral chromist might have had a rostral microtubular skeleton related to that of Myzozoa, retained for the haptonema.

If the ancestral chromist had cortical alveoli, they need only have been lost once in the ancestral cryptist. Could the smooth cortical membranes of haptophytes that also extend into the haptonema have been inherited from cortical alveoli? They are no more radically different from the archetypal amphiesmal pattern than those of perkinsids, usually similarly restricted to the anterior end of the cell. The primary role of this haptonemal smooth reticulum is probably control of calcium concentrations by active uptake and passive release. Calcium regulation is probably a general function of cortical alveoli in addition to their structural role: ciliate cortical alveoli actively accumulate calcium (Stelly et al., 1991, 1995), as does the sporozoan inner-membrane complex (Bonhomme et al., 1993). Although consistent with an origin from cortical alveoli, this is not evidence for such an origin as any ER could potentially evolve calcium regulation.

4.4.3 Heterokont Heterogeneity

Heterokonts are the most diverse chromists, divided into two purely heterotrophic phyla (Sagenista, Bigyra) and the mostly phototrophic Ochrophyta, typically with brown fucoxanthin-containing chloroplasts. Their primary divergence is between Sagenista (Labyrinthulea, Bicoecea, Placididea) and Ochrophyta/Bigyra. All 12 ochrophyte classes are ancestrally photosynthetic; all except Chrysophyceae and Actinochrysea (pedinellids, silicoflagellates, actinophryids) probably lost phagotrophy early on and are obligate phototrophs, except for a few colorless diatoms; both phagotrophic classes lost photosynthesis more than once. Among chrysophytes, the abundant active phagotrophs *Paraphysomonas* and *Spumella* retain leucoplasts bearing eyespots associated with the smooth posterior cilium (like chloroplasts in photosynthetic chrysophytes); whether *Oikomonas*, which lost the posterior cilium, retains a plastid is unknown. Pedinellids lost the posterior cilium independently; the colorless species, once thought to lack plastids (Cavalier Smith et al., 1995), have leucoplasts (Sekiguchi et al., 2002). Their retention suggests that the ancestral ochrophyte became dependent on plastid fatty acid synthesis, like plants, making it impossible to lose plastids even if photosynthesis is lost, predicting that *Oikomonas* and all colorless diatoms should also have leucoplasts (*Nitzschia alba* does, Schnepf, 1969). All chromists studied have replaced the host cytosolic fatty acid synthesis by the red algal plastid type II enzymes, as did most sporozoa (Ryall et al., 2003).

Bigyra comprise the phagotrophic biciliate *Developayella*, the saprotrophic or parasitic Pseudofungi and the saprotrophic gut symbionts Opalinata, which moved ciliary hairs to the cell body (*Proteromonas*), lost them (Opalinea) or lost cilia altogether (Blastocystea). Many rRNA trees suggest that Pseudofungi and Developayella are sisters to Ochrophyta, but resolution is poor, allowing the possibility that they evolved from ochrophyte ancestors. There is no evidence for plastids in Bigyra or Sagenista. They might have been overlooked, but could be absent — plastid loss is inherently easier early in a group's evolution before the host evolves dependence on them for nonphotosynthetic functions such as fatty acid synthesis, precluding later loss. Early divergence of nonphotosynthetic heterokonts was previously misinterpreted as evidence for a nonphotosynthetic ancestry (Leipe et al., 1996).

Labyrinthulea (thraustochytrids and the derived labyrinthulids) lost phagotrophy when they evolved an ectoplasmic network of anastomosing saprotrophic filaments, a unique chromistan mode of heterotrophy not found in Fungi or Protozoa. The cell body is covered in large (1 ± 0.5 μm) Golgi-derived galactose-rich scales that overlap as a wall. Cell organelles, e.g., mitochondria, ER and lysosomes, are excluded from the ectoplasmic net by an intracellular plug, the sagenetosome; this exclusion and the wall prevent phagotrophy. Labyrinthulids surround their cell body completely by the ectoplasmic net; the ectoplasmic membrane facing the wall topologically separates from the outer membrane. The cell plus this membrane become motile (by ectoplasmic actomyosin) within the labyrinth thus formed by the ectoplasmic compartment — a membrane topology and form of motility unique in the living world. If the ancestral chromist had cortical alveoli, the scales might be homologous with the organic lumenal skeleton of myzozoan alveoli and the labyrinthulean wall evolved by exocytic fusion of cortical alveoli with the cell surface.

Bicoecea are phagotrophs with no wall or cortical alveoli: bicoecids dwell attached by their smooth cilium in an organic lorica shaped like a drinking cup (Greek *bicos*); anoecids (Cavalier-Smith, 1997) include similar aloricate flagellates (e.g., *Cafeteria*), derivatives lacking a smooth cilium (*Siluania*) and others that lost the heterokont ciliary hairs (e.g., *Caecitellus,* possibly through cell miniaturization, adopting a gliding habit using the posterior cilium as a skid on surfaces, dramatically shortening the anterior one and relying for feeding on a new lateral cytostome instead of retronemal currents). Environmental DNA sequencing suggests that Sagenista are more diverse than previously realized (Diez et al., 2001); rRNA sequences show that Caecitellus is not single "species" but a deep clade and that Adriamonas lost ciliary hairs independently (TC-S and Chao, unpublished). The apparently early diverging Placididea have a double ciliary transition helix (Moriya et al., 2002) like Bigyra; this might be ancestral for all heterokonts and lost by other Sagenista and Ochrophyta, and not a synapomorphy for Bigyra alone (Cavalier-Smith, 1997).

Ochrophyta apparently split early into two subphyla: Khakista (diatoms and Bolidophyceae) and Phaeista (e.g., brown algae, chrysophytes, xanthophytes). The cortical sacs of Raphidophyceae are virtually indistinguishable from the cortical alveoli of alveolates and typically attached to the smooth ER that surrounds their numerous plastids (Ishida et al., 2000). This skeletal role of cortical alveoli in this multiplastid group might explain their retention; many Phaeista instead evolved a vegetative cell wall (Eustigmatophyceae, Xanthophyceae, Phaeophyceae, Chrysomerophyceae, Phaeothamniophyceae) so could understandably have lost alveoli, were they ancestral for heterokonts. Conceivably, the highly vacuolated cortex of silicoflagellates and Developayella and the cortical silicalemma of diatoms (within which the opaline frustule forms) are relics of cortical alveoli. Even if all these cortical membranes arose independently of those of alveolates and haptophytes, it is likely that an ability to secrete and polymerize opaline silica in cortical membranes was an ancestral ochrophyte character, present in both Khakista (diatoms) and Phaeista (silicoflagellates, chrysophytes). Although a new class was created for *Schizocladia* (Kawai et al., 2003), treating it as a subcleas within Phaeophyceae is preferable, in my view.

The fossil record suggests all three silicified heterokont groups arose in the early Mesozoic, like the calcified (coccolithophorid) prymnesiophyte haptophytes and the thecate dinoflagellates. Before the preceding Permian mass extinction, there must have been chromalveolate algae to serve as their ancestors, but they cannot be identified as fossils. The early divergences of the chromist plastid within the red algae on plastid gene trees (Yoon et al., 2002b) and of the various chromalveolate taxa on nuclear trees (Cavalier-Smith, 1995c) make it likely that the chromalveolate symbiogenesis took place close to the Cambrian protist explosion (Cavalier-Smith, 2002) and that many unidentified Palaeozoic spiny acritarchs might be early chromalveolate algae, some possibly representing lineages now purely heterotrophic.

4.5 Biogenesis of Cortical Alveoli

Cortical alveoli are probably not permanently distinct genetic membranes (cf. Chapter 15). The facts that some dinoflagellates can slough off and replace their amphiesma, and that in Sporozoa the inner-membrane complex can be assembled by life cycle stages that lack it (e.g., piroplasms) or in the endoplasm well away from the existing organelles (Coccidea, e.g., *Toxoplasma*), show that alveolar membranes need not arise by subdividing membranes of the same kind. It is unclear whether they arise from the Golgi (like thraustochytrid scales or the diatom silicalemma) or the ER (like synurid chrysophyte scales). In the dividing ciliate cortex, new alveoli interpolate between old ones, like centrioles, but as there is evidence for indirect membrane continuity between them (Stelly et al., 1991, 1995) the possibility of alveolar membrane division is not excluded; during conjugation, alveoli can apparently be formed directly from ER (Geyer and Kloetzel, 1987). In colpodellids, the inner-membrane complex appears to be present throughout the life cycle and to arise by subdividing the preexisting complex during multiple fission in the cyst (TC-S and Oates, unpublished). In *Toxoplasma*, the centrosome–centriolar complex and associated microtubules apparently direct the fusion of precursor vesicles, ensuring that the inner-membrane complexes of daughter cells are associated with them (Hu et al., 2002). Although occurring internally, not at the cell surface as in ciliates, this structural or nucleating role might be basically similar in both.

The origin of cortical alveoli probably involved the budding of a novel class of vesicles from either ER or Golgi bearing integral membrane proteins that provided a label for the new type of alveolus and sites for binding centriole-associated microtubules and the plasma membrane, together with mechanisms for excluding proteins targeted to other destinations (e.g., plasma membrane, lysosomes). Thus, even though ultrastructurally relatively simple, they must have great specificity and novelty in their membrane proteins. Study of these and their biogenesis should reveal molecular synapomorphies for alveolate cortical alveoli and whether those of glaucophytes and possibly related cortical membranes of chromists are homologous.

4.6 Evolution and Biogenesis of Chromalveolate Ciliary Hairs

Many Myzozoa have simple ciliary hairs. It has been assumed that these are unrelated to the simple hairs of haptophytes or the much more complex bipartite or tripartite tubular hairs of cryptophytes and heterokonts, which are assembled within the rough ER and secreted at the ciliary bases (Bouck, 1971). However, the tubular hairs must have evolved from something; it is more likely that they evolved from simple hairs than *de novo*. Therefore, I suggest that the monomer proteins of the simple hairs of Myzozoa might be homologous either with those of the similarly simple terminal filaments of chromist tubular hairs or with the monomers of their tubular parts, or even with both, for all might belong to a related protein family or superfamily. The terminal filament proteins are likely to be homologous between cryptophytes and heterokonts as are the tubule monomers. Biochemical data are needed to test this; comparison with simple glaucophyte hairs would be valuable. The essentially bipartite *Oikomonas* and *Proteromonas* hairs seem to be independent simplifications of the tripartite hairs of ancestral heterokonts by greatly reducing the large carrot-shaped base found in other heterokonts. Whether cryptophyte hairs are similarly reduced or were ancestrally bipartite as earlier assumed is unclear. The solid character of the spur-like *Goniomonas* hairs (Kugrens and Lee, 1991) might be secondary.

4.7 Envoi

The status of chromalveolates as a major holophyletic branch of the eukaryotic tree is clear. Their unity was formerly obscured by the high frequency of differential organellar loss, as in most microbial groups. These losses included multiple independent losses of cilia, ciliary hairs, photosynthesis, phagotrophy, nucleomorphs, periplastid membranes, cortical alveoli, thecal plates, cingular and sulcal grooves, ciliary roots (e.g., Actinochrysea, Khakista) and single losses of EM-targeting vesicles, aerobic respiration (in *Blastocystis*) and peroxisomes (Opalinata). This chapter has concentrated not on such losses but on major innovations, both those associated with the origin of novel genetic membranes during the symbiogenetic event that created chromalveolates and the purely autogenous origins of cortical alveoli, feeding and defensive organelles, macronuclei and ciliary hairs, because these cellular novelties give chromist cell biology a complexity not found in the simpler cells of animals or plants or the degenerate highly reduced ones of yeast. Diversity in cell organelle structure is related to varied feeding modes, several not fitting into standard textbook categories being unique to specific chromalveolate groups. To achieve the full potential of eukaryote cell biology, the cell biology and organelle biogenesis of several disparate chromists and alveolates need intensive study. Sequencing genomes is not enough; we need to understand the molecular interactions of the cell skeletons and membranes that actually build cells and give genomes their raison d'être and survival capabilities.

References

Baldauf, S. L., Roger, A. J., Wenk-Siefert, I. and Doolittle, W. F. (2000) A kingdom-level phylogeny of eukaryotes based on combined protein data. *Science* 290: 972–977.

Bapteste, E., Brinkmann, H., Lee, J. A., Moore, D. V., Sensen, C. W., Gordon, P., Duruflé, L., Gaasterland, T., Lopez, P., Müller, M. and Philippe, H. (2002) The analysis of 100 genes supports the grouping of three highly divergent amoebae: *Dictyostelium*, *Entamoeba*, and *Mastigamoeba*. *Proc. Natl. Acad. Sci. USA* 99: 1414–1419.

Beaton, M. J. and Cavalier-Smith, T. (1999) Eukaryotic non-coding DNA is functional: Evidence from the differential scaling of cryptomonad genomes. *Proc. R. Soc. Lond. B* 266: 2053–2059.

Bonhomme, A., Pingret, L., Bonhomme, P., Michel, J., Balossier, G., Lhotel, M., Pluot, M. and Pinon, J. M. (1993) Subcellular calcium localization in *Toxoplasma gondii* by electron microscopy and by x-ray and electron energy loss spectroscopies. *Microsc. Res. Tech.* 25: 276–285.

Bouck, G. B. (1971) Architecture and assembly of mastigonemes. In *Advances in Cell and Molecular Biology,* Vol. 2 (Dupraw, E. J., Ed.), Academic Press, New York, pp. 237–271.

Calado, A. J. and Moestrup, Ø. (1997) Feeding in *Peridiniopsis berolinensis* (Dinophyceae): New observations on tube feeding by an omnivorous, heterotrophic dinoflagellate. *Phycologia* 36: 47–59.

Cavalier-Smith, T. (1978) Nuclear volume control by nucleoskeletal DNA, selection for cell volume and cell growth rate, and the solution of the DNA C-value paradox. *J. Cell Sci.* 34: 247–278.

Cavalier-Smith, T. (1981) Eukaryote kingdoms: Seven or nine? *BioSystems* 14: 461–481.

Cavalier-Smith, T. (1982) The origins of plastids. *Biol. J. Linn. Soc.* 17: 289–306.

Cavalier-Smith, T. (1985) Cell volume and the evolution of genome size. In *The Evolution of Genome Size* (Cavalier-Smith, T., Ed.), Wiley, Chichester, pp. 105–184.

Cavalier-Smith, T. (1986) The kingdom Chromista: Origin and systematics. In *Progress in Phycological Research*, Vol. 4 (Round, F. E. and Chapman, D. J., Eds.), Biopress, Bristol, pp. 309–347.

Cavalier-Smith, T. (1987) Glaucophyceae and the origin of plants. *Evolut. Trends Plants* 2: 75–78.

Cavalier-Smith, T. (1989) The kingdom Chromista. In *The Chromophyte Algae: Problems and Perspectives* (Green, J. C., Leadbeater, B. S. C. and Diver, W. C., Eds.), Oxford University Press, Oxford, pp. 381–407.
Cavalier-Smith, T. (1991a) Cell diversification in heterotrophic flagellates. In *The Biology of Free-Living Heterotrophic Flagellates* (Patterson, D. J. and Larsen, J., Eds.), Clarendon Press, Oxford, pp. 113–131.
Cavalier-Smith, T. (1991b) Coevolution of vertebrate genome, cell and nuclear sizes. In *Symposium on the Evolution of Terrestrial Vertebrates: Selected Symposia and Monographs U.Z.I., 4.* (Ghiara, G., Ed.), Muchi, Modena, pp. 51–88.
Cavalier-Smith, T. (1993a) Evolution of the eukaryotic genome. In *The Eukaryotic Genome* (Broda, P., Oliver, S. G. and Sims, P., Eds.), Cambridge University Press, London, pp. 333–385.
Cavalier-Smith, T. (1993b) Kingdom Protozoa and its 18 phyla. *Microbiol. Rev.* 57: 953–994.
Cavalier-Smith, T. (1994) Origin and relationships of Haptophyta. In *The Haptophyte Algae* (Green, J. C. and Leadbeater, B. S. C., Eds.), Clarendon Press, Oxford, pp. 413–435.
Cavalier-Smith, T. (1995a) Cell cycles, diplokaryosis, and the archezoan origin of sex. *Arch. Protistenk.* 145: 189–207.
Cavalier-Smith, T. (1995b) Membrane heredity, symbiogenesis, and the multiple origins of algae. In *Biodiversity and Evolution* (Arai, R., Kato, M. and Doi, Y., Eds.) The National Science Museum Foundation, Tokyo, pp. 75–114.
Cavalier-Smith, T. (1995c) Zooflagellate phylogeny and classification. *Tsitologiia* 37: 1010–1029.
Cavalier-Smith, T. (1997) Sagenista and Bigyra, two phyla of heterotrophic heterokont chromists. *Arch. Protistenkd.* 148: 253–267.
Cavalier-Smith, T. (1998) A revised six-kingdom system of life. *Biol. Rev. Camb. Philos. Soc.* 73: 203–266.
Cavalier-Smith, T. (1999) Principles of protein and lipid targeting in secondary symbiogenesis: Euglenoid, dinoflagellate, and sporozoan plastid origins and the eukaryotic family tree. *J. Euk. Microbiol.* 46: 347–366.
Cavalier-Smith, T. (2000) What are Fungi? In *The Mycota*, Vol. VII, Part A (McLaughlin, D. J., and Lemke, P., Eds.), Springer-Verlag, Berlin, pp. 3–37.
Cavalier-Smith, T. (2002) The phagotrophic origin of eukaryotes and phylogenetic classification of Protozoa. *Int. J. Syst. Evol. Microbiol.* 52: 297–354.
Cavalier-Smith, T. (2003a) The excavate protozoan phyla Metamonada Grassé emend. (Anaeromonadea, Parabasalia, Eopharyngia) and Loukozoa emend. (Jakobea, *Malawimonas, Carpediemonas*): Their evolutionary affinities and new higher taxa. *Int. J. Syst. Evol. Microbiol.*, 53: 1741–1758.
Cavalier-Smith, T. (2003b) Genomic reduction and evolution of novel genetic membranes and protein-targeting machinery in eukaryote-eukaryote chimaeras (meta-algae). *Phil. Trans. R. Soc. B* 358: 109–134.
Cavalier-Smith, T. (2004) Economy, speed and size matter: Evolutionary forces driving nuclear genome miniaturisation and expansion. *Ann. Bot.*, in press.
Cavalier-Smith, T. and Beaton, M. J. (1999) The skeletal function of non-genic nuclear DNA: New evidence from ancient cell chimaeras. *Genetica* 106: 3–13.
Cavalier-Smith, T. and Chao, E. E. (2003a) Molecular phylogeny of centrohelid heliozoa, a novel lineage of bikont eukaryotes that arose by ciliary loss. *J. Mol. Evol.* 56: 387–396.
Cavalier-Smith, T. and Chao, E. E. (2003b) Phylogeny of Choanozoa, Apusozoa, and other Protozoa and early eukaryote megaevolution. *J. Mol. Evol.* 56: 540–563.
Cavalier-Smith, T. and Chao, E. E. (2004) Protalveolate phylogeny and systematics and the origins of Sporozoa and dinoflagellates (phylum myzozoa nom. nov.). *Eur. J. Protistol.* 40: in press.
Cavalier-Smith, T., Chao, E. E. and Allsopp, M. T. E. P. (1995) Ribosomal RNA evidence for chloroplast loss within Heterokonta: Pedinellid relationships and a revised classification of ochristan algae. *Arch. Protistenkd.* 145: 209–220.

Christensen, T. (1989) The Chromophyta, past and present. In *The Chromophyte Algae: Problems and Perspectives* (Green, J. C., Leadbeater, B. S. C. and Diver, W. C., Eds.), Oxford University Press, pp. 1–12.

Clay, B. and Kugrens, P. (1999) Systematics of the enigmatic Kathablepharids, including EM characterization of the type species, *Kathablepharis phoenikoston*, and new observations on *K. remigera* comb. nov. Protist 150: 43–59.

Corliss, J. O. (2000) Biodiversity, classification, and numbers of species of protists. In *Nature and Human Society: The Quest for a Sustainable World* (Raven, P. H., Ed.), National Academy Press, Washington, DC, pp. 130–155.

Crow, J. F. (1988) The importance of recombination. In *The Evolution of Sex: An Examination of Current Ideas* (Michod, R. C. and Levin, B. R., Eds.), Sunderland, MA, pp 56–73.

Deane, J., Strachan, I. M., Hill, D. R. A., Saunders, G. W. and McFadden, G. I. (2002) Cryptomonad evolution: Nuclear 18S rDNA phylogeny versus cell morphology and pigmentation. *J. Phycol.* 38: 1–10.

Delwiche, C. F. (1999) Tracing the thread of plastid diversity through the tapestry of life. *Am. Nat.* 154: S164–S177.

Diez, B., Pedros-Alio, C. and Massana, R. (2001) Study of genetic diversity of eukaryotic picoplankton in different oceanic regions by small-subunit rRNA gene cloning and sequencing. *Appl. Environ. Microbiol.* 67: 2932–2941.

Douglas, S., Zauner, S., Fraunholz, M., Beaton, M., Penny, S., Deng, L. T., Wu, X., Reith, M., Cavalier-Smith, T. and Maier, U. G. (2001) The highly reduced genome of an enslaved algal nucleus. *Nature* 410: 1091–1096.

Eichacker, L. A. and Henry, R. (2001) Function of a chloroplast SRP in thylakoid protein export. *Biochim. Biophys. Acta* 1541: 120–134.

Elbrächter, M. (1991) Food uptake mechanisms in phagotrophic dinoflagellates and classification. In *The biology of free-living heterotrophic flagellates* (Patterson, D. J. and Larsen, J., Eds.), Clarendon Press, Oxford, pp. 303–312.

Fast, N. M., Kissinger, J. C., Roos, D. S. and Keeling, P. J. (2001) Nuclear-encoded, plastid-targeted genes suggest a single common origin for apicomplexan and dinoflagellate plastids. *Mol. Biol. Evol.* 18: 418–426.

Fast, N. M., Xue, L., Bingham, S. and Keeling, P. J. (2002) Re-examining alveolate evolution using multiple protein molecular phylogenies. *J. Euk. Microbiol.* 49: 30–37.

Fenchel, T. (1987) *Ecology of Protozoa*, Springer-Verlag, Berlin.

Fensome, R. A., Taylor, F. J. R., Norris, G., Sargeant, W. A. S., Wharton, D. I. and Williams, G. L. (1993) A classification of living and fossil dinoflagellates. *Micropalaeontology*, Special Publication No. 7.

Frankel, J. (1989) *Pattern Formation: Ciliate Studies and Models*, Oxford University Press, Oxford.

Gardner, M. J., Hall, N., Fung, E., White, O., Berriman, M., Hyman, R. W., Carlton, J. M., Pain, A., Nelson, K. E., Bowman, S., Paulsen, I. T., James, K., Eisen, J. A., Rutherford, K., Salzberg, S. L., Craig, A., Kyes, S., Chan, M. S., Nene, V., Shallom, S. J., Suh, B., Peterson, J., Angiuoli, S., Pertea, M., Allen, J., Selengut, J., Haft, D., Mather, M. W., Vaidya, A. B., Martin, D. M., Fairlamb, A. H., Fraunholz, M. J., Roos, D. S., Ralph, S. A., McFadden, G. I., Cummings, L. M., Subramanian, G. M., Mungall, C., Venter, J. C., Carucci, D. J., Hoffman, S. L., Newbold, C., Davis, R. W., Fraser, C. M. and Barrell, B. (2002) Genome sequence of the human malaria parasite *Plasmodium falciparum*. *Nature* 419: 498–511.

Geyer, J. J. and Kloetzel, J. A. (1987) Cellular dynamics of conjugation in the ciliate *Euplotes aediculatus* II. Cellular membranes. *J. Morphol.* 192: 43–62.

Gibbs, S. P. (1979) The route of entry of cytoplasmically synthesized proteins into chloroplasts of algae possessing chloroplast ER. *J. Cell Sci.* 35: 253–266.

Gibbs, S. P. (1981) The chloroplasts of some algal groups may have evolved from endosymbiotic eukaryotic algae. *Ann. N.Y. Acad. Sci.* 361: 193–208.

Goldschmidt, R. B. (1955) *Theoretical Genetics*, University of California Press, Berkeley.

Graham, L. E. and Wilcox, L. W. (2000) *Algae*, Prentice Hall, Upper Saddler River, NJ.

Hackett, J. D., Yoon, H. S., Soares, M. B., Bonaldo, M. F., Casavant, T. L., Scheetz, T. E., Nosenko, T. and Bhattacharya, D. (2004) Migration of the plastid genome to the nucleus in a peridinin dinoflagellate. *Curr. Biol.* 14: 13–18.

Hammerschmidt, B., Schlegel, M., Lynn, D. H., Leipe, D. D., Sogin, M. L. and Raikov, I. B. (1996) Insights into the evolution of nuclear dualism in the ciliates revealed by phylogenetic analysis of rRNA sequences. *J. Euk. Microbiol.* 43: 225–230.

Harper, J. T. and Keeling, P. J. (2003) Nucleus-encoded, plastid-targeted glyceraldehyde-3-phosphate dehydrogenase (GAPDH) indicates a single origin for chromalveolate plastids. *Mol. Biol. Evol.* 20: 1730–1735.

Hoef-Emden, K. and Melkonian, M. (2003) Revision of the genus *Cryptomonas* (Cryptophyceae): A combination of molecular phylogeny and morphology provides insights into a long-hidden dimorphism. *Protist* 154: 371–409.

Hogan, D. J., Hewitt, E. A., Orr, K. E., Prescott, D. M. and Muller, K. M. (2001) Evolution of IESs and scrambling in the actin I gene in hypotrichous ciliates. *Proc. Natl. Acad. Sci. USA* 98: 15101–15106.

Hu, K., Mann, T., Striepen, B., Beckers, C. J., Roos, D. S. and Murray, J. M. (2002) Daughter cell assembly in the protozoan parasite *Toxoplasma gondii. Mol. Biol. Cell* 13: 593–606.

Ishida, K., Cavalier-Smith, T. and Green, B. R. (2000) Endomembrane structure and the chloroplast protein targeting pathway in *Heterosigma akashiwo* (Raphidophyceae, Chromista). *J. Phycol.* 36: 1135–1144.

Ishida, K. and Green, B. R. (2002) Second- and third-hand chloroplasts in dinoflagellates: Phylogeny of oxygen-evolving enhancer 1 (PsbO) protein reveals replacement of a nuclear-encoded plastid gene by that of a haptophyte tertiary endosymbiont. *Proc. Natl. Acad. Sci. USA* 99: 9294–9299.

Jacobson, D. M. and Anderson, D. M. (1992) Ultrastructure of the feeding apparatus and myonemal system of the heterotrophic dinoflagellate *Protoperidinium spinulosum. J. Phycol.* 28: 69–82.

Jahn, C. L. and Klobutcher, L. A. (2002) Genome remodeling in ciliated protozoa. *Annu. Rev. Microbiol.* 56: 489–520.

Joiner, K. A. and Roos, D. S. (2002) Secretory traffic in the eukaryotic parasite *Toxoplasma gondii*: Less is more. *J. Cell Biol.* 157: 557–563.

Kato, K. H., Moriyama, A., Huitorel, P., Cosson, J., Cachon, M. and Sato, H. (1997) Isolation of the major basic nuclear protein and its localization on chromosomes of the dinoflagellate, *Oxyrrhis marina. Biol. Cell* 89: 43–52.

Kawai, M. Maeba, S. Sasaki, M. Okuda, K. and Henry, E. (2003) *Schizocladia ischiensis*: A new filamentous marine chromophyte belonging to a new class, Schizocladiophyceae. *Protist* 154: 211–228.

Kloetzel, J. A. (1991) Identification and properties of plateins major proteins in the cortical alveolar plates of *Euplotes. J. Protozool.* 38: 392–401.

Kloetzel, J. A., Hill, B. F. and Kosaka, T. (1992) Genetic variants of plateins alveolar plate proteins among and within species of *Euplotes. J. Protozool.* 39: 92–101.

Kugrens, P. and Lee, R. E. (1991) Organization of cryptomonads. In *The Biology of Free-Living Heterotrophic Flagellates* (Patterson, D. J. and Larsen, J., Eds.), Clarendon Press, Oxford, pp. 219–233.

Lang, B. F., O'Kelly, C., Nerad, T., Gray, M. W. and Burger, G. (2002) The closest unicellular relatives of animals. *Curr. Biol.* 12: 1773–1778.

Latasa, M. and Bidigare, R. R. (1998) A comparison of phytoplankton populations of the Arabian Sea during the spring Intermonsoon and South West Monsoon of 1995 as described by HPLC-analyzed pigments. *Deep-Sea Res.* II 45: 2133–2170.

Lee, J. J., Leedale, G. and Bradbury, P. (2002 dated 2000) *An Illustrated Guide to the Protozoa*, Society of Protozoologists, Lawrence, KS.

Lee, R. E., Kugrens, P. and Mylnikov, A. P. (1991) Feeding apparatus of the colourless flagellate *Katablepharis* (Cryptopyceae). *J. Phycol.* 27: 725–733.

Leipe, D. D., Tong, S. M., Goggin, C. L., Slemenda, S. B., Pienizek, N. J. and Sogin, M. L. (1996) 16S–like rDNA seqeunces from *Developayella elegans*, *Labyrinthuloides haliotidis*, and *Proteromonas lacertae* confirm that the stramenopliles are a primarily heterotrophic group. *Eur. J. Protistol.* 32: 449–458.

Levine, N. D. (1970) Taxonomy of the Sporozoa. *J. Parasitol.* 56: 208–209.

Levine, N. D. (1978) *Perkinsus* gen. n. and other new taxa in the protozoan phylum Apicomplexa. *J. Parasitol.* 64: 549.

López-García, P., Rodriguez-Valera, F., Pedros-Alio, C. and Moreira, D. (2001) Unexpected diversity of small eukaryotes in deep-sea Antarctic plankton. *Nature* 409: 603–607.

Lynn, D. H. (1996) My journey in ciliate systematics. *J. Euk. Microbiol.* 43: 253–260.

Lynn, D. H. and Small, E. B. (2002 dated 2000) Phylum Ciliophora Doflein 1901. In *An Illustrated Guide to the Protozoa,* 2nd ed., Vol. 1 (Lee, J. J., Leedale, G. and Bradbury, P., Eds.), Society of Protozoologists, Lawrence, KS, pp. 371–656.

Moestrup, Ø. and Andersen, R. A. (1991) Organization of heterotrophic heterokonts. In *The Biology of Free-Living Heterotrophic Flagellates* (Patterson, D. J. and Larsen, J., Eds.), Clarendon Press, Oxford, pp. 333–360.

Moon-van der Staay, S. Y., De Wachter, R. and Vaulot, D. (2001) Oceanic 18S rDNA sequences from picoplankton reveal unsuspected eukaryotic diversity. *Nature* 409: 607–610.

Moriya, M., Nakayama, T. and Inouye, I. (2002) A new class of the stramenopiles, Placididea classis nova: Description of *Placidia cafeteriopsis* gen. et sp. nov. *Protist* 153: 143–156.

Nara, T., Hashimoto, T. and Aoki, T. (2000) Evolutionary implications of the mosaic pyrimidine-biosynthetic pathway in eukaryotes. *Gene* 257: 209–222.

Prescott, D. M. (1999a) Evolution of DNA organization in hypotrichous ciliates. *Ann. N. Y. Acad. Sci.* 870: 301–313.

Prescott, D. M. (1999b) The evolutionary scrambling and developmental unscrambling of germline genes in hypotrichous ciliates. *Nucleic Acids Res.* 27: 1243–1250.

Prescott, D. M. (2000) Genome gymnastics: Unique modes of DNA evolution and processing in ciliates. *Nat. Rev. Genet.* 1: 191–198.

Prescott, D. M., Prescott, J. D. and Prescott, R. M. (2002) Coding properties of macronuclear DNA molecules in *Sterkiella nova* (*Oxytricha nova*). *Protist* 153: 71–77.

Raikov, I. B. (1982) *The Protozoan Nucleus: Morphology and Evolution*, Springer-Verlag, Vienna.

Riley, J. L. and Katz, L. A. (2001) Widespread distribution of extensive chromosomal fragmentation in ciliates. *Mol. Biol. Evol.* 18: 1372–1377.

Roberts, K. R., Stewart, K. D. and Mattox, K. R. (1981) The flagellar apparatus of *Chilomonas paramecium* (Cryptophyceae) and its comparison with certain zooflagellates. *J. Phycol.* 17: 159–167.

Ryall, K., Harper, J. T. and Keeling, P. J. (2003) Plastid-derived Type II fatty acid biosynthetic enzymes in chromists. *Gene* 313, 139–148.

Saldarriaga, J. F., Taylor, F. J. R., Keeling, P. J. and Cavalier-Smith, T. (2001) Dinoflagellate nuclear SSU rRNA phylogeny suggests multiple plastid losses and replacements. *J. Mol. Evol.* 53: 204–213.

Saldarriaga, J. F., Taylor, F. J. R. "Max," Cavalier-Smith, T., Menden-Dener, S. and Keeling, P. J. (2004) Molecular data and the evolutionary history of dinoflagellates. *Eur. J. Protistol* 40: in press.

Sato, M. S., Suzuki, M. and Hayashi, H. (1998) The density of a homogeneous population of cells controls resetting of the program for swarmer formation in the unicellular marine microorganism *Noctiluca scintillans*. *Exp. Cell Res.* 245: 290–293.

Schnepf, E. (1969) Leukoplastin bei *Nitzschia alba*. *Österr. Bot. Z.* 116: 65–69.

Schuster, F. L. (1968) The gullet and trichocysts of *Cyathomonas truncata*. *Exp. Cell Res.* 49: 277–284.

Sekiguchi, H., Moriya, M., Nakayama, T. and Inouye, I. (2002) Vestigial chloroplasts in heterotrophic stramenopiles *Pteridomonas danica* and *Ciliophrys infusionum* (Dictyochophyceae). *Protist* 153: 157–167.
Siddall, M. E., Reece, K. S., Graves, J. E. and Burreson, E. M. (1997) Total evidence refutes the inclusion of *Perkinsus* species in the phylum Apicomplexa. *Parasitology* 115: 165–176.
Simpson, A. G. and Roger, A. J. (2002) Eukaryotic evolution: Getting to the root of the problem. *Curr. Biol.* 12: R691–R693.
Simpson, A. G. B. and Patterson, D. J. (1996) Ultrastructure and identification of the predatory flagellate *Colpodella pugnax* Cienkowski (Apicomplexa) with a description of *Colpodella turpis* n. sp. and a review of the genus. *Syst. Parasitol.* 33: 187–198.
Sournia, A. (1984) Classification et nomenclature de divers dinoflagellés marins (Dinophyceae). *Phycologia* 23: 345–355.
Stechmann, A. and Cavalier-Smith, T. (2002) Rooting the eukaryote tree by using a derived gene fusion. *Science* 297: 89–91.
Stechmann, A. and Cavalier-Smith, T. (2003a) Phylogenetic analysis of eukaryotes using heat shock protein Hsp90. *J. Mol. Evol.* 57: 408–419.
Stechmann, A. and Cavalier-Smith, T. (2003b) The root of the eukaryote tree pinpointed, *Curr. Biol.* 13: R665–R666.
Stelly, N., Halpern, S., Nicolas, G., Fragu, P. and Adoutte, A. (1995) Direct visualization of a vast cortical calcium compartment in *Paramecium* by secondary ion mass spectrometry (SIMS) microscopy: Possible involvement in exocytosis. *J. Cell Sci.* 108: 1895–1909.
Stelly, N., Mauger, J. P., Claret, M. and Adoutte, A. (1991) Cortical alveoli of *Paramecium*: A vast submembranous calcium storage compartment. *J. Cell Biol.* 113: 103–112.
Takishita, K., Koike, K., Maruyama, T. and Ogata, T. (2002) Molecular evidence for plastid robbery (kleptoplastidy) in *Dinophysis*, a dinoflagellate causing diarrhetic shellfish poisoning. *Protist* 153: 293–302.
Tappan, H. (1980) *The Palaeobiology of Plant Protists*, Freeman, San Francisco.
Taverna, S. D., Coyne, R. S. and Allis, C. D. (2002) Methylation of histone H3 at lysine 9 targets programmed DNA elimination in *Tetrahymena*. *Cell* 110: 701–711.
Vivier, E. (1982) Reflexions et suggestions à propos de la systématique des Sporozoaires: Création d'une classe des Hematozoa. *Protistologica* 18: 449–453.
Vørs, N. (1992) Ultrastructure and autecology of the marine, heterotrophic flagellate *Leucocryptos marina* (Braarud) Butcher 1967 (Katablepharidaceae/Kathablepharidae), with a discussion of the genera *Leucocryptos* and *Katablepharis/Kathablepharis*. *Eur. J. Protistol.* 28: 369–389.
Wastl, J. and Maier, U. G. (2000) Transport of proteins into cryptomonads complex plastids. *J. Biol. Chem.* 275: 23194–23198.
Whatley, J. M., John, P. and Whatley, F. R. (1979) From extracellular to intracellular: The establishment of mitochondria and chloroplasts. *Proc. R. Soc. Lond. B* 204: 165–187.
Yoon, H. S., Hackett, J. D. and Bhattacharya, D. (2002a) A single origin of the peridinin- and fucoxanthin-containing plastids in dinoflagellates through tertiary endosymbiosis. *Proc. Natl. Acad. Sci. USA* 99: 11724–11729.
Yoon, H. S., Hackett, J. D., Pinto, G. and Bhattacharya, D. (2002b) The single, ancient origin of chromist plastids. *Proc. Natl. Acad. Sci. USA* 99: 15507–15512.
Zhang, Z., Green, B. R. and Cavalier-Smith, T. (1999) Single gene circles in dinoflagellate chloroplast genomes. *Nature* 400: 155–159.
Zhang, Z., Green, B. R. and Cavalier-Smith, T. (2000) Phylogeny of ultra-rapidly evolving dinoflagellate chloroplast genes: A possible common origin for sporozoan and dinoflagellate plastids. *J. Mol. Evol.* 51: 26–40.
Zhang, Z., Cavalier-Smith, T. and Green, B. R. (2001) A family of selfish minicircular chromosomes with jumbled chloroplast gene fragments from a dinoflagellate. *Mol. Biol. Evol.* 18: 1558–1565.

Zhang, Z., Cavalier-Smith, T. and Green, B. R. (2002) Evolution of dinoflagellate unigenic minicircles and the partially concerted divergence of their putative replicon origins. *Mol. Biol. Evol.* 19: 489–500.

Zingmark, R. G. (1970) Sexual reproduction in the dinoflagellate *Noctiluca miliaris* Suriray. *J. Phycol.* 6: 122–126.

Chapter 5

Origin and Evolution of Animals, Fungi and Their Unicellular Allies (Opisthokonta)

Emma T. Steenkamp and Sandra L. Baldauf

CONTENTS

Abstract

The well-established exclusive grouping of animals and fungi (Opisthokonta) has recently been expanded to include a diverse collection of protistan taxa. These single-celled opisthokonts or Choanozoa include choanoflagellates (uniflagellated filter feeders), ichthyosporeans (parasites of aquatic animals), corallochytreans (free-living saprotrophs) and cristidiscoideans (nucleariid and ministeriid amoebae). The exact relationships of the choanozoans to each other and to animals and fungi are unknown, because most studies of

0-415-29904-7/04/$0.00+$1.50

these taxa are based on single-gene trees and taxonomically nonoverlapping datasets. From the limited data available, the taxon Choanozoa appears to be paraphyletic. However, to substantiate this and to reconstruct the early evolution of opisthokonts, broad sampling of the basal animal and fungal lineages, various choanozoans and appropriate outgroups (i.e., Amoebozoa and Apusozoa) is required. As with all deep phylogenetic questions, resolving the relationships among multi- and unicellular opisthokonts might be difficult, if not impossible, with single-gene trees. This is possibly because they lack sufficient phylogenetic information or are burdened by too many substitutions at individual sites. Thus, the origin of opisthokonts and their early divergence will most probably only be resolved with the analysis of taxonomically well-represented multisequence datasets.

5.1 Introduction

An exclusive grouping of animals and fungi, the Opisthokonta (Cavalier-Smith, 1987), is now well established (e.g., Wainright et al., 1993; Baldauf and Palmer, 1993; Baldauf et al., 2000). Most recently, the group has been expanded to include a diverse collection of protistan taxa, also known as the Choanozoa (Cavalier-Smith, 1998b). The specific relationships of most of these Choanozoa to each other and to animals and fungi are currently unresolved, and much remains to be discovered. To provide a background for studying the evolution of the Opisthokonta, we review the relevant literature on the subject. To achieve this, we first discuss the phylogeny within both animals and fungi, with special reference to their possible protist origins. We then briefly describe the various Choanozoa and our current best guess on their relationships to each other and to the animals and fungi. Given that reconstruction of opisthokont phylogenies will undoubtedly require appropriate outgroups, we also discuss the evolution of two groups of protists, Amoebozoa (Cavalier-Smith, 1998b) and Apusozoa (Cavalier-Smith and Chao, 1995), which are strong candidates for being the sistergroup of opisthokonts.

5.2 Animals and Their Phylogeny

The kingdom Animalia (synonym Metazoa) consists of multicellular phagotrophic organisms with collagenous connective tissue between dissimilar epithelia (Cavalier-Smith, 1998a). Between 21 and 38 animal phyla are recognized (e.g., Hyman, 1940; Cavalier-Smith, 1998a; Farabee, 2002), and on the basis of body architecture all classification schemes split animals into the two fundamental groups diploblasts and triploblasts. Diploblasts (Radiata) are for the most part radially or biradially symmetric and have two germ layers, the endoderm and ectoderm. Triploblasts (Bilateria), on the other hand, are mostly bilaterally symmetric and have three germ layers, the endoderm, ectoderm and mesoderm, which might or might not form a true coelom (a mesoderm-enclosed internal body cavity).

The Radiata include the Porifera (Calcispongiae — calcareous sponges, Hexactinellida — glass sponges and Demospongiae — spongin and siliceous sponges), Cnidaria (Anthozoa — corals, Scyphozoa — jellyfish, and Hydrozoa — anemones and hydroids), Ctenophora (comb jellies) and Placozo (e.g., *Trichoplax*) (Hyman, 1940; Cavalier-Smith, 1998a; Barnes et al., 2001; Farabee, 2002). The Bilateria consists of two deeply divided groups, deuterostomes and protostomes, which are primarily distinguished by the developmental fate of the blastopore. In the deuterostomes, the blastopore forms the anus, whereas in the protostomes it forms the anus or mouth or both. Deuterostomes are also characterized by radial cleavage in the embryo, whereas protostomes display spiral embryo cleavage (Hyman, 1940; Barnes et al., 2001; Farabee, 2002).

Traditionally, animal taxonomy was based on the degree of morphological and embryological complexity. Accordingly, the simplest animals represented the more ancient lineages, whereas those with increased complexity reflected more advanced evolutionary stages (e.g., Hyman, 1940). As a result, sponges with their loose tissue organization and few differentiated cell types have long been considered as the basal lineage in the animal clade. Cnidarians then represented the next branch up in animal evolution, followed by a series of intermediate branches emerging in succession and reflecting a steady increase in complexity. Finally, the tree culminated with the chordates at its tip. Thus, in this system, the protostomes, which constituted many of the intermediate taxa, were divided into a diversity of groups of uncertain relationship, and it was unclear whether they formed a monophyletic group.

5.2.1 Molecular Phylogeny

This oversimplified view that the degree of morphological complexity reflects the level of evolution in animals has now been rejected in favor of the more contemporary one based on molecular phylogenetic data (Figure 5.1). In this new phylogeny, the traditional classification of deuterostomes as chordates, urochordates (ascidians), hemichordates (e.g., pterobranchs) and echinoderms (e.g., sea urchin, starfish, blastoids) has been upheld by molecular data (Cavalier-Smith, 1998a; Adoutte et al., 2000). However, almost all the taxa formerly classified as protostomes can be assigned to one of two major groups, the Ecdysozoa and the Lophotrochozoa, with acoelomorphs as the possible exceptions (Halanych et al., 1995; Winnepenninckx et al., 1995a,b; Aguinaldo et al., 1997; Balavoine, 1997; Abouheif et al., 1998). Ecdysozoa possess a cuticle that is moulted, whereas Lophotrochozoa possess either lophophores (structures around the mouth with rows of ciliated hollow tentacles) for feeding or trochophore larval (microscopic free-swimming larva with characteristic equatorial bands of cilia) stages.

Not surprisingly, molecular phylogenetics has transformed animal systematics and classification over the last decade (reviewed in Adoutte et al., 1999, 2000), although the old and new ideas of animal evolution agree on some points. For example, both the new and traditional nonmolecular animal phylogenies recognize the deep split between radiates and bilaterians (Figure 5.1). For the most part, however, the contemporary view of animal evolution is radically different from old ideas. The most striking difference is the absence of the so-called intermediate taxa (reviewed by Adoutte et al., 1999). Thus, traditional intermediates between radiates and protostomes such as nematodes and platyhelminthes are now placed within the ecdysozoan and lophotrochozoan clades, respectively (Figure 5.1).

Most of the molecular evolutionary studies of animals use SSUrDNA. These data resolve the relationships among the four major animal groups Deuterostomia, Radiata, Lophotrochozoa and Ecdysozoa (Figure 5.1), as well as relationships within the deuterostomes. Many of the results of these SSU-based phylogenies have also been confirmed with information from *Hox* developmental regulator genes (e.g., De Rosa et al., 1999) and mitochondrial genes (e.g., von Nickisch-Rosenegk et al., 2001). However, SSUrDNA as well as protein-coding genes such as elongation factor 1α (EF-1α) and HSP70 have proven less useful in resolving relationships within the other animal groups, especially the protostomes (Kobayashi et al., 1996; Abouheif et al., 1998; Borchiellini et al., 1998). The reasons for these limitations might be mutational saturation and unequal rates of evolution (Philippe, Chapter 6). As a result, the phylogenetic status of various animal phyla (reviewed in Collins and Valentine, 2001) is uncertain.

5.2.2 Basal Animal Lineages

Several attempts have been made to resolve relationships among the cnidarians, placozoa, ctenophores and sponges to identify the most basal animal lineage. (See Figure 5.1 for references.) Results of these studies, using either ribosomal RNA or protein-coding genes

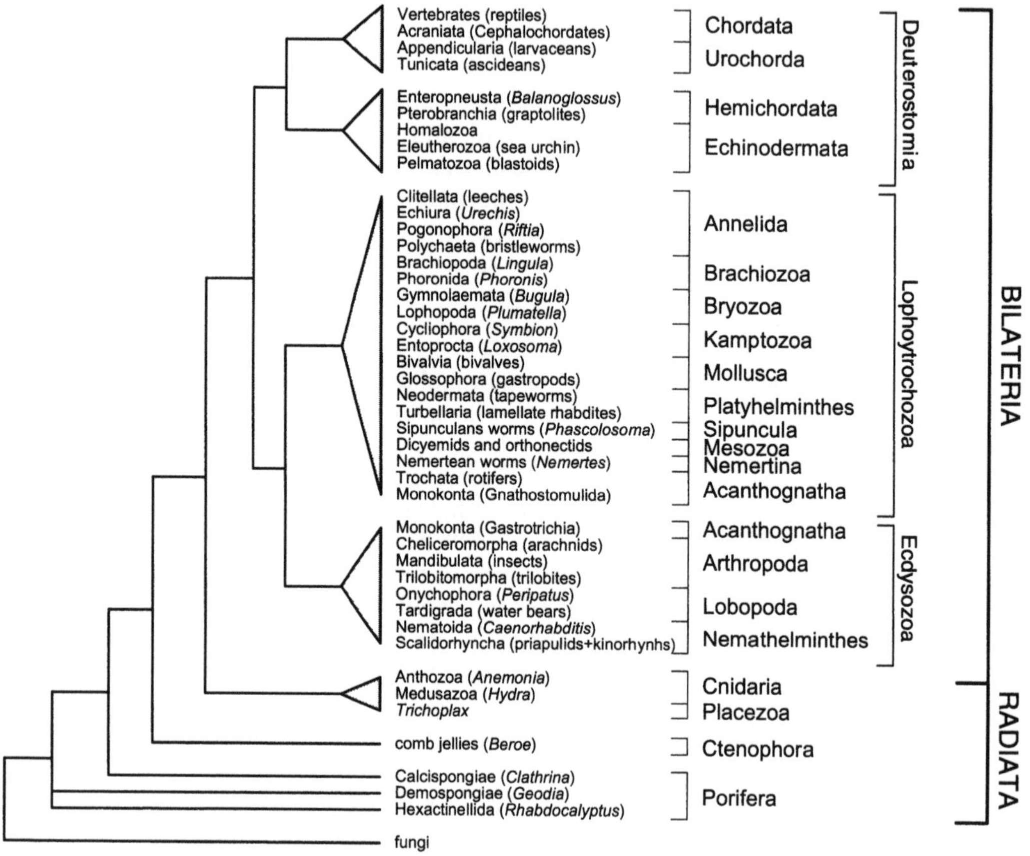

FIGURE 5.1 The phylogeny of animals based on a consensus of morphological data, protein-coding and ribosomal RNA gene trees. Only clades and branches with significant bootstrap support, as reported by the authors, are indicated. Animal subphylum, phylum and superphylum definitions (e.g., Radiata and Ecdysozoa) are mostly according to Cavalier-Smith (1998a) 203–266 and Aguinaldo et al. (1997) 489–493. Representative examples for each of the subphyla are indicated in parentheses. References on overall animal phylogeny: Wainright et al. (1993) 340–342; Kumar and Rzhetsky (1996) 183–193; Van de Peer De Wachter (1997) 619–630; Adoutte et al. (1999) 104–108; Adoutte et al. (2000) 4453–4456; Peterson and Eernisse (2001) 170–205. References on deuterostome phylogeny: Caroll, (1995) 479–485; Bromham and Degnan (1999) 166–171; Cameron, et al. (2000) 4469–4474; Wada, et al. (2002) 118–128. References on ecdysozoan and lophotrochozoan phylogeny: Caroll (1995) 479–485; Halanych et al. (1995) 1641–1642; Winnepenninckx et al. (1995a) 150–160; Aguinaldo et al. (1997) 489–493; Greiner et al. (1997) 547–553; De Rosa et al. (1999) 772–776; Kobayashi et al. (1999) 762; De Rosa, (2001) 848–859. References on radiate phylogeny: Borchiellini et al. (1998) 647–655; Borchiellini et al. (2000) 15–27; Borchiellini et al. (2001) 171–179; Cavalier-Smith et al. (1996) 2031–2045; Kobayashi et al. (1996) 414–422; Nielson et al. (1996) 385–410; Abouheif et al. (1998) 394–405; Collins, (1998) 15458–15463; Kruse et al. (1998) 721–728; Zrzavy et al. (1998) 49–285; Adams et al. (1999) 33–43; Kim et al. (1999) 423–427; Medina et al. (2001) 9707–9712; Peterson and Eernisse (2001) 170–205; Cavalier-Smith and Chao (2003) 540–563.

(EF-1α and HSP70), differ in terms of which taxon is the sistergroup of the bilaterian animals, i.e., either monophyletic ctenophores (Peterson and Eernisse, 2001) or paraphyletic cnidarians (+ placozoans) (Wainright et al., 1993; Kumar and Rzhetsky, 1996; Abouheif et al., 1998; Borchiellini et al., 2001; Medina et al., 2001). Although the question is far from resolved, these studies show consistently higher bootstrap support for cnidarians (+ placozoan) as the sistergroup to the Bilateria. Thus, the current consensus on radiate evolution is that the appearance of sponges was followed by the emergence of ctenophores, then Cnidaria (+ Placozoa), followed by the divergence of the bilaterian animals (Figure 5.1).

Sponges always appear at the base of the animal clade, and it is generally thought that the ancestor of all animals might have been a sponge-like organism. Molecular phylogenetic studies have so far not resolved the phylogeny of sponges, using either protein-coding genes such as HSP70 (Borchiellini et al., 1998) and protein kinase C (Kruse et al., 1998) or ribosomal RNA genes (Lafay et al., 1992; Cavalier-Smith et al., 1996; Adams et al., 1999; Borchiellini et al., 2001; Medina et al., 2001). Nonetheless, these studies suggest that sponges might be paraphyletic. Thus, two of the three sponge groups (hexatinellids and demosponges) tend to cluster together and separate from the calcareous sponges, which emerge as the sistergroup to the rest of the animals (Figure 5.1). This is consistent with sponge skeletal composition; calcarean skeletons are composed entirely of calcium carbonate, whereas those of the other two generally have siliceous skeletons (Cavalier-Smith, 1998a; Borchiellini et al., 2001). However, the branching order of the hexactinellid and demosponges remains unclear (Figure 5.1), because the fossil record and protein kinase C trees suggest an earlier origin for hexactinellids, whereas SSUrDNA suggests the reverse (Adams et al., 1999). Nevertheless, the nonsponge animals might have been derived from a calcareous sponge-like ancestor through the loss of poriferan characters such as choanocytes and aquiferous systems (Borchiellini et al., 2001).

5.3 Fungi and Their Phylogeny

The kingdom Fungi (also known as the Eumycota or True Fungi) consists of absorptive heterotrophs that typically have β-glucan and chitin in their cell walls and usually produce and live inside a network of apically extending branched multinucleate tubes (hyphae) (Tehler, 1988; Cavalier-Smith, 1998a, 2001; Kirk et al., 2001). Based on the presence or absence of a dikaryon (a nuclear phenomenon in which compatible nuclei pair off and cohabit without fusing), all fungi are accommodated into one of two subkingdoms, Neomycota (also known as Dikaryomycota) and Eomycota (Cavalier-Smith, 1998a).

Cavalier-Smith (1998a) separates the Eomycota into the Microsporidia (e.g., *Glugea*, *Encephalitozoon*, *Nosema*) and Archemycota, the latter including the phyla Zygomycotina (with the classes Zygomycetes and Trichomycetes) and Chytridiomycotina (Kirk et al., 2001). The neomycotan or dikaryomycotan fungi include the phyla Basidiomycotina and Ascomycotina (Cavalier-Smith, 1998a). Ascomycotina is further separated into the classes Archiascomycetes (e.g., *Pneumocystis*, *Schizosaccharomyces*, *Protomyces*), yeast Hemiascomycetes (e.g., *Saccharomyces*, *Yarrowia*, *Endomyces*) and filamentous Euascomycetes (e.g., *Neurospora*, *Fusarium*, *Aspergillus*) (Nishida and Sugiyama, 1994; Kirk et al., 2001; Eriksson and Winka, 1997). Three Basidiomycotina classes are recognized, based on the nature of septal pores and spindle pole bodies (Swann and Taylor, 1995; Kirk et al., 2001): Hymenomycetes (also known as Basidiomycetes, e.g., mushrooms, bracket fungi), Urediniomycetes (e.g., rust fungi) and Ustilaginomycetes (e.g., smut and bunt fungi).

5.3.1 Molecular Phylogeny

All molecular trees support the fungi as a monophyletic group (Baldauf et al., 2000 and references therein), and their overall phylogeny is also relatively well resolved by molecular data (Figure 5.2; e.g., Bruns et al., 1992; Tehler et al., 2000). Similar to the situation in animals, most fungal molecular phylogenies were inferred from SSUrDNA, which show the higher fungi, Ascomycotina and Basidiomycotina, to form a clade, with the glomalean fungi as their sister taxon. (See Figure 5.2 for references.) SSUrDNA also supports the morphology-based subdivisions within each of these phyla (Figure 5.2). However, SSUrDNA and protein-based analyses have revealed many previously unknown intraclass and intraorder relationships within these groups, which has led to reevaluation and revision of many classification schemes (e.g., Swann and Taylor, 1995).

Relationships among the more basal branching or lower fungi (i.e., chytridiomycetes, nonglomalean zygomycetes, trichomycetes and microsporidia) have been less easily resolved with molecular data (Figure 5.2). The exact nature of the relationship between lower and higher fungi, i.e., which lower fungal taxon is the sistergroup to glomales + higher fungi, is also unresolved. However, most SSUrDNA and protein phylogenies suggest that lower fungi are probably paraphyletic, because neither the nonglomalean zygomycetes nor the trichomycetes are monophyletic (Jensen et al., 1998; Voigt and Wöstemeyer, 2001; O'Donnell et al., 2001; Tehler et al., 2000). The same is also true for the Chytridiomycotina (Jensen et al., 1998; James et al., 2000; Tehler et al., 2000). Consequently, the base of the fungal tree, whether inferred from ribosomal RNA or protein-coding sequences, is characterized by a mixture of chytridiomycetes, trichomycetes and nonglomalean zygomycetes (Figure 5.2).

5.3.2 Basal Fungal Lineages

Among the lower fungi, those in the phylum Chytridiomycotina probably represent the oldest fungal lineages. These fungi are typically found in aquatic habitats and are the only true fungi that have flagellated stages (Alexopoulus, 1962; Alexopoulus et al., 1996; Cavalier-Smith, 1998a). The fact that chytrid zoospores swim with a single posteriorly directed flagellum is particularly important as a single basal flagellum on reproductive cells is one of the few morphological diagnostics for Opisthokonta (Cavalier-Smith, 1987). In addition to the presence of a flagellum, the morphology of most chytridiomycetes differs quite markedly from other fungi; they have thalli that are differentiated into one or more globular nucleated portions and slender branched enucleated rhizoids (Alexopoulus, 1962; Cavalier-Smith, 1998a). Five orders of Chytridiomycotina are recognized: Blastocladialles, Chytridiales, Monoblepharidales, Neocallimastigales and Spizellomycetales (Hawksworth et al., 1995). Whether these orders are all monophyletic has not been tested, but the limited data available suggest that Chytridiomycetes are broadly paraphyletic (Tehler et al., 2000).

The possibility that one of the chytridiomycete orders might constitute the deepest extant branch of Fungi has been repeatedly suggested (e.g., Alexopoulus et al., 1996; James et al., 2000; Cavalier-Smith, 2001). Although this is a widely held assumption, there is some evidence to the contrary. For example, certain members of the zygomycota appear to represent more ancient lineages than any chytridiomycetes, i.e., they branch off earlier than chytrids in fungal SSUrDNA trees (Berbee and Taylor, 2001). However, these authors suggest that all contemporary lower fungi probably still evolved from chytridiomycete-like ancestors. They argue that zygomycetes convergently lost their flagella during land colonization, and chytridiomycetes retained their flagella to persist in their aquatic habitats (Berbee and Taylor, 2001; Redecker, 2002). If this is indeed the case, relatives of these ancestral chytrid lineages might be extant but still awaiting molecular characterization (Berbee and Taylor, 2001).

5.4 The Protistan Animal–Fungal Allies (Choanozoa)

Currently, the protistan animal–fungal allies are classified in Cavalier-Smith's Choanozoa (Cavalier-Smith, 1998b). Within this protist phylum, four classes are recognized: Choanoflagellatea, Ichthyosporea, Corallochytrea and Cristidiscoidae (Cavalier-Smith, 1998b).

5.4.1 Choanoflagellatea

Choanoflagellates are single-celled heterotrophic uniflagellates that inhabit marine and freshwater systems and feed on bacteria. Morphological features that distinguish them from other protists include the presence of a single posteriorly directed flagellum in the motile stage, a collar of filter-feeding microvilli and the presence of siliceous (lorica) or organic (theca) structures that envelope the cell (Sleigh, 1989; Buck, 1990; Hausmann and Hülsmann, 1996). Despite the choanoflagellates being a highly distinct taxon, there has been considerable disagreement on their classification. For example, they have also been classified in the phylum Choanoflagellata, Kingdom Mastigota (Hausmann and Hülsmann, 1996) and as eukaryotes without known sister taxa (Patterson, 1999).

All classifications recognize three families: Codonosigidae (e.g., *Sphaeroeca*), Salpingoecidae (e.g., *Salpingoeca*) and Acanthoecidae (e.g., *Diaphanoeca*) (Buck, 1990; Hausmann and Hülsmann, 1996). The Acanthoecidae differs from the other two in that the basket-like loricae of its members are composed of silica. At some stage of their life cycles, most Salpingoecidae choanoflagellates have thecas composed of organic compounds such as chitin, whereas Codonosigidae cells are sometimes embedded in mucilaginous matrices. A number of choanoflagellates are solitary (e.g., *Monosiga*), whereas others live in colonial systems (e.g., *Codonocladium*) (Hausmann and Hülsmann, 1996).

The choanoflagellates are famous for the possibly pivotal position they hold in the ancestry of animals and fungi, and they share similarities with both the basal animals and the fungi (Cavalier-Smith, 1987, 1998b; Wainright et al., 1993). Morphological features shared with basal animals (sponges) include structural, organizational and functional similarities of the microvilli collars of choanoflagellates and sponge choanocytes, contractile vacuoles for discharging water, the ability to metabolize silica, the presence of a single flagellum and similar flagellar vanes and root systems (Buck, 1990; Wainright et al., 1993; Hausmann and Hülsmann, 1996). It is suggested that colonial, athecate choanoflagellates, such as those in the family Codonosigidae (e.g., *Proterospongia*), could have given rise to the first sponges (Cavalier-Smith, 1987, 1998b; Buck, 1990; Hausmann and Hülsmann, 1996). Morphological features shared by Choanoflagellates and the likely most basal fungi (chytridiomycetes) include similarities between chytrid zoospores and choanoflagellate flagella (Cavalier-Smith, 1998b, 2001). It is speculated that fungi could have evolved from chitinous-thecate choanoflagellates such as those from the family Salpingoecidae (Cavalier-Smith, 1998b). In this view, the choanoflagellate collar of microvilli gave rise to the rhizoids of the first chytrid-like fungi (Cavalier-Smith, 2001, 1998b).

5.4.2 Ichthyosporea

Ichthyosporea are also known as mesomycetozoa (Herr et al., 1999) and DRIPs after the initial four identified members (*Dermocystidium*, rosette agent, *Ichthyophonus* and *Psorospermium*) of this group (Ragan et al., 1996). Ichthyosporea is composed of heterotrophic walled parasites of aquatic organisms such as fish and crustaceans. Approximately 10 genera are described and placed into two orders, Ichthyophonida (e.g., *Ichthyophonus*) and Dermocystida (e.g., *Dermocystidium*) (Ragan et al., 1996; Cavalier-Smith, 1998b; Baker et al., 1999; Herr et al., 1999; Benny and O'Donnell, 2000). Because of fungus-like characteristics such as the presence of septated hyphae in some species, ichthyosporea have been previously classified as fungi (reviewed in Ragan et al., 1996). For example, *Amoebidium parasiticum*, initially classified as a trichomycete, based on morphology (Lichtwardt and Williams, 1992), is now clearly assigned to the Ichthyophonida, based on SSUrDNA trees (Benny and O'Donnell, 2000). Other important morphological and ultrastructural ichthyosporean features are uniflagellate stages in certain species (e.g., *Dermocystidium* spp.), as well as

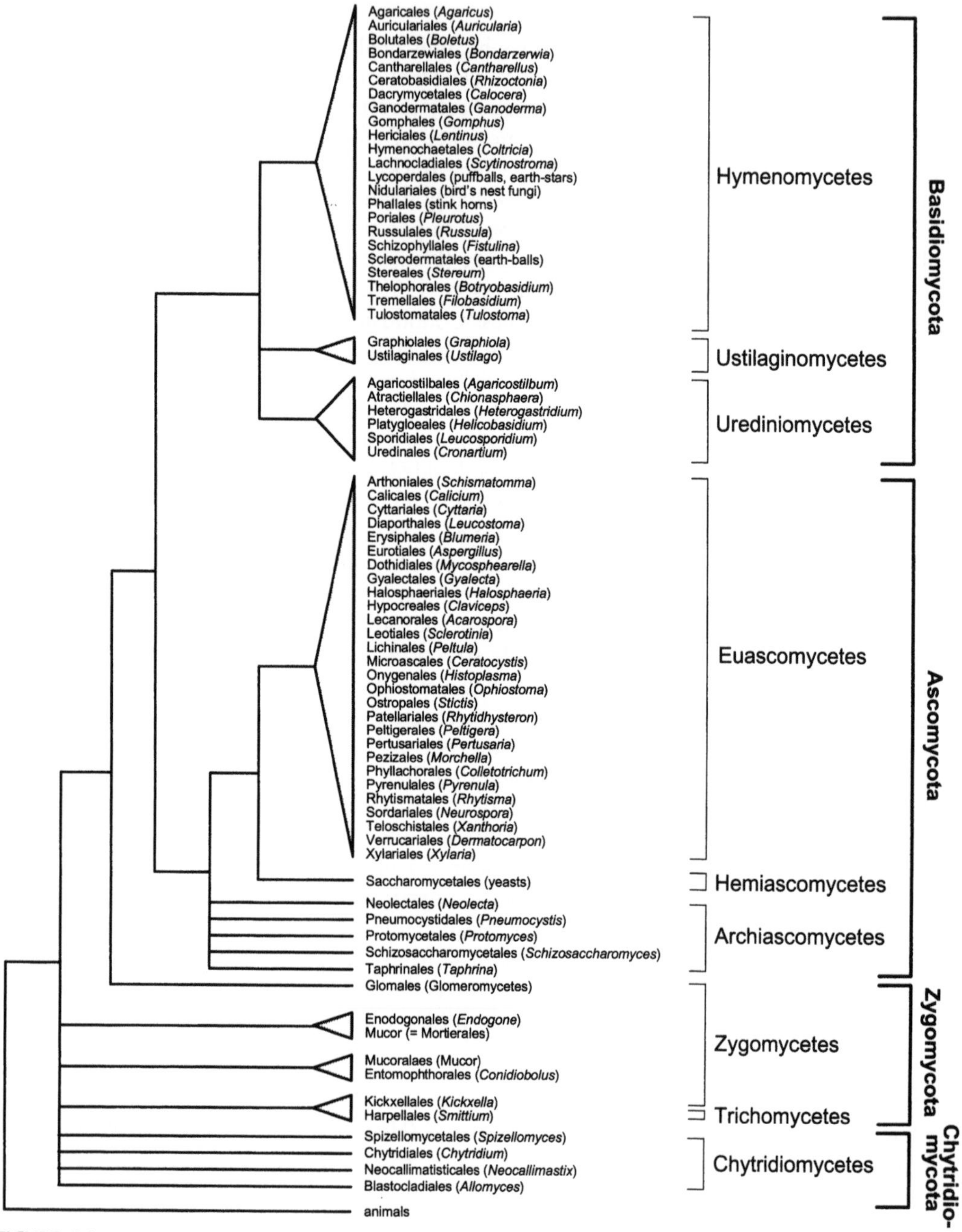

FIGURE 5.2

flattened mitochondrial cristae in some (e.g., *Rhinosporidium seeberi*) and tubulvesculate cristae in other species (e.g., *Ichthyophonus hoferi*) (Ragan et al., 1996; Herr et al., 1999).

5.4.3 *Corallochytrea*

Corallochytrea (e.g., *Corallochytrium limacisporum*) are perhaps the most enigmatic of the choanozoa because of their very simple morphology; they are known only as free-living unicellular saprophytes that apparently lack flagella. Before their designation as choanozoa, corallochytrea could not be classified into any of the existing fungal or protistan classes, because they lacked the necessary diagnostic morphological characters (Cavalier-Smith and

FIGURE 5.2 The phylogeny of fungi based on a consensus of SSUrDNA and protein-coding gene trees. Only clades and branches with significant bootstrap support, as reported by the authors, are indicated. Fungal orders (-ales), classes (-mycetes) and phyla (-mycotina) are mostly according to Hawksworth et al. (1995), except for the basidiomycotan classes that were defined by Swann and Taylor (1995) S862–S868. Representative examples for each of the orders are indicated in parentheses. The placement of the zygomycetous group, Glomales, as sister to the Ascomycotina + Basidiomycotina-clade prompted the erection of a separate phylum (Glomeromycota) for these fungi [Schüßler et al. (2001) 1414–1421]. References on the overall phylogeny of fungi: Bruns et al. (1992) 231–241; Tehler, et al. (2000) 459–474; Cavalier-Smith (2001) 3–37. References on the phylogeny of the Ascomycotina: Spatafora and Blackwell (1994) 1–9; Taylor et al. (1994) 201–212; Berbee (1996) 462–470; Sugiyama (1998) 487–511; Liu (1999) 1799–1808; Tehler et al. (2000) 459–474; Lutzoni et al. (2001) 937–940. References on the phylogeny of the Basidiomycotina: Bruns et al. (1992) 231–241; Swann and Taylor (1995) S862–S868; Hibbett et al. (1997) 12002–12006; Moncalvo et al. (2000) 278–305; Tehler et al. (2000) 459–474. References on the phylogeny of Zygomycotina: Cavalier-Smith (1998a) 203–266; Jensen (1998) 325–334; O'Donnell et al. (1998) 624–639; O'Donnell et al. (2001) 286–296; Voigt and Wöstemeyer (2001) 113–120; Tehler et al. (2000) 459–474. References on the phylogeny of the Chytridiomycotina: James et al. (2000) 336–350; Tehler et al. (2000) 459–474. Although ribosomal RNA analysis initially placed the microsporidians deep within the eukaryotic tree [Vossbrinck et al. (1987) 411–424], subsequent analysis of several protein-coding genes support their inclusion (not indicated here) in the fungal kingdom [Edlind et al. (1996) 359–367; Germot (1997) 159–168; Fast et al. (1999) 1415–1419; Hirt et al. (1999) 580–585; Weiss et al. (1999) 17S–18S; Baldauf et al. (2000) 972–977; Keeling et al. (2000) 23–31; Van de Peer et al. (2000b) 1–8].

Allsopp, 1996). Based on SSUrDNA trees, corallochytrea appear to be closely related to both ichthyosporea and choanoflagellates (Cavalier-Smith and Allsopp, 1996; Cavalier-Smith, 2001). *Corallochytrium* also has plate-like or flat mitochondrial cristae, a feature that it shares with choanoflagellates and some of the ichthyosporea, as well as animals and fungi (Cavalier-Smith and Allsopp, 1996).

5.4.4 *Cristidiscoidea*

Three orders of Cristidiscoidea are recognized: Nucleariida, Fonticulida and Ministeriida (Cavalier-Smith, 1998b). Of these, the nucleariid and fonticulid amoebae were previously classified as Rhizopoda (Page, 1987; Schuster, 1990; Cavalier-Smith, 1993), and all three have also been listed as eukaryotic taxa without known sistergroups (Patterson, 1999). Although the name Cristidiscoidea implies that its members should have disc-shaped mitochondrial cristae (Page, 1987; Schuster, 1990; Cavalier-Smith, 1993), ministeriids and some nucleariids appear to have flattened mitochondrial cristae (Patterson, 1999; Amaral-Zettler et al., 2001).

Nucleariids lack flagella and are typically filose amoebae with extremely fine pseudopodia that lack microtubules (Schuster, 1990; Cavalier-Smith, 1993; Patterson, 1999). Fonticulids also lack flagella and were previously considered as a member of the Acrasea (Blanton, 1990), because their amoebae aggregate together to form sporeforming structures similar to those of the cellular slime molds (Blanton, 1990). Unlike *Nuclearia* species, fonticulids have wider pseudopodia with subpseudopodia (Blanton, 1990). Ministeriids are characterized by the presence of 20 symmetrically distributed, stiff radiating pseudopodia (Patterson, 1999). Members of this order apparently stick to substrates with a vibratile stalk, which is thought to represent a modified flagellum (Patterson, 1999; Cavalier-Smith and Chao, 2003).

5.5 Choanozoa Phylogeny and the Origin of Animals and Fungi

5.5.1 Evidence for an Animal–Fungal Clade (Opisthokonta)

The relationship between fungi and animals has long been a contentious issue, not only because of fungi originally being classified as plants but also because of disagreement among contemporary molecular phylogeneticists. Since Whittaker's (1969) separation of the plants, fungi and animals, each into their own kingdom, it has been argued that (1) fungi are more closely related to plants than to animals (see, e.g., Löytynoja and Milinkovitch, 2001), (2) animals are more closely related to plants than to fungi (see, e.g., Vossbrinck et al., 1987) and (3) fungi and animals are more closely related to each other than to plants (see, e.g., Wainright et al., 1993; Baldauf et al., 2000). In addition, fungi and animals have also been associated with protistan eukaryotes such as heterokonts and alveolates (see, e.g., Van der Auwera and De Wachter, 1996; Löytynoja and Milinkovitch, 2001).

The current consensus on the origin of animals and fungi is that these kingdoms are more closely related to each other than either is to any of the other major eukaryote groups. The primary lines of evidence for kinship of animals and fungi are derived from SSUrDNA and protein trees. Some ultrastructural, morphological and physiological similarities also support this relationship. For example, animals and fungi both display flagellated zoospores stages, i.e., the sperm cells of animals or spores of chytridiomycetes (Cavalier-Smith, 1987). Significantly, the single flagellum on these motile cells is posteriorly directed (Cavalier-Smith, 1987). The mitochondria of both animals and fungi also display well-developed, flattened plate-like cristae (Cavalier-Smith, 1987). Various authors also point out that animals and fungi utilize and synthesize compounds such as chitin in a similar way (Corliss, 1984; Cavalier-Smith, 1987; Ragan, 1989; Sleigh, 1989; Kumar and Rzhetsky, 1996). Virtually no such morphological, ultrastructural or physiological characters link plants with either fungi or animals (Kumar and Rzhetsky, 1996).

In the past, SSUrDNA phylogenies often did not cluster animals and fungi together, but this was artifactual because among-site variation, lineage specific variation and uneven taxonomic sampling were not taken into account (reviewed in Van de Peer and De Wachter, 1997; Hillis, 1998; Philippe, Chapter 6). Subsequent reanalyses of taxonomically broadly sampled SSUrDNA sequences generate phylogenies that consistently group animals and fungi together with high confidence (Kumar and Rzhetsky, 1996; Van de Peer and De Wachter, 1997; Van de Peer et al., 2000a). This affinity is also evident in phylogenies inferred from combined (Nikoh et al., 1994; Baldauf et al., 2000; Lang et al., 2002) or single protein-coding sequences (reviewed in Baldauf et al., 2000). Additional molecular evidence for the kinship of fungi and animals is the presence of a uniquely shared ~12 amino acid insertion in their EF-1αs (Baldauf and Palmer, 1993).

5.5.2 Phylogenetic Position of the Choanozoa

Molecular phylogenetic studies have revealed that the evolution of the protistan animal–fungal allies is interwoven with the origin of animals and fungi. Unfortunately, compared with the large number of animals and fungi that have been studied, the molecular evolution of very few animal–fungal allies has been addressed. Five animal–fungal allies from the Cristidiscoidea, one corallochytrean, eight choanoflagellates and ca. 12 ichthyosporea have been used in various phylogenetic analyses. (See Figure 5.3 for references.) These results, mostly based on SSUrDNA, suggest that the taxon Choanozoa is paraphyletic and that the animal–fungal allies share a common ancestor with animals and fungi (Figure 5.3). Cavalier-Smith (1998b) subsequently proposed inclusion of these single-celled animal-fungal allies in the Opisthokonta, citing uniflagellated stages and flattened mitochondrial cristae as possible synapomorphies.

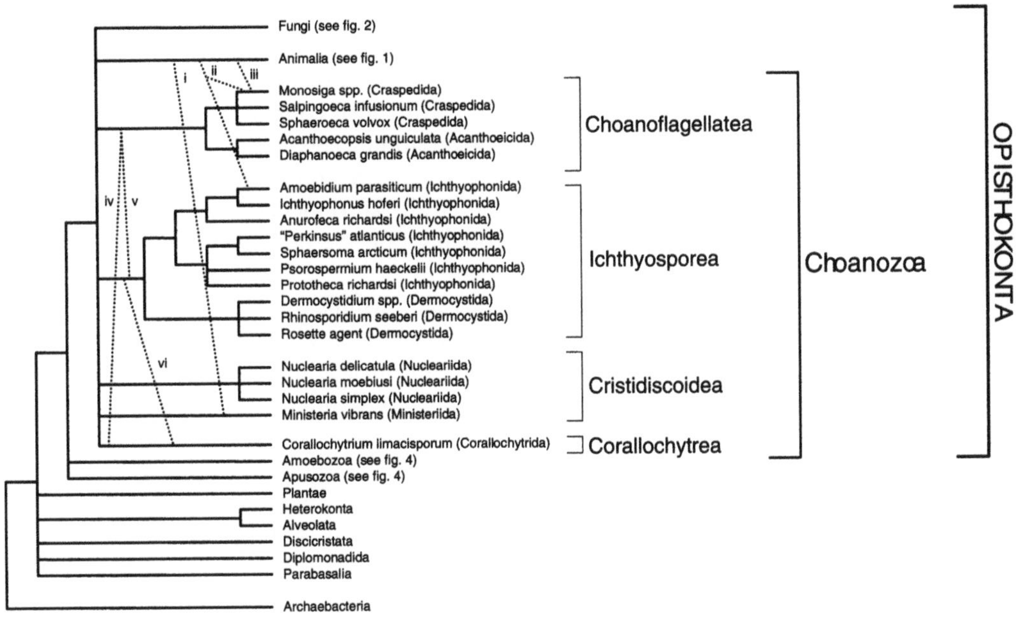

FIGURE 5.3 A preliminary phylogeny of choanozoa based on limited SSUrDNA and protein-gene trees. References for choanoflagellate phylogeny: Cavalier-Smith and Chao (1995) 1–6; Cavalier-Smith and Chao (2003) 540–563; Cavalier-Smith, (1998b) 375–407; Cavalier-Smith (2000) 361–390; Atkins et al. (2000) 278–285; Medina et al. (2001) 9707–9712. References for ichthyosporean phylogeny: Ragan et al. (1996) 11907–11912; Baker et al. (1999) 1777–1784; Herr et al. (1999) 2750–2454; Benny and O'Donnell (2000) 1133–1137. References for corallochytrean and cristidiscoidean phylogenies: Cavalier-Smith and Allsopp (1996) 306–310; Cavalier-Smith (2000) 361–390; Amaral-Zettler et al. (2001) 275–282; Cavalier-Smith and Chao (2003) 540–563. Choanozoan orders (in parentheses), class and superphylum designations follow those of Cavalier-Smith (1998b) 375–407 and Cavalier-Smith (2000) 361–390. Dotted lines show individual phylogenetic relationships with significant bootstrap support as reported by different authors. Relationship (i) was reported by Cavalier-Smith (2000) 361–390 and Cavalier-Smith, and Chao, (2003) 540–563; relationship (ii) was reported by Lang et al. (2002); relationship (iii) was reported by King and Carroll (2001) 15032–15037 and Snell et al. (2001) 967–970; relationship (iv) was reported by Amaral-Zettler et al. (2001) 293–297 and Cavalier-Smith, (1998b) 375–407; Cavalier-Smith and Chao (1995) 1–6; relationship (v) was reported by Ragan et al. (1996) 11907–11912 and Cavalier-Smith (2000) 361–390; and relationship (vi) was reported by Cavalier-Smith (2000) 361–390. Only clades and branches with significant bootstrap confidences, as reported by the authors, are indicated.

The exact branching order of animals, fungi and their various protistan allies within the Opisthokonta is unclear (Figure 5.3). For example, depending on the taxa included, ichthyosporea have been reported as a sistergroup of the choanoflagellates + animals (Ragan et al., 1996; Lang et al., 2002), choanoflagellates + corallochytrea (Amaral-Zettler et al., 2001; Cavalier-Smith, 1998b) and choanoflagellates + animals + fungi (Ragan et al., 1996). A SSUrDNA phylogeny by Cavalier-Smith (2000) also suggested a sistergroup relationship between a choanoflagellate + ichthyosporea + corallochytrea and animals + fungi clade. However, some of these associations were not supported by high bootstrap values. A similar trend has also been observed for choanoflagellates: SSUrDNA and protein sequences place them as a sistergroup of animals + fungi (Cavalier-Smith, 2000; Snell et al., 2001), animals (Snell et al., 2001; King and Carroll, 2001; Lang et al., 2002) or fungi (Snell et al., 2001).

The phylogenetic instability of the animal–fungal allies might be because they are not a monophyletic group or because their placement with respect to the animals and fungi is not easily resolved by current available data, or both. The possibility that lower animals (sponges) and fungi (chytrids) are paraphyletic probably also contributes to this instability. The fact that many of these datasets include nonoverlapping sets of taxa also does not help. For example, almost all the studies on choanoflagellates and ichthyosporea have used different taxa (e.g., Cavalier-Smith, 1998b; Snell et al., 2001; King and Carroll, 2001). A stable phylogeny of opisthokonts will most probably only be obtained through a broader taxonomic sampling.

Even with more taxa, such deep phylogenetic relationships might be difficult to resolve in analyses based on single genes (Baldauf et al., 2000). Individual genes tend to lack sufficient phylogenetic information or be plagued with misleading multiple mutations that obscure the true phylogenetic signal in deep branches (Baldauf et al., 2000). In these situations, concatenated multigene datasets might be required to resolve relationships. So far, the only analyses of combined data to address deep opisthokont phylogeny have been based on combined mitochondrial gene sequences (Lang et al., 2002). These strongly support placement of the choanoflagellate *Monosiga brevicollis* as the sistergroup of the animals, with the ichthyosporean *Amoebidium* sister to the animals + *Monosiga* clade. Thus, if these taxa are truly representative of their respective groups, the combined analysis of mitochondrial genes suggests the monophyly of animals + choanoflagellates + ichthyosporea, at least, but the paraphyly of choanoflagellates + ichthyosporea. However, inclusion of a broader representation of animal taxa and animal–fungal allies is required to substantiate this.

5.6 Possible Sistergroups of the Opisthokonta

The most likely closest relatives of the Opisthokonta (Animals + Fungi + Choanozoa) are the Amoebozoa and the Apusozoa (Figure 5.4). Only SSUrDNA data are available for Apusozoa, which place them as the immediate sistergroup to Opisthokonta, with the lobose amoebae (Amoebozoa) on the next branch out (Cavalier-Smith, 2000). However, there is no significant bootstrap support for either of these groupings (Figure 5.4). Protein sequence data strongly support the placement of Amoebozoa as the closest sistergroup to Opisthokonta (Baldauf et al., 2000; Lang et al., 2002), but no apusozoan data were included in these analyses.

The Apusozoa have one recognized subdivision, the Thecomonadea (Cavalier-Smith 2000, 2002), and include the Ancyromonadida (e.g., *Ancyromonas*), Apusomonadida (e.g., *Apusomonas* and *Amastigomonas*) and Hemimastigida (e.g., *Hemimastix*). Members of the Apusozoa are mostly biflagellate (i.e., Ancyromonadida and Apusomonadida) or multiflagellated in the case of Hemimastigida (Patterson, 1999). Apusomonadida and Hemimastigida are characterized by tubular mitochondrial cristae, whereas Ancyromonadida are characterized by flat mitochondrial cristae (Patterson, 1999). SSUrDNA trees consistently show a close relationship between Opisthokonts and the Apusozoa (Cavalier-Smith and Chao, 1995, 2003; Atkins et al., 2000; Cavalier-Smith, 2000). No other molecular data are, however, available on these taxa.

The Amoebozoa appear to include the Lobosa, Conosa and Phalansterea (*Phalansterium*) (Cavalier-Smith, 1998a, 2000). Lobosa are amoebae with tubular mitochondrial cristae, lobose pseudopodia and move in a slow, noneruptive manner (Hausmann and Hülsmann, 1996). Three subgroups are recognized: Amoebaea (e.g., *Acanthamoeba* and *Leptomyxa*), Testacealobosea (e.g., *Arcella*) and Holomastigea (*Multicilia*) (Cavalier-Smith, 2000). The Conosa of Cavalier-Smith (2000, 2002) consists of the so-called Archamoebae and the Mycetozoa. Archamoebae are all amitochondriate amoebae, flagellates or amoeboflagellates and include the Pelobiontea (e.g., *Pelomyxa* and *Mastigamoeba*) and Entamoebea (e.g., *Entamoeba*). The Mycetozoa or true slime molds include the Dictyostelea (cellular slime

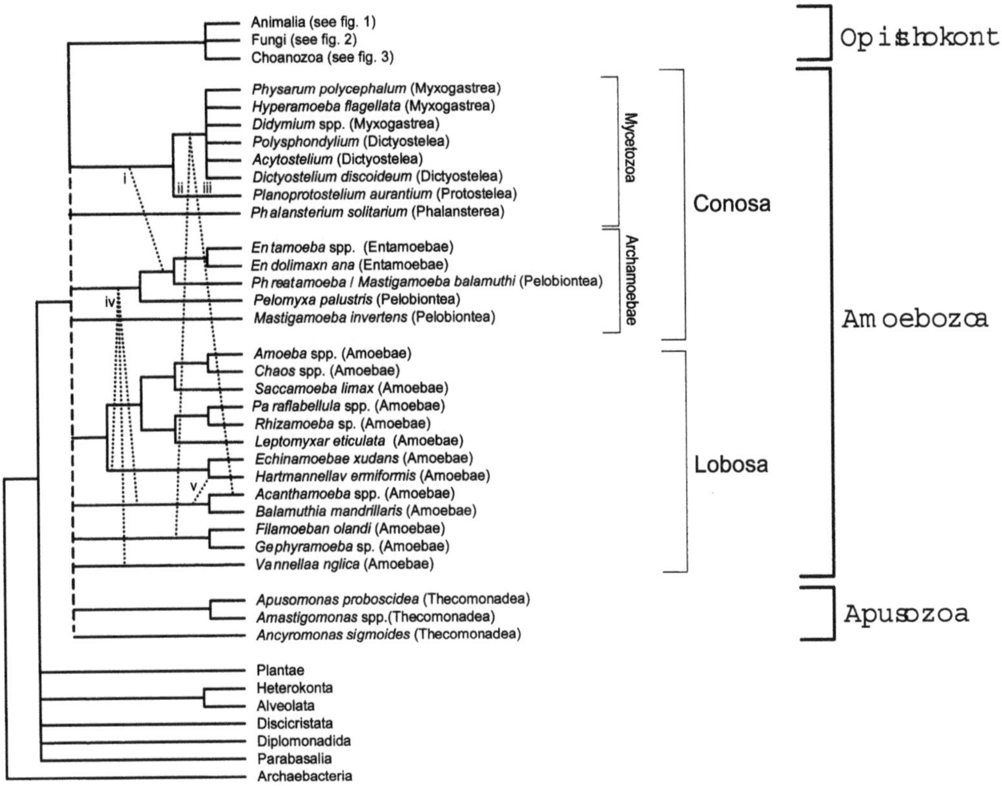

FIGURE 5.4 The phylogeny of Amoebozoa and Apusozoa based on a consensus of SSUrDNA and protein-coding gene trees. Only clades and branches with significant bootstrap confidences, as reported by the authors, are indicated. A vertical broken line in the cladogram indicates the taxa that are hypothetically part of either the Amoebozoa or Apusozoa. Class (in parentheses) and subphylum designations follow those of Cavalier-Smith (2000) 361–390 and Cavalier-Smith (2002) 297–354. References for phylogeny of the Mycetozoa: Baldauf and Doolittle (1997) 12007–12012; Philippe and Adoutte (1998) 25–56; Baldauf et al. (2000) 972–977; Cavalier-Smith (2000) 361–390; Archibald et al. (2002) 422–431; Arisue et al. (2002) 1–10. References for the phylogeny of the Archamoebae: Silberman et al. (1999) 1740–1751; Bolivar et al. (2001) 2306–2314; Milyutina et al. (2001) 131–139. References for the Lobosa phylogeny: Amaral-Zettler et al. (2000) 275–282; Bolivar et al. (2001) 2306–2314; Milyutina et al. (2001) 131–139. References for the Apusozoa: Cavalier-Smith and Chao (1995) 1–6, Cavalier-Smith and Chao (2003) 540–563; Atkins et al. (2000) 278–285; Cavalier-Smith (2000)361–390. Additionally, the dotted lines show individual phylogenetic relationships that were reported by (i) Arisue et al. (2002) 1–10; Bapteste et al. (2002) 1414–1419; and Horner and Embley (2001) 1970–1975; (ii) Bolivar et al. (2001) 2306–2314; (iii) Baldauf et al. (2000) 972–977; (iv) Milyutina et al. (2001) 131–139; and (v) Weekers et al. (1994) 684–690, Amaral-Zettler et al. (2000) 275–282; Sims, et al. (1999) 740–749; and Milyutina et al. (2001)131–139.

molds, e.g., *Dictyostelium*), Myxogastrea (plasmodial slime molds, e.g., *Physarum* and *Hyperamoeba*) and Protostelea (protostelid slime molds, e.g., *Protostelium*) (Baldauf and Doolittle, 1997).

5.6.1 Amoebozoa Phylogeny

The phylogeny of Amoebozoa is poorly understood. Most available amoebozoan sequences are from only a few taxa. For example, the genomes of the model organisms *Dictyostelium discoideum* and *Entamoeba histolytica* are nearly completely sequenced, whereas only SSUrDNA (with a few exceptions) is available for a limited number of other Amoebozoa. The best molecularly characterized group of Amoebozoa are the Mycetozoa (reviewed in Baldauf, 1999), which were long thought to be a nonmonophyletic and early emerging eukaryotic lineage, an idea strongly supported by ribosomal RNA data. [For references see Baldauf and Doolittle (1997), Philippe and Adoutte (1998) and Philippe and Germot (2000).] However, this now appears to be an artifact of accelerated SSUrDNA evolution in *Physarum*, and extreme and opposite nucleotide compositional bias in this taxon (GC-rich) vs. *Dictyostelium* (AT-rich) (Baldauf and Doolittle, 1997).

Numerous protein phylogenies unequivocally show the Mycetozoa to be very closely related to each other (Baldauf and Doolittle, 1997; Baldauf et al., 2000) and to form a strong grouping with the lobose amoeba *Acanthamoeba* (Baldauf et al., 2000). Several single-protein trees also support a close connection between Lobosa and Mycetozoa (e.g., Kelleher et al., 1995; Iwamoto et al., 1998; Philippe and Adoutte, 1998; Figure 5.4), as do unique shared characters such as fused *cox1* and *cox2* mitochondrial genes (Lang et al., 1999). Additionally, analyses of large concatenated datasets recover a strong group of *Dictyostelium* + archamoebae (i.e., *Mastigamoeba* and *Entamoeba*) (Arisue et al., 2002; Bapteste et al., 2002), a relationship that has been particularly difficult to resolve because of the highly accelerated rate of sequence evolution in the obligate parasite *E. histolytica*.

Most recently, new evidence suggests that the root of the eukaryote tree might lie close to the branch leading to the Opisthokonta (Stechman and Cavalier-Smith, 2002; Cavalier-Smith and Chao, 2003). Amoebozoa could therefore occupy a critical position either as the sistergroup to Opisthokonta, the sistergroup to the rest of eukaryotes or as a paraphyletic group at the root of the eukaryote tree [See Martin and Müller (1998), Moreira and López-Garcia (1998), Margulis et al. (2000) and Cavalier-Smith (2000, 2002) for alternative hypotheses on the nature of the first eukaryote.] Cavalier-Smith (2000, 2002) suggests that the first eukaryotes were simple single-celled, uniflagellated, amoeboid organisms, much like the Phalansterea. In this scenario, these ancient uniflagellated amoebae then diverged into the two major eukaryotic lineages, one giving rise to the Opisthokonta and the other leading to modern Amoebozoa, Apusozoa and the rest of the eukaryotes (Cavalier-Smith, 2000, 2002; Cavalier-Smith and Chao, 2003). If this is true, Amoebozoa would bear important clues about the nature of the last common ancestor of extant eukaryotes.

5.7 Conclusions

At present, a large body of data supports a grouping of animals, fungi and their protistan allies. Likewise, a large body of data places the Amoebozoa as a close sistergroup to the opisthokonts, although the exact nature of this relationship depends on where the root of the tree is found to lie. There are many remaining questions, among which four stand out: (1) Are the animal–fungal allies a monophyletic group? (2) What is the exact nature of the relationships among animals, fungi and their protistan allies? (3) Where do the Apusozoa fit in? (4) How are these relationships affected by the position of the root of the tree?

To address these questions, we will need to use a much broader taxonomic representation than we have seen so far, particularly of Choanozoa, Apusozoa, Amoebozoa and Porifera. This is particularly important as one, more or even all of these taxa might be paraphyletic. We will probably also need to develop concatenated multigene datasets, as single genes seem so far to resolve these relationships particularly poorly (e.g., Snell et al. 2001; King and Caroll, 2001). However, care must be taken to avoid paralogous genes, as animals probably experienced major genome duplications early in their evolution (Suga et al., 1999). Finally, we need to keep looking for new opisthokont taxa among the ca. 200 currently unclassified eukaryotes (Patterson, 1999) and the many, novel and diverse pico- and nanoeukaryote lineages currently being discovered (e.g., Moriera and López-Garcia, 2002; Dawson and Pace, 2002; Amaral-Zettler et al., 2002). If animals and fungi are indeed as old as the new rooting of the eukaryote tree suggests, then it is also to be expected that many taxa have branched off these lines before multicellularity evolved, some of which should still be extant and awaiting discovery.

Acknowledgments

We thank Jane Wright and Kate Perkins for technical assistance. This work was supported by the BBSRC grant #G13911.

References

Abouheif, E., Zardoya, R. and Meyer, A. (1998) Limitations of metazoan 18S rRNA sequence data: Implications for reconstructing a phylogeny of the animal kingdom and inferring the reality of the Cambrian explosion. *J. Mol. Evol.* 47: 394–405.

Adams, C.L., McInerney, J.O. and Kelly, M. (1999) Indications of relationships between poriferan classes using full-length 18S rRNA gene sequences. *Mem. Queensl. Mus.* 44: 33–43.

Adoutte, A., Balavoine, G., Lartillot, N. and de Rosa, R. (1999) Animal evolution: The end of intermediate taxa? *Trends Genet.* 15: 104–108.

Adoutte, A, Balavoine, G, Lartillot, N, Lespinet, O., Prudhomme, B. and de Rosa R. (2000) The new animal phylogeny: Reliability and implications. *Proc. Natl. Acad. Sci. USA* 97: 4453–4456.

Aguinaldo, A.M., Turbeville, J.M., Linford, L.S., Rivera, M.C., Garey, J.R., Raff, R.A. and Lake, J.A. (1997) Evidence for a clade of nematodes, arthropods and other moulting animals. *Nature* 387: 489–493.

Alexopoulus, C.J. (1962) *Introductory Mycology*, 2nd ed., John Wiley & Sons, London.

Alexopoulus, C.J., Mims, C.W. and Blackwell, M. (1996) *Introductory Microbiology*, 4th ed., John Wiley & Sons, New York.

Amaral-Zettler, L.A., Gómez, F., Zettler, E., Keenan, B.G., Amils, R. and Sogin, M.L. (2002) Eukaryotic diversity in Spains river of fire. *Nature* 417: 137.

Amaral-Zettler, L.A., Nerad, T.A., O'Kelly, C.J., Peglar, M.T., Gillevet, P.M., Silberman, J.D. and Sogin, M.L. (2000) A molecular reassessment of the Leptomixid amoebae. *Protist* 151: 275–282.

Amaral-Zettler, L.A., Nerad, T.A., O'Kelly, C.J. and Sogin, M.L. (2001) The nucleariid amoebae: More protists at the animal-fungal boundary. *J. Euk. Microbiol.* 48: 293–297.

Archibald, J.M., O'Kelly, J.C. and Doolittle, W.F. (2002) The chaperonin genes of jakobid and jakobid-like flagellates: Implications for eukaryotic evolution. *Mol. Biol. Evol.* 19: 422–431.

Arisue, N., Hashimoto, T., Lee, J.A., Moore, D.V., Gordon, P., Sensen, C.W., Gaasterland, T., Hasegawa, M. and Müller, M. (2002) The phylogenetic position of the pelobiont *Mastigamoeba balamuthi* based on sequences of rDNA and translation elongation factors EF-1α and EF-2. *J Euk. Microbiol.* 49: 1–10.

Atkins, M.S., McArthur, A.G. and Teske, A.P. (2000) Ancyromonadida: A new phylogenetic lineage among the protozoa closely related to the common ancestor of metazoa, fungi and choanoflagellates (Opisthokonta). *J. Mol. Evol.* 51: 278–285.

Baker, G.C., Beebee, T.J.C. and Ragan, M.A. (1999) *Prototheca richardsi*, a pathogen of anuran larvae, is related to a clade of protistan parasites near the animal-fungal divergence. *Microbiology* 145: 1777–1784.

Balavoine, G. (1997) The early emergence of platyhelminths is contradicted by the agreement between 18S rRNA and Hox gene data. *C. R. Acad. Sci. III — Life Sci.*, 320: 83–94.

Baldauf, S.L. (1999) A search for the origins of animals and fungi: Comparing and combining molecular data. *Am. Nat.* 154: S178–S188.

Baldauf, S.L. and Doolittle, W.F. (1997) Origin and evolution of slime moulds (Mycetozoa). *Proc. Natl. Acad. Sci. USA* 94: 12007–12012.

Baldauf, S.L. and Palmer, J.D. (1993) Animals and fungi are each others closest relatives: Congruent evidence from multiple proteins. *Proc. Natl. Acad. Sci. USA* 90: 11558–11562.

Baldauf, S.L., Roger, A.J., Wenk-Sierfert, I. and Doolittle, W.F. (2000) A kingdom-level phylogeny of eukaryotes based on combined protein data. *Science* 290: 972–977.

Bapteste, E., Brinkmann, H., Lee, J.A., Moore, D.V., Sensen, C.W., Gordon, P., Duruflé, L., Gaasterland, T., Lopez, P., Müller, M. and Philippe, H. (2002) The analysis of 100 genes support the grouping of three highly divergent amoebae: *Dictyostelium*, *Entamoeba*, and *Mastigamoeba*. *Proc. Natl. Acad. Sci. USA* 99: 1414–1419.

Barnes, R., Calow, P., Olive, P., Golding, D. and Spicer, J.I. (2001) *The Invertebrates: A Synthesis*, 3rd ed., Blackwell Science, Oxford.

Benny, G.L. and O'Donnell, K. (2000) *Amoebidium parasiticum* is a protozoan, not a Trichomycete. *Mycologia* 92: 1133–1137.

Berbee, M.L. (1996) Loculoascomycete origins and evolution of filamentous ascomycete morphology based on 18S rRNA gene sequence data. *Mol. Biol. Evol.* 13: 462–470.

Berbee, M.L., and Taylor, J.W. (2001) Fungal molecular evolution: Gene trees and geologic time. In *The Mycota* (McLaughlin, J.W. and Lemke, P.A., Eds.), Springer Verlag, New York.

Blanton, R.L. (1990) Phylum Acrasea. In *Handbook of Protoctista* (Margulis, L. Corliss, J.O., Melkonian, M. and Chapman, D.J., Eds.) Jones and Bartlett, Boston, pp. 75–87.

Bolivar, I., Fahrni, J.F., Smirnov, A. and Pawlowski, J. (2001) SSU rRNA-based phylogenetic position of Amoeba and Chaos (Lobosa, Gymnamoebia): The origin of gymnamoebae revisited. *Mol.Biol. Evol.* 8: 2306–2314.

Borchiellini, C., Boury-Esnault, N., Vacelet, J. and Le Parco, Y. (1998) Phylogenetic analysis of the HSP70 sequences reveals the monophyly of Metazoa and specific phylogenetic relationships between animals and fungi. *Mol. Biol. Evol.* 15: 647–655.

Borchiellini, C., Chombard, C., Lafay, B. and Boury-Esnault, N. (2000) Molecular systematics of sponges (Porifera). *Hydrobiologia* 420: 15–27.

Borchiellini, C., Manuel, M., Alivon, E., Boury-Esnault, N., Vacelet, J. and Le Parco, Y. (2001) Sponge paraphyly and the origin of Metazoa. *J. Evol. Biol.* 14: 171–179.

Bromham, L.D. and Degnan, B.M. (1999) Hemichordates and deuterostome evolution: Robust molecular phylogenetic support for a hemichordate + echinoderm clade. *Evol. Dev.* 1: 166–171.

Bruns, T.D., Vilgalys, R., Barns, S.M., Gonzales, D., Hibbett, D.S., Lane, D.J., Simon, L., Szaro, T.M., Wesburg, W.G. and Sogin, M.L. (1992) Evolutionary relationships within the fungi: Analyses of nuclear small subunit rRNA sequences. *Mol. Biol. Evol.* 1: 231–241.

Buck, K.R. (1990) Phylum Zoomastigina Class Choanomastigotes (Choanoflagellates). In *Handbook of Protoctista* (Margulis, L. Corliss, J.O., Melkonian, M. and Chapman, D.J., Eds.) Jones and Bartlett, Boston, pp. 194–199.

Cameron, C.B., Garey, J.R. and Swalla, B.J. (2000) Evolution of the chordate body plan: New insights from phylogenetic analyses of deuterostome phyla. *Proc. Natl. Acad. Sci. USA* 97: 4469–4474.

Caroll, S.B. (1995) Homeotic genes and the evolution of arthropods and chordates. *Nature* 376: 479–485.

Cavalier-Smith, T. (1987) The origin of fungi and pseudofungi. In *Evolutionary Biology of Fungi* (Rayner, A.D.M., Brasier, C.M. and Moore, D., Eds.) Cambridge University Press, Cambridge, pp. 339–353.

Cavalier-Smith, T. (1993) Kingdom Protozoa and its 18 phyla. *Microbiol. Rev.* 57: 953–994.

Cavalier-Smith, T. (1998a) A revised six kingdom system of life. *Biol. Rev.* 73: 203–266.

Cavalier-Smith, T. (1998b) Neomonada and the origin of animals and fungi. In *Evolutionary Relationships among Protozoa* (Coombs, G.H., Vickerman, K., Sleigh, M.A. and Warren, A., Eds.), Kluwer, London, pp. 375–407.

Cavalier-Smith, T. (2000) Flagellate megaevolution: The basis for eukaryote diversification. In *The Flagellates* (Green, J.R. and Leadbeater, B.S.G., Eds.), Taylor & Francis, London, pp. 361–390.

Cavalier-Smith, T. (2001) What are fungi? In *The Mycota VII, Part A: Systematics and Evolution* (McLaughlin, J.W. McLaughlin, J.W. and Lemke, P.A., Eds.) Springer Verlag, Berlin, pp. 3–37.

Cavalier-Smith, T. (2002) The phagotrophic origin of eukaryotes and phylogenetic classification of protozoa. *Int. J. Syst. Evol. Microbiol.* 52: 297–354.

Cavalier-Smith, T. and Allsopp, M.T.E.P. (1996) Corallochytrium, an enigmatic non-flagellate protozoan related to choanoflagellates. *Eur. J. Protistol.* 32: 306–310.

Cavalier-Smith, T., Allsopp, M.T.E.P., Chao, E.E., Boury-Esnault, N. and Vacelet, J. (1996) Sponge phylogeny, animal monophyly, and the origin of the nervous system: 18S rRNA evidence. *Can. J. Zool.* 74: 2031–2045.

Cavalier-Smith, T. and Chao E.E. (1995) The opalozoan Apusomonas is related to the common ancestor of animals, fungi, and choanoflagellates. *Proc. R. Soc. Lond. B* 261: 1–6.

Cavalier-Smith, T. and Chao E.E. (2003) Phylogeny of Choanozoa, Apusozoa, and other protozoa and early eukaryote megaevolution. *J. Mol. Evol.* 56: 540–563.

Collins, A.G. (1998) Evaluating multiple alternative hypotheses for the origin of Bilateria: An analysis of 18S rRNA molecular evidence. *Proc. Natl. Acad. Sci. USA* 95: 15458–15463.

Collins, A.G. and Valentine, J.W. (2001) Defining phyla: Evolutionary pathways to metazoan body plans. *Evol. Dev.* 3: 432–442.

Corliss, J.O. (1984) The kingdom Protista and its 45 phyla. *Biosystems* 17: 87–126.

Dawson, S.C. and Pace, N.R. (2002) Novel kingdom-level eukaryotic diversity in anoxic environments. *Proc. Natl. Acad. Sci. USA* 99: 8324–8329.

De Rosa, R. (2001) Molecular data indicate the protostome affinity of Brachipods. *Syst. Biol.* 50: 848–859.

De Rosa, R., Grenier, J.K., Andreeva, T., Cook, C.E., Adoutte, A., Akam, M., Carroll, S.B. and Balavoine, G. (1999) Hox genes in brachiopods and priapulids and protostome evolution. *Nature* 399: 772–776.

Edlind, T.D., Li, J., Visvesvara, G.S., Vodkin, M.H., McLaughlin, G.L. and Katiyar, S.K. (1996) Phylogenetic analysis of β-tubulin sequences from amitochondrial protozoa. *Mol. Phylogenet. Evol.* 5: 359–367.

Eriksson, O. and Winka, K. (1997) Superordinal taxa of Ascomycota. *Myconet* 1: 1–16.

Farabee, M.J. (2002). Biological Diversity: Animals I, II and III in Online Biology Book. Available at http://www.emc.maricopa.edu/faculty/farabee/BIOBK/BioBookTOC.HTML, accessed February16, 2004.

Fast, N.M., Logsdon J.M., Jr. and Doolittle, W.F. (1999) Phylogenetic analysis of the TATA box binding protein (TBP) gene from *Nosema locustae*: Evidence for a microsporidia-fungi relationship and spliceosomal intronloss. *Mol. Biol. Evol.* 16: 1415–1419.

Germot, A., Philippe, H. and Le Guyader, H. (1997) Evidence for loss of mitochondria in Microsporidia from a mitochondrial-type HSP70 in *Nosema locustae. Mol. Biochem. Parasitol.* 87: 159–168.

Graham, A. (2000) Animal phylogeny: Root and branch surgery. *Curr. Biol.* 10: R36–R38.

Greiner, J.K., Garber, T.L., Warren, R., Whitington, P.M. and Caroll, S. (1997) Evolution of the entire arthropod Hox gene set predated the origin and radiation of the onycophoran/arthropod clade. *Curr. Biol.* 7: 547–553.

Halanych, K.M., Bacheller, J.D., Aguinaldo, A.M.A., Liva, S.M., Hillis, D.M. and Lake, J.A. (1995) Evidence from 18S ribosomal DNA that the lophophorates are protostome animals. *Science* 267: 1641–1642.

Hausmann, K. and Hülsmann, N. (1996) *Protozoology*, 2nd ed., Thieme Medical Publishers, New York.

Hawksworth, D.L., Kirk, P.M., Sutton, B.C. and Pelger, D.N. (1995) *Ainsworths & Bisbys Dictionary of the Fungi*, 8th ed., Cambridge University Press, Cambridge.

Herr, R.A., Ajello, L., Taylor. J.W., Arseculeratne, S.N. and Mendoza, L. (1999) Phylogenetic analysis of *Rhinosporidium seeberis* 18S small-subunit ribosomal DNA groups this pathogen among members of the protoctistan Mesomycetozoa clade. *J. Clin. Microbiol.* 37: 2750–2754.

Hibbett, D.S., Pine, E.M., Langer, G. and Donoghue, M.J. (1997) Evolution of gilled mushrooms and puffballs inferred from ribosomal DNA sequences. *Proc. Natl. Acad. Sci. USA* 94: 12002–12006.

Hillis, D.M. (1998) Taxonomic sampling, phylogenetic accuracy, and investigator bias. *Syst. Biol.* 47: 3–8.

Hirt, R.P., Logsdon J.M., Jr., Healy, B., Dorey, M.W., Doolittle, W.F. and Embley, T.M. (1999) Microsporidia are related to fungi: Evidence from the largest subunit of RNA polymerase II and other proteins. *Proc. Natl. Acad. Sci. USA* 96: 580–585.

Horner, D.S. and Embley, T.M. (2001) Chaperonin 60 phylogeny provides further evidence for secondary loss of mitochondria among putative early-branching eukaryotes. *Mol. Biol. Evol.* 18: 1970–1975.

Hyman, L.H. (1940) *The Invertebrates*, Vol. 1, McGraw-Hill, New York.

Iwamoto, M., Pi, M., Kurihara, M., Morio, T. and Tanaka, Y. (1998). A ribosomal protein gene cluster is encoded in the mitochondrial DNA of *Dictyostelium discoideum*: UGA termination codons and similarity of gene order to *Acanthamoeba castellanii*. *Curr. Genet.* 33: 304–310.

James, T.Y., Porter, D., Leander, C.A., Vilgalys, R. and Longcore, J.E. (2000) Molecular phylogenetics of the Chytridiomycota supports the utility of ultrastructural data in chytrid systematics. *Can. J. Bot.* 78: 336–350.

Jensen, A.B., Gargas, A., Eilenberg, J. and Rosendahl, S. (1998) Relationships of the insect-pathogenic order Entomophthorales (Zygomycota, Fungi) based on phylogenetic analysis of nuclear small subunit ribosomal DNA sequences (SSU rDNA). *Fung. Genet. Biol.* 24: 325–334.

Keeling, P.J., Luker, M.A. and Palmer, J.D. (2000) Evidence from beta-tubulin phylogeny that microsporidia evolved from within the fungi. *Mol. Biol. Evol.* 17: 23–31.

Kelleher, J.F., Atkinson, S.J. and Pollard, T.D. (1995) Sequences, structural models, and cellular localization of the actin-related proteins Arp2 and Arp3 from Acanthamoeba. *J. Cell Biol.* 131: 385–397.

Kim, J., Kim, W. and Cunningham, C.W. (1999) A new perspective on lower metazoan relationships from 18S rDNA sequences. *Mol. Biol. Evol.* 16: 423–427.

King, N. and Carroll, S.B. (2001) A receptor tyrosine kinase from choanoflagellate: Molecular insights into early animal evolution. *Proc. Natl. Acad. Sci. USA* 98: 15032–15037.

Kirk, P.M., Cannon, P.F., David, J.C. and Stalpers, J.A. (2001) *Ainsworths & Bisbys Dictionary of the Fungi*, 9th ed., CABI Publishing, Wallingford.

Kobayashi, M., Furuya, H. and Holland, P.W.H. (1999) Dicyemids are higher animals. *Nature* 401: 762.

Kobayashi, M., Wada, H. and Satoh, N. (1996) Early evolution of the Metazoa and phylogenetic status of Diploblasts as inferred from amino acid sequence of elongation factor-1α. *Mol. Phylogenet. Evol.* 5: 414–422.

Kruse, M., Leys, S.P., Müller, I.M. and Müller, W.E.G. (1998) Phylogenetic position of the Hexactinellida within the phylum Porifera based on the amino acid sequence of the protein-kinase C from *Rhabdocalyptus dawsoni*. *J. Mol. Evol.* 46: 721–728.

Kumar, S. and Rzhetsky, A. (1996) Evolutionary relationships of eukaryotic kingdoms. *J. Mol. Evol.* 42: 183–193.

Lafay, B., Boury-Esnault, N., Vacelet, J. and Christen, R. (1992) An analysis of partial 28S ribosomal RNA sequences suggests early radiations of sponges. *Biosystems* 28: 139–151.

Lang, B.F., O'Kelly, C., Nerad, T., Gray, M.W. and Burger, G. (2002) The closest unicellular relatives of animals. *Curr. Biol.* 12: 1773–1778.

Lang, B.F., Seif, E., Gray, M.W., O'Kelly, C.J. and Burger, G. (1999) A comparative genomics approach to the evolution of eukaryotes and their mitochondria. *J. Euk. Microbiol.* 46: 320–326.

Lichtwardt, R.W. and Williams, M.C. (1992) Two new Australasian species of Amoebidiales associated with aquatic insect larvae, and comments on their biogeography. *Mycologia* 84: 376–383.

Liu, Y.J., Whelen, S. and Hall, B.D. (1999) Phylogenetic relationships among Ascomycetes: Evidence from an RNA polymerase II subunit. *Mol. Biol. Evol.* 16: 1799–1808.

Löytynoja, A. and Milinkovitch, M.C. (2001) Molecular phylogenetic analysis of the mitochondrial ADP-ATP carriers: The Plantae/Fungi/Metazoa trichotomy revisited. *Proc. Natl. Acad. Sci. USA* 98: 10202–10207.

Lutzoni, F., Pagel, M. and Reeb, V. (2001) Major fungal lineages are derived from lichen symbiotic ancestors. *Nature* 411:937–940.

Margulis, L., Dolan, M.F. and Guerrero, R. (2000) The chimeric eukaryote: Origin of the nucleus from the karyomastigont in amitochondriate protists. *Proc. Natl. Acad. Sci. USA* 97: 6954–6959.

Martin, W. and Müller, M. (1998) The hydrogen hypothesis for the first eukaryote. *Nature* 392: 37–41.

Medina, M., Collins, A.G., Silberman, J.D. and Sogin, M.L. (2001) Evaluating hypotheses of basal animal phylogeny using complete sequences of large and small subunit rRNA. *Proc. Natl. Acad. Sci. USA* 98: 9707–9712.

Milyutina, I.A., Aleshin, V.V., Mikrjukov, K.A., Kedrova, O.S. and Petrov, N.B. (2001) The unusually long small subunit ribosomal RNA gene found in amitochondriate amoeboflagellate *Pelomyxa palastris*: Its rRNA predicted secondary structure and phylogenetic implication. *Gene* 272: 131–139.

Moncalvo, J.-M., Lutzoni, F.M., Rehner, S.A., Johnson, J. and Vilgalys, R. (2000) Phylogenetic relationships of agaric fungi based on nuclear large subunit ribosomal DNA sequences. *Syst. Biol.* 49: 278–305.

Moreira, D. and López-Garcia, P. (1998) Symbiosis between methanogenic archaea and δ-proteobacteria as the origin of eukaryotes: The syntrophic hypothesis. *J. Mol. Evol.* 47: 517–530.

Moreira, D. and López-Garcia, P. (2002) The molecular ecology of microbial eukaryotes unveils a hidden world. *Trends Microbiol.* 10: 266–267.

Nikoh, N., Hayase, N., Iwabe, N., Kuma, K. and Miyata, T. (1994) Phylogenetic relationships of the kingdoms Animalia, Plantae, and Fungi inferred from 23 different protein species. *Mol. Biol. Evol.* 11: 762–768.

Nielson, C., Scharff, N. and Eibye-Jacobsen, D. (1996) Cladistic analysis of the animal kingdom. *Biol. J. Linn. Soc.* 57: 385–410.

Nishida, H. and Sugiyama, J. (1994) Archiascomycetes: Detection of a major lineage within the Ascomycota. *Mycoscience* 35: 361–366.

O'Donnell, K., Cigelnik, E. and Benny, G.L. (1998) Phylogenetic relationships among the Harpellales and Kickxellales. *Mycologia* 90: 624–639.

O'Donnell, K., Lutzoni, F.M., Ward, T.J. and Benny, G.L. (2001) Evolutionary relationships among mucoralean fungi (Zygomycota): Evidence for family polyphyly on a large scale. *Mycologia* 93: 286–296.

Page, F.C. (1987) The classification of naked amoebae (Phylum Rhizopoda). *Arch. Protistenk.* 133: 199–217.

Patterson, D.J. (1999) The diversity of eukaryotes. *Am. Nat.* 154: S96–S124.

Peterson, K.J. and Eernisse, D.J. (2001) Animal phylogeny and the ancestry of bilaterians: Inference from morphology and 18S rDNA gene sequences. *Evol. Dev.* 3: 170–205.

Philippe, H. and Adoutte, A. (1998) The molecular phylogeny of Eukaryota: Solid facts and uncertainties. In *Evolutionary Relationships among Protozoa* (Coombs, G.H., Vickerman, K. Sleigh, M.A. and Warren, A., Eds.), The Systematics Association Special Volume Series 56, Kluwer Academic Publishers, London, pp. 25–56.

Philippe, H. and Germot, A. (2000) Phylogeny of eukaryotes based on ribosomal RNA: Long branch attraction and models of sequence evolution. *Mol. Biol. Evol.* 17: 830–834.

Ragan, M.A. (1989) Biochemical pathways and the phylogeny of the eukaryotes. In *The Hierarchy of Life* (Fernholm, B., Bremer, K. and Jornvall, H., Eds.), Elsevier, pp. 145–160.

Ragan, M.A., Goggins, C.L., Cawthorn, R.J., Cerenius, L., Jamieson, A.V.C., Plourde, S.M., Rand, T.G., Söderhäll, K. and Gutell, R.R. (1996) A novel clade of protistan parasites near the animal-fungal divergence. *Proc. Natl. Acad. Sci. USA* 93: 11907–11912.

Redecker, D. (2002) New views on fungal evolution based on DNA markers and the fossil record. *Res. Microbiol.* 153: 125–130.

Schüßler, A., Scwarzott, D. and Walker, C. (2001) A new fungal phylum, the Glomeromycota: Phylogeny and evolution. *Mycol. Res.* 105: 1414–1421.

Schuster F.L. (1990) Phylum Rhizopoda. In *Handbook of Protoctista* (Margulis, L. Corliss, J.O., Melkonian, M. and Chapman, D.J., Eds.) Jones and Bartlett, Boston, pp. 3–18.

Silberman, J.D., Clark, C.G., Diamond, L.S. and Sogin, M.L. (1999) Phylogeny of the genera *Entamoeba* and *Endolimax* as deduced from small-subunit ribosomal RNA sequences. *Mol. Biol. Evol.* 16: 1740–1751.

Sims, G.P., Rogerson, A. and Aitken, R. (1999) Primary and secondary structure of the small-subunit ribosomal RNA of the naked, marine amoeba *Vannella anglica*: Phylogenetic implications. *J. Mol. Evol.* 48: 740–749.

Sleigh, M. (1989) *Protozoa and Other Protists*, 2nd ed., Edward Arnold, London.

Snell, E.A., Furlong, R.F. and Holland, P.W.H. (2001) HSP70 sequences indicate that choanoflagellates are closely related to animals. *Curr. Biol.* 11: 967–970.

Spatafora, J.W. and Blackwell, M. (1994) The polyphyletic origins of ophiostomatoid fungi. *Mycol. Res.* 98: 1–9.

Stechman, A. and Cavalier-Smith, T. (2002) Rooting the eukaryote tree by using a derived gene fusion. *Science* 297: 89–91.

Suga, H., Koyanagi, M., Hoshiyama, D., Ono, K., Iwabe, N., Kuma, K. and Miyata, T. (1999) Extensive gene duplication in the early evolution of animals before the parazoan-eumetazoan split demonstrated by G proteins and protein tyrosine kinases from sponge and hydra. *J. Mol. Evol.* 48: 646–653.

Sugiyama, J. (1998) Relatedness, phylogeny, and evolution of the fungi. *Mycoscience* 39: 487–511.

Swann, E.C. and Taylor, J.W. (1995) Phylogenetic perspectives on Basidiomycete systematics: Evidence from the 18S rRNA gene. *Can. J. Bot.* 73: S862–S868.

Taylor, J.W., Swann, E.C. and Berbee, M.L. (1994) Molecular evolution of ascomycete fungi: Phylogeny and conflict. In *Ascomycete Systematics: Problems and Perspectives in the Nineties* (Hawksworth, D.L., Ed.), Plenum Press, New York, pp. 201–212.

Tehler, A. (1988) A cladistic outline of the Eumycota. *Cladistics* 4: 227–277.

Tehler, A., Farris, J.S., Lipscomb, D.L. and Källersjö, M. (2000) Phylogenetic analyses of the fungi based on large rDNA data sets. *Mycologia* 92: 459–474.

Van de Peer, Y., Baldauf, S.L., Doolittle, W.F. and Meyer, A. (2000a) An updated and comprehensive rRNA phylogeny of (crown) eukaryotes based on rate-calibrated evolutionary distances. *J. Mol. Evol.* 51: 565–576.

Van de Peer, Y., Ben Ali, A. and Meyer, A. (2000b) Microsporidia: Accumulating molecular evidence that a group of amitochondriate and suspectedly eukaryotes are just curious fungi. *Gene* 246: 1–8.

Van de Peer, Y. and De Wachter, R. (1997) Evolutionary relationships among the eukaryotic crown taxa taking into account site-to-site rate variation in 18S rRNA. *J. Mol. Evol.* 45: 619–630.

Van der Auwera, G. and De Wachter, R. (1996) Large-subunit rRNA sequence of the chytridiomycete *Blastocladiella emersonii*, and implications for the evolution of zoosporing fungi. *J. Mol. Evol.* 43: 467–483.

Voigt, K. and Wöstemeyer, J. (2001) Phylogeny and origin of 82 zygomycetes from all 54 genera of the Mucorales and Mortierellales based on combined analysis of actin and translation elongation factor EF-1α genes. *Gene* 270: 113–120.

von Nickisch-Rosenegk, M., Brown, W.M. and Boore, J.L. (2001) Complete sequence of the mitochondrial genome of the tapeworm *Hymenolepis diminuta*: Gene arrangements indicate that Platyhelminths are eutrochozoans. *Mol. Biol. Evol.* 18: 721–730.

Vossbrinck, C.L., Maddox, J.V., Friedman, B.A., Debrunner-Vossbrinck, B.A. and Woese, C.R. (1987) Ribosomal RNA sequence suggests microsporidia are extremely ancient eukaryotes. *Nature* 326: 411–414.

Wada, H., Kobayashi, M., Sato, R., Satoh, N., Miyasaka, H. and Shirayama, Y. (2002) Dynamic insertion-deletion of introns in deuterostome EF-α genes. *J. Mol. Evol.* 54: 118–128.

Wainright, P.O., Hinkle, G., Sogin, M.L. and Stickel, S.K. (1993) Monophyletic origins of the Metazoa: An evolutionary link with fungi. *Science* 260: 340–342.

Weekers, P.H.H., Gast, R.J., Fuerst, P.A. and Byers, T.J. (1994) Sequence variations in small-subunit ribosomal RNAs of *Hartmannella vermiformis* and their phylogenetic implications. *Mol. Biol. Evol.* 11: 684–690.

Weiss, L.M., Edlind, T.D., Vossbrinck, C.R. and Hashimoto, T. (1999) Microsporidian molecular phylogeny: The fungal connection. *J. Euk. Microbiol.* 46:17S–18S.

Whittaker, R.H. (1969) New concepts of kingdoms of organisms. *Science* 163: 150–160.

Winnepenninckx, B., Backeljau, T. and Dewachter, R. (1995a) Phylogeny of protostome worms derived from 18S ribosomal-RNA sequences. *Mol. Biol. Evol.* 12: 641–649.

Winnepenninckx, B., Backeljau, T., Mackey, L.Y., Brooks, J.M., De Wachter, R., Kumar, S. and Garey, J.R. (1995b) 18S rRNA data indicate that Aschelminthes are polyphyletic in origin and consist of at least three distinct clades. *Mol. Biol. Evol.* 12: 1132–1137.

Zrzavy, J., Mihulka, S., Kepka, P. and Bezdek, A. (1998) Phylogeny of the metazoa based on morphological and 18S ribosomal DNA evidence. *Cladistics* 14: 49–285.

Section II

Phylogenetics and Comparative Genomics

Chapter 6

Pitfalls in Tree Reconstruction and the Phylogeny of Eukaryotes*

Simonetta Gribaldo and Hervé Philippe

CONTENTS

Abstract

Traditional views on deep evolutionary events have been seriously challenged over the last few years, following the identification of major pitfalls affecting molecular phylogeny reconstruction. This chapter describes the principally encountered artifacts, notably long-branch attraction, and their causes (i.e., difference in evolutionary rates, mutational saturation, compositional biases). Additional difficulties due to phenomena of biological nature (i.e., lateral gene transfer, recombination, hidden paralogy) are also discussed. Contrary to common beliefs, we show that the use of rare genomic events can also be misleading and should be treated with the same caution as used for standard molecular phylogeny. The universal tree of life, as described in most textbooks, is partly affected by tree reconstruction artifacts, e.g., (1) the bacterial rooting of the universal tree of life, (2) the early emergence of

* Reprinted with modifications from Gribaldo, S. et al. (2000) Ancient phylogenetic relationships. *Theor. Popul. Biol.* 61(4), 2002: 391–408. With permission from Elsevier Science, Amsterdam.

0-415-29904-7/04/$0.00+$1.50

amitochondriate lineages in eukaryotic phylogenies; and (3) the position of hyperthermophilic taxa in bacterial phylogenies. We present an alternative view of this tree, based on recent evidence obtained from reanalyses of ancient datasets and from novel analyses of large combination of genes.

6.1 Introduction

At present, the SSUrRNA phylogeny rooted between Bacteria and Archaea/Eukarya has become the universal tree of life represented in all textbooks (Figure 6.1). The three-domain classification of life (Woese et al., 1990) and the inference that the last universal common ancestor (LUCA) was a bacterial-like organism (Martin and Müller, 1998) constitute a paradigm widely accepted by the scientific community, with a few exceptions (Gupta, 1998; Margulis, 1996; Mayr, 1998).

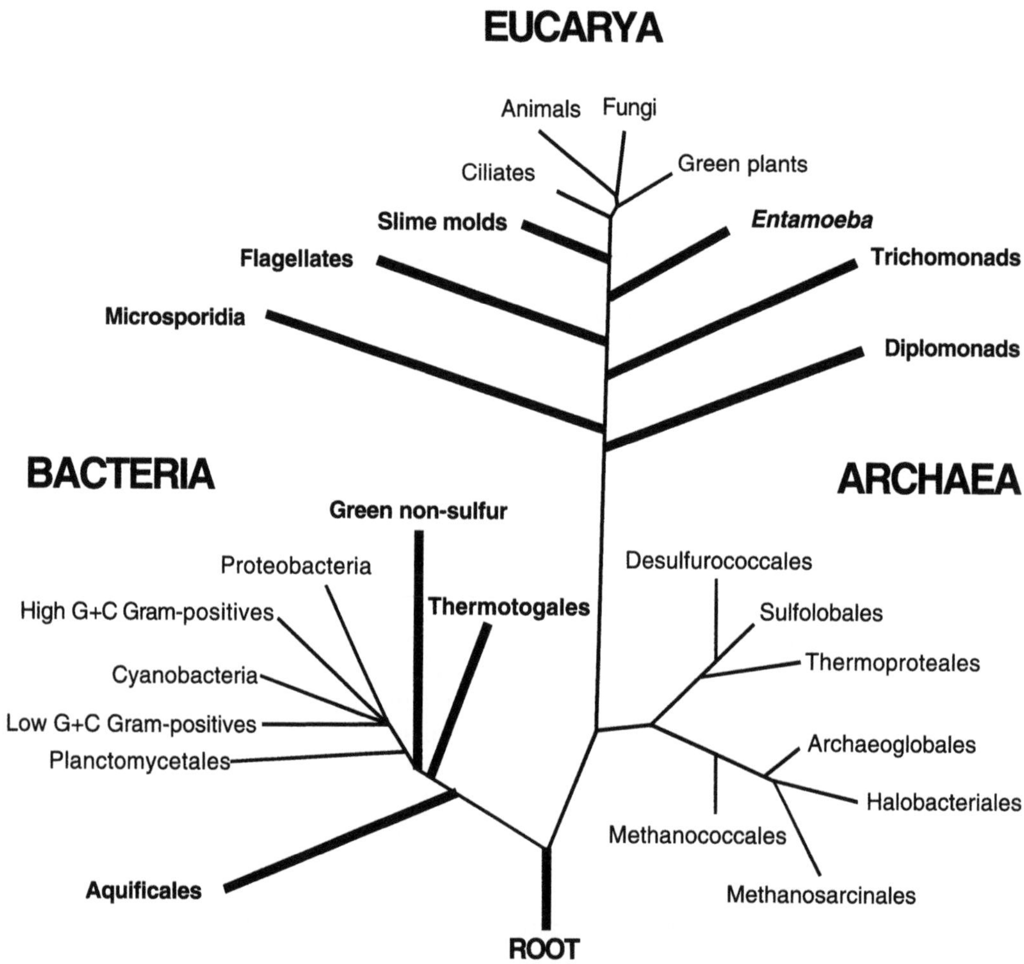

FIGURE 6.1 The classical view of the universal tree of life. The topology is inspired from Stetter, K. O. (1996) *Ciba Found. Symp.* 202: 1–10. It is mainly based on rRNA comparison. The branches that could be affected by long branch attraction artifacts (e.g., the placement of the root in the bacterial branch or the early emergence of hyperthermophilic taxa among bacteria) are given as thick lines.

Nonetheless, this view is based on a very small sample of the genome (ca. 1000 nucleotides for SSUrRNA and ca. 1000 amino acids for ancient paralogues of a few protein coding genes; (Woese, 1987; Brown and Doolittle, 1997). With the completion of genome sequences from representatives of the three domains, it has become evident that trees based on alternative markers are largely in contradiction with the SSUrRNA phylogeny as well as with each other, weakening the general consensus (Pennisi, 1998). The causes for these incongruencies often reside in phenomena of biological nature, such as lateral genetic transfer (LGT), or misidentification of paralogous from orthologous genes (hidden paralogy; Doolittle, 1999b). However, phenomena of mathematical nature (tree-reconstruction artifacts) are increasingly recognized as another important source of misleading results in molecular phylogenies (Philippe and Laurent, 1998). We first focus on pitfalls of current tree-making methods and the way they can be identified and possibly tackled. Then we discuss the phenomena of biological nature, notably LGT, affecting ancient phylogeny reconstruction. Finally, we synthesize the joint impact of these phenomena on the universal tree of life, with a focus on eukaryotic phylogeny.

6.2 Pitfalls in Tree Reconstruction Methodologies

6.2.1 Mutational Saturation vs. Resolving Power: A Central Issue

Ideally, genuine signatures for inferring bifurcations in phylogenetic trees would be provided by positions that underwent a single mutation event (e.g., the K to Y change at Site 1 in Figure 6.2). Given enough such positions, any evolutionary history may be easily reconstructed (Swofford et al., 1996). In practice, to solve a phylogeny of ca. 50 taxa, more than 150 such positions would be needed (i.e., at least three changes for each node). Unfortunately, mutations mostly involve a subset of positions throughout time (Fitch and Markowitz, 1970) and generate more noise than signal for molecular phylogeny inference. Indeed, the sites that contain important information (i.e., those having undergone substitutions on internal branches) generally reach a point of mutational saturation, and the signal they carry is lost and becomes random noise (e.g., Site 2 in Figure 6.2, harboring 25 changes). Consequently, although the resolving power of molecular phylogeny is directly proportional to both the length of the data set and its evolutionary rate, it dramatically decreases when rates become too high at a large fraction of sites, i.e., with the increase of mutational saturation. This leads to a paradox — an abundance of information blurs the authentic signal.

To illustrate this point, we analyzed an alignment of 97 Ile/Val tRNA synthetases (RS) with a maximum parsimony approach (Figure 6.3). Only ca. 20% of positions presented less than five changes (i.e., 68 out of 344, 15 of which were constant). The information carried by these positions (144 steps) was modest with respect to that (567 steps) brought by the fraction of sites with the highest evolutionary rates (i.e., those harboring more than 51 changes), and it became negligible when compared with the information carried by the positions with more than 30 changes (3970 steps). Thus, selection of the best tree was largely contributed by noisy positions (3970 steps) rather than by the more reliable ones (144 steps). This example highlights the need to develop efficient methods permitting to discriminate signal from noise (see later). Saturation can also be identified as a plateau on a diagram wherein, for each couple of species, the number of inferred multiple substitutions are plotted against the number of observed differences (Philippe et al., 1994). Mutational saturation represents a major concern in ancient phylogenetic reconstruction. For example, all the markers used to infer universal trees appear to be saturated mutationally (Philippe and Adoutte, 1998; Philippe and Forterre, 1999; Roger et al., 1999).

Although saturated data should harbor no signal and would, in principle, produce a star phylogeny, biases of different nature might misleadingly lead to a resolved tree, because

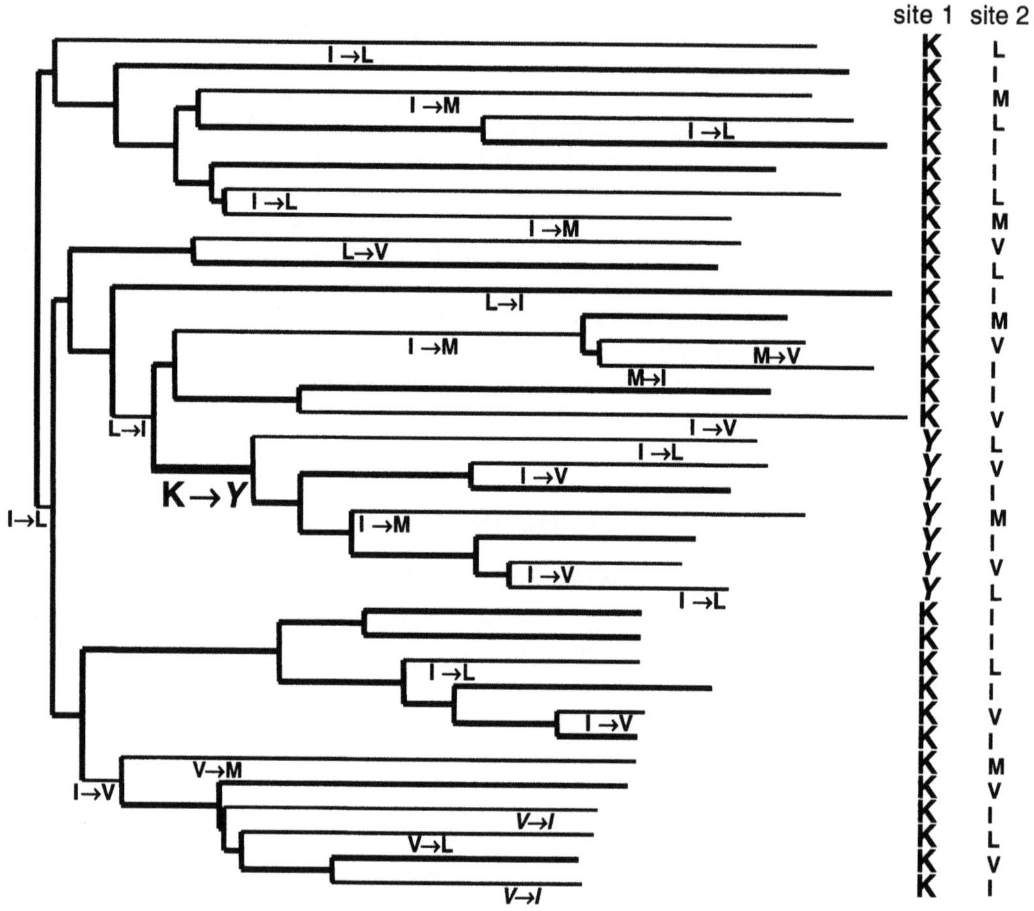

FIGURE 6.2 Archetypes of "good" and "bad" positions. Site 1 has undergone a single mutation event (lysine to tyrosine) and it thus retains a genuine signal. Conversely, Site 2 has experienced multiple substitutions involving hydrophobic residues that have completely blurred its original signal, that is, it has reached a point of mutational saturation and it will contribute more noise than reliable information to phylogenetic reconstruction.

they are not properly handled by current tree reconstruction methodologies. Three major biases can be highlighted: (1) compositional biases, (2) variable evolutionary rates and (3) heterotachy (covarion structure).

6.2.2 Compositional Biases

As underlined in early studies, the G+C content of SSUrRNA is substantially heterogeneous (ranging from 30 to 70%) and might lead to the artifactual grouping of unrelated species that share a similar G+C content (Loomis and Smith, 1990; Weisburg et al., 1989). This is the case for thermophilic organisms, all of which harbor a high G+C content (Galtier and Lobry, 1997) that produces a systematic bias. Modifications of the three major tree reconstruction methods have been developed to correct for this problem, e.g., transversion parsimony (Woese et al., 1991), log-det distances (Lockhart et al., 1994), and nonstationary maximum likelihood (Galtier and Gouy, 1995). Although some improvements were observed (Lockhart et al., 1994; Woese et al., 1991), these methods generally did not lead to substantial changes in the resulting phylogeny. Corrections sometimes produced even worse results,

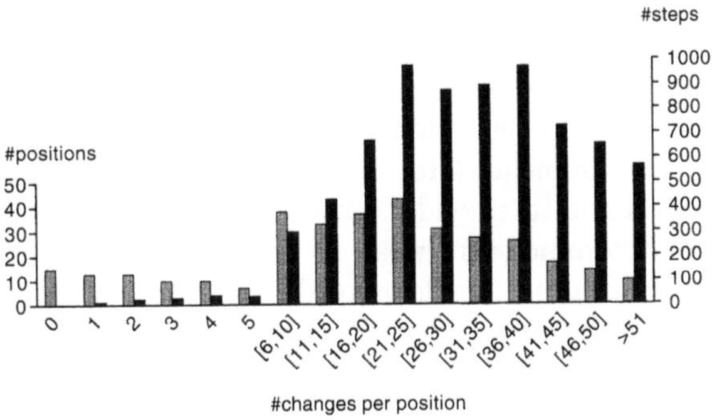

FIGURE 6.3 Mutational saturation of Ile/Val-tRNA synthetase sequences. On the x-axis, positions are grouped according to their number of changes, as inferred by maximum parsimony. On the y-axis, the number of different positions belonging to the different classes of substitutions is given on left columns (in light gray) and the number of steps they contribute to the most parsimonious tree is given on right columns (in dark gray). The information carried by positions that have undergone a few substitutions (e.g., sites with <5 changes provide 144 steps) is overwhelmed by that carried by positions that have undergone many more changes (e.g., sites with >30 changes provide 3970 steps). The more noisy a position, the more important is its contribution to tree selection.

such as the increased support for an early branching of microsporidia when G+C content biases were accounted for (Galtier and Gouy, 1995). Moreover, contrary to early claims (Hasegawa and Hashimoto, 1993), protein sequences too might be prone to compositional bias effects (Foster and Hickey, 1999). For example, in a recent phylogeny based on concatenated ribosomal proteins, taxa with G+C rich genomes (e.g., *Mycobacterium* and *Deinococcus*) were artifactually grouped (Brochier et al., 2002). Nonetheless, being rather easily identified, compositional biases cannot be considered as a major problem of phylogenetic reconstruction.

6.2.3 Long-Branch Attraction

When the genuine phylogenetic signal is compromised, a false resolution of organismal relationships can be provided by the variation in evolutionary speed among taxa. For example, SSUrRNA evolutionary rates vary by a factor of 100 among planktonic foraminifers (Pawlowski et al., 1997). The grouping of the longest branches representing the fastest-evolving taxa, irrespective of their true phylogenetic affiliations, constitutes the renowned long-branch attraction (LBA) artifact (Felsenstein, 1978). In phylogenies rooted by a distant outgroup, unrelated fast-evolving taxa will emerge independently as the deepest offshoots, because of being attracted by the long branch of the outgroup (Philippe and Laurent, 1998). The LBA artifact is at present a major concern of molecular phylogenetics, as it is believed to affect the position of virtually every deep-branching lineage. As a result, many organismal relationships in the universal tree should be regarded as highly suspect (see bold lines in Figure 6.1), especially concerning eukaryotic phylogeny, in which a number of protist groups emerge in a ladder-like fashion as the earliest branches. Indeed, one of these groups was recently reassigned to a radically different placement. The Microsporidia have become an infamous example of LBA. These intracellular parasites, originally considered to be among the first emerging eukaryotes (Vossbrinck et al., 1987), are now unmistakably identified as derived fungi (Keeling and Fast, 2002; Chapter 10) based on a number lines of evidence, such as indel sharing and several phylogenetic reconstructions accounting for among-site rate variation (see later).

Three main approaches have been used to detect or reduce the impact of LBA:

1. *Increased or modified taxonomic sampling.* Early in the development of molecular systematics, it was proposed that addition of species would reduce the impact of LBA by the breaking of long branches (Hendy and Penny, 1989). The eukaryotic phylogeny based on elongation factor 1α provides a good example of this approach, although the resolution of the tree becomes very weak as species number increases (Moreira et al., 1999). Despite concerns about its theoretical basis (Poe and Swofford, 1999; Zwickl and Hillis, 2002), practice shows that enlargement of the taxonomic spectrum is an efficient approach. Alternatively, when several representatives are available for a given group, it is sensible, rather than just increasing the sample size, to perform a more accurate choice of taxa, by retaining only the slowest-evolving ones. For example, the monophyly of molting animals was established by focusing on a slowly evolving nematode species, *Trichinella* (Aguinaldo et al., 1997).
2. *Restriction of analysis to slowly evolving positions.* The LBA artifact can also be circumvented by using the slowest-evolving positions (Olsen, 1987), as they are less prone to contain noise (i.e., multiple substitutions, see Figure 6.2) and therefore more likely to retain the ancient phylogenetic signal (Felsenstein, 2001). Our group has recently developed the slow–fast (SF) method to detect these sites (Brinkmann and Philippe, 1999). Briefly, parsimony analysis is employed to calculate the number of changes for each position within predefined monophyletic groups. Then, the evolutionary rate of a position is estimated as the sum of the number of substitutions it harbors within each group, and is thus independent from intergroup relationships. As the number of substitutions per site for which the noise exceeds the signal is not known *a priori*, several alignments are made consisting of sites covering a complete range of substitution thresholds, and changes in phylogenetic inference between datasets are followed accordingly. If a fast-evolving group is misplaced as an early branch with standard reconstruction methods because of LBA, its true position can be recovered by the slowest sites, and it will progressively emerge earlier and earlier as faster-evolving positions are used. Although the reliable signal is scarce and the resolving power often limited, the SF method is very efficient in detecting groups that are very likely misplaced because of LBA (Brinkmann and Philippe, 1999; Philippe et al., 2000).
3. *Use of improved models of sequence evolution.* The primary cause of the LBA artifact is the underestimation of the true number of substitutions, especially for fast-evolving sequences (Olsen, 1987). In probabilistic approaches, this underestimation occurs when the model of sequence evolution underlying the reconstruction process is unrealistic. The first model that was used to correct for multiple substitutions treated all types of substitutions as equivalent and assumed that all sites have the same probability to accept mutations (Jukes and Cantor, 1969). Some progress has been made since, by differentiating the types of substitutions (e.g., transitions vs. transversions; Swofford et al., 1996). Yet the most significant improvement was obtained when the unrealistic assumption of an equal substitution rate for all sites was bypassed (Yang, 1996a). For example, because the presence of invariant sites, although uninformative, violates the equal-rate-across-site assumption, their identification and removal greatly improves phylogenetic reconstruction by distance and likelihood methods (Lockhart et al., 1996). For instance, the support for an early emergence of Microsporidia in phylogenies based on elongation factors decreased when invariant sites were removed (Hirt et al., 1999). Currently, the most widely used method is the implementation of a rate across site (RAS) model, which approximates a continuous model of substitution rate heterogeneity on a gamma distribution. This approach has proved successful in confirming the reassignment of Microsporidia as highly derived Fungi (Van de Peer et al., 2000).

6.2.4 Heterotachy

RAS models handle substitution rate variation within the molecule, but assume that the rate of a position remains the same throughout time (i.e., on all branches of a phylogenetic tree). We called this assumption homotachy, for same speed (Lopez et al., 2002). Homotachy was shown to be unrealistic as early as 30 years ago, with the pioneering work of Fitch, who demonstrated that in cytochrome *c*, substitution rates at positions are differently distributed across the molecule between Fungi and Metazoa (Fitch, 1971). A single model of sequence evolution may explain this observation, i.e., the so-called covarion model. This model assumes that, at a given time, only a fraction of positions (the *co*ncomitantly *var*iable cod*ons*, or covarions) are subjected to variation, and this fraction might change over time. As covarion changes occur independently in different lineages, the covarion model allows sequence sites to display different rates of substitutions in separate parts of a phylogeny (Fitch and Markowitz, 1970). Rejection of homotachy is often referred to as a covarion-like behavior (Lockhart et al., 1996; Penny et al., 2001; Galtier, 2001).

Large taxonomic samples are needed to reject the assumption of homotachy (Lopez et al., 1999) by accurately inferring evolutionary rate variations at each site of a protein. The ongoing enlargement of sequence databases permits more efficient testing. We have recently analyzed ca. 2000 sequences of vertebrate cytochrome *b* and shown that for 95% of the variable positions, substitution rates over time are significantly heterogeneous (Lopez et al., 2002). To describe this phenomenon, we used the term heterotachy instead of covarion-like, because the covarion model does not fit the evolutionary pattern of cytochrome *b* sequences (Lopez et al., 2002). As it is only one among many possible heterotachous models, this is not unexpected.

Although heterotachous substitution patterns can give rise to phylogenetically informative signals in sequence data (Lopez et al., 1999), they can sometimes represent a major source of homoplasy. For instance, the analysis of 16S rDNA and tufA sequences from a range of nonphotosynthetic prokaryotes and oxygenic photosynthetic prokaryotes and eukaryotes showed that the sharing of similar covarion sets can bias the correct assessment of chloroplast origins (Lockhart et al., 1998). Because programs able to handle the heterotachy process are not yet available, a simple method to increase the fit of the RAS model to the data is to remove heterotachous positions from the alignment. When applied to fused SSU and LSU rRNAs, this approach showed that a profoundly revised phylogeny of eukaryotes (see Figure 6.5) could not be rejected (Philippe and Germot, 2000). Reconstruction of ancient phylogenies can especially benefit from the implementation of this method, because covarion-like processes could be responsible for the preservation of a deep phylogenetic signal even when mutational saturation has almost completely erased the evolutionary trace (e.g., when a substitution has occurred in an early branching lineage and has remained unchanged since; Lopez et al., 1999; Penny et al., 2001). Heterotachy is very complex to model, but its implementation in phylogeny reconstruction techniques is the focus of active investigation (Penny et al., 2001; Galtier, 2001; Huelsenbeck, 2002).

6.2.5 Rare Genomic Events as an Alternative Approach?

Faults in phylogeny reconstruction methodologies might be eschewed by employing a Hennigian approach (i.e., the use of characters not sensitive to saturation; Philippe and Laurent, 1998). Rare genomic changes (e.g., insertions or deletions, or indels) might represent good markers of common descent (Rokas and Holland, 2000; see also Chapter 4 and Chapter 9). For instance, some promising results were obtained in mammalian phylogenies by using patterns of retroposon integrations (Nikaido et al., 1999). Similarly, intron positions were employed for vertebrates (Venkatesh et al., 1999) and gene order for animals (Boore and Brown, 1998). In the case of ancient phylogenies, four indels were used to support the monophyly of Opisthokonta (animals + fungi; Baldauf and Palmer, 1993). One of these indels was also present in Microsporidia, in agreement with the proximity of this group to Fungi (Hashimoto and Hasegawa, 1996). Nevertheless, short indels can be misleading phylogenetic markers, as they are highly prone to convergence. They usually occur on protein regions (i.e., surface loops) that can easily accommodate insertion or deletion of stretches of a few hydrophilic amino acids. On these grounds, the reliability of a two-amino-acids indel in enolase, claimed to support the early emergence of trichomonads in eukaryote evolution (Keeling and Palmer, 2000), has been questioned (Hannaert et al., 2000; Bapteste and Philippe, 2002). Is this reasoning also applicable to large homologous indels?

The eukaryotic gene for Valyl-tRNA synthetase (Val RS) (which is targeted both to the cytoplasm and mitochondria in fungi) is considered to be of likely mitochondrial origin because, in Val RS phylogenies, eukaryotic sequences formed a monophyletic sister group to proteobacteria (Brown and Doolittle, 1995). This affiliation was further supported by a very conserved large insertion (37 amino acids) shared by eukaryotes and proteobacteria of the gamma subdivision, consistently with the phylogeny based on this gene (Hashimoto et al., 1998). A lateral gene transfer with replacement from Archaea to Rickettsiaceae was later inferred from the first alpha-proteobacterial sequence (*Rickettsia*) that became publicly available (Woese et al., 2000). While waiting for genuine Val RS representatives from other members of the alpha subdivision, a mitochondrial origin for this gene remained the most reasonable interpretation. To test this, we built an updated Val RS data set containing 185 sequences, comprising many alpha-Proteobacteria currently available. A phylogenetic analysis of Val RS (Figure 6.4A) indicates that alpha- and beta/gamma-Proteobacteria form two sister monophyletic groups, as expected. However, eukaryotes do not show any specific affiliation to alpha-Proteobacteria, but emerge earlier, possibly because of an LBA artifact. Moreover, although present in all beta/gamma-Proteobacteria and eukaryotes, the large insertion is surprisingly absent in alpha-Proteobacteria (Figure 6.4B). At first sight, this fact would discard a mitochondrial origin of eukaryotic Val-RS and rather suggest an LGT from beta/gamma-Proteobacteria. However, another well-conserved insertion of six amino acids is shared by a monophyletic group consisting of four gamma-Proteobacteria and two representatives of the alpha subdivision (*Sinorhizobium* and *Agrobacterium*), which are nonetheless grouped with high statistical confidence together with their insertless kin. Although longer and less conserved, the same insertion is also present in one member of high G+C Gram-positives (*Mycobacterium*). Therefore, because these two large indels in Val RS sequences contradict both each other as well as the phylogeny, their reliability is dubious and they leave the question of the origin of eukaryotic Val RS unanswered.

Finally, rare genomic events can be as sensitive to LGT as standard phylogenetic inference. For example, a specific N-terminal indel in HSP70 has long been taken as evidence for the grouping of Archaea and Gram-positive bacteria and the hypothesis of a chimerical origin of eukaryotes, all the more so that HSP70 phylogenies were claimed to support this view (Gupta and Singh, 1994). However, with the use of a larger taxonomic spectrum, it was shown that (1) several Gram-negative bacteria harbor a copy of HSP70 with the deletion specific to Gram-positive bacteria (Philippe et al., 1999) and (2) most archaeal species do not possess any HSP70, suggesting that Archaea are originally devoid of this gene (Gribaldo et al., 1999). Indeed, from currently available complete genomes, the taxonomic distribution of HSP70 (i.e., its absence in all Crenarchaeota, Pyrococci and *Methanococcus*) strongly suggests a single LGT from Gram-positive Bacteria, subsequent to the emergence of *Methanococcus*. In conclusion, rare genomic changes can be helpful Hennigian markers in reconstructing organismal phylogenies only when they are used to complement sequence data (Rokas and Holland, 2000), but their use as per se kinship indicators (Gupta and Johari, 1998) should be treated with great caution.

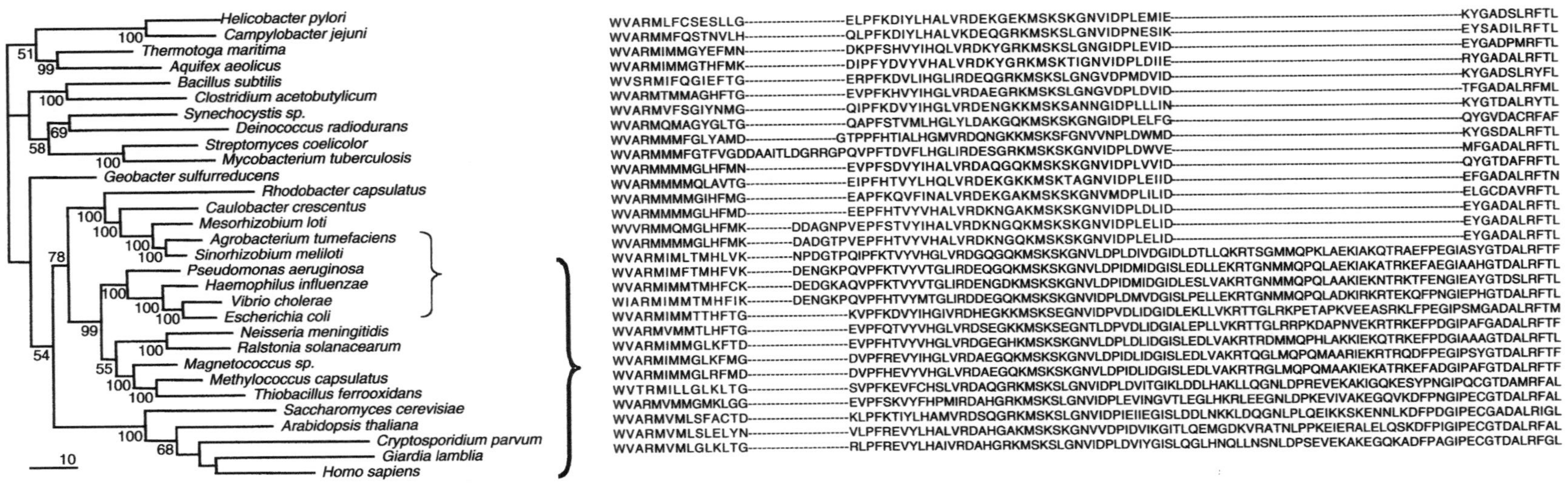

FIGURE 6.4 Rare genomic events and the evolution of Val tRNA synthetase. (A) A representative sample of 30 sequences was selected from a dataset comprising 185 ValRS sequences encompassing a broad taxonomic sampling from the three domains. The tree was inferred with the maximum likelihood method with protml and bootstrap values (on the left on the nodes) were computed with the RELL method [Adachi and Hasegawa (1996) 1–150]. Proteobacteria of the alpha- and beta/gamma subdivisions form two sister monophyletic groups, as expected. However, eukaryotes do not show any specific affiliation to alpha-Proteobacteria. (B) An excerpt of the ValRS alignment focusing on the region encompassing two highly conserved insertions. The large insertion of 37 residues thought to be a hallmark of the mitochondrial origin of eukaryotic ValRS [Hashimoto et al. (1998)], although present in all beta/gamma-Proteobacteria and eukaryotes, is surprisingly absent from alpha-Proteobacteria. Moreover, the distribution of another well-conserved insertion of six amino acids is evidently incongruent with the phylogeny presented in Figure 4A. In fact, it is shared by four gamma-Proteobacteria and two beta-Proteobacteria, which do not cluster at all. Moreover, the taxonomic distribution of this indel is incongruent with that of the large indel. Although longer and less conserved, an insertion at the same position of the insertion of the six residues is also unevenly harbored by members of high G+C Gram-positives (i.e., present in *Mycobacterium* but absent in *Streptomyces*).

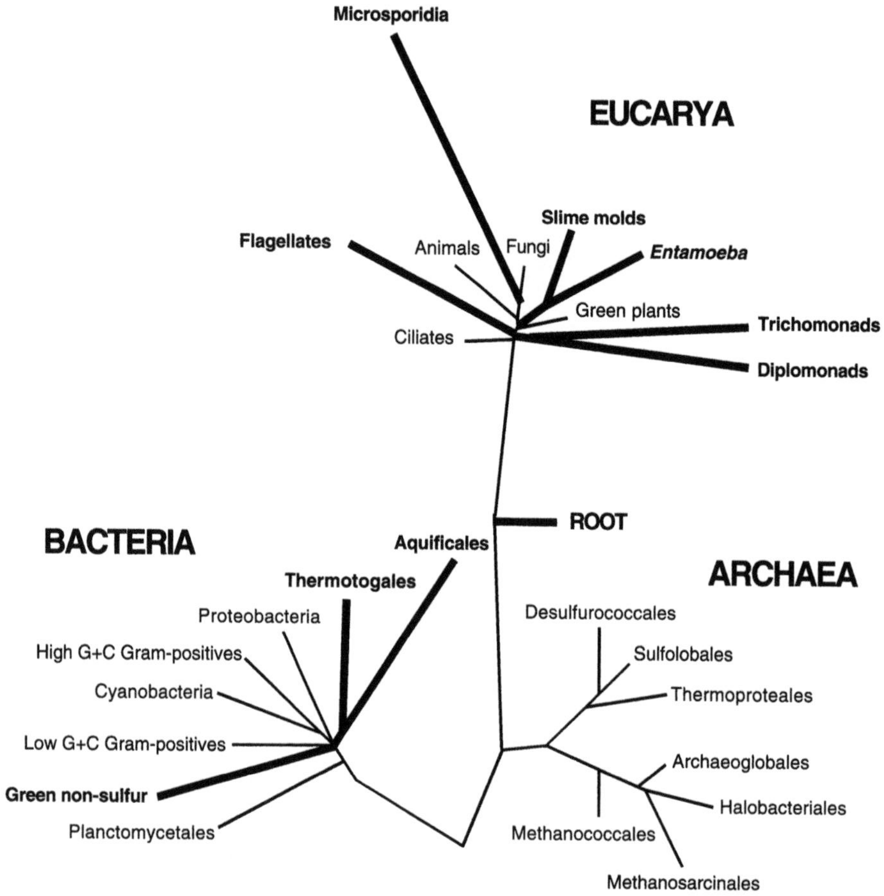

FIGURE 6.5 A revisited view of the eukaryotic tree. The taxa that are misplaced because of LBA on Figure 6.1 are relocated on the phylogeny, based either on the reanalysis of either rRNA (e.g., hyperthermophilic bacteria) or of anciently duplicated genes (e.g., the root) or on the analysis of new markers (e.g., the grouping of slime molds with *Entamoeba*). Although some of these revisions are robust (e.g., microsporidia within fungi), some others remain less settled.

6.3 LGTs and the Quest for a Species Phylogeny

The sequencing of complete genomes has shown that phylogenies inferred from different proteins are often contradictory both between themselves and with the SSUrRNA tree (Pennisi, 1998). LGTs, rather than tree-reconstruction artifacts, could explain these incongruencies. Although aberrant nucleotide composition and unexpected BLAST hits, on which the main methods employed to detect LGTs are based, are poor indicators (Koski and Golding, 2001; Koski et al., 2001, Wang, 2001; Guindon and Perriere, 2001; Ragan, 2001), it is indisputable that LGTs appear to be very common and to affect all sorts of genes (Ochman et al., 2000; Koonin et al., 2001). It has even been suggested that the history of life cannot be properly represented as a tree (Doolittle, 1999b).

To test the very existence of an organismal phylogeny, several groups have turned to genomic approaches (Wolf et al., 2001; Fitz-Gibbon and House, 1999; Tekaia et al., 1999; Lin and Gerstein, 2000; Korbel et al., 2002; also see Chapter 9). Generally based on gene content (presence or absence) or on gene order, these methods produce phylogenies that are more or less consistent with the SSUrRNA tree. For example, the monophyly of the three domains and of several other major groups (e.g., animals, spirochaetes, alpha-beta- Proteobacteria) are generally recovered. Yet, unexpected relationships, such as that of *Thermoplasma* with Crenarchaeotes (Korbel et al., 2002), testify to the biases that might be introduced in such approaches by extensive LGTs among phylogenetically unrelated taxa thriving in the same ecological niches (Ruepp et al., 2000). A striking example of such massive LTGs is given by hyperthermophilic bacteria, which are proposed to have acquired up to ca. 24% of their genes from Archaea (Aravind et al., 1998; Nelson et al., 1999). Additionally, convergent loss or acquisition of similar sets of genes because of close physiological requirements, such as adaptation to intracellular parasitism, could also generate biases in genomic approaches. In conclusion, whole genome-based methodologies should be regarded more as phenetic than phylogenetic approaches (Wolf et al., 2001; Doolittle, 1999a).

Even if LGTs are frequent, it is possible that only a subset of the gene pool might be affected by this phenomenon, to the exclusion of a bona fide nontransferable genetic core, which would mirror the organismal phylogeny. The complexity hypothesis states that informational gene products (i.e., those involved in transcription, translation and related processes) are typically members of large, complex systems, making horizontal transfer of informational genes less probable (Jain et al., 1999). Our group recently developed a method for evaluating the congruence among genes without the use of any predefined starting phylogeny and applied it to the proteins involved in translation from 45 complete bacterial genomes (Brochier et al., 2002) and from 14 archaeal genomes (Matte-Tailliez et al., 2002). The identification by this approach of a genuine core of 52 genes that appeared nontransferred among the 45 species under analysis argues for the existence of an organismal phylogeny of Bacteria. Notably, the phylogenies based on the concatenation of the nontransferred proteins are very congruent with the one based on the two rRNA genes for both Bacteria (Brochier et al., 2002) and Archaea (Matte-Tailliez et al., 2002). Therefore, although LGTs are frequent and can affect rRNA genes (Yap et al., 1999), they do not imply in themselves that a reevaluation of the universal tree of life based on rRNA is necessary (Figure 6.1).

Rather than the numerously detected LGTs, the identification of tree reconstruction artifacts has had more impact on the rRNA tree. Two reappraisals of the tree of Figure 6.1 have potential major impact on our view of ancient evolution (Figure 6.5). First, while the deepest branches in the bacterial domains are occupied by hyperthermophilic species (Aquificales and Thermotogales), concurring with the hypothesis that the LUCA was a hyperthermophile (Stetter, 1996), the analysis of slowly evolving positions of SSUrRNA demonstrates that the early branching of Aquificales and Thermotogales is due to LBA (Brochier and Philippe, 2002). This strongly argues for a secondary adaptation to high temperature in these lineages (Forterre et al., 2000) and for a nonhyperthemophilic LUCA, in agreement with other evidence (Galtier et al., 1999). Interestingly, the same analysis suggests that Planctomycetales might be the earliest emerging bacterial lineage (Brochier and Philippe, 2002). This phylum is a major division of Bacteria, whose members share several interesting characteristics, such as lack of peptidoglycan in their cell walls and a budding mode of reproduction (Fuerst, 1995). Another very intriguing feature is the existence of a single or double membrane around the bacterial chromosome in *Gemmata* and *Pirellula* species, respectively, which has been compared to the eukaryotic nuclear envelope (Fuerst, 1995). If the early emergence of Planctomycetales is confirmed by the analysis of additional markers and if their nucleus-like structure appears homologous to the eukaryotic nucleus, the origin of Bacteria should be seriously reconsidered. Second, there is good evidence that the rooting of the universal tree of life in the bacterial branch is mainly due to LBA (Brinkmann and Philippe, 1999; Lopez et al., 1999). However, the evidence for alternative rooting (in particular, an eukaryotic rooting, i.e., a monophyly of Prokaryotes, Figure 6.5) remains weak, as slow-evolving positions — that might generate a strong statistical signal — are too few. The implications of a eukaryotic rooting are thoroughly discussed elsewhere (Philippe and Forterre, 1999; Penny and Poole, 1999; Forterre and Philippe, 1999a,b) and will not be repeated here. However, it seems now that the traditional rooting of the universal tree of life should at least be considered as an open question.

6.4 Toward a Resolution of the Eukaryotic Phylogeny?

Similar to the problems described with molecular phylogenetic reconstruction and of their impact on prokaryotic phylogeny, there is particular concern regarding the eukaryotic section of the tree of life, because virtually all basal branches might be erroneous (Figure 6.1). LGTs, even if they might be more frequent than generally thought (Andersson et al., 2003), have not been shown to constitute a major issue for eukaryotes and we will not discuss this point further. The impact of multiple paralogous genes is a more important issue, especially for the most commonly used markers (e.g., actin and tubulins). Even if the use of paralogous instead of orthologous genes likely explains some unexpected results [e.g., the paraphyly of Edysozoa (Baldauf et al., 2000)], this problem is difficult to handle and will not be discussed here.

6.4.1 The Archezoa Hypothesis

The phylogeny of eukaryotes based on SSUrRNA (Figure 6.1) can be roughly divided into a lower section, displaying the successive emergence of several amitochondriate lineages (such as microsporidia, diplomonads, trichomonads), followed by mitochondriate ones (such as Euglenozoa, slime molds), and an upper section, encompassing all later-emerging organisms organized in a vast unresolved radiation, which was called the crown (Sogin, 1991; Knoll, 1992). The early emergence of amitochondriate protists agrees with the Archezoa hypothesis (Cavalier-Smith, 1987), stating that several lineages (microsporidia, diplomonads, trichomonads and Archaemoeba) emerged before mitochondrial endosymbiosis. Yet the discovery of genes of mitochondrial origin in all these lineages (for a review see Embley and Hirt, 1998) has rejected this hypothesis. Interestingly, antibodies raised against these proteins detect the presence of tiny organelles with double membranes in *Entamoeba* (Tovar et al., 1999; Mai et al., 1999) and microsporidia (Williams et al., 2002; also see Chapter 10 and Chapter 13). This finding of relict mitochondria provides evidence for the reluctance of eukaryotes to lose the mitochondrial organelle, even when its canonical function of aerobic respiration has been apparently lost. However, the presence of a relict mitochondrion remains an open question for diplomonads (Lloyd et al., 2002). Moreover, the early emergence of the Archezoa lineages, which constitutes the second basis of the premitochondrial hypothesis, is highly suspect, given the very long branch of the prokaryotic outgroup (Figure 6.1). Many independent analyses have demonstrated that all basal branches in the SSUrRNA phylogeny are likely misplaced because of LBA artifacts. [For a review, see Philippe (2000) and Chapter 2]. Consequently, finding the correct locations of all the lineages for which SSUrRNA sequences evolve fast represents a long-standing challenge to the resolution of the eukaryotic phylogeny.

6.4.2 Single-Gene Approaches

All eukaryotic single-gene phylogenies suffer from a very poor resolution among the major eukaryotic lineages (Budin and Philippe, 1998). The SSUrRNA tree provides the best-known example, leading to the hypothesis of a rapid burst of evolution (Knoll, 1992). The uncertainty in the branching orders of major eukaryotic groups increases with the adding of taxa and with the use of more sophisticated tree reconstruction methods. For example, in a phylogeny based on 55 species harboring a slow-evolving SSUrRNA, bootstrap supports for the internal branches are all but one less than 30% (Brugerolle et al., 2002). The situation is even worse when fast-evolving sequences are included (Simpson et al., 2002). Indeed, even the concatenation of the small and large subunits of rRNA does not allow discrimination between quite different eukaryotic phylogenies, such as the ones shown in Figure 6.1 and Figure 6.5 (Philippe and Germot, 2000).

However, single-gene analyses have shed some light on the resolution of the eukaryotic phylogeny. For example, microsporidia have been grouped with fungi on the basis of tubulin (Edlind et al., 1996), mitochondrial HSP70 (Germot et al., 1997) and RNA polymerase (Hirt et al., 1999) phylogenies. Trees based on actin have suggested a grouping of foraminiferans with Cercozoa (Keeling, 2001). An insertion of one or two amino acids in the highly conserved ubiquitin provides further evidence in favor of this grouping (Archibald et al., 2003). To circumvent the limited resolving power of phylogenies based on single genes, an obvious measure is the simultaneous use of several genes.

6.4.3 Multigene Approaches

The first application of the use of large datasets to eukaryotic phylogenies dates back to the mid-1990s (Nikoh et al., 1994; Kuma et al., 1995). However, these early analyses suffered from a very constrained taxonomic sampling (four or five species), leading to limited phylogenetic inference. The use of a larger number of species necessitated a decrease in the number of genes used, leading to further badly resolved phylogenies (Budin and Philippe, 1998; Germot and Philippe, 1999). A notable exception is a phylogeny based on concatenated mitochondrial genes (cob, cox1, cox2 and cox3), which provided strong support for the monophyly of Opisthokonta (animals + fungi) and Plantae (red algae + green plants; Burger et al., 1999). The recent burst of sequence data from a much wider taxonomic spectrum has led to a proliferation of multigene analyses, in particular of angiosperms (Chase and Fay, 2001), mammals (Madsen et al., 2001; Murphy et al., 2001) and eukaryotes (Moreira et al., 2000; Baldauf et al., 2000; Arisue et al., 2002a; Fast et al., 2002; Arisue et al., 2002b; Bapteste et al., 2002; Lang et al., 2002).

For instance, strong support for the grouping of three highly divergent amoebae (*Dictyostelium*, *Entamoeba* and *Mastigamoeba*; Arisue et al., 2002a; Bapteste et al., 2002) is in sharp contrast with their pronounced polyphyly observed in early SSUrRNA-based phylogenies (Sogin, 1991). This evidence suggests that the origin of the amoebic phenotype should be reconsidered, although detailed molecular characterization of many more amoeboid lineages (e.g., *Acanthamoeba*, *Pelomyxa*, *Amoeba*) is still lacking.

Important questions on eukaryotic evolution (e.g., acquisition or loss of organelles) have been tackled by several recent analyses. The use of large gene fusions (Moreira et al., 2000; Burger et al., 1999) has provided significant support for a single primary cyanobacterial endosymbiosis at the origin of chloroplasts, by showing a monophyly of Plantae (green plants, red algae and glaucophytes; see Chapter 3 and Chapter 4). Similarly, the clustering of the two major groups that harbor a chloroplast surrounded by four membranes, i.e., alveolates and stramenopiles (Baldauf et al., 2000; Arisue et al., 2002b; Bapteste et al., 2002; Fast et al., 2001), strongly suggests that secondary endosymbioses (i.e., engulfment of a photosynthetic eukaryote by another eukaryote) are much less frequent than previously thought. This way, only two endosymbiotic events would have occurred, as opposed to the seven previously hypothesized (Cavalier-Smith, 1999). More data on haptophytes, cryptophytes, chloroarachniophytes and euglenids are needed to settle this question.

The groupings of some amitochondriate eukaryotes, i.e., *Entamoeba* and *Mastigamoeba*, (Arisue et al., 2002a; Bapteste et al., 2002), or trichomonads and diplomonads (Baldauf et al., 2000; Henze et al., 2001; Embley and Hirt, 1998), have similarly reduced the number of secondary losses of typical mitochondria. Thus, independent losses of mitochondria and of chloroplasts are very likely more frequent than their acquisition through endosymbiosis.

When constructing phylogenies from concatenated sequences, the implicit assumption is made that branch lengths are homogeneous for all genes (Yang, 1996b). This assumption is clearly incorrect, especially for the most frequently used markers (Moreira et al., 2002). One sensible approach (Yang, 1996b) was recently indicated to circumvent this problem. In brief, the likelihood for each gene is independently computed, allowing different sets of parameters (i.e., branch lengths and alpha parameter of the gamma law), and the best topology is the one that maximizes the sum of all likelihood values. With this method, for a data set of 123 genes (totaling up to 25,000 amino acid positions), the fit to the data was significantly improved as compared with that by traditional concatenation, despite the large number of free parameters (ca. 7000; Bapteste et al., 2002). This approach represents a first step toward improving information extraction from large amount of data.

6.5 Perspective

Since its first theoretical enunciation (Zuckerkandl and Pauling, 1965) and its consecration as a novel and promising discipline (Woese and Fox, 1977), molecular phylogenetics has experienced a difficult stage of growth at the end of the 1990s, when most of the problems described in this chapter, and their significant impact on ancient phylogenies, were progressively highlighted. Several erroneous inferences (e.g., the placements of amitochondriate eukaryotes, of hyperthermophilic bacteria and potentially of the root of the universal tree of life) have been acknowledged, leading to the challenging of some well-settled hypotheses. Nonetheless, this has opened new avenues of research for the origin of life and the early diversification of organisms. Molecular phylogeneticists can now look at the future with optimism, thanks to the high throughput of sequences (i.e., genomes from model organisms and cDNA EST from interesting, but less tractable, taxa) and to progresses in the implementation of tree reconstruction methods (e.g., by accounting for heterotachy). This will ultimately lead to the inference of a well-resolved universal tree, to the characterization of the dynamics of LGTs and to a better understanding of the molecular mechanisms at the origin of the major evolutionary transitions.

Acknowledgment

SG was supported by a *poste de chercheur associé* from CNRS.

References

Adachi, J. and Hasegawa, M. (1996) MOLPHY version 2.3: Programs for molecular phylogenetics based on maximum likelihood. *Comput. Sci. Monogr.* 28: 1–150.

Aguinaldo, A. M., Turbeville, J. M., Linford, L. S., Rivera, M. C., Garey, J. R., Raff, R. A. and Lake, J. A. (1997) Evidence for a clade of nematodes, arthropods and other moulting animals. *Nature* 387: 489–493.

Andersson, J. O., Sjogren, A. M., Davis, L. A., Embley, T. M. and Roger, A. J. (2003) Phylogenetic analyses of diplomonad genes reveal frequent lateral gene transfers affecting eukaryotes. *Curr. Biol.* 13: 94–104.

Aravind, L., Tatusov, R. L., Wolf, Y. I., Walker, D. R. and Koonin, E. V. (1998) Evidence for massive gene exchange between archaeal and bacterial hyperthermophiles. *Trends Genet.* 14: 442–444.

Archibald, J. M., Longet, D., Pawlowski, J. and Keeling, P. J. (2003) A novel polyubiquitin structure in cercozoa and foraminifera: Evidence for a new eukaryotic supergroup. *Mol. Biol. Evol.* 20: 62–66.
Arisue, N., Hashimoto, T., Lee, J. A., Moore, D. V., Gordon, P., Sensen, C. W., Gaasterland, T., Hasegawa, M. and Müller, M. (2002a) The phylogenetic position of the pelobiont *Mastigamoeba balamuthi* based on sequences of rDNA and translation elongation factors EF-1a and EF-2. *J. Euk. Microbiol.* 49: 1–10.
Arisue, N., Hashimoto, T., Yoshikawa, H., Nakamura, Y., Nakamura, G., Nakamura, F., Yano, T.-A. and Hasegawa, M. (2002b) Phylogenetic position of *Blastocystis hominis* and of stramenopiles inferred from multiple molecular sequence data. *J. Euk. Microbiol.* 49: 42–53.
Baldauf, S. L. and Palmer, J. D. (1993) Animals and fungi are each others closest relatives: Congruent evidence from multiple proteins. *Proc. Natl. Acad. Sci. USA* 90: 11558–11562.
Baldauf, S. L., Roger, A. J., Wenk-Siefert, I. and Doolittle, W. F. (2000) A kingdom-level phylogeny of eukaryotes based on combined protein data. *Science* 290: 972-977.
Bapteste, E., Brinkmann, H., Lee, J. A., Moore, D. V., Sensen, C. W., Gordon, P., Durufle, L., Gaasterland, T., Lopez, P., Muller, M. and Philippe, H. (2002) The analysis of 100 genes supports the grouping of three highly divergent amoebae: *Dictyostelium*, *Entamoeba*, and *Mastigamoeba*. *Proc. Natl. Acad. Sci. USA* 99: 1414–1419.
Bapteste, E. and Philippe, H. (2002) The potential value of indels as phylogenetic markers: Position of trichomonads as a case study. *Mol. Biol. Evol.* 19: 972–977.
Boore, J. L. and Brown, W. M. (1998) Big trees from little genomes: Mitochondrial gene order as a phylogenetic tool. *Curr. Opin. Genet. Dev.* 8: 668–674.
Brinkmann, H. and Philippe, H. (1999) Archaea sister group of Bacteria? Indications from tree reconstruction artifacts in ancient phylogenies. *Mol. Biol. Evol.* 16: 817–825.
Brochier, C., Bapteste, E., Moreira, D. and Philippe, H. (2002) Eubacterial phylogeny based on translational apparatus proteins. *Trends Genet.* 18: 1–5.
Brochier, C. and Philippe, H. (2002) Phylogeny: A non-hyperthermophilic ancestor for bacteria. *Nature* 417: 244.
Brown, J. R. and Doolittle, W. F. (1995) Root of the universal tree of life based on ancient aminoacyl-tRNA synthetase gene duplications. *Proc. Natl. Acad. Sci. USA* 92: 2441–2445.
Brown, J. R. and Doolittle, W. F. (1997) Archaea and the prokaryote-to-eukaryote transition. *Microbiol. Mol. Biol. Rev.* 61: 456–502.
Brugerolle, G., Bricheux, G., Philippe, H. and Coffe, G. (2002) *Collodictyon triciliatum* and *Diphylleia rotans* (= *Aulacomonas submarina*) form a new family of flagellates (Collodictyonidae) with tubular mitochondrial cristae that is phylogenetically distant from other flagellate groups. *Protist* 153: 59–70.
Budin, K. and Philippe, H. (1998) New insights into the phylogeny of eukaryotes based on ciliate Hsp70 sequences. *Mol. Biol. Evol.* 15: 943–956.
Burger, G., Saint-Louis, D., Gray, M. W. and Lang, B. F. (1999) Complete sequence of the mitochondrial DNA of the red alga *Porphyra purpurea*. Cyanobacterial introns and shared ancestry of red and green algae. *Plant Cell* 11: 1675–1694.
Cavalier-Smith, T. (1987) Eukaryotes with no mitochondria. *Nature* 326: 332–333.
Cavalier-Smith, T. (1999) Principles of protein and lipid targeting in secondary symbiogenesis: Euglenoid, dinoflagellate, and sporozoan plastid origins and the eukaryote familly tree. *J. Euk. Microbiol.* 46: 347–366.
Chase, M. W. and Fay, M. F. (2001) Ancient flowering plants: DNA sequences and angiosperm classification. *Gen. Biol.* 2: REVIEWS1012.
Doolittle, W. F. (1999a) Lateral gene transfer, genome surveys, and the phylogeny of prokaryotes. Technical comment on M. Huynen et al. *Science* 286: 1443.
Doolittle, W. F. (1999b) Phylogenetic classification and the universal tree. *Science* 284: 2124–2129.

Edlind, T. D., Li, J., Visvesvara, G. S., Vodkin, M. H., McLaughlin, G. L. and Katiyar, S. K. (1996) Phylogenetic analysis of beta-tubulin sequences from amitochondrial protozoa. *Mol. Phylogenet. Evol.* 5: 359–367.

Embley, T. M. and Hirt, R. P. (1998) Early branching eukaryotes? *Curr. Opin. Genet. Dev.* 8: 624–629.

Fast, N. M., Kissinger, J. C., Roos, D. S. and Keeling, P. J. (2001) Nuclear-encoded, plastid-targeted genes suggest a single common origin for apicomplexan and dinoflagellate plastids. *Mol. Biol. Evol.* 18: 418–426.

Fast, N. M., XUE, L., Bingham, S. and Keeling, P. J. (2002) Re-examining alveolate evolution using multiple protein molecular phylogenies. *J. Euk. Microbiol.* 49: 30–37.

Felsenstein, J. (1978) Cases in which parsimony or compatibility methods will be positively misleading. *Syst. Zool.* 27: 401–410.

Felsenstein, J. (2001) Taking variation of evolutionary rates between sites into account in inferring phylogenies. *J. Mol. Evol.* 53: 447–55.

Fitch, W. M. (1971) The nonidentity of invariable positions in the cytochromes c of different species. *Biochem. Genet.* 5: 231–241.

Fitch, W. M. and Markowitz, E. (1970) An improved method for determining codon variability in a gene and its application to the rate of fixation of mutations in evolution. *Biochem. Genet.* 4: 579–593.

Fitz-Gibbon, S. T. and House, C. H. (1999) Whole genome-based phylogenetic analysis of free-living microorganisms. *Nucleic Acids Res.* 27: 4218–4222.

Forterre, P., Bouthier De La Tour, C., Philippe, H. and Duguet, M. (2000) Reverse gyrase from hyperthermophiles: Probable transfer of a thermoadaptation trait from archaea to bacteria. *Trends Genet.* 16: 152–154.

Forterre, P. and Philippe, H. (1999a) The last universal common ancestor (LUCA), simple or complex? *Biol. Bull.* 196: 373–375; discussion 375–377.

Forterre, P. and Philippe, H. (1999b) Where is the root of the universal tree of life? *BioEssays* 21: 871–879.

Foster, P. G. and Hickey, D. A. (1999) Compositional bias may affect both DNA-based and protein-based phylogenetic reconstructions. *J. Mol. Evol.* 48: 284–290.

Fuerst, J. A. (1995) The planctomycetes: Emerging models for microbial ecology, evolution and cell biology. *Microbiology* 141: 1493–1506.

Galtier, N. (2001) Maximum-likelihood phylogenetic analysis under a covarion-like model. *Mol. Biol. Evol.* 18: 866–873.

Galtier, N. and Gouy, M. (1995) Inferring phylogenies from DNA sequences of unequal base compositions. *Proc. Natl. Acad. Sci. USA* 92: 11317–11321.

Galtier, N. and Lobry, J. R. (1997) Relationships between genomic G+C content, RNA secondary structures, and optimal growth temperature in prokaryotes. *J. Mol. Evol.* 44: 632–636.

Galtier, N., Tourasse, N. and Gouy, M. (1999) A nonhyperthermophilic common ancestor to extant life forms. *Science* 283: 220–221.

Germot, A. and Philippe, H. (1999) Critical analysis of eukaryotic phylogeny: A case study based on the HSP70 family. *J. Euk. Microbiol.* 46: 116–124.

Germot, A., Philippe, H. and Le Guyader, H. (1997) Evidence for loss of mitochondria in Microsporidia from a mitochondrial- type HSP70 in *Nosema locustae*. *Mol. Biochem. Parasitol.* 87: 159–168.

Gribaldo, S., Lumia, V., Creti, R., de Macario, E. C., Sanangelantoni, A. and Cammarano, P. (1999) Discontinuous occurrence of the hsp70 (dnaK) gene among Archaea and sequence features of HSP70 suggest a novel outlook on phylogenies inferred from this protein. *J. Bacteriol.* 181: 434–443.

Guindon, S. and Perriere, G. (2001) Intragenomic base content variation is a potential source of biases when searching for horizontally transferred genes. *Mol. Biol. Evol.* 18: 1838–1840.

Gupta, R. S. (1998) What are archaebacteria: Life's third domain or monoderm prokaryotes related to Gram-positive bacteria? A new proposal for the classification of prokaryotic organisms. *Mol. Microbiol.* 229: 695–708.

Gupta, R. S. and Johari, V. (1998) Signature sequences in diverse proteins provide evidence of a close evolutionary relationship between the deinococcus-thermus group and cyanobacteria. *J. Mol. Evol.* 46: 716–720.

Gupta, R. S. and Singh, B. (1994) Phylogenetic analysis of 70 kD heat shock protein sequences suggests a chimeric origin for the eukaryotic cell nucleus. *Curr. Biol.* 4: 1104–1114.

Hannaert, V., Brinkmann, H., Nowitzki, U., Lee, J. A., Albert, M. -A., Sensen, C. W., Gaasterland, T., Müller, M., Michels, P. and Martin, W. (2000) Enolase from *Trypanosoma brucei*, from the amitochondriate protist *Mastigamoeba balamuthi*, and from the chloroplast and cytosol of *Euglena gracilis*: Pieces in the evolutionary puzzle of the eukaryotic glycolytic pathway. *Mol. Biol. Evol.* 17: 989–1000.

Hasegawa, M. and Hashimoto, T. (1993) Ribosomal RNA trees misleading? *Nature* 361: 23.

Hashimoto, T. and Hasegawa, M. (1996) Origin and early evolution of eukaryotes inferred from the amino acid sequences of translation elongation factors 1alpha/Tu and 2/G. *Adv. Biophys.* 32: 73–120.

Hashimoto, T., Sanchez, L. B., Shirakura, T., Muller, M. and Hasegawa, M. (1998) Secondary absence of mitochondria in *Giardia lamblia* and *Trichomonas vaginalis* revealed by valyl-tRNA synthetase phylogeny. *Proc. Natl. Acad. Sci. USA* 95: 6860–6865.

Hendy, M. and Penny, D. (1989) A framework for the quantitative study of evolutionary trees. *Syst. Zool.* 38: 297–309.

Henze, K., Horner, D. S., Suguri, S., Moore, D. V., Sanchez, L. B., Muller, M. and Embley, T. M. (2001) Unique phylogenetic relationships of glucokinase and glucosephosphate isomerase of the amitochondriate eukaryotes Giardia intestinalis, Spironucleus barkhanus and Trichomonas vaginalis. *Gene* 281: 123–131.

Hirt, R. P., Logsdon, J. M., Jr., Healy, B., Dorey, M. W., Doolittle, W. F. and Embley, T. M. (1999) Microsporidia are related to fungi: Evidence from the largest subunit of RNA polymerase II and other proteins. *Proc. Natl. Acad. Sci. USA* 96: 580–585.

Huelsenbeck, J. P. (2002) Testing a covariotide model of DNA substitution. *Mol. Biol. Evol.* 19: 698–707.

Jain, R., Rivera, M. C. and Lake, J. A. (1999) Horizontal gene transfer among genomes: The complexity hypothesis. *Proc. Natl. Acad. Sci. USA* 96: 3801–3806.

Jukes, T. H. and Cantor, C. R. (1969) Evolution of protein molecules. In *Mammalian Protein Metabolism* (Munro, H. N., Ed.), Academic Press, New York, pp. 21–132.

Keeling, P. J. (2001) Foraminifera and Cercozoa are related in actin phylogeny: Two orphans find a home? *Mol. Biol. Evol.* 18: 1551–1557.

Keeling, P. J. and Fast, N. M. (2002) Microsporidia: Biology and evolution of highly reduced intracellular parasites. *Annu. Rev. Microbiol.* 56: 93–116.

Keeling, P. J. and Palmer, J. D. (2000) Parabasalian flagellates are ancient eukaryotes. *Nature* 405: 635–637.

Knoll, A. H. (1992) The early evolution of eukaryotes: A geological perspective. *Science* 256: 622–627.

Koonin, E. V., Makarova, K. S. and Aravind, L. (2001) Horizontal gene transfer in prokaryotes: Quantification and classification. *Annu. Rev. Microbiol.* 55: 709–42.

Korbel, J. O., Snel, B., Huynen, M. A. and Bork, P. (2002) SHOT: A web server for the construction of genome phylogenies. *Trends Genet.* 18: 158–162.

Koski, L. B. and Golding, G. B. (2001) The closest BLAST hit is often not the nearest neighbor. *J. Mol. Evol.* 52: 540–542.

Koski, L. B., Morton, R. A. and Golding, G. B. (2001) Codon bias and base composition are poor indicators of horizontally transferred genes. *Mol. Biol. Evol.* 18: 404–412.

Kuma, K., Nikoh, N., Iwabe, N. and Miyata, T. (1995) Phylogenetic position of *Dictyostelium* inferred from multiple protein datasets. *J. Mol. Evol.* 41: 238–246.

Lang, B. F., OKelly, C., Nerad, T., Gray, M. W. and Burger, G. (2002) The closest unicellular relatives of animals. *Curr. Biol.* 12: 1773–1778.

Lin, J. and Gerstein, M. (2000) Whole-genome trees based on the occurrence of folds and orthologs: Implications for comparing genomes on different levels. *Gen. Res.* 10: 808–818.

Lloyd, D., Harris, J. C., Maroulis, S., Wadley, R., Ralphs, J. R., Hann, A. C., Turner, M. P. and Edwards, M. R. (2002) The "primitive" microaerophile *Giardia intestinalis* (syn. *lamblia, duodenalis*) has specialized membranes with electron transport and membrane-potential-generating functions. *Microbiology* 148: 1349–1354.

Lockhart, P., Steel, M., Hendy, M. and Penny, D. (1994) Recovering evolutionary trees under a more realistic model of sequence evolution. *Mol. Biol. Evol.* 11: 605–612.

Lockhart, P. J., Larkum, A. W., Steel, M., Waddell, P. J. and Penny, D. (1996) Evolution of chlorophyll and bacteriochlorophyll: The problem of invariant sites in sequence analysis. *Proc Natl. Acad. Sci. USA* 93: 1930–1934.

Lockhart, P. J., Steel, M. A., Barbrook, A. C., Huson, D., Charleston, M. A. and Howe, C. J. (1998) A covariotide model explains apparent phylogenetic structure of oxygenic photosynthetic lineages. *Mol. Biol. Evol.* 15: 1183–1188.

Loomis, W. F. and Smith, D. W. (1990) Molecular phylogeny of *Dictyostelium discoideum* by protein sequence comparison. *Proc. Natl. Acad. Sci. USA* 87: 9093–9097.

Lopez, P., Casane, D. and Philippe, H. (2002) Heterotachy, an important process of protein evolution. *Mol. Biol. Evol.* 19: 1–7.

Lopez, P., Forterre, P. and Philippe, H. (1999) The root of the tree of life in the light of the covarion model. *J. Mol. Evol.* 49: 496–508.

Madsen, O., Scally, M., Douady, C. J., Kao, D. J., DeBry, R. W., Adkins, R., Amrine, H. M., Stanhope, M. J., de Jong, W. W. and Springer, M. S. (2001) Parallel adaptive radiations in two major clades of placental mammals. *Nature* 409: 610–614.

Mai, Z., Ghosh, S., Frisardi, M., Rosenthal, B., Rogers, R. and Samuelson, J. (1999) Hsp60 is targeted to a cryptic mitochondrion-derived organelle ("crypton") in the microaerophilic protozoan parasite *Entamoeba histolytica. Mol. Cell. Biol.* 19: 2198–2205.

Margulis, L. (1996) Archaeal-eubacterial mergers in the origin of Eukarya: Phylogenetic classification of life. *Proc. Natl. Acad. Sci. USA* 93: 1071–1076.

Martin, W. and Müller, M. (1998) The hydrogen hypothesis for the first eukaryote. *Nature* 392: 37–41.

Matte-Tailliez, O., Brochier, C., Forterre, P. and Philippe, H. (2002) Archaeal phylogeny based on ribosomal proteins. *Mol. Biol. Evol.* 19: 631–639.

Mayr, E. (1998) Two empires or three? *Proc. Natl. Acad. Sci. USA* 95: 9720–9723.

Moreira, D., Kervestin, S., Jean-Jean, O. and Philippe, H. (2002) Evolution of eukaryotic translation elongation and termination factors: Variations of evolutionary rate and genetic code deviations. *Mol. Biol. Evol.* 19: 189–200.

Moreira, D., Le Guyader, H. and Philippe, H. (1999) Unusually high evolutionary rate of the elongation factor 1 alpha genes from the Ciliophora and its impact on the phylogeny of eukaryotes. *Mol. Biol. Evol.* 16: 234–245.

Moreira, D., Le Guyader, H. and Philippe, H. (2000) The origin of red algae: Implications for the evolution of chloroplasts. *Nature* 405: 69–72.

Murphy, W. J., Eizirik, E., Johnson, W. E., Zhang, Y. P., Ryder, O. A. and OBrien, S. J. (2001) Molecular phylogenetics and the origins of placental mammals. *Nature* 409: 614–618.

Nelson, K. E., Clayton, R. A., Gill, S. R., Gwinn, M. L., Dodson, R. J., Haft, D. H., Hickey, E. K., Peterson, J. D., Nelson, W. C., Ketchum, K. A., McDonald, L., Utterback, T. R., Malek, J. A., Linher, K. D., Garrett, M. M., Stewart, A. M., Cotton, M. D., Pratt, M. S., Phillips, C. A., Richardson, D., Heidelberg, J., Sutton, G. G., Fleischmann, R. D., Eisen, J. A., Fraser, C. M. et al. (1999) Evidence for lateral gene transfer between Archaea and bacteria from genome sequence of *Thermotoga maritima. Nature* 399: 323–329.

Nikaido, M., Rooney, A. P. and Okada, N. (1999) Phylogenetic relationships among cetartiodactyls based on insertions of short and long interpersed elements: Hippopotamuses are the closest extant relatives of whales [see comments]. *Proc. Natl. Acad. Sci. USA* 96: 10261–10266.

Nikoh, N., Hayase, N., Iwabe, N., Kuma, K. and Miyata, T. (1994) Phylogenetic relationship of the kingdoms Animalia, Plantae, and Fungi, inferred from 23 different protein species. *Mol. Biol. Evol.* 11: 762–768.

Ochman, H., Lawrence, J. G. and Groisman, E. A. (2000) Lateral gene transfer and the nature of bacterial innovation. *Nature* 405: 299–304.

Olsen, G. (1987) Earliest phylogenetic branching: Comparing rRNA-based evolutionary trees inferred with various techniques. *Cold Spring Harb. Symp. Quant. Biol.* LII: 825–837.

Pawlowski, J., Bolivar, I., Fahrni, J. F., de Vargas, C., Gouy, M. and Zaninetti, L. (1997) Extreme differences in rates of molecular evolution of foraminifera revealed by comparison of ribosomal DNA sequences and the fossil record. *Mol. Biol. Evol.* 14: 498–505.

Pennisi, E. (1998) Genome data shake tree of life. *Science* 280: 672–674.

Penny, D., McComish, B. J., Charleston, M. A. and Hendy, M. D. (2001) Mathematical elegance with biochemical realism: The covarion model of molecular evolution. *J. Mol. Evol.* 53: 711–723.

Penny, D. and Poole, A. (1999) The nature of the last universal common ancestor. *Curr. Opin. Genet. Dev.* 9: 672–677.

Philippe, H. (2000) Long branch attraction and protist phylogeny. *Protist* 51: 307–316.

Philippe, H. and Adoutte, A. (1998) The molecular phylogeny of Eukaryota: Solid facts and uncertainties. In *Evolutionary Relationships among Protozoa* (Coombs, G., Vickerman, K., Sleigh, M. and Warren, A., Eds.) Kluwer, Dordrecht, pp. 25–56.

Philippe, H., Budin, K. and Moreira, D. (1999) Horizontal transfers confuse the prokaryotic phylogeny based on the HSP70 protein family. *Mol. Microbiol.* 31: 1007–1009.

Philippe, H. and Forterre, P. (1999) The rooting of the universal tree of life is not reliable. *J. Mol. Evol.* 49: 509–523.

Philippe, H. and Germot, A. (2000) Phylogeny of eukaryotes based on ribosomal RNA: Long-branch attraction and models of sequence evolution. *Mol. Biol. Evol.* 17: 830–834.

Philippe, H. and Laurent, J. (1998) How good are deep phylogenetic trees? *Curr. Opin. Genet. Dev.* 8: 616–623.

Philippe, H., Lopez, P., Brinkmann, H., Budin, K., Germot, A., Laurent, J., Moreira, D., Müller, M. and Le Guyader, H. (2000) Early branching or fast evolving eukaryotes? An answer based on slowly evolving positions. *Philos. Trans. R. Soc. Lond. B* 267: 1213–1221.

Philippe, H., Sörhannus, U., Baroin, A., Perasso, R., Gasse, F. and Adoutte, A. (1994) Comparison of molecular and paleontological data in diatoms suggests a major gap in the fossil record. *J. Evol. Biol.* 7: 247–265.

Poe, S. and Swofford, D. L. (1999) Taxon sampling revisited. *Nature* 398: 299–300.

Ragan, M. A. (2001) On surrogate methods for detecting lateral gene transfer. *FEMS Microbiol. Lett.* 201: 187–191.

Roger, A. J., Sandblom, O., Doolittle, W. F. and Philippe, H. (1999) An evaluation of elongation factor 1 alpha as a phylogenetic marker for eukaryotes. *Mol. Biol. Evol.* 16: 218–233.

Rokas, A. and Holland, P. W. H. (2000) Rare genomic changes as a tool for phylogenetics. *Trends Ecol. Evol.* 15: 454–459.

Ruepp, A., Graml, W., Santos-Martinez, M. L., Koretke, K. K., Volker, C., Mewes, H. W., Frishman, D., Stocker, S., Lupas, A. N. and Baumeister, W. (2000) The genome sequence of the thermoacidophilic scavenger *Thermoplasma acidophilum*. *Nature* 407: 508–513.

Simpson, A. G., Roger, A. J., Silberman, J. D., Leipe, D. D., Edgcomb, V. P., Jermiin, L. S., Patterson, D. J. and Sogin, M. L. (2002) Evolutionary history of "early-diverging" eukaryotes: The excavate taxon Carpediemonas is a close relative of *Giardia*. *Mol. Biol. Evol.* 19: 1782–1791.

Sogin, M. L. (1991) Early evolution and the origin of eukaryotes. *Curr. Opin. Genet. Dev.* 1: 457–463.

Stetter, K. O. (1996) Hyperthermophiles in the history of life. *Ciba Found. Symp.* 202: 1–10.

Swofford, D. L., Olsen, G. J., Waddell, P. J. and Hillis, D. M. (1996) Phylogenetic inference. In *Molecular Systematics* (Hillis, D. M., Moritz, C. and Mable, B. K.) Sinauer Associates, Sunderland, pp. 407–514.

Tekaia, F., Lazcano, A. and Dujon, B. (1999) The genomic tree as revealed from whole proteome comparisons. *Genome Res.* 9: 550–557.

Tovar, J., Fischer, A. and Clark, C. G. (1999) The mitosome, a novel organelle related to mitochondria in the amitochondrial parasite *Entamoeba histolytica. Mol. Microbiol.* 32: 1013–1021.

Van de Peer, Y., Ben Ali, A. and Meyer, A. (2000) Microsporidia: Accumulating molecular evidence that a group of amitochondriate and suspectedly primitive eukaryotes are just curious fungi. *Gene* 246: 1–8.

Venkatesh, B., Ning, Y. and Brenner, S. (1999) Late changes in spliceosomal introns define clades in vertebrate evolution. *Proc. Natl. Acad. Sci. USA* 96: 10267–10271.

Vossbrinck, C. R., Maddox, J. V., Friedman, S., Debrunner-Vossbrinck, B. A. and Woese, C. R. (1987) Ribosomal RNA sequence suggests microsporidia are extremely ancient eukaryotes. *Nature* 326: 411–414.

Wang, B. (2001) Limitations of compositional approach to identifying horizontally transferred genes. *J. Mol. Evol.* 53: 244–50.

Weisburg, W. G., Giovannoni, S. J. and Woese, C. R. (1989) The Deinococcus-Thermus phylum and the effect of rRNA composition on phylogenetic tree construction. *Syst. Appl. Microbiol.* 11: 128–34.

Williams, B. A., Hirt, R. P., Lucocq, J. M. and Embley, T. M. (2002) A mitochondrial remnant in the microsporidian *Trachipleistophora hominis. Nature* 418: 865–869.

Woese, C. R. (1987) Bacterial evolution. *Microbiol. Rev.* 51: 221–271.

Woese, C. R., Achenbach, L., Rouviere, P. and Mandelco, L. (1991) Archaeal phylogeny: Reexamination of the phylogenetic position of *Archaeoglobus fulgidus* in light of certain composition-induced artifacts. *Syst. Appl. Microbiol.* 14: 364–371.

Woese, C. R. and Fox, G. E. (1977) Phylogenetic structure of the prokaryotic domain: The primary kingdoms. *Proc. Natl. Acad. Sci. USA* 74: 5088–5090.

Woese, C. R., Kandler, O. and Wheelis, M. L. (1990) Towards a natural system of organisms: Proposal for the domains Archaea, Bacteria, and Eucarya. *Proc. Natl. Acad. Sci. USA* 87: 4576–4579.

Woese, C. R., Olsen, G. J., Ibba, M. and Soll, D. (2000) Aminoacyl-tRNA synthetases, the genetic code, and the evolutionary process. *Microbiol. Mol. Biol. Rev.* 64: 202–236.

Wolf, Y. I., Rogozin, I. B., Grishin, N. V., Tatusov, R. L. and Koonin, E. V. (2001) Genome trees constructed using five different approaches suggest new major bacterial clades. *BMC Evol. Biol.* 1: 8.

Yang, Z. (1996a) Among-site rate variation and its impact on phylogenetic analyses. *Trends Ecol. Evol.* 11: 367–370.

Yang, Z. (1996b) Maximum-likelihood models for combined analyses of multiple sequence data. *J. Mol. Evol.* 42: 587–596.

Yap, W. H., Zhang, Z. and Wang, Y. (1999) Distinct types of rRNA operons exist in the genome of the actinomycete *Thermomonospora chromogena* and evidence for horizontal transfer of an entire rRNA operon. *J. Bacteriol.* 181: 5201–5209.

Zuckerkandl, E. and Pauling, L. (1965) Evolutionary divergence and convergence in proteins. In *Evolving Genes and Proteins* (Bryson, V. and Vogel, H. J., Eds.) Academic Press, New York, pp. 97–166.

Zwickl, D. J. and Hillis, D. M. (2002) Increased taxon sampling greatly reduces phylogenetic error. *Syst. Biol.* 51: 588–598.

Chapter 7

The Importance of Evolutionary Biology to the Analysis of Genome Data

Deborah Charlesworth

CONTENTS

Abstract

Evolution is very important in modern biology, and with the appearance of complete genome sequences it is becoming more important than ever before. An understanding of the concepts and findings of important molecular evolution is now essential for molecular biologists, and evolutionary arguments are increasingly being used as part of functional studies of individual genes and of genomes. Population genetics concepts are involved in most aspects of genome analysis, such as cataloging genes and other sequences, understanding gene families and their behavior, examining genome data for evidence of the action of natural selection (including detecting molecular adaptation) and understanding patterns that are emerging in levels of sequence diversity within species and divergence between species. This chapter reviews some of these concepts and describe some of the diversity of studies by using these approaches.

7.1 Introduction

Evolution is central to modern biology as a whole. Biology did not begin to develop into its modern form until organisms were classified in an orderly manner, and progress became

0-415-29904-7/04/$0.00+$1.50

steady once the theory of evolution had shown that this classification reflects relationships through descent from common ancestors. A direct consequence of this understanding is the modern concept of model organisms, which rests on the assumption that people can learn about organisms that particularly interest them, such as crops, pests and pathogens, and themselves, by studying more convenient organisms. The model plant *Arabidopsis thaliana*, although a weed, is hardly a problem one. It is studied not because of the need to learn how to eradicate it, but in the hope that all plants work in much the same way and that the set of genes in the *A. thaliana* genome will be similar to the set in maize or rice, even though *A. thaliana* is a dicotyledon species whereas maize and rice are monocotyledons. Similarly, it is hoped that important developmental processes in vertebrates, including humans, and the answers to questions such as the number of genes in the human genome, will become clear from knowledge of the genomes of zebra fish (Crollius et al., 2000).

The use of model organisms is largely aimed at discovering how organisms work, and genome analysis is within this tradition. Starting from sequences, genes will be deduced, the expression of their products and any interactions with the products of other genes described and their functions worked out. It is becoming increasingly clear that this enterprise requires the necessary involvement of evolutionary approaches, with close collaboration between evolutionary biologists and scientists, such as cell biologists, who study organismal functions directly. Such collaborations are already advanced in developmental biology.

In addition, molecular evolution is very closely involved with genome analysis, at every level, from detecting genes to understanding the evolution of genomes (see Chapter 9). At the present stage of genome sequencing, advances in our understanding are almost as much in terms of learning the extent of our ignorance as in terms of answering the questions that biologists set out to answer. When genome sequencing was first thought to be practicable, the questions to be asked and the analyses needed to answer them were not entirely clear (Miller, 2001), and work on genomes entered a stage of natural history. Unexpectedly, for the molecular biologists who promoted genome sequencing, many of the most interesting emerging questions are evolutionary ones. In the analysis of genome sequences, evolutionary biology is therefore fundamental, and concepts derived from population genetics, on which such comparisons are based, need to be more widely understood. Molecular evolutionary studies of single genes, and sometimes of gene families, have led to the development of approaches that will be useful in analyzing complete genome sequences. Genome sequences will also provide raw data on evolutionary changes in genomes, which will be used to understand processes of genome evolution. In turn, this knowledge will lead to refinement of comparative genome analysis. Evolutionary questions are thus arising in more and more studies within molecular biology. Despite the widespread neglect of evolutionary concepts in biology education and research compared with functional studies, its importance and essential role in understanding organisms is now recognized and evolutionary approaches are in daily use in molecular biology.

7.2 Cataloging Genes and Other Sequences and Understanding Their Placement in Gene Families

The first goal in analyzing a genome sequence is to recognize coding sequences and other sequences with functional importance, such as regulatory sequences, i.e., to ask how many genes it contains, what kinds of genes they are and what the rest of the sequence consists of. It has become clear that this task is extremely difficult by analysis of only a single genome (Hogenesch et al., 2001; Schmid and Aquadro, 2001). Our current ignorance of the range and diversity of genes is considerable. At present, we know only the kinds of genes that have been studied in the past decades, and various biases might limit our views. For instance, a large open reading frame is convincing, but without complete information we do not know what length is small enough to dismiss (Oshiro et al., 2002). Very short plant peptides with important functions were unrecognized until recently (Lindsey, 2001), and even with the complete sequence of the *A. thaliana* genome, the large pollen coat protein superfamily was missed (Vanoosthuyse et al., 2001). Other kinds of genes also exist in genomes, e.g., genes for noncoding RNAs, such as transfer RNA and RNA species involved in signaling processes, such as *Xist* (Brown et al., 1991; Nesterova et al., 2001; Willard and Carrel, 2001), and unless these are already known they might be missed in complete genome annotations, which use properties of known genes to guide the analysis; in tests for transcription, many more transcripts are detected from human chromosomes than the known mRNA numbers (Kapranov et al., 2002). The same is true for other genes with unusual properties, such as very large genes, genes with very large or numerous introns (e.g., Carvalho et al., 2001; Celotto and Graveley, 2001) or genes with extreme or unusual codon usage (Schmid and Aquadro, 2001). Genome sequences provide the first large unbiased set of data, and increasing numbers of small genes are being found, e.g., genes for noncoding RNAs (Rivas et al., 2001), but at present accurate estimates of numbers of genes are unobtainable for many organisms and the numbers of genes missed by current approaches is unknown (Hogenesch et al., 2001).

Even this first step of detecting coding sequences is thus greatly helped if comparative data are available from two or more genomes at a suitable evolutionary distance from one another (Miller, 2001). Counting genes in a model species as a way to estimate the number in the human genome is not a new idea (Crollius et al., 2000). However, gene numbers change and turnover is greater than was guessed before genome sequences were available (e.g., Aravind et al., 2000). Furthermore, some genes are highly conserved, whereas other sequences, even some coding sequences, might evolve rapidly. Many examples have recently been discovered (e.g., Schmid and Tautz, 1997; Ting et al., 1998; Hellberg et al., 2000; Wyckoff et al., 2000; Nesterova et al., 2001). Homologous sequences in different species can thus differ in unexpected ways. Any highly conserved sequences can provide a rough map of a region, making it possible to recognize less conserved genes and sequences that, on their own, could not have been recognized or even aligned in sequences from different species. For noncoding sequences, which might evolve differently from, and perhaps faster than, coding sequences, sequences from more than two genomes are almost essential (Sumiyama et al., 2001). At present, this kind of analysis is often done by using intuitive approaches, such as searching for sequences with 100% conservation in a set of organisms, or visualizing percent identity (Hardison, 2000), but methods are needed that can recognize the kinds of changes that are more likely or less so, like those used in analyzing coding sequences (see later). Just as alignment of sequences of a single gene takes account of evolutionary principles and avoids evolutionarily implausible features, such as numerous amino acid replacements between sequences from related species, or an unlikely frequency of gaps, recognizing conserved features of genomes will become possible once patterns of genome evolution are better understood.

7.2.1 Gene Families

Before the availability of complete genome sequences, gene families presented enormous difficulties. For any analysis, even to catalog the genes in a single species, it is necessary to determine which sequences represent the same genes, but this is not as simple as it sounds. Gene sequences are often polymorphic, and sequences from diploid organisms might therefore often differ. It is essential to distinguish between allelic differences (polymorphism) and different genes (paralogy, see Figure 7.1), but this cannot be done from only the sequences, particularly for large gene families, as is illustrated by the complexity of sequences such as plant pollen coat protein genes (Vanoosthuyse et al., 2001). Such issues can, in principle,

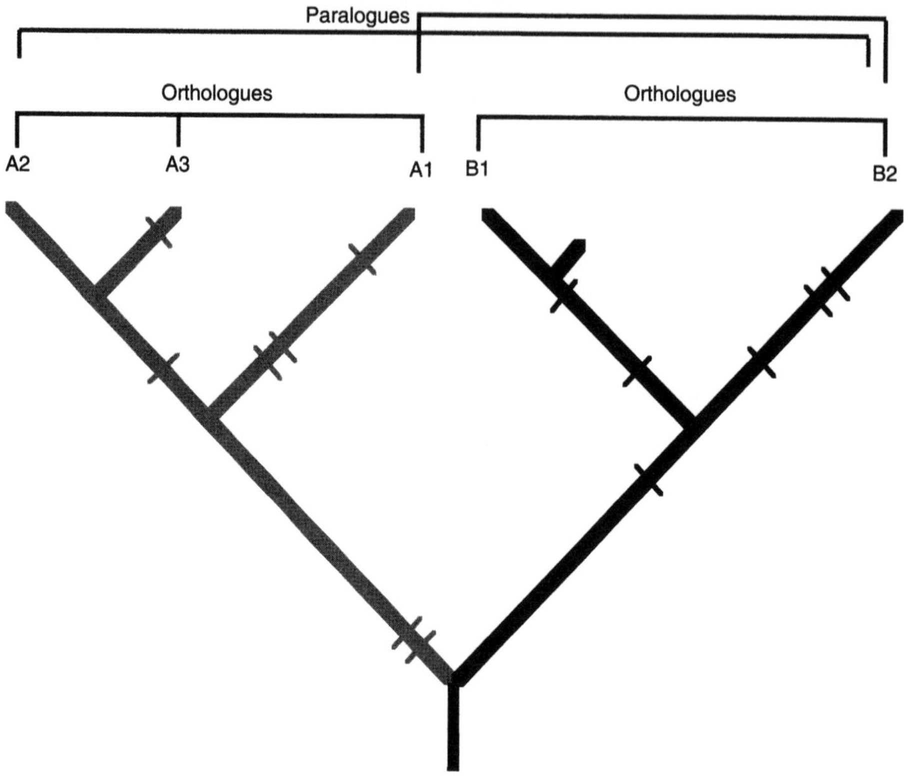

FIGURE 7.1 Orthologous and paralogous genes. The figure shows a gene duplication, creating a new gene A (grey), from an ancestral gene B (black). Substitutions, indicated by short bars crossing the lines of descent, lead to the sequences of the genes diverging independently in these two gene lineages. When species split from one another (successively creating Species 1, 2 and 3), the sequences continue diverging, or one duplicate might be lost (as shown for Species 3). If comparisons are made between species, it is important to be sure that the sequences are not paralogues, because this will suggest longer divergence times than the true ones. Within species, it is also important to avoid including paralogues, because this will overestimate diversity. If the speciation occurs long after the gene duplication, comparisons of the sequences should allow the true ancestry to be inferred, because orthologues in the different species should form clusters, even if genes have been lost in some species. However, large gene families can be difficult to interpret in the absence of all members, and rapid sequence evolution can obscure the clusters and make it difficult to know which genes should be included.

be resolved by using the polymorphisms as genetic markers; lack of linkage between different sequence variants establishes that they are different genes, but this is extremely laborious and unhelpful when genes in a family are in a tandem array.

Approaches are now being developed that make use of information about numbers of copies present in genomes, estimated from their representation in whole-genome shotgun sequences, to detect duplicate genes (Bailey et al., 2002). But such analyses must be confirmed by physical map data. This is the basis for our knowledge of the homologies of the best-understood gene families, such as the different HOX genes in the duplicated clusters found in vertebrates (Duboule, 1994). After gene duplication, the flanking sequences of genes, and even their coding sequences, evolve differences that, given enough time, uniquely identify the different copies in the genome. Even when different copies exchange information by gene conversion, which can cause concerted evolution or differences in divergence between two species (Wang et al., 1999), a physical map can help show what has happened. A complete genome sequence provides such a map for all the organism's gene families and can be viewed as merely the fullest development of earlier genetic mapping approaches. Just as genetics requires both variability (to detect phenotypes) and conservation (to recognize alleles in families and crosses), genome analysis benefits from both similarities maintained by evolutionary constraints, and from evolved differences, but even with complete genome sequences great care is needed to infer numbers of duplicate genes (Gu et al., 2002).

Physical map information is also extremely helpful in identifying orthologous genes in different species, which is essential before one can do any analyses to detect the action of natural selection (see later). It is impossible accurately to compare genes between species without knowing which genes are the same in the different species. Duplications involve both the nuclear genome and the organelle genomes (Blanchard and Lynch, 1999). For example, mitochondrial pseudogenes are found in animal nuclear genomes (Bensasson et al., 2001; Mourier et al., 2001; Woischnik and Moraes, 2002), and transfers from plant mitochondrial to nuclear genomes are well documented (Adams et al., 2002). Because sequences evolve at different rates, it is sometimes unclear which loci in different species are orthologues (true homologues, which have been diverging only since the two species became separated) and difficult to distinguish these from paralogous loci (derived from gene duplications, and thus potentially having diverged since long before the split between the species in question; see Figure 7.1). This distinction is made even more difficult by lineage-specific loss of members of gene families (Theissen, 2002). Although the rate of gene turnover will not be clear until genomes of closely related species have been compared, these processes (sometimes called gene birth and death, Nei et al., 1997) have been detected in the genomes of all taxa studied, even prokaryotes (Jordan et al., 2001).

Evidence from physical map locations has been used for several years to help detect genome rearrangements and duplications of small and large segments of genomes (Hardison and Miller, 1993; Wolfe and Shields, 1997; Semple and Wolfe, 1999; Vision et al., 2000), and it is becoming increasingly clear that these are important aspects of genome evolution. Without physical map information, one can only resort to methods such as the clusters of orthologous genes (COG) approach. This assumes that pairs of sequences in two genomes that are each other's most similar matches are true orthologues (Tatusov et al., 2000). However, this will sometimes be wrong for members of gene families, particularly when birth and death of genes occurs, because duplicated gene sequences often diverge (Theissen, 2002). Although some duplications lead to loss of function of one of the duplicates (Wolfe and Shields, 1997), they might instead acquire new functions. This is expected on theoretical grounds (Walsh, 1995), and many real instances are well documented (e.g., Wallis, 1996; Wu et al., 1997).

Additional information, such as data on the expression of genes, can help suggest which of a set of genes are true homologues (Eisen, 1998), but changes in function certainly occur, sometimes very rapidly in evolutionary time, and so this approach cannot always succeed. Similarity of physical maps is therefore of great importance. It is very helpful that chromosome rearrangements are infrequent evolutionary events (Burt et al., 1999; Schoen, 2000; Kumar et al., 2001). When two organisms, such as humans and mice, have sets of genes in common, one can therefore hope to find regions that contain sets of genes in the same order. Information about the homology of genes can then be acquired for an entire set of genes that is found in a similar physical map. This is much more convincing evidence than mere sequence similarity between pairs of genes from the two species. However, uncertainties sometimes remain, especially when the numbers of genes differ between species, as in the case of the mammalian globin gene clusters (Hardison and Miller, 1993).

Once orthologues can be determined between species and distinguished from paralogues, one can ask how much the sets of genes of different species overlap and what proportions are unique to individual species or taxa (Castresana, 2002). This question can be answered for the HOX genes, for example, because the combination of physical map and sequence information allows each gene to be identified, and this approach has also contributed to recent progress in understanding mammalian odorant receptor gene families (Mombaerts, 2001). For the plant MADS-box genes, in contrast, there is a baffling diversity of sequences (Münster et al., 1997; Kramer et al., 1999), but assignment of sequences as orthologues of identifiable *A. thaliana* genes should be very helpful, as it has been for plant pathogen resistance genes (Bergelson et al., 2001).

7.3 Comparative Genomics: Molecular Evolution and Natural Selection

One of the most important contributions to genome studies of using molecular evolutionary approaches is that it can provide evidence about the action of natural selection and its nature. Merely discovering the genes and other functional sequences is only the first step in genome analysis. Within the realm of functional studies, genes must be tested to discover when and where in organisms they are expressed, and genes and gene products must be tested to discover interactions between them, such as involvement in multimeric proteins or in regulation of signal transduction systems. Complementing such studies are tests and analyses within the realm of molecular evolution that can help determine whether a sequence is functional rather than being a pseudogene or a region of junk DNA. To help do this, one needs to test whether natural selection is acting on the sequence and distinguish between different forms of selection. It is widely realized that an evolutionary approach might help detect sequences with functional importance, because such sequences are likely to be conserved in different species.

It is starting to be understood that finding evidence that selection has led to evolutionary change in a protein sequence is also a good way to discover candidate sequences that might have important functions (e.g., Bielawski and Yang, 2001a; Knudsen and Miyamoto, 2001) This is closely related to another goal of genome sequencing, to discover the causes of differences between species. It is evident that this question will not be answered merely by looking at the differences between species; approaches are needed that can help analyze differences and test whether they have been driven by natural selection, i.e., are adaptive. This is essential because neutral differences must also exist between species. The neutral theory of molecular evolution shows that genetic drift will lead to steady divergence of neutral sites in the genome sequences of different species, so that differences alone are meaningless without a way to test which of them are adaptive. Part of gaining an understanding of genome evolution will be working to get evidence about the distribution of selective effects on different sequence elements. Selection might preserve amino acid sequences, but might also lead to changed sequences, and both kinds of selection on genome are of great interest.

7.3.1 Coding Sequences

Conservation of amino acid sequences in coding regions is a valuable type of evidence for function and can help show which regions of the sequence are most and least subject to selective constraints, by comparing different genes and different regions of the same gene. Given the value of this information, it is surprising how frequently coding sequences are compared in terms of a simple divergence measure using all base positions. Much greater understanding can be achieved by taking into account the relationship of the genomic sequence to the coding sequence (reviewed by Bielawski and Yang, 2001a; Kreitman, 2001; Nekrutenko et al., 2002).

The general approach used in molecular evolution is based on using the neutral theory of molecular evolution as a null hypothesis and testing for differences in the behavior of different sequences that are unlikely to arise under neutrality. A sign that natural selection has acted to keep a coding sequence from changing over evolutionary time is the familiar observation that when the sequences of a given gene are compared in two species, amino acid replacements are less frequent, per site, than silent ones (synonymous changes). These two measures of divergence per nucleotide site are often symbolized by Ka and Ks. If there were no selection, their values would be expected to be equal, as is found in pseudogenes (Nei et al., 1981), because the silent site divergence (Ks) takes account of the amount of time during which the species being compared occurred have been diverging, and allows for differences in mutation rates between different loci. When amino acid changes are disadvantageous (selective constraints severe), the Ka value will be low, and thus a low Ka/Ks ratio suggests purifying selection (preserving the amino acid sequence). The range of Ka/Ks values when sequences of genes from different species are compared is wide, because some genes are highly conserved whereas other sequences might be more permissive of amino acid changes, but there is widespread evidence for purifying selection (Li, 1993; Lagercrantz and Axelsson, 2000). It is thus possible that genes that can mutate to lethal alleles will encode more highly conserved amino acids than will genes whose presence is nonessential and which are nonlethal when knocked out experimentally (Jordan et al., 2002).

Although the logic of such tests is straightforward, estimating these values is not a simple matter of counting the numbers of amino acid and synonymous changes between the sequences. It is also necessary to estimate the numbers of synonymous and nonsynonymous (replacement) sites in the sequence to be studied in the two species being compared; only fourfold degenerate sites experience just synonymous mutations (Li, 1993). Estimation is therefore based on modeling sequence evolution, which also permits testing the significance of differences. Clearly, this requires data on how sequences evolve, which, in turn, depends on having good knowledge of relationships between taxa. Methods continue to be developed to include as much as possible of the realities of sequence evolution (Thorne, 2000; Bielawski and Yang, 2001b; McVean, 2001), but different models yield different estimates, and the possibilities are so varied that no model is appropriate for all situations. For instance, mutations can be affected by their sequence context (Bulmer, 1986), a possibility that is usually ignored, and mutation rates can differ in different parts of the genome (e.g., Lercher et al., 2001; Smith et al., 2002).

A related kind of analysis depends on the fact that all amino acids are not equivalent. Some amino acid replacements between species are more frequent than others. It is well known, for example, that cysteine residues are often highly conserved in protein sequences. The frequencies of different amino acid replacements between species depend on the codons used, the mutation rates of the bases and the selective effects of the amino acids. When replacements are tabulated, the commonest changes are, as expected, those involving amino acids with similar size and charge (Li, 1997). However, it is again important to take into account the composition of sequences studied and possible mutational biases; without such information, a false appearance of selection might be created (Dagan et al., 2002).

7.3.2 Directional Selection, Molecular Adaptation and Evolution of New Functions

Another goal of genome sequencing is to discover the causes of differences between species (particularly our own species, and how and why we differ from our closest ape relatives). High Ka/Ks values in coding sequences have been used to detect selectively driven amino acid change, such as changes occurring after gene duplication (Walsh, 1995), and many examples of molecular adaptation have recently been found in gene families (e.g., Zhang et al., 1998; Merritt and Quattro, 2001; Wallis, 2001). These analyses use information from phylogenetic relationships inferred from previous evolutionary studies to define when the changes in evolutionary rates occurred and how long the periods of change lasted. When sequences from several species are compared, rather than just two, it is possible to estimate the numbers of changes in individual lineages and to infer the direction of changes (Li, 1997). These kinds of information are much richer and more informative than merely knowing that orthologous sequences from two species have diverged.

Probably many more genes have undergone adaptive changes than can be detected from high Ka/Ks ratios averaged over all codons. Important functional differences can be caused by changes in only a few, perhaps even single, amino acid residues of a protein, while much of its sequence remains under selective constraint, so that an analysis that does not differentiate between regions of the coding sequence would conclude that the sequence encodes a protein of little importance. Methods are now being developed to differentiate which regions and residues are evolving in a manner that suggests selection (Yang et al., 2000; Gu, 2001; Knudsen and Miyamoto, 2001). It is sometimes possible to detect selection affecting parts of a sequence even when overall nonsynonymous and synonymous substitution rates of a gene are similar to one another, and to the rate in introns, as in the case of the DAZ gene family (Bielawski and Yang, 2001a).

It is very helpful to combine analyses of sequence divergence between species with data on polymorphism within species. In addition to low Ka/Ks values estimated from divergence between species, coding sequences that are under selective constraints tend to have less replacement site variability within populations than neutral sequences do. Purifying selection acting on a sequence can be inferred when nonsynonymous sites have both lower diversity and and also lower divergence than synonymous sites. The MacDonald–Kreitman test combines diversity and divergence data (McDonald and Kreitman, 1991). This approach detected rapid amino acid evolution after the formation of a new gene, *jingwei*, a chimaera of an alcohol dehydrogenase and a different gene, in a *Drosophila* species; *jingwei* evolved by a duplication and an insertion of a portion of the alcohol dehydrogenase coding sequence (Long and Langley, 1993). Many other examples of molecular adaptation have been found in recent years, and the approach is now starting to be used to test hypotheses of adaptive change. For example, to study centromere evolution of sequence differences between *Drosophila* species, a centromere-specific histone, *Cid*, was studied. Evidence for adaptive evolution was found by this approach (Malik and Henikoff, 2001), though selection on this protein could not have been detected by direct functional studies. Evidence for molecular adaptation and estimates of the fraction of loci affected, using MacDonald–Kreitman tests, might even be possible (Fay et al., 2002; Smith and Eyre-Walker, 2002).

The frequency distribution of variants also provides information about selection. If amino acid variants that occur within species are mostly disadvantageous variants, they will arise by mutation, but will generally be eliminated by natural selection over the course of a few generations, and so they should generally be at low frequencies. In contrast, neutral variants are subject to genetic drift in finite populations, and can occasionally by chance rise to high frequencies, and sometimes to fixation, causing a substitution at the sites. These differences are the basis of tests for selection, again using the distinction between synonymous and nonsynonymous sites. Purifying selection can be detected from excess low-frequency nonsynonymous variants, detected using Tajima (1989)'s test to compare the number of polymorphic sites in a sample of sequences with the frequency of sites differing between pairs of sequences in the sample (i.e., the nucleotide diversity, sometimes called heterozygosity). Higher than expected frequencies of amino acid variants might suggest molecular adaptation (Fay et al., 2001). Finally, balancing selection with long-term maintenance of variability is detectable from excess of variants at intermediate frequencies. Balancing selection leads to high variability at silent sites linked to the selected site, because haplotypes are held in populations long enough to diverge in sequence, so the effects might often be readily detectable. Well-studied examples include plant self-incompatibility genes (Charlesworth and Awadalla, 1998), disease-resistance genes (Bergelson et al., 2001) mammalian

MHC loci (Hughes et al., 1990) and the variability of organelle genomes in plants with cytoplasmic male-sterility (Ingvarsson and Taylor, 2002; Städler and Delph, 2002).

However, great care is needed in inferring selection, because other situations can affect variant frequencies. For instance, population subdivision increases numbers of polymorphisms and can give the appearance of balancing selection even when within-population polymorphism is not maintained, whereas population growth can lead to excess low-frequency polymorphisms. There is currently intense interest in modeling neutral variability in subdivided populations to help understand when selection can and cannot be inferred (Wakeley and Aliacar, 2001). Moreover, it must not be forgotten that silent changes including synonymous changes in coding regions are not always neutral, but can often be under selection (Akashi et al., 1998), albeit weak compared with selection on changes in the amino acid sequences of proteins (and probably on sequences regulating gene expression).

7.3.3 Noncoding Sequences

The human genome sequence is at present incomplete. Initial estimates that noncoding sequences are a major component might not be correct if many human genes are very large and contain large introns (Wong et al., 2001). Whatever the proportion, the important task of understanding the importance of these sequences will involve both functional and molecular evolutionary studies. Classification of the diversity of noncoding sequences is incomplete. Some of these sequences are parasitic DNA, whose presence in the genome is due to natural selection acting on the sequences themselves. This category includes the many families of transposable elements found in most, if not all, genomes, which are selected to transpose to new locations and thus to spread into new carrier individuals (Charlesworth and Langley, 1989) as well as incomplete copies of transposable elements, some of which can still transpose by relying on the transposase function of complete (active) elements, but some of which lack sequences essential for transposition.

How can one tell whether sequences are present in genomes because they serve organismal functions? Similar logic to that used for coding sequences has been used to show that other kinds of changes in genomes are disadvantageous. The most important regions should diverge less than the average and should also be less likely to accept transpositions of repetitive sequences (Chiaromonte et al., 2001). For instance, transposable element are rarely inserted within coding sequences or at the ends of introns (Duret and Hurst, 2001), but are largely found in regions between genes, and this makes sense in terms of purifying selection. These elements avoid genes, not because they cannot transpose into genes, but because those that do are largely eliminated from populations by natural selection. Consistent with their distribution, transposable elements are also generally at low frequencies at the sites where they are found; that is, an occupied site is usually occupied in only a minority of the members of a species or population (Charlesworth et al., 1994; Wright et al., 2001). This example, like the detection of selection on silent sites already mentioned (Akashi et al., 1998) illustrates the importance of evolutionary evidence. The evolutionary approach can infer selection even when it is too weak to detect by direct studies of gene functions or expression patterns. A small proportion of the huge number of such elements in the human genome are, nevertheless, found within genes, mostly in introns (Nekrutenko and Li, 2001), and integration of elements might occasionally contribute to forming new genes (Long, 2001).

It is unclear how best to study noncoding sequences to detect regions with functional importance and evolutionary approaches need to be developed for this. It is frequently suggested that morphological evolution might involve mainly changes in flanking regulatory regions of genes, and only to a lesser extent changes in amino acid sequences of proteins (Purugganan, 1998; White and Doebley, 1999; Carroll, 2000). Some changes in gene regulation can be due to changed sequences of protein transcription factors and signaling proteins

(Löhr et al., 2001), but at least some of them could involve changes in enhancer or other DNA sequences (Jenkins et al., 1995; Ludwig et al., 2000). It will be important to understand how each of the kinds of sequences evolve and how changes in DNA sequences involved in signaling are related to the evolution of the proteins that recognize them (Stern, 2000; Galant and Carroll, 2002).

Intron sites can also sometimes be subject to selection. There is evidence for selection that maintains intron and other sequences involved in base pairing that determines the three-dimensional structure of RNA. As might be expected, such sequences show compensatory changes to restore base pairing whenever mutations are fixed (Chen et al., 1999; Innan and Stephan, 2001), consistent with the hypothesis that the slow evolution of such regions is due to selective constraint, even though they are noncoding.

Analyses of aligned noncoding sequences from two or more species will be very important aids to discovering conserved, and potentially functionally important, noncoding sequences (Sumiyama et al., 2001). In the genomes and regions sampled so far, even the best alignments of conserved sequence tend mostly to be very short (Bergman and Kreitman, 2001; Dermitzakis, 2002; Frazer et al., 2001), consistent with what is known about enhancer sequences (Ludwig et al., 2000), and the degree of conservation between sequences from a given pair of species varies greatly from one region to another. It can, nevertheless, be possible to develop measures of the proportion of constrained sequence (Shabalina et al., 2001).

7.4 Patterns in Genomes

As we attempt to understand genomes and their functions, the results of evolutionary analyses must also be incorporated into our thinking. Regional differences and some patterns in genomes are starting to be observed, and these might complicate some of the simple analyses outlined previously. It is emerging that neither mutation rates nor the effects of a given force of selection are the same throughout a genome, and that local differences in sequence can sometimes affect sequence evolution, and must therefore be taken into account (Galtier et al., 2001; Lercher et al., 2001). A recent advance in the understanding of genome evolution is the realization that selection is expected to be less able to act when recombination is infrequent, so that variability within species, and codon usage, might depend on which genome region is studied. Lower diversity in genome regions, such as near the centromeres, where recombination is infrequent is consistent with these theories, and other explanations such as different mutation rates are unlikely because synonymous site divergence between related species is not unusually low in these regions (Charlesworth et al., 1993; Aquadro et al., 1994; Braverman et al., 1995). Differences in amino acid divergence between genomes of different species can, however, sometimes be due to differences in the ability of selection to constrain sequence evolution, rather than reflecting differences in the strength of selection itself. For example, amino acid divergence is expected to be fast for sequences on nonrecombining Y chromosomes, compared with orthologous X-linked genes, even if the Y-linked genes are expressed. This effect has been detected in the W-linked *CHD1* gene in birds (Fridolfsson and Ellegren, 2000) and in genes on the recently evolved neo-Y chromosome of *Drosophila miranda* (Bachtrog and Charlesworth, 2002).

Different efficacy of selection in different regions of genomes can also sometimes affect the evolution of synonymous sites. Weak selection, such as selection for preferred codons, might dominate the evolution of such sites in recombining sequences, but will be less likely to do so when recombination is infrequent, and genetic drift at such sites might lead to substitutions at a rate approaching that for neutral sites. This is one possible explanation for the lower bias in codon usage in low-recombination regions of chromosomes of *Drosophila* and other species (Kliman and Hey, 1993). The rate of the silent molecular clock, as well as the rate of nonsynonymous substitutions, would then differ between genome regions, being lowest when recombination allows the greatest efficacy of selection. Patterns such as these cannot be ignored in future analyses using evolutionary approaches.

However, recombination differences are only one of several correlated differences between genome regions, and it is difficult to test between different possible explanations for patterns such as this. In the few eukaryote genomes so far sequenced, data from genome sequences are consistent with earlier evidence that the number of genes per unit of DNA differs between regions with different recombination frequencies. In *D. melanogaster*, humans and in the plant *A. thaliana*, genes are more densely packed in high-recombination regions and sparser in the centromeric regions of the chromosomes where recombination occurs less often. In *C. elegans*, however, defined centromeres do not exist, and genes are most abundant in regions with lower recombination frequencies (in the centres of chromosomes). Nucleotide composition also differs in different genome regions, often associated with differences in recombination frequencies (Galtier et al., 2001). Many transposable elements are more abundant per unit of DNA in regions where recombination frequencies are low (Charlesworth et al., 1994; Duret et al., 2000). It is therefore important to devise tests of different possibilities and to go beyond merely showing that a pattern correlates with a difference in some property such as recombination frequency. Regional codon usage patterns in the *D. melanogaster* and *C. elegans* genomes, and in base composition in yeast genomes, might be caused by gene conversion biased toward GC at mismatched sites in DNA heteroduplexes, which is correlated with recombination but does not involve selection or any regional differences in the efficacy of selection (Marais et al., 2001; Birdsell, 2002).

7.5 Conclusions

As genomes are sequenced and analyzed, there is a continuous process of revising our ideas. More is learnt about the kinds of genes that are present, their relationships with other genes in the genomes, the physical organization of the members of gene families and about patterns in the sequences, such as patterns of differences in nucleotide composition or of precision in usage of synonymous codons. These observations, in turn, suggest processes that must be understood in evolutionary terms. We want to know the fundamental reasons for differences. For example, one should be able to say whether a pattern within a genome could be an effect of different mutation processes in different regions of genomes and chromosomes, or whether genes in different species have different sequences because of natural selection causing different adaptation. Evolutionary approaches provide ways to ask such questions. Instead of being restricted to intuitive ideas, such as suggestions of adaptive significance of a genome feature because it is widespread, tests are possible, allowing mistakes to be avoided. For instance, it is often suggested that the large representation of transposable elements in mammalian genomes implies that such a prominent genome feature must contribute some valuable genome function. This is not a valid inference. The parasitic DNA hypothesis for transposable elements is a simple alternative explanation for the representation of such elements in genomes. The fact that transposable elements cause genome rearrangements and are sometimes involved in formation of new genes, and thus undoubtedly have importance in genome evolution, does not imply that they are other than occasional chance consequences of element activities (Edwards and Brookfield, 2003). Only events that have advantageous effects are retained in genomes, and there is clear evidence for generally disadvantageous effects of transposable element insertions.

To take another example, gene duplications should not be viewed as allowing evolutionary change, as is often suggested, even though they have this effect. Arrangement of genes in a cluster almost certainly does not evolve to facilitate allelic diversity (e.g., Mayfield et al., 2001), even though this allows gene conversion to generate diversity at individual loci. Greater understanding of evolution is needed to help prevent such misunderstandings and thereby focus on testable and potentially productive hypotheses. Features such as duplications (which cause redundancy) can be a favorable condition for evolutionary changes, but evolution has no foresight and natural selection cannot lead to features that will promote evolution in the future unless the features have benefits when they arise. In modern evolutionary thinking, there is a clear conceptual distinction between adaptations that are themselves selectively advantageous and situations that permit disadvantageous changes to occur. Given our increasing knowledge of genome structure and function, it is important to distinguish sequence changes caused by permissive situations from changes directly driven by selection, i.e., adaptive change. The first kind of change is important in genome evolution, and the second includes functional changes in genes and other sequences. Because these occur in the context of the genome, differences in the local sequence and recombinational environment cannot be ignored as we try to understand functions.

Understanding functions and understanding evolution go hand in hand. Without a basis in evolutionary understanding, unhindered speculation is possible, and ideas such as developmental macromutations are untestable. With the detailed information from genome sequences, and even with only sequence data from a few genes, we are starting to be able to test such hypotheses. It is becoming increasingly clear that much evolution is gradual. Although it is true that new genes can form from sudden events, such as transposition of sequence to a new location, only insertions without highly detrimental consequences are preserved in the face of natural selection, and further gradual evolution often occurs once a new gene is formed (Long and Langley, 1993; Long, 2001). *Drosophila fushi tarazu* evolved from a gene that, in other insects, still has homeotic functions, and this has gradually been lost and new functions aquired (Löhr et al., 2001). The same type of process can occur between paralogues within a genome. Some plant MYB genes encode proteins that when tested in *A. thaliana* are functionally equivalent, whereas others differ functionally and are unable to replace one another (Lee and Schiefelbein, 2001).

It is often questioned whether genomics is more than the study of many genes. The answer is certainly yes, if we are trying to understand how genomes function and evolve. If we hope to use sequence data in research, or for other applications, this understanding is important. We cannot ignore the way evolution happens in genomes or assume simple properties that might not always be correct.

References

Adams, K. L., Qiu, Y. L., Stoutemyer, M. and Palmer, J. D. (2002) Punctuated evolution of mitochondrial gene content: High and variable rates of mitochondrial gene loss and transfer to thenucleus during angiosperm evolution. *Proc. Natl. Acad. Sci. USA* 99: 9905–9912.

Akashi, H., Kliman, R. M. and Eyre-Walker, A. (1998) Mutation pressure, natural selection, and the evolution of base composition in *Drosophila. Genetica* 103: 49–60.

Aquadro, C. F., Begun, D. J. and Kindahl, E. C. (1994) Selection, recombination, and DNA polymorphism in Drosophila. In *Non-Neutral Evolution: Theories and Molecular Data* (Golding, B., Ed.), Chapman & Hall, London, pp. 46–56.

Aravind, L., Watanabe, H., Lipman, D. J. and Koonin, E. V. (2000) Lineage-specific loss and divergence of functionally linked genes in eukaryotes. *Proc. Natl. Acad. Sci. USA* 97: 11319–11324.

Bachtrog, D. and Charlesworth, B. (2002) Reduced adaptation of a non-recombining neo-Y chromosome. *Nature* 416: 323–326.

Bailey, J. A., Gu, Z., Clark, R. A., Reinert, K., Samonte, R. V., Schwartz, S., Adams, M. D., Myers, E. W., Li, P. W. and Eichler, E. E. (2002) Recent segmental duplications in the human genome. *Science* 99: 1003–1007.

Bensasson, D., Zhang, D. X., Hartl, D. L. and Hewitt, G. M. (2001) Mitochondrial pseudogenes: Evolution's misplaced witnesses. *Trends Ecol. Evol.* 16: 314–321.

Bergelson, J., Kreitman, M., Stahl, E. A. and Tian, D. (2001) Evolutionary dynamics of plant R-genes. *Science* 292: 2281–2285.

Bergman, C. M. and Kreitman, M. (2001) Analysis of conserved noncoding DNA in Drosophila reveals similar constraints in intergenic and intronic sequences. *Genome Res.* 11: 1335–1345.

Bielawski, J. P. and Yang, Z. H. (2001a) Positive and negative selection in the *DAZ* gene family. *Mol. Biol. Evol.* 18: 523–529.

Bielawski, J. P. and Yang, Z. H. (2001b) Statistical methods for detecting molecular adaptation. *Trends Ecol. Evol.* 15: 496–503.

Birdsell, J. A. (2002) Integrating genomics, bioinformatics, and classical genetics to study the effects of recombination on genome evolution. *Mol. Biol. Evol.* 19: 1181–1197.

Blanchard, J. L. and Lynch, M. (1999) Organellar genes: Why do they end up in the nucleus? *Trends Genet.* 16: 315–320.

Braverman, J. M., Hudson, R. N., Kaplan, N. L., Langley, C. H. and Stephan, W. (1995) The hitchhiking effect on the site frequency spectrum of DNA polymorphisms. *Genetics* 140: 783–796.

Brown, C. J., Ballabio, A., Rupert, J. K., Lafreniere, R. G., Grompe, M., Tonlorenzi, R. and Willard, H. F. (1991) A gene from the region of the human X-inactivation center is expressed exclusively from the inactive ZX-chromosome. *Nature* 349: 38–44.

Bulmer, M. G. (1986) Neighboring base effects on substitution rates in pseudogenes. *Mol. Biol. Evol.* 3: 322–329.

Burt, D. W., Bruley, C., Dunn, I. C., Jones, C. T., Law, A. S., Morrice, D. R., Paton, I. R., Smith, J., Windsor, D., Sazanov, A., Fries, R. and Waddington, D. (1999) The dynamics of chromosome evolution in birds and mammals. *Nature* 402: 411–413.

Carroll, S. B. (2000) Endless forms: The evolution of gene regulation and morphological diversity. *Cell* 101: 557–580.

Carvalho, A. B., Lazzaro, B. P. and Clark, A. G. (2001) Y chromosomal fertility factors kl-2 and kl-3 of *Drosophila melanogaster* encode dynein heavy chain polypeptides. *Proc. Natl. Acad. Sci. USA* 97: 13239–13244.

Castresana, J. (2002) Estimation of genetic distances from human and mouse introns. *Genome Biol.* 3: research 0028 .1–7.

Celotto, A. M. and Graveley, B. R. (2001) Alternative splicing of the *Drosophila Dscam* pre-mRNA is both temporally and spatially regulated. *Genetics* 159: 599–608.

Charlesworth, B. and Langley, C. H. (1989) The population genetics of *Drosophila* transposable elements. *Annu. Rev. Genet.* 23: 251–287.

Charlesworth, B., Morgan, M. T. and Charlesworth, D. (1993) The effect of deleterious mutations on neutral molecular variation. *Genetics* 134: 1289–1303.

Charlesworth, B., Sniegowski, P. and Stephan, W. (1994) The evolutionary dynamics of repetitive DNA in eukaryotes. *Nature* 371: 215–220.

Charlesworth, D. and Awadalla, P. (1998) The molecular population genetics of flowering plant self-incompatibility polymorphisms. *Heredity* 81: 1–9.

Chen, Y., Carlini, D. B., Baines, J. F., Parsch, J., Braverman, J. M., Tanda, S. and Stephan, W. (1999) RNA secondary structure and compensatory evolution. *Genes Genet. Syst.* 74: 271–286.

Chiaromonte, F., Yang, S., Elnitski, L., Yap, V. B., Miller, W. and Hardison, R. C. (2001) Association between divergence and interspersed repeats in mammalian noncoding genomic DNA. *Proc. Natl. Acad. Sci. USA* 98: 14503–14508.

Crollius, H. R., Jaillon, O., Bernot, A., Dasilva, C., Bouneau, L., Fischer, C., Fizames, C., Wincker, P., Brottie, P., Quetier, F., Saurin, W. and Weissenbach, J. (2000) Estimate of human gene number provided by genome-wide analysis using *Tetraodon nigroviridis* DNA sequence. *Nature Genet.* 25: 235–238.

Dagan, T., Talmor, Y. and Graur, D. (2002) Ratios of radical to conservative amino acid replacement are affected by mutational and compositional factors and may not be indicative of positive darwinian selection. *Mol. Biol. Evol.* 19: 1022–1025.

Duboule, D. (1994) How to make a limb. *Science* 266: 575–576.

Duret, L. and Hurst, L. D. (2001) The elevated GC content at exonic third sites is not evidence against neutralist models of isochore evolution. *Mol. Biol. Evol.* 18: 757–762.

Duret, L., Marais, G. and Biémont, C. (2000) Transposons but not retrotransposons are located preferentially in regions of high recombination rate in *Caenorhabditis elegans*. *Genetics* 156: 1661–1669.

Edwards, R. J. and Brookfield, J.F.Y. (2003) Transiently beneficial insertions could maintain mobile DNA sequences in variable environments. *Mol. Biol. Evol.* 20: 30–37.

Eisen, J. A. (1998) Phylogenomics: Improving functional predictions for uncharacterized genes by evolutionary analysis. *Genome Res.* 8: 163–167.

Fay, J.C., Wyckoff, G.J. and Wu, C.I. (2001) Positive and negative selection on the human genome. *Genetics* 158: 1227–1234.

Fay, J. C., Wyckoff, G. J. and Wu, C. I. (2002) Testing the neutral theory of molecular evolution with genomic data from *Drosophila*. *Nature* 415: 1024–1026.

Frazer, K. A., Sheehan, J. B., Stokowski, R. P., Chen, X., Hosseini, R., Cheng, J. -F., Fodor, S. P. A., Cox, D. R. and Patil, N. (2001) Evolutionarily conserved sequences on human chromosome 21. *Genome Res.* 11: 1651–1659.

Fridolfsson, A. -K. and Ellegren, H. (2000) Molecular evolution of the avian *CHD1* genes on the Z and W sex chromosomes. *Genetics* 155: 1903–1912.

Galant, R. and Carroll, S. B. (2002) Evolution of a transcriptional repression domain in an insect Hox protein. *Nature* 415: 910–913.

Galtier, N., Piganeau, G., Mouchiroud, D. and Duret, L. (2001) GC-content evolution in mammalian genomes: The biased gene conversion hypothesis. *Genetics* 159: 907–911.

Gu, X. (2001) Maximum-likelihood approach for gene family evolution under functional divergence. *Mol. Biol. Evol.* 18: 453–464.

Gu, Z., Cavalcanti, A., Chen, F. -C., Bouman, P. and Li, W. -H. (2002) Extent of gene duplication in the genomes of *Drosophila*, nematode, and yeast. *Mol. Biol. Evol.* 19: 256–262.

Hardison, R.C. (2000) Conserved noncoding sequences are reliable guides to regulatory elements. *Tr. Genet.* 16: 369–372.

Hardison, R. C. and Miller, W. (1993) Use of long sequence alignment to study the evolution and regulation of mammalian globin clusters. *Mol. Biol. Evol.* 10: 73–102.

Hellberg, M. E., Moy, G. W. and Vacquier, V. D. (2000) Positive selection and propeptide repeats promote rapid interspecific divergence of a gastropod sperm protein. *Mol. Biol. Evol.* 17: 458–466.

Hogenesch, J. B., Ching, K. A., Batalov, S., Su, A. I., Walker, J. R., Zhou, Y., Kay, S. A., Schultz, P. G. and Cooke M. P. (2001) A comparison of the Celera and Ensembl predicted gene sets reveals little overlap in novel genes. *Cell* 106: 413–415.

Hughes, A., Ota, T. and Nei, M. (1990) Positive Darwinian selection promotes charge profile diversity in the antigen-binding cleft of class I major-histocompatibility-complex molecules. *Mol. Biol. Evol.* 76: 515–524.

Ingvarsson, P. K. and Taylor, D. R. (2002) Genealogical evidence for epidemics of selfish genes. *Proc. Natl. Acad. Sci. USA* 99: 11265–11269.

Innan, H. and Stephan, W. (2001) Selection intensity against deleterious mutations in RNA secondary structures and rate of compensatory nucleotide substitutions. *Genetics* 159: 389–399.

Jenkins, D. L., Ortori, C. A. and Brookfield, J. F. Y. (1995) A test for adaptive change in DNA-sequences controlling transcription. *Proc. R. Soc. Lond. B* 261: 203–207.

Jordan, I. K., Makarova, K. S., Spouge, J. L., Wolf, Y. I. and Koonin, E. V. (2001) Lineage-specific gene expansions in bacterial and archeal genomes. *Genome Res* 11: 555–565.

Jordan, I. K., Rogozin, I. B., Wolf, Y. I. and Koonin, E. V. (2002) Essential genes are more evolutionarily conserved than are nonessential genes in bacteria. *Genome Res.* 12: 962–968.

Kapranov, P., Cawley, S. E., Drenkow, J., Bekiranov, S., Strausberg, R. L., Fodor, S. P. A. and T. Gingeras, R. (2002) Large-scale transcriptional activity in chromosomes 21 and 22. *Science* 296: 916–919.

Kliman, R. M. and Hey, J. (1993) Reduced natural selection associated with low recombination in *Drosophila melanogaster. Mol. Biol. Evol.* 10: 1239–1258.

Knudsen, B. and Miyamoto, M. M. (2001) A likelihood ratio test for evolutionary rate shifts and functional divergence among proteins. *Proc. Natl. Acad. Sci. USA* 98: 14512–14517.

Kramer, E. M., Dorit, R. L. and Irish, V. F. (1998) Molecular evolution of genes controlling petal and stamen development: Duplication and divergence within the APETALA3 and PISTILLATA MADS-box gene lineages. *Genetics* 149: 765–783.

Kreitman, M. (2001) Methods to detect natural selection in populations with applications to the humans. *Annu. Rev. Genom. Human Genet.* 1: 539–559.

Kumar, S., Gadagkar, S. R. and Filipski, A. (2001) Determination of the number of conserved chromosomal segments between species. *Genetics* 157: 1387–1395.

Lagercrantz, U. and Axelsson, T. (2000) Rapid evolution of the CONSTANS LIKE genes in plants. *Mol. Biol. Evol.* 17: 1499–1507.

Lee, M. M. and Schiefelbein, J. (2001) Developmentally distinct MYB genes encode functionally equivalent proteins in *Arabidopsis. Development* 128: 1539–1546.

Lercher, M. J., Williams, E. J. B. and Hurst, L. D. (2001) Local similarity in evolutionary rates extends over whole chromosomes in human-rodent and mouse-rat comparisons: Implications for understanding the mechanistic basis of the male mutation bias. *Mol. Biol. Evol.* 18: 2032–2039.

Li, W. -H. (1993) Unbiased estimation of the rates of synonymous and nonsynonymous substitution. *J. Mol. Evol.* 36: 96–99.

Li, W. -H. (1997) *Molecular Evolution*, Sinauer, Sunderland, MA.

Lindsey, K. (2001) Plant peptide hormones: The long and the short of it. *Curr. Biol.* 11: R741–743.

Löhr, U., Yussa, M. and L. Pick (2001) *Drosophila fushi tarazu*: A gene on the border of homeotic function. *Curr. Biol.* 11: 1403–1412.

Long, M. (2001) Evolution of novel genes. *Curr. Opin. Genet. Dev.* 11: 673–680.

Long, M. and C. H. Langley (1993) Natural selection and the origin of *Jingwei*, a chimeric processed functional gene in *Drosophila. Science* 260: 91–95.

Ludwig, M. Z., Bergman, C., Patel, N. H. and M. Kreitman (2000) Evidence for stabilizing selection in a eukaryotic enhancer element. *Nature* 403: 564–567.

Malik, H. S. and S. Henikoff (2001) Adaptive evolution of Cid, a centromere-specific histone in *Drosophila. Genetics* 157: 1293–1298.

Marais, G., Mouchiroud, D. and L. Duret (2001) Does recombination improve selection on codon usage? Lessons from nematode and fly complete genomes. *Proc. Natl. Acad. Sci. USA* 156: 1661–1669.

Mayfield, J., Fiebig, A., Johnstone, S. E. and D. Preuss (2001) Gene families from the *Arabidopsis thaliana* pollen coat proteome *Science* 292: 2482–2485.

McDonald, J. H. and M. Kreitman (1991) Accelerated protein evolution at the Adh locus in Drosophila. *Nature* 351: 652–654.

McVean, G. A. T. (2001) What do patterns of genetic variability reveal about mitochondrial recombination? *Heredity* 87: 613–620.

Merritt, T. J. S. and Quattro, J. M. (2001) Evidence for a period of directional selection following gene duplication in a neurally expressed locus of triosephosphate isomerase. *Genetics* 159: 689–697.

Miller, W. (2001) Comparison of genomic DNA sequences: Solved and unsolved problems. *Bioinformatics* 17: 391–397.

Mombaerts, P. (2001) The human repertoire of odorant receptor genes and pseudogenes. *Annu. Rev. Genomics Human Genet.* 2: 493–510.

Mourier, T., Hansen, A. J., Willerslev, P. and Arctander, P. (2001) The human genome project reveals a continuous, transfer of large mitochondrial fragments to the nucleus. *Mol. Biol. Evol.* 18: 1833–1837.

Münster, T., Pahnke, J., DiRosa, A., Kim, J. T., Martin, W., Saedler, H. and Theissen, G. (1997) Floral homeotic genes were recruited from homologous MADS-box genes preexisting in the common ancestor of ferns and seed plants. *Proc. Natl. Acad. Sci. USA* 94: 2415–2420.

Nei, M., Gojobori, T. and Li, W. H. (1981) Pseudogenes as a paradigm of neutral evolution. *Nature* 292: 237–239.

Nei, M., Gu, X. and Sitnikova, T. (1997) Evolution by the birth-and-death process in multigene families of the vertebrate immune system. *Proc. Natl. Acad. Sci. USA* 94: 7799–7806.

Nekrutenko, A. and Li, W. -H. (2001) Transposable elements are found in a large number of human protein-coding genes. *Trends Genet.* 17: 619–621.

Nekrutenko, A., Makova, K. D. and Li, W. -H. (2002) The Ka/Ks ratio test for assessing the protein-coding potential of genomic regions: An empirical and simulation study. *Gen. Res.* 12: 198–202.

Nesterova, T. B., Slobodyanyuk, S. Y., Elisaphenko, E. A., Shevchenko, A. I., Johnston, C., Pavlova, M. E., Rogozin, I. B., Kolesnikov, N. N., Brockdorff, N. and Zakian, S. M. (2001) Characterization of the genomic Xist locus in rodents reveals conservation of overall gene structure and tandem repeats but rapid evolution of unique sequence. *Genome Res.* 11: 833–849.

Oshiro, G., Wodicka, L. M., Washburn, M. P., Yates, J. R., Lockhart, D. J. and Winzeler, E. A. (2002) Parallel identification of new genes in *Saccharomyces cerevisiae. Genome Res.* 12: 1210–1220.

Purugganan, M. D. (1998) The molecular evolution of development. *BioEssays* 20: 700–711.

Rivas, E., Klein, R. J., Jones, T. A. and Eddy, S. R. (2001) Computational identification of noncoding RNAs in *E. coli* by comparative genomics. *Curr. Biol.* 11: 1369–1373.

Schmid, K. J. and Aquadro, C. F. (2001) The evolutionary analysis of "orphans" from the *Drosophila* genome identifies rapidly diverging and incorrectly annotated genes. *Genetics* 159: 589–598.

Schmid, K. J. and Tautz, D. (1997) A screen for fast evolving genes from *Drosophila. Proc. Natl. Acad. Sci. USA* 94: 9746–9750.

Schoen, D. J. (2000) Comparative genomics, marker density and statistical analysis of chromosome rearrangements. *Genetics* 154: 381–393.

Semple, C. and Wolfe, K. H. (1999) Gene duplication and gene conversion in the *Caenorhabditis elegans* genome. *J. Mol. Evol.* 48: 555–564.

Shabalina, S. A., Ogurtsov, A. Y., Kondrashov, V. A. and Kondrashov, A. S. (2001) Selective constraint in intergenic regions of human and mouse genomes. *Trends Genet.* 17: 373–376.

Smith, N. G. and Eyre-Walker, A. (2002) Adaptive protein evolution in *Drosophila. Nature* 415: 1022–1024.

Smith, N. G. C., Webster, M. T. and Ellegren, H. (2002) Deterministic mutation rate variation in the human genome. *Genome Res.* 12: 1350–1356.

Städler, T. and Delph, L. F. (2002) Ancient mitochondrial haplotypes and evidence for intragenic recombination in a gynodioecious plant. *Proc. Natl. Acad. Sci. USA* 99: 11730–11735.

Stern, D. L. (2000) Perspective: Evolutionary developmental biology and the problem of variation. *Evolution* 54: 1079–1091.

Sumiyama, K., Kim, C. B. and Ruddle, F. H. (2001) An efficient cis-element discovery method using multiple sequence comparisons based on evolutionary relationships. *Genomics* 71: 260–262.

Tajima, F. (1989) Statistical method for testing the neutral mutation hypothesis. *Genetics* 123: 585–595.

Tatusov, R. L., Galperin, M. Y., Natale, D. A. and Koonin, E. V. (2000) The COG database: A tool for genome-scale analysis of protein functions and evolution. *Nucleic Acids Res.* 28: 33–36.

Theissen, G. (2002) Secret life of genes. *Nature* 415: 741.

Thorne, J. L. (2000) Models of protein sequence evolution and their applications. *Curr. Opin. Genet. Dev.* 10: 602–606.

Ting, C. T., Tsaur, S. C., Wu, M. L. and Wu, C. I. (1998) A rapidly evolving homeobox at the site of a hybrid sterility gene. *Science* 282: 1501–1504.

Vanoosthuyse, V., Miege, C., Dumas, C. and Cock, J. M. (2001) Two large *Arabidopsis thaliana* gene families are homologous to the Brassica gene superfamily that encodes pollen coat proteins and the male component of the self-incompatibility response. *Plant Mol. Biol.* 46: 17–34.

Vision, T. J., Brown, D. G. and Tanksley, S. D. (2000) The origin of genomic duplications in *Arabidopsis. Science* 290: 2114–2117.

Wakeley, J. and Aliacar, N. (2001) Gene genealogies in a metapopulation. *Genetics* 159: 893–905.

Wallis, M. (1996) The molecular evolution of vertebrate growth hormones: A pattern of near-stasis interrupted by sustained bursts of rapid change. *J. Mol. Evol.* 43: 93–100.

Wallis, M. (2001) Episodic evolution of protein hormones in mammals. *J. Mol. Evol.* 53: 10–18.

Walsh, J. B. (1995) How often do duplicated genes evolve new functions? *Genetics* 139: 421–428.

Wang, S. J., Magoulas, C. and Hickey, D. (1999) Concerted evolution within a trypsin gene cluster in *Drosophila. Mol. Biol. Evol.* 16: 1117–1124.

White, S. E. and Doebley, J. F. (1999) The molecular evolution of *terminal ear1*, a regulatory gene in the genus *Zea. Genetics* 153: 1455–1462.

Willard, H. F. and Carrel, L. (2001) Making sense (and antisense) of the X inactivation center. *Proc. Natl. Acad. Sci. USA* 98: 10025–10027.

Woischnik, M., and Moraes, C. T. (2002) Compositional gradients in Gramineae genes. *Genome Res.* 12: 885–893.

Wolfe, K. H. and Shields, D. C. (1997) Molecular evidence for an ancient duplication of the entire yeast genome. *Nature* 387: 708–713.

Wong, G. K., Passey, D. A. and Wu, J. (2001) Most of the human genome is transcribed. *Genome Res.* 11: 1975–1977.

Wright, S. I., Le, Q. H., Schoen, D. J. and Bureau, T. E. (2001) Population dynamics of an Ac-like transposable element in self- and cross-fertilizing *Arabidopsis. Genetics* 158: 1279–1288.

Wu, W., Goodman, M., Lomax, M. I. and Grossman, L. I. (1997) Molecular evolution of cytochrome c oxidase subunit.4. Evidence for positive selection in simian primates. *J. Mol. Evol.* 44: 477–491.

Wyckoff, G. J., Wang, W. and Wu, C. I. (2000) Rapid evolution of male reproductive genes in the descent of man. *Nature* 403: 304–309.

Yang, Z., Swanson, W. J. and Vacquier, V. D. (2000) Maximum-likelihood analysis of molecular adaptation in abalone sperm lysin reveals variable selective pressures among lineages and sites. *Mol. Biol. Evol.* 17: 1446–1455.

Zhang, J., Rosenberg, H. F. and Nei, M. (1998) Positive Darwinian selection after gene duplication in primate ribonuclease genes. *Proc. Natl. Acad. Sci. USA* 95: 3708–3713.

Chapter 8

Eukaryotic Phylogeny in the Age of Genomics: Evolutionary Implications of Functional Differences

John W. Stiller

CONTENTS

Abstract

The ongoing revolutions in biotechnology and genomics are producing an ever-growing body of molecular sequence data for comparative evolutionary study. As more and more genes become available from diverse organisms, trees constructed from sequence-based phylogenetic analyses hold the promise of strongly supported clades of major eukaryotic taxa, and thereby the hope of solving a number of long-standing controversies over ancient evolutionary relationships. This chapter examines some of the assumptions and potential problems inherent to broad-scale sequence-based phylogenetic analyses, particularly the potential for circularity when these analyses are used both to formulate and to test a given evolutionary hypothesis. In addition to direct sequence comparisons, enormous quantities of molecular information are generated from genomic studies, and evolutionary researchers face challenges in developing new methodologies that can complement and test the results of traditional phylogenetic investigations. The exploration of complex, multigenic and coadapted functional systems is one such approach. For example, differences have begun

0-415-29904-7/04/$0.00+$1.50

to emerge among eukaryotes in mechanisms for controlling gene expression, the cell cycle and cellular differentiation during ontogenetic development. Such complicated systems, although not immediately apparent at the phenotypic level, are neither likely to be the result of evolutionary convergence nor are they easily lost once fully integrated into the molecular machinery of a given lineage. As such, they hold great promise as clearly defined, shared-derived characters that can be used to infer monophyletic relationships among major eukaryotic taxa.

8.1 Introduction

The genomics era has generated an enormous body of molecular data from a growing number of organisms for resolving questions of comparative broad-scale eukaryotic evolution. The vast amount of evidence from whole genome comparisons holds great promise for resolving long-standing controversies over taxonomic relationships and unlocking the secrets of the early history of life. How genomic data are analyzed, however, will have an enormous impact on how questions about ancient evolutionary events are answered and on what new evolutionary hypotheses are proposed. The dominant methodology currently used to investigate questions of comparative evolution is sequence-based phylogenetic analysis. Unlike classical studies of morphological and cytological characters, which yielded few apparently shared-derived features useful for grouping major eukaryotic taxa together, sequence-based analyses usually produce a favored set of relationships in phylogenetic reconstruction. Moreover, as the size of molecular datasets grow, there generally is a comparable increase in internal support for a given favored tree topology. Thus, analyses of ever-larger, concatenated alignments are likely to provide a progressively more clear definition of inferred relationships among major eukaryotic groups.

This apparent increase in phylogenetic clarity is desirable if it is derived from historical signal, which strengthens and comes to dominate over random homoplasy, or noise, as datasets increase in size. A problem arises, however, if the signal used to infer relationships does not reflect evolutionary history but is rather due to biases in the data, which lead to mistaken phylogenetic inferences (Felsentstein, 1978; Hendy and Penny, 1989). Various biases that produce phylogenetic artifacts have been documented in alignments of molecular sequences from diverse eukaryotes (Hasegawa and Hashimoto, 1993; Philippe and Adoutte, 1998; Naylor and Brown, 1998; Hirt et al., 1999; Stiller and Hall, 1999; Philippe and Germot, 2000; Stiller et al., 2001). Moreover, these biases can result in statistical inconsistency in sequence analyses and can therefore be expected to worsen as the number of molecular characters increases (Felsenstein, 1978). Consequently, addressing potential artifacts is of paramount importance as phylogenetic analyses are based on increasingly large, concatenated datasets. If a favored tree topology is consistent with biases in the sequences, evolutionary inferences are compromised and internal support values are not reliable measures of support for the evolutionary hypothesis depicted. Even if biases appear to be negligible or are too small to be detected easily, they might be responsible for the recovered tree topology if historical signal is weak in very large datasets.

Chapter 6 reviews various problems related to phylogenetic artifacts in broad-scale molecular analyses; therefore, the general issue is not addressed here at length. Rather, this chapter attempts to step back and examine the underlying problem of methodological circularity when the same phylogenetic methods that can recover incorrect relationships in the presence of sequence biases are used both to generate and also to test phylogenetic hypotheses of ancient evolutionary relationships. Because of this potential for circularity, sequence-based phylogenetic analyses are most powerful as a tool for the study of evolutionary patterns when they are used to test previous suppositions derived from alternative data and methods. Unfortunately, for relationships among many major eukaryotic taxa, no such prior hypotheses are available from classical studies; thus, many of the current and most intriguing proposals about broad-scale eukaryotic relationships are based largely, or even exclusively, on molecular sequence analyses. With the promise of a virtually limitless supply of sequence data, it is of paramount importance to develop new and independent methodological approaches to complement sequence-based phylogenetic hypotheses. In particular, this chapter focuses on the opportunity afforded by the genomics revolution to examine potential homologies in complex molecular and biochemical processes.

8.2 Red Algal and Green Plant Origins as a Case Study for Genomic Analyses

An investigation into the origins of green plants and red algae is a timely example of both the great potential and inherent challenges associated with broad-scale sequence-based phylogenies in the genomics era. Many of the difficulties encountered when interpreting the results of molecular analyses of these two groups are relevant to other investigations of ancient evolution as well. The puzzle of red and green origins also provides an opportunity to examine the possible evolutionary implications of complex and coadapted functional processes. Therefore, this review applies the theme of rhodophyte and chlorophyte origins as a framework for discussing genome-level eukaryotic phylogenetics.

Historically, classical investigations of morphological and cytological features produced highly reliable characters for circumscribing both red algae and green plants as distinct monophyletic lineages; however, little evidence emerged from these studies that indicated specific evolutionary relatives for either group. [See Ragan and Gutell (1995) for a thorough review.] In contrast, molecular phylogenetic analyses have led a number of researchers to hypothesize a sister relationship between plants and red algae. Analyses of concatenated plastid (Martin et al., 1998), mitochondrial (Burger et al., 1999), and, most recently, nuclear genes (Moreira et al., 2000) have produced a monophyletic association of rhodophyte and chlorophyte sequences. Thus, it has been argued that a consensus is emerging from all three genomes in support of a common red–green ancestor (Moreira et al., 2000; Palmer, 2000). It is clear that sequence-based phylogenetic investigations have led to a provocative new hypothesis about eukaryotic relationships. What is not yet clear, however, is whether the clustering of red and green sequences in these analyses is based on a historical signal present in the sequences and really represents support for the proposed hypothesis.

8.3 Do Genomic Studies Provide Clear Support for a Relationship between Green Plants and Red Algae?

The phylogenetic signal favoring one or another hypothesis about ancient evolution is seldom without contradiction. Generally there is some conflicting evidence that favors an alternative set of relationships. Assumptions must be introduced to reconcile these conflicting data with the favored hypothesis, and the validity of those assumptions can be difficult to interpret. This appears to be the case for analyses of molecular data from chloroplast genomes.

There have been a number of recent and extensive reviews of molecular phylogenetic and comparative genomic analyses of plastids from diverse photosynthetic eukaryotes. Many of these analyses indicate a monophyletic relationship among plastids, which has been argued to be most consistent with a single origin of all plastids in the common ancestor of green plants and red algae (Delwiche and Palmer, 1997; Turner, 1997; Delwiche, 1999; McFadden, 2001; Moreira and Philippe, 2001; Archibald and Keeling in Chapter 4). Therefore, although serious questions remain regarding evolutionary history and analyses of plastids and their genomes (Barbrook et al., 1998 ; Moreira and Philippe, 2001), including whether the available plastid data actually favor a single origin of plastids (Stiller et al., 2003), these issues are not addressed here. What is germane to this discussion is whether, as suggested by most reviewers, even clear evidence of a single origin of plastids necessarily favors a sister relationship between red algae and green plants.

Whether plastids have a single origin is not directly related to the question of a red–green relationship, because the genomes of the host cell and organelle need not share the same evolutionary history. For example, because of the secondary transfer of a red plastid to a stramenopile, red and brown algal plastids are monophyletic despite the vast evolutionary divergence between their host cell lineages (Delwiche, 1999). It remains possible that the apparent monophyly of red and green plastid is a reflection of this same phenomenon (Stiller and Hall, 1997). Alternatively, the recent discovery of genes of apparent cyanobacterial ancestry in the genomes of a number of nonphotosynthetic eukaryotic groups suggests that plastids might have been lost from many eukaryotic taxa (Andersson and Roger, 2002). A single origin of plastids appears to be equally consistent with at least three different viable hypotheses about red and green origins, none of which can be dismissed based on the evidence available. Therefore, whether genome-level analyses of plastids favor a red–green relationship depends on which largely untested assumptions are made about the early evolution of plastids rather than on the results of the analyses themselves.

Interpreting plastid-derived evidence with respect to host cell relationships highlights the long-recognized problem that gene, and, in the case of organelles, even whole genome phylogenies are not necessarily congruent with the evolutionary history of the organisms under investigation. As a result, reconciling conflicts between the inferred histories of different sets of genes requires the introduction of assumptions about the source of the conflict. Suppositions that most accurately reflect the process and pattern of eukaryotic evolution are not always understood. The importance of assumption choice in the interpretation of large molecular datasets is illustrated plainly by the results of phylogenetic analyses of mitochondrial genomes.

8.4 Do Mitochondrial Genome Investigations Support a Red–Green Relationship?

Like plastid investigations, phylogenetic analyses of mitochondrial genes also have been cited widely as support for a sister relationship between red algae and plants (Delwiche and Palmer, 1997; Burger et al., 1999; Moreira et al., 2000; Palmer, 2000; Moreira and Philippe, 2001). Also like plastids, the possibility of lateral transfer among eukaryotes exists with mitochondria. Nevertheless, broad-scale phylogenetic analyses of genes from this organelle raise important caveats regarding phylogenetic investigations of large, concatenated datasets and are therefore particularly important for this discussion.

Combined analyses of cytochrome oxidase and apocytochrome *b* genes recover a sister relationship between rhodophyte and green plant mitochondria with reasonably strong bootstrap support, provided certain problematical sequences are removed from the analyses (Burger et al., 1999). Unfortunately, these problematical sequences include a number of chlorophycean algae that do not group with other members of the green plant clade (Figure 8.1). On the other hand, concatenated mitochondrial NADH dehydrogenase genes provide even stronger support against a monophyletic relationship between reds and greens (Stiller et al., 2001), whether or not all green plants and algae are included in the analyses. Thus, mitochondrial genes support a red–green relationship only if a number of *a priori* assumptions are made as to which organisms and genes are more likely to produce phylogenetic artifacts. In addition, variations in evolutionary rates among sequences in both sets of genes result in biases that are consistent with an artificial clustering of red and green mitochondria (Stiller et al., 2001).

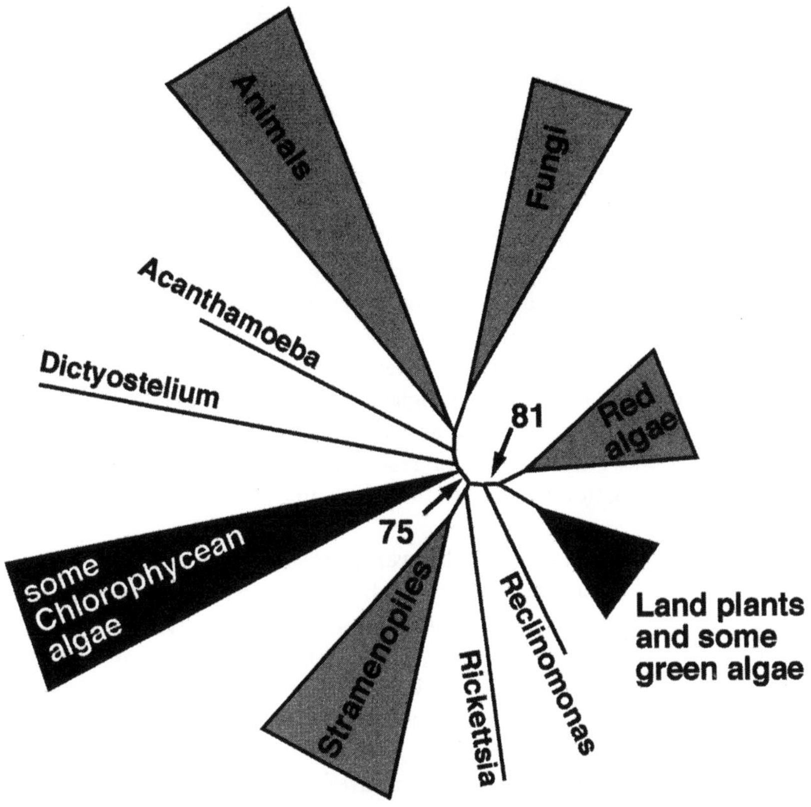

FIGURE 8.1 Neighbor-joining (N-J) tree based on mitochondrial encoded cytochrome oxidase (COX) and apocytochrome *b* (COB) sequences. The tree was recovered in PHYLIP 3.5 [Felsenstein. (1989) 164–165] from a distance matrix recovered in PROTDIST under the categories model of amino acid substitution and using a concatenated alignment of 1352 residues inferred from COX/COB genes. Bootstrap values, based on 1000 PROTDIST/N-J replicates, show support values for the relative branching positions of two different groups of the Viridaeplantae. Land plants and some green algae group with red algae, whereas other green algae associate with different eukaryotes but with comparable bootstrap support. This discrepancy in branching positions between different members of the same major taxa highlights the problem of conflicts in tree-building signal inherent to genomic-level sequence based phylogenetic analyses. In addition, the variations in branch lengths that are evident among different eukaryotic groups reflect dramatic differences in apparent evolutionary rates that are consistent with the clustering of slow- and fast-evolving sequences, respectively [Stiller et al. (2001) 527–539]. Because red algal and (some) green plant mitochondrial genomes are among the slowest evolving, their grouping together is consistent with biases in the data and therefore remains suspect.

This example from genome-wide investigations of mitochondrial evolution raises important issues for consideration in large-scale molecular phylogenetic analyses in general. Even when there are valid reasons to assume that all sequences under analysis share the same evolutionary history, as is likely true for NADH and cytochrome genes in mitochondrial genomes, they might not indicate the same history in phylogenetic analyses. In nearly all or perhaps all cases, there are conflicts in tree-building signal among the different genes from the same genome that are used to create large, concatenated alignments. For example, Moreira et al. (2000) found that an alignment of 13 nuclear genes supported a monophyletic relationship of red algae and green plants; however, the majority of genes included did not recover a red–green clade when analyzed individually.

To overcome these inconsistencies, analyses of multigene concatenated alignments introduce an essential and explicit assumption: as sequences are combined into large datasets, the dominant tree-building signal that emerges from any given set of genomes comes from their historical pattern of relationships. Conflicting signals found in many individual genes are then assumed to result from random noise or phylogenetic artifacts associated with smaller datasets. The actual source of the dominant signal, however, could be consistent biases that can come to dominate tree-building algorithms as datasets increase in size (Felsenstein, 1978), whereas certain individual genes might, in fact, retain enough historical signal from shared-derived characters to recover the true evolutionary history. This uncertainty is less of a concern if molecular phylogenies are used to test prior hypotheses based on independent characters and methodologies. Unfortunately, in the case of a putative sisterhood of red algae and green plants, as with many ancient evolutionary relationships, the hypothesis itself is based entirely on the results of molecular phylogenetic analyses. [See Lipscombe (1989) and Ragan and Gutell (1995) for analyses and discussions of nonmolecular characters.]

Although a clear phylogenetic pattern has yet to emerge from nuclear genomes, expanded analyses of concatenated datasets will very likely reveal a strong global genomic signal in favor of one or another hypothesis regarding red–green origins. That is, a dominant tree-building signal will support one particular tree topology and conflicting signals in the data will be increasingly overwhelmed. In more general terms, larger sample sizes will increase statistical confidence in the precision of phylogenetic analyses; however, they cannot in themselves increase confidence in the accuracy of those analyses (Figure 8.2). For that reason, phylogenetic analyses of larger datasets that are used to test conclusions derived previously from smaller sequence-based analyses could be argued to be tautological.

It seems unlikely that independent confirmation of a sequence-based hypothesis will emerge from additional investigations of traditional cytological and morphological characters. Fortunately, the burgeoning field of comparative genomics offers alternative approaches for addressing questions of ancient evolutionary relationships. A variety of promising approaches are being explored, including examination of conserved, shared gene insertions and deletions (Gupta, 1998), patterns of syntany in gene arrangement (Stoebe and Kowalik, 1999) and whole-genome correlations of functionally orthologous sequences (De Las Rivas et al., 2002; House and Fitz-Gibbon, 2002). This discussion focuses on how genomic data afford evolutionary researchers an opportunity to examine complex biochemical and molecular processes within a phylogenetic framework. It is analogous to the approach used so successfully by evolutionary biologists and systematists in the premolecular era to describe most major eukaryotic taxa.

8.5 Molecular and Biochemical Processes as Shared-Derived Characters

Although classical morphological and cytological characters offered little evidence for polarizing relationships among a number of major eukaryotic groups, the circumscriptions of these taxa themselves have largely stood the tests of time and molecular scrutiny. Molecular phylogenetic analyses have upheld the monophyletic nature of animals, true fungi, green algae and plants, red algae and a number of other eukaryotic groups described in the premolecular era. Molecular investigations have also provided strong and independent confirmation of some classical inferences about relationships among those taxa. For example, as early as the mid-19th century it was proposed that choanoflagellate protists were related specifically to metazoans, based on a number of key cellular features. [See Wainright et al.

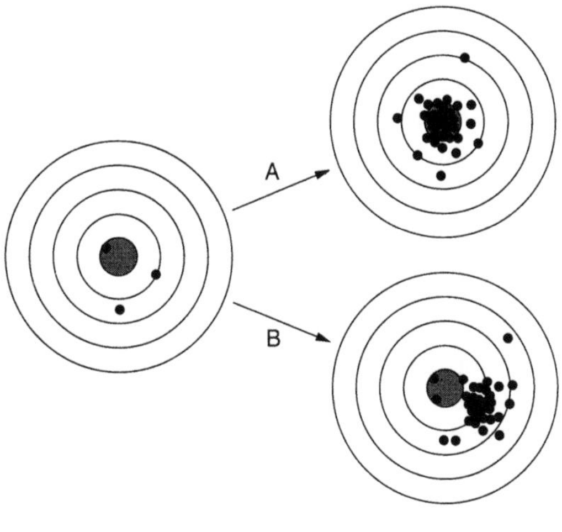

FIGURE 8.2 The effect of adding more sequences to phylogenetic analyses on errors of statistical precision and accuracy. On the left is a schematic target diagram with the tree-building signal from each of three genes indicated by a black dot. The gray circle at the center of the target represents the set of trees containing the true relationship under investigation, in this case the evolutionary positions of red algae and green plants. In Scenario A, the dominant tree-building signal present in the sampled genomes is historical in nature. Thus, as the results of phylogenetic analyses become more precise, they also become more accurate. In contrast, Scenario B demonstrates the effect of an inconsistent data set in which the dominant signal that leads to phylogenetic artifact. In this case the results of phylogenetic analyses become as precise but increasingly inaccurate. Statistical tests derived from the sequence data themselves, such as bootstrap resampling, can assess the precision of the analysis, but do not indicate accuracy unless, as in Scenario A, the two are coincidental.

(1994) for a review.] Likewise, the aquatic phycomycetes (oomycete fungi) and various heterokontic protists were argued to be relatives of chromophytic algae, based on features of their flagella (Sparrow, 1958). Both the hypotheses have been strongly supported by molecular data and have received wide acceptance (see Chapters 3, 4 and 5).

Those classical evolutionary inferences were made successfully because they were based on shared-derived phenotypic features produced by complex, coadapted, multigenic processes. Such complex processes are exceedingly unlikely to be products of convergent evolution and, once fully integrated into cell or physiological function, were not easily lost by most members of a descendent lineage. Thus, they provided clear synapomorphic characters for unifying many major eukaryotic taxa. Broad-scale genomic comparisons of diverse eukaryotes offer an opportunity to uncover similar coadapted molecular and biochemical processes that might not be immediately apparent at a phenotypic level. Evolutionary inferences based on these processes, combined with sequence-based phylogenetic methods to test those inferences, can provide evolutionary biologists with a powerful new approach to genome-level analyses.

8.6 Major Innovations in How the Genome Is Expressed

Major innovations in the mechanisms of control over gene expression have had a profound impact on the evolution of eukaryotic organisms. This is particularly true of complex, multicellular forms. The relatively small differences in gene content between, for example, mouse and human (Mural et al., 2002) suggest that complex organisms are distinguished not so much by the gene complements in their respective genomes as by how those genes are expressed. One of the most striking evolutionary transformations in the history of the

Eukaryota was the acquisition of an ability to differentiate cells in a developmentally coordinated manner.

Only a few groups of organisms are capable of such elaborate development, most notably animals and green plants. There is growing evidence from genomic and mechanistic investigations that this capacity might be a shared-derived feature. A number of the central genetic and biochemical mechanisms underlying developmental complexity are homologous in these two groups; some of them have not been found in any other eukaryotic organisms.

Neither are all green plants (e.g., many green algae) complex, multicellular organisms, nor are the nearest suspected relatives of metazoans (see Chapter 5). Many factors determine which adaptations are favored during the course of evolution, and having the capacity to differentiate cell lines does not mean that all, or even any, members of a given lineage will have done so. Nevertheless, it is possible that one of the most obvious and outwardly profound characteristics found in more than one major eukaryotic taxa, the *capacity* to differentiate cells into functionally specific tissues and ontogenetic developmental stages, also represents a shared-derived feature that unites those taxa as close relatives. In contrast to plants, and despite their remarkable age (Butterfield, 2000) and diversity in morphological form and life history patterns, red alga never managed to differentiate true parenchymatous tissue (Coomans and Hommersand, 1990). The reasons for an absence of true tissues in rhodophytes remain a mystery, but a reasonable hypothesis is that they lack comparable mechanisms for controlling gene expression. Although little is known about the molecular biology of red algae, there are tantalizing new data that support this idea.

8.7 An Overhaul of RNA Polymerase II Transcription?

Lying at the heart of the cellular machinery that determines how eukaryotic organisms express their protein-encoding genes is DNA-dependent RNA polymerase II (RNA pol II). The largest subunit of RNA pol II contains an additional C-terminal domain (CTD) that is not present in other eukaryotic or prokaryotic polymerases (Corden, 1990). The CTD is composed of tandemly repeated heptapeptides, of varying numbers depending on the organism, with the consensus sequence Tyr_1-Ser_2-Pro_3-Thr_4-Ser_5-Pro_6-Ser_7 (Corden, 1990). In animals, and yeast, in which the mechanics of RNA pol II transcription are most well characterized, the CTD has been shown to interact with a variety of transcription-related proteins that are essential for the control of initiation, elongation and processing of mRNA (Figure 8.3; Steinmetz, 1997; Shilatifard, 1998; Myer and Young, 1998; Hirose and Manley, 2000; Proudfoot, 2000).

A typical CTD has been found in all animal, plant and fungal RPB1 sequences examined, as well as those from a number of related protistan taxa (Stiller et al., 2001; Stiller and Hall, 2002) . It has been conserved even in microsporidian parasites (Hirt et al., 1999), which have the most severely reduced genomes and molecular machinery of all eukaryotes (Biderre et al., 1995; Peyretaillade et al., 1998). Given its essential role in RNA pol II transcription, it is not surprising that the CTD has been maintained by intense stabilizing selection in all of these groups. What is surprising is that the CTD has not been conserved in a number of eukaryotic lineages, including the Rhodophyta (Figure 8.4).

The absence of complex, tissue-specific developmental patterns in red algae and the lack of a canonical CTD are probably not coincidental. Among its diverse functions (Figure 8.3), the CTD is essential for regulated gene expression, because it binds the mediator, a multi-subunit complex that transduces control signals to the pol II–promoter complex (Nonet and Young, 1989; Kim et al., 1994; Svejstrup et al., 1997; Myers and Kornberg, 2000). Further, phosphorylated CTD heptads recruit SR proteins and other splicing factors to the elongating message (Yuryev et al., 1996; Corden and Patturajan, 1997; Hirose et al., 1999; Misteli and Spector, 1999), thereby mediating alternative-splicing of exon junctions to produce different

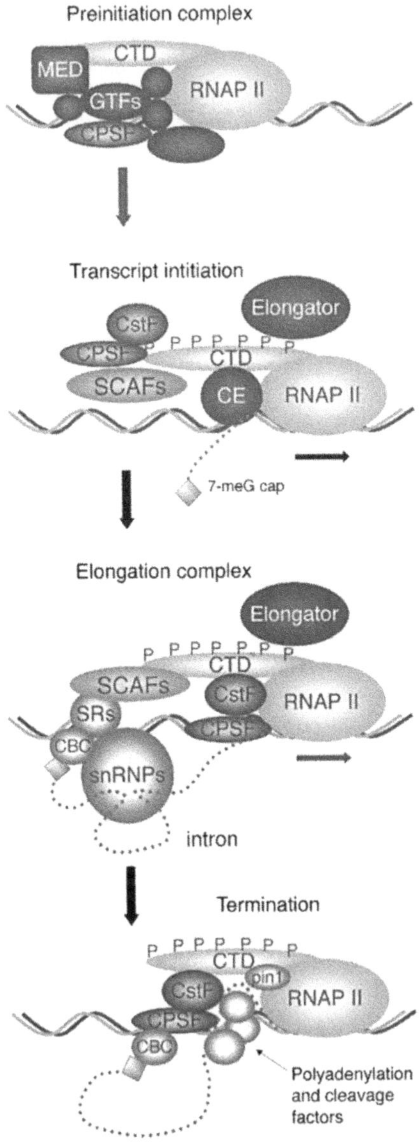

FIGURE 8.3 An overview of CTD-protein interactions during the RNA polymerase II transcription cycle, after descriptions of CTD functions in reviews by Hirose and Manley (2000) 1234–1239 and Proudfoot (2000) 290–293. During initiation of the cycle, the CTD associates with the multi-subunit srb-mediator (MED), which transduces signals from upstream regulatory elements to RNA pol II complexed with general transcription factors (GTFs) at the gene promoter. Phosphorylation of the CTD accompanies promoter clearance and release of the mediator. Once phosphorylated, the CTD binds factors (elongator) that stabilize the elongation complex, as well as the capping enzyme (CE) that catalyzes the addition of a 7-methyl guanine to the 5′ end of the nascent pre-mRNA. The phospho-CTD also binds to SR-like CTD-associated factors (SCAFs) that promote recruitment of the spliceosome to the elongation complex, as well as cleavage polyadenylation specificity factor (CPSF) and cleavage-stimulating factor (CstF), which also attract additional factors related to polyadenylation and cleavage of the mRNA at transcription termination. Cap-binding complex (CBC) stimulates both splicing and polyadenylation and Pin1 affects CTD conformation to increase the efficiency of 3′ mRNA processing and termination.

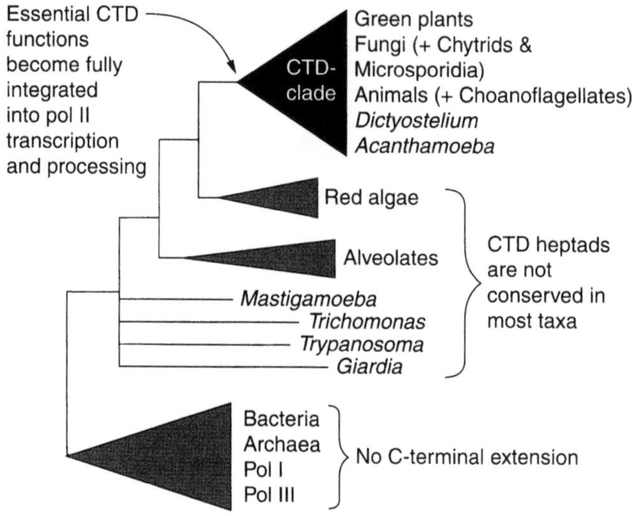

FIGURE 8.4 Evolutionary distribution of a canonical RNA pol II CTD-based on phylogenetic analyses of sequences encoding RPB1, the pol II largest subunit, after Stiller and Hall (2002). All members of the CTD clade have retained a set of tandemly repeated heptapeptides with the canonical sequence Y-S-P-T-S-P-S. In contrast, no RPB1 gene isolated from an organism outside the CTD clade has a typical CTD and most have no semblance of tandemly repeated heptads whatsoever. Regardless of whether the specific set of relations depicted is accurate, the restricted distribution of canonical heptads suggests that they are under intense stabilizing selection in members of the CTD clade, but relatively free to degenerate or vary in other eukaryotes. Thus far, the apparent phylogenetic distribution of CTD heptads is in agreement with the hypothesis that CTD-based protein–protein interactions represent an important synapomporphy that unifies certain eukaryotic groups.

products from the same gene. Recent characterization of a number of additional phospho-CTD associated proteins (PCAPs) suggests that the CTD acts as a kind of nuclear organizing center and plays important roles in coordinating functions far beyond simply transcription and RNA processing (Carty and Greenleaf, 2002).

Presumably, without a CTD red algae have evolved alternative mechanisms for the delicate regulation of gene expression, mechanisms that do not rely on these same protein–protein interactions. Functional analyses of the constraints on CTD structure, through *in vivo* complementation analyses in yeast, support the inference that the differences in RPB1 C-terminal sequences found in red algae are incompatible with CTD-based RNA pol II function (West and Corden, 1995; Pei et al., 2001; JWS, unpublished data). Thus, the processes by which red algae regulate mRNA synthesis and processing must differ substantially from those of organisms that use a CTD-based RNA pol II transcription cycle. The mechanisms present in red algae might not permit the kind of tissue- and developmental-specific differentiation that is the hallmark of multicellular green plants.

The shared complexity of RNA polymerase II transcription and related factors that are present in animals, fungi and plants is not likely to be the product of convergent evolution and once acquired has not been lost easily. Thus, the CTD-based transcription system might well represent the kind of character, at a molecular level, that was used so successfully to circumscribe major eukaryotic taxa in classical studies. Moreover, the elaborate set of interactions between the pol II CTD and transcription-related proteins is but one of a number of complicated regulatory and developmental mechanisms that appear to be shared between green plants and animals (see discussion later). Many of the mechanisms for controlling the cell cycle and cell differentiation in these groups are tied to RNA pol II transcription through CTD-based interactions (Bregman et al., 2000).

To date, phylogenetic analyses of *RPB*1 sequences (Stiller and Hall, 2002) support a so-called CTD clade of all organisms in which a canonical CTD has been absolutely conserved (Figure 8.4). Green plants are members of this clade whereas red algae are not, a result that does not agree with some nuclear gene phylogenies (Moreira et al., 2000). It should be noted that this CTD clade, and in particular statistical support for it, is not always recovered in *RPB*1-based analyses, depending on the methods employed and sequences included (Moreira et al., 2000; Stiller et al., 2001; Dacks et al., 2002). Moreover, one enigmatic and poorly characterized protist, *Mastigamoeba invertens*, that does not group with the CTD clade (Figure 8.4) has 25 tandemly repeated heptapeptides, albeit with an alternative consensus sequence (Stiller et al., 1998). Almost nothing is known about the molecular biology of this organism, much less the role of tandem heptads in its RNA pol II transcription machinery. Should *M. invertens* turn out to share the same CTD-based mechanisms present in green plants, animals and fungi, the interpretation of these coadapted mechanisms as a synapomorphy would be in conflict with the *RPB*1 tree topology. In this context it is important to recognize that a hypothesis of relationships based on RNA pol II functional constraints is independent of hypotheses derived from phylogenetic analyses of *RPB*1 sequences (although it is reassuring that the two are not in conflict thus far). In microcosm, this exemplifies the potential interplay between function and phylogeny and the capacity of one to provide an independent test of the accuracy of hypotheses based on the other.

Although many questions remain, evolutionary and functional investigations of the RNA polymerase II CTD have raised a provocative new idea about deep eukaryotic evolution. Before its true evolutionary significance can be determined, a more complete circumscription of the phylogenetic distribution of the CTD is needed, along with a better understanding of RNA pol II transcription from red algae and other poorly studied eukaryotes. More mechanistic investigations and genome-wide bioinformatic comparisons are needed to determine which CTD–protein interactions result in strong stabilizing on tandem heptad structure in certain eukaryotes but not in others. Nevertheless, even if investigations continue to support the CTD as a key evolutionary synapomorphy useful for polarizing eukaryotic relationships, no single character, no matter how complex, can provide confirmation of specific eukaryotic relationships. Congruent evidence will be required from comparative analyses of other molecular and biochemical processes. The revolutions in genomics and proteomics are beginning to make such comparisons possible.

8.8 Control of the Cell Cycle and Cellular Differentiation

Mechanisms for regulating RNA polymerase II function, specifically when, where and how specific genes are transcribed, lay at the heart of cellular differentiation and ontogenic development in animals and green plants. There is growing evidence that a number of homologous mechanisms serve as master controls over complex gene expression in these two groups. One of the most striking examples is a key checkpoint in animal development that is exerted by the retinoblastoma protein (Rb), which is important for regulating both cell division and differentiation (Murray, 1997). An Rb homologue (Xie et al., 1996; Grafi et al., 1996) has been shown to play comparable roles in regulation of the green plant cell cycle (Huntley et al., 1998). Even more remarkable is the discovery that the Rb protein in plants is at the center of a complex set of interactions that occur during the G_1 phase of cell division, involving E2F transcription regulators (Ramirez-Parra et al., 1999) and D-type cyclin kinases (Huntley et al., 1998). All these proteins have homologues in animals and perform the same roles in controlling cell division, coordinating growth and differentiation of tissue specific cell type (Huntley and Murray, 1999).

Although absent from the yeast and other fungal genomes examined thus far, a retinoblastoma-like protein has been reported from the cellular slime mold *Physarum polycephalum* (Loidl and Loidl, 1996) and might play some role in controlling cell division in many eukaryotes. However, the suite of G_1 interactions that are central to developmental and tissue specific cell differentiation have been described only in green plants and animals (Huntley et al., 1998; Oakenfull et al., 2002). Thus, it is possible that they represent derived processes that convey more intricate control over this cell cycle checkpoint and represent another key synapomorphy that unites certain major eukaryotic taxa.

As with the canalization of a CTD-based RNA pol II cycle, it has been argued that the invention of the G_1 pathway was a defining moment in the evolution of eukaryotes, which permitted the elaboration of complex organisms with multicellular tissues (Huntley and Murray, 1999). Given the inability of red algae to differentiate cells into tissues, they may well lack the G_1 pathway that is present in both green plants and animals.

8.9 Homologies in the Mechanisms of Homeotic Development

In animals, plants and fungi, homeotic genes encode central transcriptional regulators of key biological processes and developmental programs (Coen and Meyerowitz, 1991; Shore and Sharrocks, 1995). The specific homeotic pathways, which control overall spacial pattern formation, appear to have evolved independently in plants and metazoans (Meyerowitz, 2002). In both vertebrate and invertebrate animals, homeobox (HOX) genes act as core transcription regulators for establishing the central body plan (Lander et al., 2001), whereas this role is played by MADS box genes in plants (Jack, 2001; Meyerowitz, 2002). The fact that multicellular animals and plants have evolved different mechanisms for establishing their basic body plans is not surprising, given that their last common ancestor is believed to have been a unicellular or, at most, a simple colonial form (Meyerowitz. 2002). What might be of great significance for understanding eukaryotic evolution, however, is the fact that the raw material for attaining homeotic development was present in that last common ancestor.

Homeobox genes are present in plants, although thus far they have not been implicated as master regulators of morphological form (Meyerowitz, 2002); rather, flower development in plants is controlled by a different class of homeotic genes belonging to the MADS box family of transcription factors (Jack, 2001). MADS box genes also are present in animals, and several have been shown to be important regulators of cell and tissue differentiation (Taylor et al., 1995; Montegne et al., 1996). Moreover, the two major MADS gene families (MADS type I/SRF-like and MADS type II/MEF2-like) resulted from a duplication that occurred before the divergence of plants, animals and fungi (Alvarez-Buylla et al., 2000).

Although the specific genes chosen as master homeotic regulators differ between plants and animals, the mechanisms for establishing and maintaining expression of those genes, through regulation of chromatin structure, appears to be homologous in the two groups (Meyerowitz, 2002). Both SWI2/SNF2 chromatin remodeling proteins (Verbsky and Richards, 2001) and members of the polycomb gene family (Jones and Gelbart, 1993; Goodrich et al., 1997) perform homologous functions in activation and repression of homeotic genes in plants and animals. The chief difference seems to be that they act to regulate MADS box as opposed to HOX genes in the two respective systems (Meyerowitz, 2002). Polycomb genes also have similar functions in early development, determining the anterior–posterior polar axis in both animal embryos and plants seeds (Sorenson et al., 2001).

There are additional key animal homeotic regulators for which homologues have been found in plants, such as the Trithorax gene family (Alvarez-Venegas and Avramova, 2001) and SHAGGY-like protein kinases that are important for establishing polarity and cell differentiation in animals and for cell elongation in plants (Pérez-Pérez et al., 2001). One of the most interesting examples involves the mechanisms used to maintain undifferentiated stem cells in animals and meristem cells in plants. The ZWILLE gene, which encodes a protein responsible for preventing differentiation of meristematic cells in *Arabidopsis* (Moussian et al., 1998), is homologous with the gene (*piwi/hiwi/prg-1*) that maintains stem cell lines in animals (Cox et al., 1998). Moreover, in both systems the regulator is expressed in somatic tissues and results in an intercellular signal that prevents nearby stem or meristem cells from differentiating (Benfey, 1999). A possible *ZWILLE* homologue has been reported in the ciliate *Paramecium* (Obara et al., 2000), suggesting that the gene might have a relatively ancient origin. It remains to be seen whether *ZWILLE* plays a role in cell differentiation in diverse eukaryotes and whether its function in multicellular development is evolutionarily analogous or homologous in animals and plants. Nevertheless, the apparent similarity in the overall modes of action of *ZWILLE* and *PIWI* suggests yet another example of a homology in mechanisms of developmental control. More detailed investigation of both plant and animal genomes should soon clarify this point.

At present, it is unknown whether red algae contain any of this growing list of mechanisms that exert control over gene regulation, and through it over cell and tissue differentiation and development. That so many of the complicated and apparently derived systems are shared by green plants, animals and sometimes fungi can have profound implications for understanding the evolutionary relationships among these groups. Whole genome sequences should provide a clear picture of when in the course of eukaryotic evolution the different key innovations for control of cell differentiation originated. If red algae do not share these integrated mechanisms, it would be strong evidence that they also do not share a common ancestor with green plants to the exclusion of animals and fungi. It would further support the argument, based initially on differences in RNA polymerase II C-terminal structures, that red algae lack true tissue differentiation because they diverged from a lineage of eukaryotic ancestors that had not evolved the prerequisite innovations in genome regulation (Stiller and Hall, 1998). On the other hand, rhodophyte genomes might turn out to encode the same mechanisms that are present in green plants, and perhaps some will display specific evolutionary homologies that are absent from animals and other eukaryotes. If so, these homologies will provide powerful support for the proposition, based initially on gene phylogenies (Ragan and Gutell, 1995; Moreira et al., 2000), that red algae really are red plants after all.

8.10 Conclusions

As is true for a number of other long-standing evolutionary problems, molecular phylogenetic data remain inconclusive with respect to a possible relationship between green plants and red algae. As the signal from ever-larger datasets increases, this most likely will change. Regardless of whether that signal eventually supports a sister relationship between the two groups or indicates that they have independent origins, alternative approaches are needed to determine the ultimate reliability of the trees produced. Genomic-level comparisons of molecular and biochemical processes provide such an approach. Unfortunately, such wholesale comparisons between rhodophytes and chlorophytes will have to await the emergence of full-scale red algal genomics.

Homologies in mechanisms of cell cycle and developmental regulation have important implications for other eukaryotic relationships as well. Complete genome sequences from several animals, as well as *Saccharomyces* and *Arabidopsis*, have been examined closely. It is interesting that a number of important regulatory genes and pathways (e.g., retinoblastoma protein, polycomb genes, *zwille/piwi*) that are homologous in plants and animals are absent from yeast. Whether more complex multicellular fungi share the developmental controls common to green plants and animals has yet to be determined. A more comprehensive evaluation of coadapted regulatory processes from the three groups might well confirm the dominant view that has emerged from molecular phylogenetic analyses that metazoans and fungi are most closely related (see Chapter 5). On the other hand, a determination that, like yeast, multicellular fungi also lack key processes shared between plants and animals could lead to new and provocative ideas about the relationships among the three groups.

The regulatory pathways discussed here are only a few examples of the kinds of complex processes that can be investigated using genome sequences from diverse eukaryotic organisms. When molecular phylogenetic analyses are used to investigate classical evolutionary models, they do not always yield results congruent with those models. Likewise, some genome-wide investigations of coadapted functional processes will probably lead to dramatic new ideas about ancient evolution, whereas others will add support to conclusions drawn previously from molecular phylogenies. In some cases, a more thorough understanding of biochemical and molecular processes will undoubtedly provide the kind of clearly shared-derived characters that permit unequivocal determinations of relationships among major eukaryotic groups.

Acknowledgment

I thank D. Bhattacharya for review and helpful suggestions on the initial manuscript.

References

Alvarez-Buylla, E.R., Pelaz, S., Lilegren, S.J., Gold, S.E., Burgeff, C., Ditta, G.S., de Pouplana, L.R., Martinez-Castilla, L. and Yanofsky, M.F. (2000) An ancestral MADS-box gene duplication occurred before the divergence of plants and animals. *Proc. Natl. Acad. Sci. USA* 97: 5328–5333.

Alvarez-Venegas, R. and Avramova, Z. (2001) Two *Arabidopsis* homologs of the animal trithorax genes: A new structural domain is a signature feature of the *trithorax* gene family. *Gene* 271: 215–221.

Andersson, J.O. and Roger, A.J. (2002) A cyanobacterial gene nonphotosynthetic protists: An early chloroplast acquisition in eukaryotes. *Curr. Biol.* 12: 115–119.

Baldauf, S.L., Roger, A.J., Wenk-Siefert, I. and Doolittle W.F. (2000) A kingdom level phylogeny of eukaryotes based on combined protein data. *Science* 290: 972–977.

Barbrook, A.C., Lockhart, P.J. and Howe, C.J. (1998) Phylogenetic analysis of plastid origins based on *secA* sequences. *Curr. Genet.* 34: 336–341.

Benfey, P.N. (1999) Stem cells: A tale of two kingdoms. *Curr. Biol.* 9: R171–R172.

Biderre, C., Pages, M., Metenier, G., Canning E.U. and Vivares C. P. (1995) Evidence for the smallest nuclear genome (2.9 Mb) in the microsporidian *Encephalitozoon cuniculi. Mol. Biochem. Parasitol.* 74: 229–231.

Bregman, D.B., Pestell, R.G. and Kidd, V.J. (2000) Cell cycle regulation and RNA polymerase II. *Front. Biosci.* 5: D244–D257.

Burger, G., Saint-Louis, D., Gray, M.W. and Lang, B.F. (1999) Complete sequence of the mitochondrial DNA of the red alga *Porphyra purpurea*: Cyanobacterial introns and shared ancestry of red and green algae. *Plant Cell* 11: 1675–1694.

Butterfield, N.J. (2000) *Bangiomorpha pubescens* n gen, n sp: Implications for the evolution of sex, multicellularity and the Mesoproterozoic–Neoproterozoic radiation of eukaryotes. *Paleobiology* 26: 386–404.

Carty, S.M. and Greenleaf, A.L. (2002) PhosphoCTD-associated proteins in the nuclear proteome link transcription to DNA/chromatin modification and RNA processing. *Mol. Cell. Proteoml.* !: 598–610.

Cavalier-Smith, T. (2000) Membrane heredity and early chloroplast evolution. *Trends Plant Sci.* 5: 174–182.

Coen, E.S. and Meyerowitz, E.M. (1991) The war of the whorls: Genetic interactions controlling flower development. *Nature* 353: 31–37.

Coomans, R.J. and Hommersand, M.H. (1990) Vegetative growth and organization. In *Biology of the Red Algae* (Cole, K.M. and Sheath, R.G., Eds.), Cambridge University Press, New York, 517 pp.

Corden, J.L. (1990) Tails of polymerase II. *Trends Biol. Sci.* 15: 383–387.

Corden, J.L. and Patturajan, M. (1997) A CTD function linking transcription to splicing. *Trends Biol. Sci.* 22: 413–416.

Cox, D.N., Chao, A., Baker, J., Chang, L., Qiao, D. and Lin, H. (1998) A novel class of evolutionarily conserved genes defined by *piwi* are essential for stem cell self-renewal. *Genes Dev.* 12: 3715–3727.

Dacks, J.B., Marinets, A., Doolittle, W.F., Cavalier-Smith, T. and Logsdon Jr., J.M. (2002) Analyses of RNA polymerase II genes from free-living protists: Phylogeny, long branch attraction, and the eukaryotic big bang. *Mol. Biol. Evol.* 19: 830–840.

de Jager S.M. and Murray J.A.H. (1999) Retinoblastoma proteins in plants. *Plant Mol. Biol.* 41: 295–299.

De Las Rivas, J., Lozano, J.J. and Ortiz, A.R. (2002) Comparative analysis of chloroplast genomes: Functional annotation, genone-based phylogeny, and deduced evolutionary patterns. *Genome Res.* 12: 567–583.

Delwiche, C.F. (1999) Tracing the thread of plastid diversity through the tapestry of life. *Am. Nat.* 154: S164–S177.

Delwiche, C.F. and Palmer, J.D. (1997) The origin of plastids and their spread via secondary symbiosis. *Plant Syst. Evol. (Suppl.)* 11: 53–86.

Felsenstein J. (1978) Cases in which parsimony or compatibility methods will be positively misleading. *Syst. Zool.* 25: 401–410.

Felsenstein, J. (1989) PHYLIP – phylogenetic inference package (vers. 3.5). *Cladistics* 5:164–165.

Giesecke, H., Barale, J.C., Langsley, G., and Cornelissen, A.W.C.A. (1991) The C-terminal domain of RNA polymerase II of the malaria parasite *Plasmodium berghei. Biochem. Biophys. Res. Comm.* 180: 1350–1355.

Goodrich, J., Puangsomlee, P. Martin, M., Long, D. Meyerowitz, E. and Coupland, G. (1997) A Polycomb-group gene regulates homeotic gene expression in *Arabidopsis. Nature* 386: 44–51.

Grafi, G., Burnett, R.J., Helentjaris, T., Larkins, B.A., DeCaprio, J.A., Sellers, W.R. and Kaelin Jr., W.G. (1996) A maize cDNA encoding a member of the retinoblastoma family: Involvment in endoreduplication. *Proc. Natl. Acad. Sci. USA* 93: 8962–8967.

Greenleaf, A. (1993) Positive patches and negative noodles: Linking RNA processing to transcription. *Trends Biochem. Sci.* 18: 117–119.

Gupta, R.S. (1998) Protein phylogenies and signatures sequences: A reappraisal of evolutionary relationships among Archaebacteria, Eubacteria and eukaryotes. *Microbiol. Mol. Biol. Rev.* 62: 1435–1491.

Hasegawa, M. and Hashimoto, T. (1993) Ribosomal RNA trees misleading? *Nature* 326: 411–414.

Hendy, M.D. and Penny, D. (1989) A framework for the quantitative study of evolutionary trees. *Syst. Zool.* 38: 297–309.

Hirose, Y. and Manley, J.L. (2000) RNA polymerase II and the integration of nuclear events. *Genes Dev.* 14: 1415–1429.

Hirose, Y., Tacke, R. and Manley, J.L. (1999) Phosphorylated RNA polymerase II stimulates pre-RNA splicing. *Genes Dev.* 13: 1234–1239.

Hirt, R.P., Logsdon, J.M., Healy, B., Dorey, M.W., Doolittle, W.F. and Embley, T.M. (1999) Microsporidia are related to fungi: Evidence from the largest subunit of RNA polymerase II and other proteins. *Proc. Natl. Acad. Sci. USA* 96: 580–585.

House, C.H. and Fitz-Gibbon, S.T. (2002) Using homolog groups to create a whole-genome tree of free-living organisms: An update. *J. Mol. Evol.* 54: 539–547.

Huntley, R., Healy, S., Freeman, D., Lavender, P., de Jager, S, Greenwood, J., Makker, J., Walker, E., Jackman, M. Xie, Q., Bannister, A.J., Kourarides, T., Guitiérrez, C., Doonan, J.H. and Murray, J.A.H. (1998) The maize retinoblastoma protein homologue ZmRb-I is regulated during leaf development and displays conserved interactions with G1/S regulators and plant cyclin D (CycC) proteins. *Plant Mol. Biol.* 37: 155–169.

Huntley, R.P. and Murray, J.A.H. (1999) The plant cell cycle. *Curr. Opin. Plant Biol.* 2: 440–446.

Kang, M.E. and Dahmus, M.E. (1995) The unique C-terminal domain of RNA polymerase II and its role in transcription. In *Advances in Enzymology and Related Areas of Molecular Biology* (Meister, A., Ed.), John Wiley & Sons, pp. 41–77.

Kim, Y.-J., Björklund, S., Li, Y., Sayre, M.H. and Kornberg, R.D. (1994) A multiprotein mediator of transcriptional activation and its interaction with the C-terminal repeat domain of RNA polymerase II. *Cell* 77: 599–608.

Jack, T. (2001) Plant development going MADS. *Plant Mol. Biol.* 46: 515–520.

Jones, R. and Gelbart, W. (1993) The *Drosophila* polycomb-group gene enhancer of zeste contains a region of sequence similar to *trithorax*. *Mol. Cell. Biol.* 13: 6357–6366.

Lander, E.S. et al. (2001) Initial sequencing and analysis of the human genome. *Nature* 409: 860–921.

Lipscombe, D. L. (1989) Relationships among eukaryotes. In *Hierarchy of Life (*Fermholm, B., Bremer, K. and Jornvall, H., Eds.), Excerpta Medica, Amsterdam, pp. 161–178.

Loidl, A. and Loidl, P. (1996) Oncogene- and tumor-suppressor gene-related proteins in plants and fungi. *Crit. Rev. Oncogen.* 7: 49–64.

Martin, W.S., Stoebe, B., Goremykin, V., Hansmann, S., Hasegawa, M. and Kowallik, K.V. (1998) Gene transfer to the nucleus and the evolution of chloroplasts. *Nature* 393: 162–165.

McFadden, G.I. (2001) Primary and secondary endosymbiosis and the origin of plastids. *J. Phycol.* 37: 951–959.

Meyerowitz, E.M. (2002) Plants compared to animals: The broadest comparative study of development. *Science* 295: 1482–1485.

Minvielle-Sebastia, L. and Keller W. (1999) mRNA polyadenylation and its coupling to other RNA processing reactions and to transcription. *Curr. Opin. Cell Biol.* 11: 352–357.

Misteli, T. and Spector, D.L. (1999) RNA polymerase II targets pre-mRNA splicing factors to transcription sites in vivo. *Mol. Cell* 3: 697–705.

Montagne, J., Groppe, J., Guillemin, K., Krasnow, M.A., Gehring, W.J. and Affolter, M. (1996) The *Drosophila* serum response factor gene is required for the formation of intervein tissue of the wing and is allelic to blistered. *Development* 122: 2589–2597.

Moreira, D., Le Guyader, H. and Philippe, H. (2000) The origin of red algae and the evolution of chloroplasts. *Nature* 405: 69–72.

Moreira, D. and Philippe, H. (2001) Sure facts and open questions about the origin and evolution of photosynthetic plastids. *Res. Microbiol.* 152: 771–780.

Moussian,B., Schoof, H., Haecker, A. Jergens, G. and Laux, T. (1998) Role of the *ZWILLE* gene in the regulation of central shoot meristem cell fate during *Arabidopsis* embryogenesis. *EMBO J.* 17: 1799–1809.

Mural et al . (2002) A comparison of whole-genome shotgun-derived mouse chromosome 16 and the human genome. *Science* 296: 1661–1671.

Murray, J.A.H. (1997) The retinoblastoma protein is in plants! *Trends Plant Sci.* 2: 82–84.

Myer, V.E. and Young, R.A. (1998) RNA polymerase II holoenzymes and subcomplexes. *J. Biol. Chem.* 273: 27757–27760.

Myers, L.C. and Kornberg, R.D. (2000) Mediator of transcriptional regulation. *Annu. Rev. Biochem.* 69: 729–749.

Naylor, G.J.P. and Brown, W.M. (1998) Amphioxus mitochondrial DNA, chordate phylogeny, and the limits of inference based on comparison of sequences. *Syst. Biol.* 47: 61–76.

Nonet, M.L. and Young, R.A. (1989) Intragenic and extragenic suppressers of mutations in the heptapeptide repeat domain of *Saccharomyces cerevisiae* RNA polymerase II. *Genetics* 123: 714–715.

Oakenfull, E.A., Riou-Khamlichi, C. and Murray, J.A.H. (2002) Plant D-type cyclins and the control of G1 progression. *Philos. Trans. R. Soc. Lond. B* 357: 749–760.

Obara, S., Iwataki, Y. and Mikami, K. (2000) Identification of a possible stem-cell maintenance homologue in the unicellular eukaryote *Paramecium caudatum. Proc. Natl. Acad. Sci. USA* 76: 57–62.

Palmer, J.D. (2000) Molecular evolution: A single birth of all plastids? *Nature* 405: 32–33.

Pei, Y., Hausmann, S., Ho, C.K., Schwer, B. and Shuman, S. (2001) The length, phosphorylation state, and primary structure of the RNA polymerase II carboxyl-terminal domain dictate interactions with mRNA capping enzymes. *J. Biol. Chem.* 276: 28075–28082.

Pérez- Pérez, J.M., Ponce, M.R. and Micol, J.L. (2001) *ULTRACURATA 1*, a *SHAGGY*-like *Arabidopsis* gene required for cell elongation. *Int. J. Dev. Biol.* 45: S51–S52.

Peyretaillade, E., Biderre, C., Peyret, P., Duffieux, F., Metenier, G., Gouy, M., Michot, B. and Vivares, C.P. (1998) Microsporidian *Encephalitozoon cuniculi,* a unicellular eukaryote with an unusual chromosomal dispersion of ribosomal genes and a LSU rRNA reduced to the universal core. *Nucleic Acids Res.* 26: 3513-3520.

Philippe, H. and Adoutte, A. (1998) The molecular phylogeny of Eukaryota: Solid facts and uncertainties. In *Evolutionary Relationships among Protozoa* (Coombs, G, Vickerman, K., Sleigh, M. and Warren, A., Eds.), Kluwer Academic, Dordrecht, pp. 25–56.

Philippe, H. and Germot, A. (2000) Phylogeny of eukaryotes based on ribosomal RNA: Long-branch attraction and models of sequence evolution. *Mol. Biol. Evol.* 17: 830–834.

Proudfoot, N. (2000) Connecting transcription to messenger RNA processing. *Trends Biol. Sci.* 25: 290–293.

Ragan, M. and Gutell, R. (1995) Are red algae plants? *Bot. J. Linn. Soc.* 118: 81–105.

Ramirez-Parra, E., Xie, Q., Boniotti, M.B. and Guiterrez, C. (1999) The cloning of plant E2F, a retinoblastoma binding protein, reveals unique and conserved features with animal G(1)/S regulators. *Nucleic Acids Res.* 27: 3527–3533.

Shilatifard, A. (1998) The RNA polymerase II general elongation complex. *J. Biol. Chem.* 379: 27–31.

Shore, P. and Sharrocks, A.D. (1995) The MADS-box family of transcription factors. *Eur. J. Biochem.* 229: 1–13.

Sorensen, M.B., Chaudhury, A.M., Robert, H., Bancharel, E. and Berger, F. (2001) Polycomb group genes control pattern formation in plant seeds. *Curr. Biol.* 11: 277–281.

Sparrow, F.K. (1958) Interrelationships and phylogeny of the aquatic phycomycetes. *Mycologia* 6: 797–813.

Steinmetz, E.J. (1997) Pre-mRNA processing and the CTD of RNA polymerase II: The tail that wags the dog? *Cell* 89: 491–494.

Stiller, J.W., Duffield, E.C.D. and Hall, B.D. (1998) Amitochondriate amoebae and the evolution of DNA-dependent RNA polymerase II. *Proc. Natl. Acad. Sci. USA* 95: 11760–11774.

Stiller, J.W. and Hall B.D. (1997) The origin of red algae: Implications for plastid evolution. *Proc. Natl. Acad. Sci. USA* 94: 4520–4525.

Stiller, J.W. and Hall, B.D. (1998) Sequences of the largest subunit of RNA polymerase II from two red algae and their implications for rhodophyte evolution. *J. Phycol.* 34: 857–864.

Stiller, J.W. and Hall, B.D. (1999) Long-branch attraction and the rDNA model of early eukaryotic evolution. *Mol. Biol. Evol.* 16: 1270–1279.

Stiller, J.W. and Hall, B.D. (2002) Evolution of the RNA polymerase II C terminal domain. *Proc. Natl. Acad. Sci. USA* 99: 6091–6096.

Stiller, J.W., Reel, D.C. Johnson, J.C. (2003) The case for a single plastid origin revisited: Convergent evolution in organellar genome content. *J. Phycol.* 39: 95–105.

Stiller, J.W., Riley, J. and Hall, B.D. (2001) Are red algae plants? A critical evaluation of three key molecular datasets. *J. Mol. Evol.* 52: 527–539.

Stroebe, B. and Kowallik, D.V. (1999) Gene-cluster analysis of chloroplast genomics. *Trends Genet.* 15: 344–347.

Svejstrup, J.Q., Li, Y., Fellows, J., Gnatt, A., Bjorklund, S. and Kornberg, R.D. (1997) Evidence for a mediator cycle at the initiation of transcription. *Proc. Natl. Acad. Sci. USA* 94: 6075–6078.

Taylor, M.V., Beatty, K.E., Hunter, H.K. and Baylies, M.K. (1995) *Drosophila* mef2 is regulated by twist and is expressed in both the primordial and differentiated cells of the embryonic somatic, visceral and heart musculature. *Mech. Develop.* 50: 29–41.

Turner, S. (1997) Molecular systematics of oxygenic photosynthetic bacteria. *Plant Syst. Evol. (Suppl.)* 11: 13–52. [Chapter 2 in *The Origins of Algae and their Plastids* (Bhattacharya, D., Ed.), Springer Wien, New York.]

Verbsky, M.L. and Richards, E.J. (2001) Chromatin remodeling in plants. *Curr. Opin. Plant Biol.* 4: 494–500.

Wainright, P.O., Patterson, D.J. and Sogin, M.L. (1994) Monophyletic origin of animals: A shared ancestry with fungi. In: *Molecular Evolution of Physiological Processes*, Society of General Physiologists Series No. 49 (Farmborough, D.M., Ed.), Rockefeller University Press, New York, pp. 39–53.

West, M.L. and Corden, J.L. (1995) Construction and analysis of yeast RNA polymerase II CTD deletion and substitution mutants. *Genetics* 140: 1223–1233.

Yuryev, A., Patturajan, M., Litingtung, Y., Joshi, R.V., Gentile, C., Gebara, M. and Corden, J.L. (1996) The C-terminal domain of the largest subunit of RNA polymerase II interacts with a novel set of serine/arginine-rich proteins. *Proc. Natl. Acad. Sci. USA* 93: 6975–6980.

Chapter 9

Genome Phylogenies

Robert L. Charlebois, Robert G. Beiko and Mark A. Ragan

CONTENTS

Abstract

The classification of organisms has value beyond taxonomy as it can reveal their natural relationships and evolutionary history. Morphological and physiological traits have largely given way to genetic characterization in determining phylogeny, though the gene might not always represent the organism. The problem goes beyond statistical concerns as real biological processes can muddle or shuffle inferred relationships. The collective of genes within a genome might better represent the organism if the goings-on of individual genes do no more than add unbiased noise to an otherwise resilient signal. We explore such genome-based phylogenic methods and find that genome trees largely resemble the standard small subunit ribosomal RNA (SSUrRNA) tree, suggesting that genomes are suitable proxies for the organisms that contain them.

9.1 Introduction

Genomes, like the genes they contain and the organisms in which they occur, retain evidence of the evolutionary processes and historical path by which they have arisen. This chapter considers genomes from a phylogenetic perspective. We discuss what a genome phylogeny might mean, take note of processes that could diminish the usefulness of genome phylogenies and present methods for inferring phylogenetic trees from genomic data. It is instructive to begin by briefly considering the historical development of relevant concepts.

0-415-29904-7/04/$0.00+$1.50

9.1.1 Organism, Gene and Genome

The classification of organisms into abstract types is as old as human language (Greene, 1909). Plato introduced a more formalized classification based no longer on convenience but instead on idealized concepts (Platonic forms). Cesalpino, Ray, Linnaeus and de Jussieu replaced these forms with shared morphological and ecological characters, and later von Baer, Cuvier and Geoffroy arranged animals according to idealized developmental and mechanical principles. By the early 19th century the true scale of geological (hence potentially biological) time had become known, giving Darwin scope to recast natural groupings as genealogical lineages shaped by gradual selection. Genealogies require temporal continuity of genetic determinants, although what these might be and how they might be transmitted was quite unknown in 1859. In the succeeding decade, Mendel discovered that phenotypic characters could be transmitted in statistically describable ways that required the underlying genetic determinants to behave as discrete units, although his work was ignored in the debate between Nägeli's mechanophysiological and Haeckel's ontogenic–phylogenetic explanations for organismal development and diversity. By the end of the 19th century, idealistic formulations of organism had largely been replaced with concepts based on group identity, phenotype, developmental principles and continuity through time.

Genes came into their own in the 20th century. Animals and plants had been bred for desirable traits since time immemorial, but the rediscovery of Mendel's laws (independently by Correns, von Tschermak and de Vries) refocused attention on the mechanistic basis of heredity. Garrod (1909) showed that mutant genes generate specific metabolic errors, a fundamental relationship later generalized by Beadle and Tatum (1941) as the one gene – one protein hypothesis. Avery et al. (1944) and Hershey and Chase (1952) localized these genes in DNA. When the structural principles of DNA became known (Watson and Crick, 1953), it was obvious how DNA both provides physical continuity of information through time and specifies the full range of cellular proteins via parsing rules (the genetic code). By the 1960s, the gene was fundamentally understood, although important details (e.g., introns, reverse transcription, alternative splicing) would be added later. With the rise of sequencing, functionally important regions within genes were found to be relatively conserved in structure and sequence across even large phyletic distances, whereas intergenic regions and pseudogenes are much more variable. If (to a first approximation) information inherent in DNA must be either used or lost, it is not too great an extrapolation to imagine that the genetic material (genome) in each organism must be significantly coupled with the currently realized suite of phenotypes within its total phenotype space, actual and potential (phenome).

Genomes thus face two ways: on the one hand they represent the collection of all genes, and on the other the organism *in potentio*. In the past few years, genomes of diverse bacteria, archaea and eukaryotes have been sequenced. As expected, these sequences provide unprecedented insight into the basis and development of organismal form and function and are changing the face of biotechnology, medicine, agriculture, environmental sciences and our appreciation of biodiversity. Genomics likewise promises to revolutionize the understanding of how present-day biological systems from molecules to ecosystems have originated and diversified. Our expectations for genome phylogenetics are necessarily based (rightly or wrongly) on methods first developed in application to organisms and genes.

9.1.2 Taxonomy and Phylogeny Based on Organisms and Genes

Genes are relevant to taxonomy because only heritable characters are taxonomically useful. Nonetheless, genes themselves have until recently been inaccessible, and taxonomy has grown from a phenetic base. Phenetic taxonomy and phylogenetics have been recognized for at least a century to be distinct undertakings (e.g., Bather, 1927), but in the early 20th-century modern synthesis they tended to be muddled together. By contrast, the numerical taxonomy arising in the 1950s (Sokal and Sneath, 1963) offered clean statistical methods, and later, phylogenetics arising from the popularization of Hennig's work (Hennig, 1950) focused on character analysis and reconstruction of genealogical history. Both approaches yielded important (albeit sometimes conflicting) insights when applied to morphologically complex organisms, yet neither was greatly helpful for prokaryotes or protists with their simple, variable or unfamiliar phenotypes.

Zuckerkandl and Pauling (1965) recognized that genes, and the proteins they encode, necessarily retain in their sequences evidence of their phylogenetic histories. If genes attain their distribution among organisms predominantly via transmission along genealogical lineages from parents to offspring, then (statistical issues aside) individual molecular phylogenies would be expected to be congruent with each other and with that of their host organisms. Of course, certain processes — notably gene transfer associated with the endosymbiotic origins of organelles (Sagan, 1967) epigenetics (Chapter 15 and Chapter 16), and cryptic paralogy arising from gene duplication and loss (Fitch, 1970) — can complicate this extrapolation from molecular sequence to organism. Nonetheless, much optimism remained that genes associated with ancient, universally distributed, biochemically stable and mechanistically complex processes (e.g., translation) had been little affected by such processes and would yield gene trees that could be interpreted at face value and extrapolated to the organism level. Chief among these were genes for ribosomal RNAs (Woese, 1987). The rDNA tree soon became not only the reference molecular tree but also the proxy universal tree of life (Woese, 2000). Each microbial species probably contains a species-specific set of core genes that are genealogically ancestral within and functionally indispensable to that species, are maintained primarily by vertical descent and thus show phylogenies congruent with the rDNA tree (or, more precisely, with subtrees within the rDNA tree, because core gene sets can differ among species, and few if any genes are core to all organisms; Lan and Reeves, 2000). The core is likely enriched for informational (central dogma) genes, owing to their various interactions and interdependencies (Jain et al., 1999), though no genes are immune to phylogenetic misbehavior (Gogarten et al., 2002).

Phylogenetic methods are statistically based, and individual gene or protein sequences are often too short (or, more accurately, contain too few informative patterns) to support the inference of stable trees. With the rapid expansion of gene-sequence data from diverse organisms: interest has grown in methods that address these statistical limitations by drawing more data into the analysis. These fall into two broad classes: methods that combine data (e.g., sequence concatenation) and those that combine trees (e.g., consensus methods and supertrees). However, it is first important to ask whether individual data sets (or trees inferred from individual data sets) *should* be combined (Sanderson et al., 1998). If two genes truly have incongruent evolutionary histories, what would a combined data set or a single tree *mean*? Would it not be better to prescreen each data set to ensure mutual congruence? Why should congruence tests be restricted to genes — can they not also be applied to find domains, or smaller regions, of incompatible sequence *within* genes? At what point might such a process become a single-minded pursuit of an unrealistic ideal? These issues are under active debate in fields far outside genomics (Bininda-Emonds et al., 2002), but are immediately relevant to the issue at hand — genome phylogenies.

9.1.3 Genomes as a Basis for Phylogenetic Inference

Genomes have been shaped by diverse processes, such as endogenous origin, exogenous gain, catastrophic loss and sequence divergence of genes and other regions; reassortment of domains into large modular genes; expansion and contraction of gene and repeat families; gene fusion; operon formation and dissolution; rearrangement of gene order; and integration of introns, extranuclear elements and viruses. As a consequence, genomes are replete with semantically rich characters — at least in principle a phylogeneticist's dream.

Homology is the basis of all comparative biology. Homology is established in organismal phylogenetics by careful delineation of characters and for molecular sequences via multiple sequence (or structure) alignment or model fitting. *A priori*, one might imagine that genes (and perhaps a few other well-defined sequence-based features) would be the characters for which orthology would be easiest to establish in genome phylogenetics, as genes can, without too great a degree of arbitrariness, be assessed as present or absent or assigned similarity scores. Not surprisingly, most inferences of genome phylogenies so far (Huynen and Bork, 1998; Ragan and Gaasterland, 1998; Fitz-Gibbon and House, 1999; Snel et al., 1999; Tekaia et al., 1999; Grishin et al., 2000; Lin and Gerstein, 2000; Natale et al., 2000; Wolf et al., 2001; Clarke et al., 2002; Korbel et al., 2002; Wolf et al., 2002) have used genes as primary characters, computing or inferring trees by well-established methods based on distance measures. If rearrangements have been infrequent, gene order (breakpoint analysis) offers an alternative approach (Sankoff and Blanchette, 1998). As experimental and comparative genomics progressively reveal the dynamics of genomes, models of genome change can be made richer and increasingly realistic. Statistically based (e.g., likelihood and Bayesian) methods can accommodate complex models (see Chapter 6). Methodological issues are discussed later.

Genomics has brought surprises too, not least about the degree to which genomes of highly similar, closely related taxa can (and do) differ in gene content. (By *similar* we refer to phenetically based grouping into the same species or genus, and by *closely related* we refer to phylogenetic relatedness inferred from patterns of sequence variation in genes held in common.) On both grounds, for example, *Escherichia coli* K12 and *E. coli* O157:H7 — strains that differ in pathogenicity but otherwise share the suite of morphological, chemical and metabolic properties considered diagnostic of the *E. coli* phenotype — might be expected to have almost identical gene contents. But strain K12 encodes (by our reckoning) 253 genes not found in strain O157:H7, whereas O157:H7 encodes 1076 genes absent from the genome of strain K12.

How could a difference of such magnitude have arisen? Not purely by gene loss, unless we accept that the immediate ancestor of K12 (4289 ORFs) and O157:H7 (5361 ORFs) encoded at least 5614 genes, and similarly back through ever-more-gargantuan genomes of sizes unknown among present-day bacteria and unlikely to be maintainable by selection, nor purely by expansion and contraction of gene families, as this would be readily detected by comparative analysis of sequences, motifs or structural elements. [We estimate the number of nonredundant orthologous genes in the first 67 bacterial genomes to be greater than 21,500 (Charlebois, unpublished), and this number can only increase as more genomes are sequenced.] There is no reason to suspect that all reduced genomes are on their way to extinction. Thus, it seems inescapable that differences in gene content have come about in no small part via transfers from other genomes (Lawrence and Ochman, 1998; Nelson et al., 1999; Paul, 1999; Doolittle, 1999a; Ochman et al., 2000). Gene transfer is mechanistically well characterized and forms the basis for much of modern biotechnology, although it is little understood which specific transfer mechanisms might be quantitatively important outside the laboratory. Similarly, little is known of the tempo or mode of lateral gene transfer (LGT) or of the possible contribution of LGT to genomic diversity, though we are beginning to gain an appreciation for its importance in prokaryotes (Gogarten et al., 2002) and in eukaryotes (Gogarten, 2003).

LGT is inherently subversive of genome phylogenetics. It is far from clear that a genome composed of genes with diverse noncongruent histories can be considered to have a phylogeny at all or be meaningfully represented on a tree (Doolittle, 1999a). Indeed, if the genome is the organism in potential, perhaps organisms whose genomes are susceptible to LGT cannot be said to have phylogenies. Doolittle (1999b) has asked, by analogy, whether any analysis of names listed in telephone directories could be considered to yield a meaningful phylogeny of cities.

The contribution of LGT to actual genomes remains an open question and is beyond the scope of this chapter. Assuming that LGT exists, how might it be dealt with in the context of genome phylogenetics?

1. If relatively few genes or gene regions have arisen by LGT, and especially if any such genes have originated from phyletically diverse sources, it might be possible simply to ignore LGT without incurring dire consequences. Simulations by John Logsdon, Jr. (unpublished) suggest that phylogenetic signal is preserved even against relatively large proportions of lateral transfer.
2. The genome might be equated (at least for phylogenetic purposes) with a core gene set. It would be necessary to operationalize this term — does *core* mean universally distributed, functionally irreplaceable, untouched by LGT or constituting the largest congruence clique (Mushegian and Koonin, 1996; Gaasterland and Ragan, 1998a,b; Marakova et al., 1999; Lan and Reeves, 2000; Woese, 2000; Sicheritz-Pontén and Andersson, 2001)? But whatever the criteria, genes outside this core could be identified and removed from analysis (e.g., Clarke et al., 2002).
3. Genomes might be considered to be related via a network rather than a tree (e.g., Huson, 1998), with relative contributions along different edges represented as probabilities, magnitudes or fluxes.
4. A genome's phylogeny might be considered to be the set of gene trees for all its genes individually. This solution is perhaps not easily human comprehensible but is readily machine parsable. As lateral transfer is not confined to discrete genetic units, recursive assessments will be required.
5. The problem might be mapped into a purpose-built genome space, in which the question does not arise or is somehow addressed naturally, and in which genome phylogenies are represented as vectors or trajectories within its coordinate system. Stripped of mathematical trappings, this would leave us with a concept of genome with which Plato would have felt quite comfortable. (Proponents of extensive LGT might argue that organismal phylogenies, especially of prokaryotes, currently occupy a similar space.)

Let us assume that a meaningful concept of genome phylogeny can be preserved against the challenges, real or imagined, of LGT. More specifically, let us assume either that LGT is quantitatively negligible (Option 1), or that all genes significantly affected by LGT can be identified and removed (Option 2). There then remain methodological issues. What genomic data can or should be used in phylogenetics? What phylogenetic methods are appropriate for genome data sets? We see basically four options.

First, a genome tree can be derived from discrete-state gene content data (presence or absence of orthologs) in exactly the same way as is done for phenetic data. Each genome is scored for the presence (1 or +) or absence (0 or –) of each ortholog. Pairwise similarity scores are computed, and a tree derived by distance-matrix analysis (Fitz-Gibbon and House, 1999; Snel et al., 1999; Tekaia et al., 1999; Lin and Gerstein, 2000; Clarke et al., 2002; Korbel et al., 2002). Alternatively, the presence–absence matrix might be analyzed by parsimony (Fitz-Gibbon and House, 1999; Lin and Gerstein, 2000; Wolf et al., 2001). In an obvious variant, the mean pairwise distance between any two genomes can be computed from quantitative pairwise similarities of all orthologs held in common, and a tree derived by distance analysis (e.g., Fitch–Margoliash or neighbor joining: Ragan and Gaasterland, 1998; Grishin et al., 2000; Wolf et al., 2001; Clarke et al., 2002). In place of gene content or quantitative pairwise similarities, comparisons might instead be based on instances of conserved gene order (Wolf et al., 2001; Korbel et al., 2002) or on protein-fold content (Lin and Gerstein, 2000).

Second, a genome tree can be built up from component trees, one for each orthologous gene family represented. The approach would be simple, if computationally daunting: one would first infer a tree for each orthologous family, then compute the consensus or supertree. Ortholog-by-ortholog phylogenetic analysis is in any case a prerequisite for identifying congruent subsets (Option 2). This might be the only approach by which one could test whether ortholog trees fall into a small number of mutually compatible types (cliques), e.g., because the eukaryotic cell arose via one or a few ancestral fusion events (Sogin, 1991; Martin and Müller, 1998). Because few orthologous families are represented in all genomes and because of the well-known loss of resolution in consensus analysis, the final step of this approach would probably require the application of supertree methods.

Third, one can concatenate sequences of all core genes (however defined) and infer a tree by standard molecular phylogenetic methods. This might be feasible for prokaryotic genomes, but probably not for analyses across broad phyletic distances involving genomes that contain large families of modular genes.

Finally, it is possible to reconstruct genomic histories from patterns of gene order. Rigorous methods are computationally hard (Sankoff and El-Mabrouk, 2002) and gene-order conservation is in many cases extremely limited; therefore, we do not anticipate that this approach is likely to yield a universal genome tree anytime soon. Given a data set of sufficient phyletic density, however, gene order could provide an independent view of genome phylogeny. Even in less-favorable circumstances, gene-order information will aid in identifying orthologs.

Other approaches to elucidate evolutionary information from whole genome sequences have also been proposed (see Chapter 8). With any of these approaches, it is possible to focus not only on the complete (or core) genome but also on functional or other subcategories (see later), e.g., to test hypotheses of eukaryotic genome evolution.

Organisms, genes and genomes are not necessarily three sides of the same coin. Notably, gene and organismal phylogenies might be noncongruent due in significant part to genome-level events, including chromosomal duplications and LGT. If gene content were always stable over evolutionary time, then gene, genome and organismal phylogenies should be fully harmonious. (In morphologically complex eukaryotes, this might mostly be the case, at least at the level of exons or protein-structural modules.) The example of *E. coli* K12 and O157:H7 shows that over evolutionarily shorter periods, large numbers of phenetic characters can be maintained in common and transmitted to successive generations, despite substantial variation in genome size, gene content and gene order. But over longer times, gene-content variation must tend to uncouple both genome and phenome (as perhaps idealistic entities with continuity through time) from the sort of unitary phylogenetic history enjoyed individually by the underlying genes (or regions thereof).

So, are genomes cities? Are gene, genome and organismal phylogenies tightly coupled, loosely coupled or uncoupled? Do genome trees (based here on measures of gene content that reflect quantitative pairwise similarity of putative orthologs) resemble trees of core genes, particularly of rDNA? Is there evidence that sets of genes encoding proteins of different functional classes have different phylogenetic histories? Can we find evidence from genomic comparisons that LGT presents a major obstacle to discovering the paths along which genomes have diversified?

9.2 Methods

Orthologs are defined topologically on a tree (Fitch, 1970), but to implement this rigorously in genome phylogenetics would require us to infer thousands of trees. We therefore follow Mushegian and Koonin (1996) in basing genome comparison on reciprocal best BLASTP matches (Clarke et al., 2002). For the 87 genomes covered by the NGIBWS database (www.neurogadgets.com/genomes.php) when this calculation was made, pairwise (genome-to-genome) dissimilarities were calculated as 1.0 minus the mean of normalized BLASTP similarities (bit scores *S′*) for all reciprocally best-matching pairs (BLASTP $e < 10^{-5}$) and assembled into a distance (dissimilarity) matrix (Clarke et al., 2002). Trees were constructed by using the Fitch–Margoliash (least-squares) method as implemented in the FITCH program within PHYLIP (Felsenstein, 1989).

Confidence measures on subtrees are assessed by bootstrapping (Fitz-Gibbon and House, 1999; Grishin et al., 2000; Lin and Gerstein, 2000; Wolf et al., 2001; Clarke et al., 2002). Suppose that Genome A and Genome B share *N* ORFs better than the chosen threshold. For bootstrapping, the *N* ORFs are sampled with replacement, generating a new set of *N* ORFs that will contain duplicates and omit some of the original ORFs. The mean of their normalized BLASTP scores is then one of the bootstrap distances. For each pair of genomes we constructed 100 such replicate distances.

To produce trees for specific functional subsets of genes, we first assigned each ORF within NGIBWS to one or more of the functional categories defined within NCBI's clusters of orthologous groups (COG) database (Tatusov et al., 1997; http://ncbi.nlm.nih.gov/COG/). Where data for a given gene or for a given genome were not directly available from this COG database, we extrapolated functional categories based on the best BLASTP match to genes whose functional category was available. Function-specific trees were then inferred as given from the distance matrix based on correspondingly annotated ORFs. The tree labeled *Metabolism*, for instance, is based solely on the subset of ORFs assigned to one or more of the metabolism categories. (See http://ncbi.nlm.nih.gov/COG/ for the definition of which subcategories of genes this includes.) A small proportion of ORFs are assigned to more than one category, so the function-specific trees are based on slightly overlapping subsets of genes.

9.3 Results and Discussion

Figure 9.1 to Figure 9.4 present trees for 87 genomes (67 bacteria, 16 archaea and 4 eukaryotes) without filtering for phylogenetically discordant sequences (Clarke et al., 2002) but with the requirement of reciprocal best match. Figure 9.1 shows the Fitch–Margoliash tree based on reciprocally best-matching ORFs in these genomes. The bootstrap numbers were generated by majority-rule consensus from 100 bootstrap replicates over the same data; as the topology happens to be identical with and without bootstrapping, we have labeled nodes in the standard Fitch–Margoliash tree with the corresponding bootstrap values. Figure 9.2 to Figure 9.4 are based on subsets of ORFs annotated as involved in information storage and processing (essentially NCBI functional categories J, K and L), cellular processes (D, M, N, O, P and T) and metabolism (C, E, F, G, H, I and Q). (See http://ncbi.nih.nlm.gov/COG/ for details about the classification scheme.)

The four trees are identical or closely similar in most major respects (Table 9.1). In all but the metabolism tree (Figure 9.4; see previous discussion), bacterial genomes are resolved as a unitary group vis-à-vis genomes of eukaryotes and archaea. Many groups familiar from SSUrDNA trees and phyletic classifications appear in all four trees, including Crenarchaeota, Chlamydia, Spirochaetes, Cyanobacteria, High G+C Firmicutes, Proteobacteria and its α, β and ε divisions. β-Proteobacteria are always embedded within γ-Proteobacteria, forming a unitary β + γ grouping although without γ-Proteobacteria appearing intact. Where we have data from genomes of two or more strains within one species or two or more species within a genus, expected groupings are found without exception. Branching order within major groupings is remarkably similar among trees. The extensive similarity in major features over these trees implies that there are few, if any, functionally based special processes or distributions of sequence change apparent at this granularity.

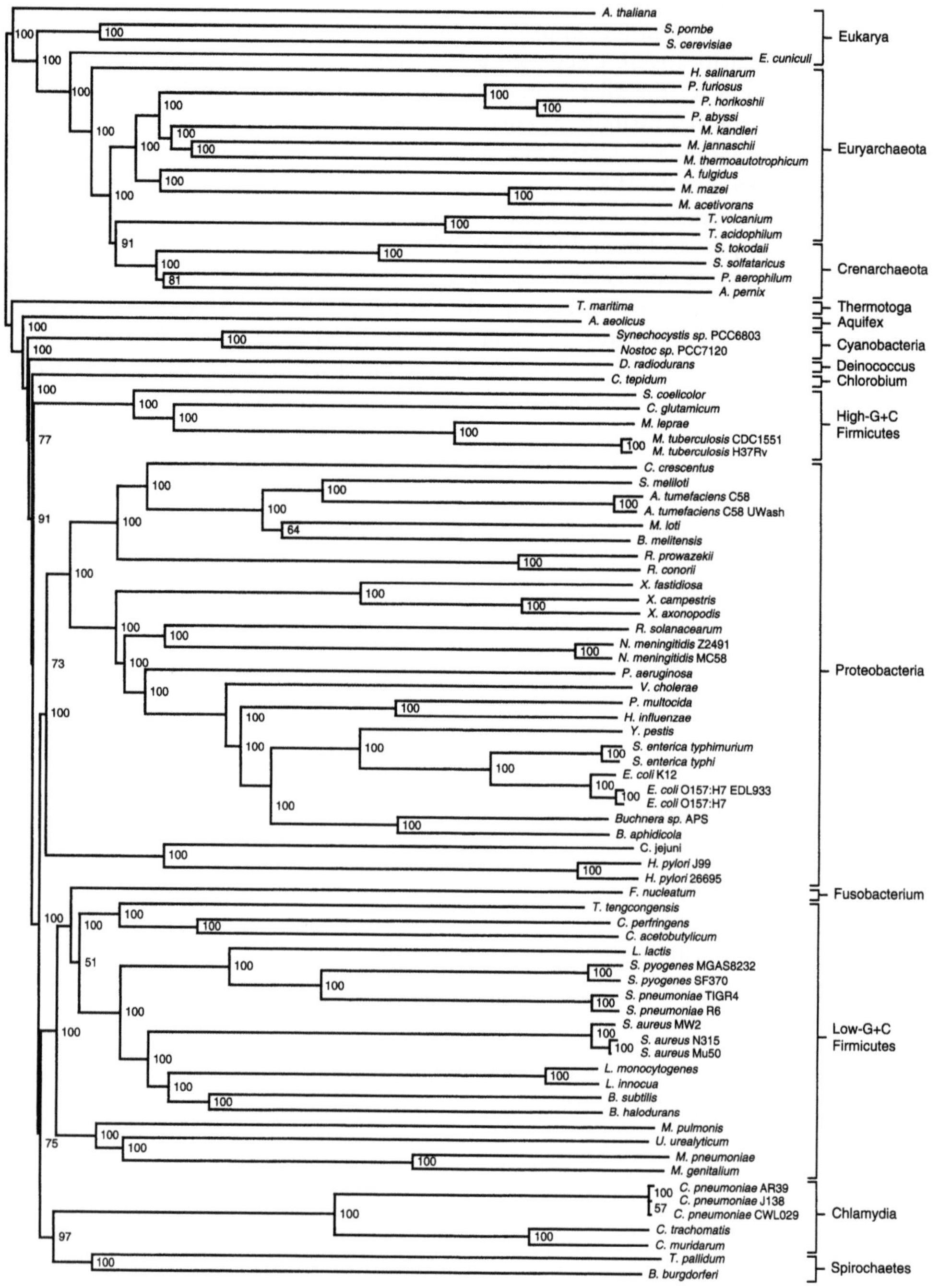

FIGURE 9.1 Fitch–Margoliash tree based on conceptually translated complete genomic ORF sets. [See text and Clarke et al. (2002) 2072–2080 for details.] Bootstrap support was determined in a separate analysis by strict consensus among 100 bootstrap replicates.

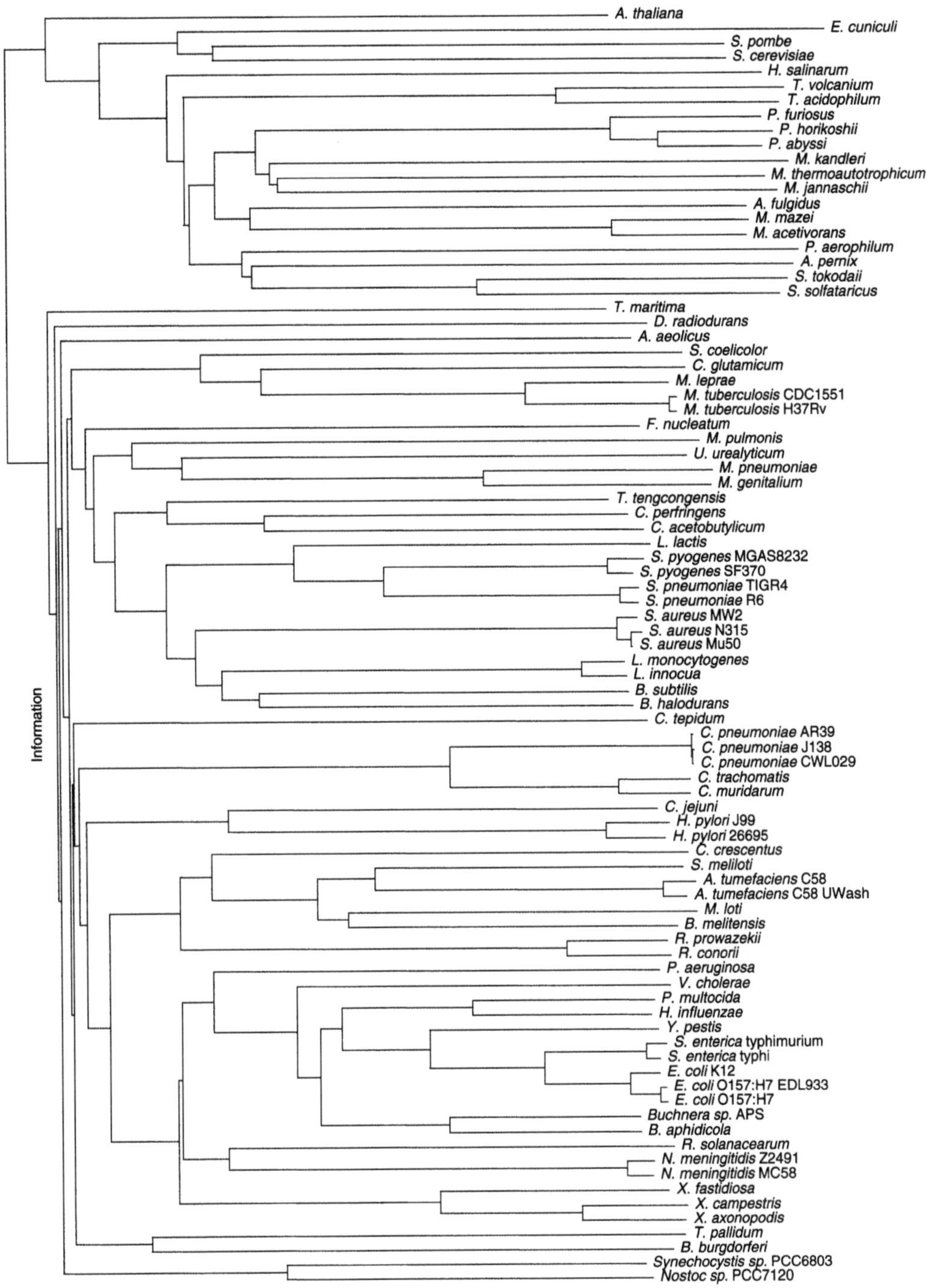

FIGURE 9.2 Fitch–Margoliash tree based on conceptually translated ORFs corresponding to genes annotated as Information Storage and Processing (essentially NCBI functional categories J, K and L).

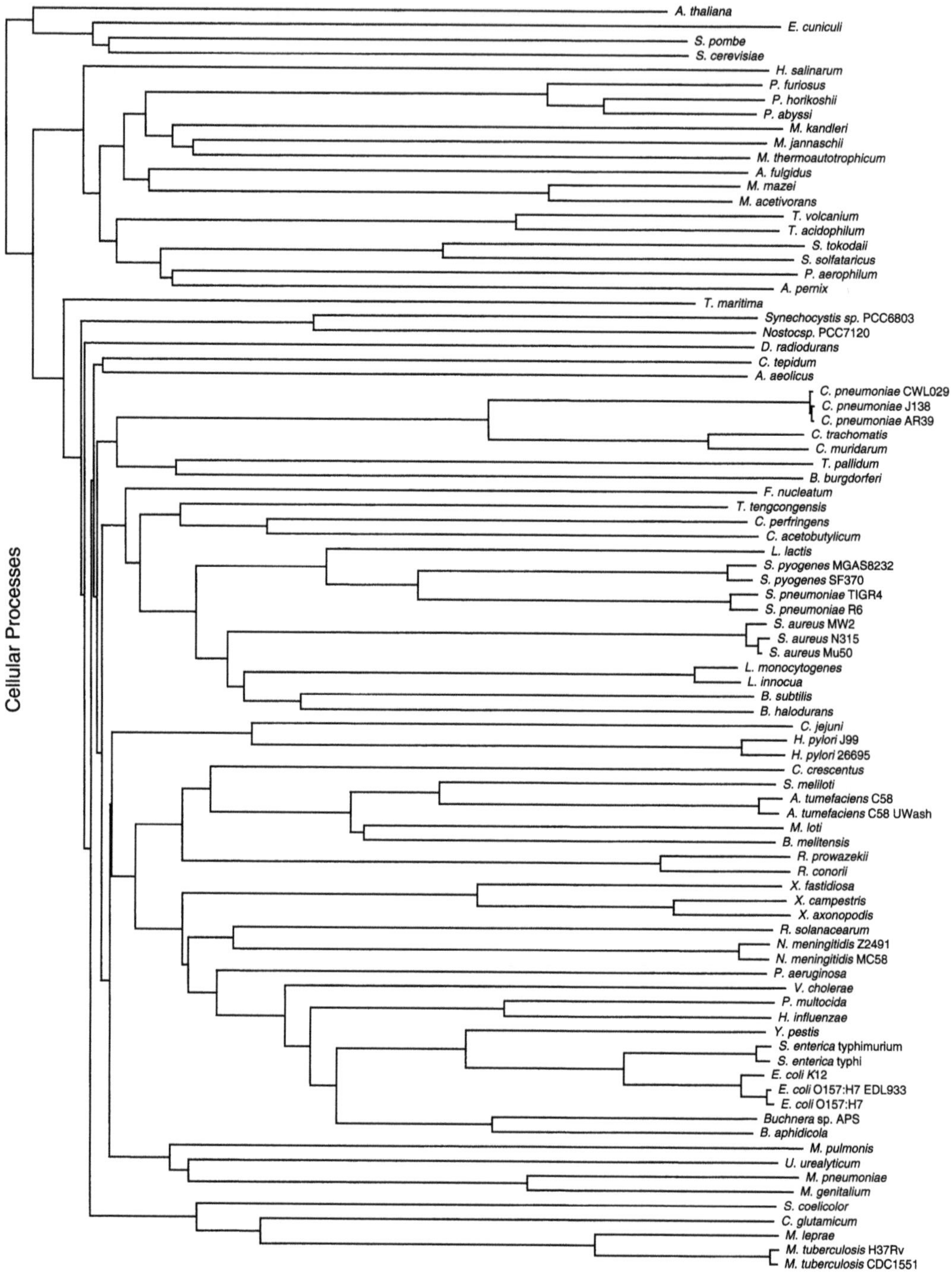

FIGURE 9.3 Fitch–Margoliash tree based on conceptually translated ORFs corresponding to genes annotated as involved in Cellular Processes (essentially NCBI functional categories D, M, N, O, P and T).

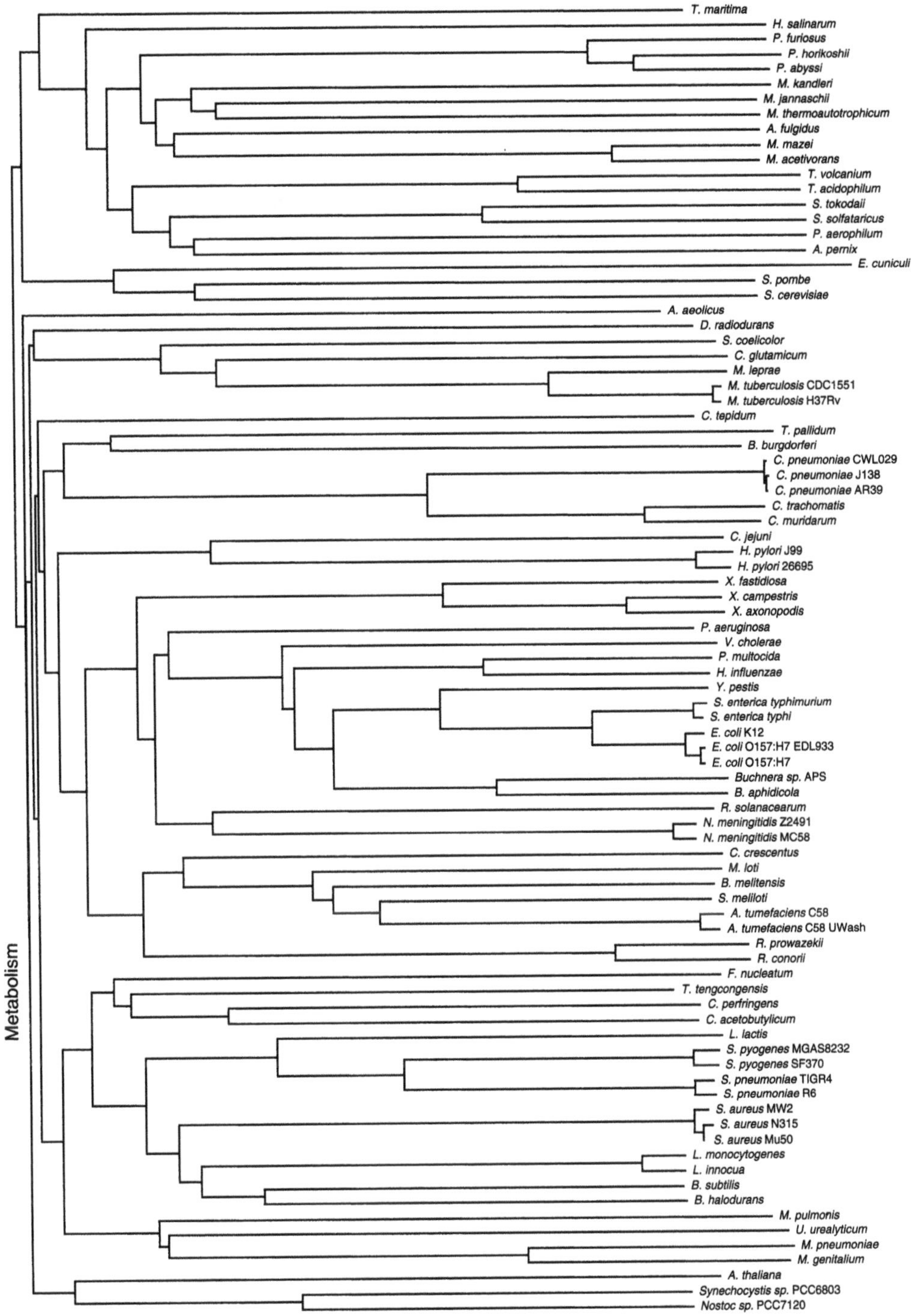

FIGURE 9.4 Fitch–Margoliash tree based on conceptually translated ORFs corresponding to genes annotated as involved in Metabolism (essentially NCBI functional categories C, E, F, G, H, I and Q).

TABLE 9.1 Groupings in Genome Trees

Grouping	*Full Tree/Boot*	*Information*	*Cell Process*	*Metabolism*	*PDS/Boot*
Eukarya	PARA	PARA	MONO	NO	(MONO)[a]
Archaea	MONO/100	MONO	MONO	MONO	MONO/100
Euryarchaeota	PARA	PARA	PARA	PARA	PARA
Crenarchaeota	MONO/100	MONO	MONO	MONO	(MONO)
Bacteria	MONO/100	MONO	MONO	NO	MONO/100
Chlamydia + Spirochaetes	MONO/97	NO	MONO	MONO	MONO/99
Chlamydia	MONO/100	MONO	MONO	MONO	MONO/100
Spirochaetes	MONO/100	MONO	MONO	MONO	MONO/100
Cyanobacteria	MONO/100	MONO	MONO	MONO	(MONO)
Cyanobacteria + Arabidopsis	NO	NO	NO	MONO	n/p
Firmicutes	NO	PARA	NO	NO	NO
High G+C	MONO/100	MONO	MONO	MONO	(MONO)
Low G+C	PARA/100+51[b]	MONO	NO	PARA	MONO/92
Fusobacteria + Firmicutes	NO	MONO	NO	NO	n/p
Fusobacteria + low-G+C Firmicutes	MONO/100	NO	NO	MONO	n/p
Proteobacteria	MONO/100	MONO	MONO	MONO	MONO/100
Alpha + beta + gamma	MONO/100	MONO	MONO	MONO	MONO/100
Alpha	MONO/100	MONO	MONO	MONO	(MONO)
Beta + gamma	MONO/100	MONO	MONO	MONO	MONO/100
Beta	MONO/100	MONO	MONO	MONO	MONO/100
Gamma	PARA/100+100[b]	PARA	PARA	PARA	PARA
Epsilon	MONO/100	MONO	MONO	MONO	MONO/100

Note: Full tree, Figure 9.1; informational subtree, Figure 9.2; cellular processes subtree, Figure 9.3; metabolism subtree, Figure 9.4. MONO, monophyletic; PARA, paraphyletic; NO, neither monophyletic nor made paraphyletic by a single extraneous sequence or group; n/p, not possible because of lack of representation in the data set at that time. Groupings in an earlier 37-genome tree [Clarke, G. D. P. et al. (2002) *J. Bacteriol.* 184: 2072–2080] in which data were filtered to remove phylogenetically discordant sequences (PDS) are shown for comparison. Bootstrap support (out of 100 replicates) is shown for topological features in the full and PDS trees.

[a] MONO in parentheses indicates that only one genome was present in that data set.

[b] The Low G+C Firmicutes and γ-Proteobacteria are made paraphyletic by an extraneous sequence or small group branching basally within the larger grouping. We report the bootstrap proportions for each of the two subsections into which the otherwise monophyletic group was thereby broken.

Maximum likelihood analysis of a set of 32 concatenated ribosomal protein sequences from each of 40 prokaryotes (Wolf et al., 2001) revealed support for three nontraditional high-level bacterial clades: Spirochaetes-Chlamydia, Thermotogales–Aquificales and Actinomycetes–Deinococcales–Cyanobacteria. All our genome trees except the one based on informational genes (i.e., Figure 9.1, Figure 9.3 and Figure 9.4) likewise support a grouping of Spirochaetes and Chlamydia, with 97% bootstrap probability where assessed (Figure 9.1); this is interesting because ribosomal proteins are classified as informational molecules (NCBI Class J). None of our trees, however, supports a specific grouping of the *T. maritima* and *A. aeolicus* genomes or of the High G+C Firmicutes with *Deinococcos radiodurans* and the two cyanobacteria. Our trees usually resolve the latter two sets of genomes as relatively basal branches within Bacteria.

As is commonly seen in SSUrDNA trees (e.g., Woese; 1987, 2000; Clarke et al., 2002), genomes branching along the backbone of Bacteria are for the most part poorly resolved. Genomes of *Thermotoga maritima* and *Aquifex aeolicus* are basal within Bacteria in the full genome tree with 100% bootstrap support (Figure 9.1). However, in the metabolism tree (Figure 9.4) the *T. maritima* genome appears on the Archaea + Eukarya branch as the specific neighbor of Archaea. Genomes of cyanobacteria, *D. radiodurans* (*Thermus/Deinococcus* group) and *Chlorobium tepidum* (Chlorobi) branch basally within Bacteria in some, but not all, of our analyses.

Fusobacterium nucleatum is an interesting case. Ultrastructurally, it is a Gram-negative bacterium, with a standard outer membrane. However, many of its genes, gene functions and gene clusters most closely resemble those of the low G+C Gram-positives, and its 16S rRNA sequence is most similar to those of streptococci (Kapatral et al., 2002). These observations and our genome trees suggest that *F. nucleatum* is a Gram-negative member of the Firmicute lineage. We speculate that (the Gram-positive members of) this lineage lost the outer membrane and took advantage of peptidoglycan's three-dimensional cross-linking potential. It will be interesting to observe where *Heliobacterium* spp. eventually appear on genome trees, as this genus is thought to be another deep-branching Gram-negative member of this group.

In all trees, Archaea form a unitary grouping, but in every case Crenarchaeota is resolved within a paraphyletic Euryarchaeota. Crenarchaeal genomes appear as specific neighbors to those of *Thermoplasma* species in the whole-genome (Figure 9.1), cell-process (Figure 9.3) and metabolism (Figure 9.4) trees, but avoid *Thermoplasma* species (and *Halobacterium salinarum*) in the informational-gene tree (Figure 9.2). Thus, informational genes of archaea collectively present a different signal in this analysis than do cell-process or metabolic genes.

Vis-à-vis bacterial genomes, those of Eukarya are resolved as either a cohesive group (Figure 9.3) or a paraphyletic assemblage (Figure 9.1 and Figure 9.2). Eukaryotic genomes are not seen to group within Archaea (or Bacteria). In the metabolic-gene tree (Figure 9.4), *Arabidopsis thaliana* groups with cyanobacteria, no doubt reflecting the contribution of cyanobacterial genes to green-plant genomes via an endosymbiotic origin of the plastid, recently estimated to be ca. 18% of protein-coding genes (Martin et al., 2002). In the informational and cell-process (Figure 9.2 and Figure 9.3) but not the whole-genome tree (Figure 9.1), the microsporidian *Encephalitozoon cuniculi* appears as the specific neighbor of the fungi *Saccharomyces cerevisiae* and *Schizosaccharomyces pombe*. (These three group together in the metabolism tree also, but perhaps by default, as *A. thaliana* is separated from the other eukaryotes; data from additional eukaryotic genomes will presumably resolve this point.) This is interesting in light of suggestions (Keeling and McFadden, 1998; Hirt et al., 1999; Van de Peer et al., 2000) that microsporidia are degenerate fungi (see also Chapter 10).

Our new results support our earlier conclusion (Clarke et al., 2002) that the topology of prokaryote genome trees is little, if at all, affected by LGT. The comparison is imperfect because we are now working with 50 additional genomes and use a less-stringent BLAST threshold, but to the extent that our new whole-genome tree (Figure 9.1) can be compared with a tree for which phylogenetically discordant sequences ($P < 0.05$) were first removed, we find no notable topological difference (the loss of monophyly for the Low G+C Firmicutes is due to the inclusion of *Fusobacterium*). LGT appears either to have been quantitatively unimportant or at least not to have greatly favored any particular pairs of donor and recipient genomes represented in our trees. This conclusion should, however, be tempered by the discovery of the *Arabidopsis*–cyanobacteria grouping in the metabolic tree (Figure

9.4 and previous discussion); although the lateral contribution of genes from cyanobacteria into photosynthetic eukaryotes is strongly supported by individual gene trees and is substantial enough to have been found in the metabolic-gene tree (Figure 9.4) and by software-based consistency filtering of a phyletic data set (Ragan and Lee, 1991), our whole-genome result (Figure 9.1) not only fails to indicate it but also groups *A. thaliana* with the other eukaryotes with a 100% bootstrap score. Simulations are urgently required to understand the sensitivity of genome trees to noncoherent signals.

9.4 Envoi

If genomes are cities, their inhabitants (genes) show little evidence of having migrated en masse from one established city to another. This is not to say that their populations have remained completely immobile, as both single-gene phylogenetics (Brown and Doolittle, 1997; Jain et al., 1999; Nesbø et al., 2001) and surrogate methods of LGT detection (Lawrence and Ochman, 1997; Hayes and Borodovsky, 1998; Karlin et al., 1998; Ragan, 2001; Clarke et al., 2002; Ragan and Charlebois, 2002) indicate that most sequenced prokaryotic genomes contain genes with phylogenetic histories that differ from each other and from that imagined for the host organism. In some genomes, the proportion of such genes can exceed 25% (Lawrence and Ochman, 2002; Ragan, 2002). Nonetheless, even when no special steps are taken to remove phylogenetically discordant genes, inferred genome trees differ little from those based on rRNA sequences and remain topologically stable as more genomes, even of parasites and extremophiles, are added. Migration per se at the levels at which it has actually occurred — whether between cities or from countryside to city and back again — seems not to prevent us from reconstructing the historical roots of civic founding populations, and in this sense of cities themselves.

A more difficult question is whether stable genome trees could exist and be topologically congruent with rRNA-sequence trees, if genomes had arisen not via a treelike process of diversification and descent, but rather (for example) by selective accretion, retention and loss of genes from a large and perhaps inhomogeneous gene pool. It might be formally possible that such processes could generate and maintain the pattern of diversity that is actually observed among microbial genomes, but it is far from obvious that a tree based on such genomes could (e.g., Figure 9.1) exhibit near-100% bootstrap support for most internal nodes or be topologically near-identical with the tree based on rRNA sequence. If rDNAs are part of a conserved core of genes largely resistant to lateral transfer (Lan and Reeves, 2000), then any LGT-based paradigm must also explain why different processes (selective LGT for transferable genes, vertical descent for the core) analyzed at different scales (overall trends and individual gene sequences, respectively) somehow yield the same tree. Alternatively, if there is no universally conserved core (or a strongly overlapping set of regionally valid cores, which would produce the same effect), and our well-supported genome tree is purely the consequence of historical patterns of individual migration among regions (including, but not limited to, cities), then the similarity of trees for different functional subsets (Figure 9.2 to Figure 9.5) seems to exclude function as a basis for differential LGT, at least at the (substantial) range of granularities covered by the existing data. And if what appears to be genomic (and organismal) phylogeny is actually an epiphenomenon of LGT, we should be prepared to explain how the well-understood processes of vertical inheritance that have formed the basis of biology for more than 100 years, and appear adequate for eukaryotes, can be so irrelevant to prokaryotic phylogenetics.

Genome trees can be simply and naturally interpreted within the paradigm of vertical descent with modification as on a bifurcating tree, with some further level of undirected (or at least not too tightly constrained) LGT. Moving beyond this qualitative description to a quantitative understanding will require the development of explicit models, parameterization, computational simulation and analysis of sensitivity to model violation and perturbation. Microbial genomics has been successful in making available a huge data set of unprecedented richness and admirable phyletic scope. We must now take the next steps toward understanding what these data collectively tell us about the microbial world.

Acknowledgment

We are indebted to many colleagues within the Canadian Institute for Advanced Research, Program in Evolutionary Biology for stimulating conversations and difficult questions.

References

Avery, O. T., MacLeod, C. M. and McCarty, M. (1944) Studies on the chemical nature of the substance inducing transformation of pneumococcal types. *J. Exp. Med.* 79: 137-158.

Bather, F. A. (1927) Biological classification: Past and future. *Q. J. Geol. Soc. Lond.* 83: lxii–civ.

Beadle, G.W. and Tatum, E.L. (1941) Genetic control of biochemical reactions in *Neurospora. Proc. Natl. Acad. Sci. USA* 27: 499–506.

Bininda-Emonds, O. R. P., Gittleman, J. L. and Steel, M. A. (2002) The (super)tree of life: Procedures, problems, and prospects. *Annu. Rev. Ecol. Syst.* 33: 265–289.

Brown, J. R. and Doolittle, W. F. (1997) *Archaea* and the prokaryote-to-eukaryote transition. *Microbiol. Mol. Biol. Rev.* 61: 456–502.

Clarke, G. D. P., Beiko, R. G., Ragan, M. A. and Charlebois, R. L. (2002) Inferring genome trees by using a filter to eliminate phylogenetically discordant sequences and a distance matrix based on mean normalized BLASTP scores. *J. Bacteriol.* 184: 2072–2080.

Doolittle, W. F. (1999a) Phylogenetic classification and the universal tree. *Science* 284: 2124–2128.

Doolittle, W. F. (1999b) Lateral gene transfer, genome surveys, and the phylogeny of prokaryotes. *Science* 286: 1443a.

Felsenstein, J. (1989) PHYLIP: Phylogeny inference package. *Cladistics* 5: 164–166.

Fitch, W. M. (1970) Distinguishing homologous from analogous proteins. *Syst. Zool.* 19: 99–113.

Fitz-Gibbon, S. T. and House, C. H. (1999) Whole genome-based phylogenetic analysis of free-living microorganisms. *Nucleic Acids Res.* 27: 4218–4222.

Gaasterland, T. and Ragan, M. A. (1998a) Constructing multi-genome views of whole microbial genomes. *Microb. Compar. Genomics* 3: 177–192.

Gaasterland, T. and Ragan, M. A. (1998b) Microbial genescapes: Phyletic and functional patterns of ORF distribution among prokaryotes. *Microb. Compar. Genomics* 3: 199–217.

Garrod, A.E. (1909) *Inborn Errors of Metabolism*, Oxford University Press, London.

Gogarten, J. P. (2003) Gene transfer: Gene swapping craze reaches eukaryotes. *Curr. Biol.* 13: R53–R54.

Gogarten, J. P., Doolittle, W. F. and Lawrence, J. G. (2002) Prokaryotic evolution in light of gene transfer. *Mol. Biol. Evol.* 19: 2226–2238.

Greene, E. L. (1909, reprinted 1983) *Landmarks of Botanical History*, Vol. 1, Stanford University Press, Stanford, CA, pp. 177, 187, 192.

Grishin, N. V., Wolf, Y. I. and Koonin, E. V. (2000) From complete genomes to measures of substitution rate variability within and between proteins. *Genome Res.* 10: 991–1000.

Hayes, W. S. and Borodovsky, M. (1998) How to interpret an anonymous bacterial genome: Machine learning approach to gene identification. *Genome Res.* 8: 1154–1171.

Hennig, W. (1950) *Grundzüge einer Theorie der phylogenetischen Systematik*, Deutscher Zentralverlag, Berlin.

Hershey, A.D. and Chase, M. (1952) Independent functions of viral protein and nucleic acid in growth of bacteriophage. *J. Gen. Physiol.* 36: 39–56.

Hirt, R. P., Logsdon. J. M. Jr., Healy, B., Dorey, M. W., Doolittle, W. F. and Embley, T. M. (1999) Microsporidia are related to Fungi: Evidence from the largest subunit of RNA polymerase II and other proteins. *Proc. Natl. Acad. Sci. USA* 96: 580–585.

Huson, D. H. (1998) SplitsTree: Analyzing and visualizing evolutionary data. *Bioinformatics* 14: 68–73.

Huynen, M. A. and Bork, P. (1998) Measuring genome evolution. *Proc. Natl. Acad. Sci. USA* 95: 5849–5856.

Jain, R., Rivera, M. C. and Lake, J. A. (1999) Horizontal gene transfer among genomes: The complexity hypothesis. *Proc. Natl. Acad. Sci. USA* 96: 3801–3806.

Kapatral, V., Anderson, I., Ivanova, N., Reznik, G., Los, T., Lykidis, A., Bhattacharya, A., Bartman, A., Gardner, W., Grechkin, G., Zhu, L., Vasieva, O., Chu, L., Kogan, Y., Chaga, O., Goltsman, E., Bernal, A., Larsen, N., D'Souza, M., Walunas, T., Pusch, G., Haselkorn, R., Fonstein, M., Kyrpides, N. and Overbeek, R. (2002) Genome sequence and analysis of the oral bacterium *Fusobacterium nucleatum* strain ATCC 25586. *J. Bacteriol.* 184: 2015–2018.

Karlin, S., Mrázek, J. and Campbell, A. M. (1998) Codon usages in different gene classes of the *Escherichia coli* genome. *Mol. Microbiol.* 29: 1341–1355.

Keeling, P. J. and McFadden, G. I. (1998) Origins of Microsporidia. *Trends Microbiol.* 6: 19–23.

Korbel, J. O., Snel, B., Huynen, M. A. and Bork, P. (2002) SHOT: A web server for the construction of genome phylogenies. *Trends Genet.* 18: 158–162.

Lan, R. and Reeves, P. R. (2000) Intraspecies variation in bacterial genomes: The need for a species genome concept. *Trends Microbiol.* 8: 396–401.

Lawrence, J. G. and Ochman, H. (1997) Amelioration of bacterial genomes: Rates of change and exchange. *J. Mol. Evol.* 44: 383–397.

Lawrence, J. G. and Ochman, H. (1998) Molecular archaeology of the *Escherichia coli* genome. *Proc. Natl. Acad. Sci. USA* 95: 9413–9417.

Lawrence, J. G. and Ochman, H. (2002) Reconciling the many faces of lateral gene transfer. *Trends Microbiol.* 20: 1–4.

Lin, J. and Gerstein, M. (2000) Whole-genome trees based on the occurrence of folds and orthologs: Implications for comparing genomes on different levels. *Genome Res.* 10: 808–818.

Makarova, K. S., Aravind, L., Galperin, M. Y., Grishin, N. V., Tatusov, R. L., Wolf, Y. I. and Koonin, E. V. (1999) Comparative genomics of the Archaea (Euryarchaeota): Evolution of conserved protein families, the stable core, and the variable shell. *Genome Res.* 9: 608–628.

Martin, W. and Müller, M. (1998) The hydrogen hypothesis for the first eukaryote. *Nature* 392: 37–41.

Martin, W., Rujan, T., Richly, E., Hansen, A., Cornelson, S., Lins, T., Leister, D., Streobe, B., Hasegawa, M. and Penny, D. (2002) Evolutionary analysis of *Arabidopsis*, cyanobacterial, and chloroplast genomes reveals plastid phylogeny and thousands of cyanobacterial genes in the nucleus. *Proc. Natl. Acad. Sci. USA* 99: 12246–12251.

Mushegian, A. R. and Koonin, E. V. (1996) A miminal gene set for cellular life derived by comparison of complete bacterial genomes. *Proc. Natl. Acad. Sci. USA* 93: 10268–10273.

Natale, D. A., Shankavaram, U. T., Galperin, M. Y., Wolf, Y. I., Aravind, L. and Koonin, E. V. (2000) Towards understanding the first genome sequence of a crenarchaeon by genome annotation using clusters of orthologous groups of proteins (COGs). *Genome Biol.* 1: RESEARCH0009.

Nelson, K. E., Clayton, R. A., Gill, S. R., Gwinn, M. L., Dodson, R. J., Haft, D. H., Hickey, E. K., Peterson, J. D., Nelson, W. C., Ketchum, K. A., McDonald, L., Utterback, T. R., Malek, J. A., Linher, K. D., Garrett, M. M., Stewart, A. M., Cotton, M. D., Pratt, M. S., Phillips, C. A., Richardson, D., Heidelberg, J., Sutton, G. G., Fleischmann, R. D., Eisen, J. A., White, O., Salzberg, S. L., Smith, H. O., Venter, J. C. and Fraser, C. M. (1999) Evidence for lateral gene transfer between Archaea and Bacteria from genome sequence of *Thermotoga maritima. Nature* 399: 323–329.

Nesbø, C. L., Boucher, Y. and Doolittle, W. F. (2001) Defining the core of nontransferable prokaryotic genes: The euryarchaeal core. *J. Mol. Evol.* 53: 340–350.

Ochman, H., Lawrence, J. G. and Groisman, E. A. (2000) Lateral gene transfer and the nature of bacterial innovation. *Nature* 405: 299–304.

Paul, J. H. (1999) Microbial gene transfer: An ecological perspective. *J. Mol. Microbiol. Biotechnol.* 1: 45–50.

Ragan, M. A. (2001) On surrogate methods for detecting lateral gene transfer. *FEMS Microbiol. Lett.* 201: 187–191.

Ragan, M. A. (2002) Reconciling the many faces of lateral gene transfer: Response. *Trends Microbiol.* 20: 4.

Ragan, M. A. and Charlebois, R. L. (2002) Distributional profiles of homologous open reading frames among bacterial phyla: Implications for vertical and lateral transmission. *Int. J. Syst. Evol. Microbiol.* 52: 777–787.

Ragan, M. A. and Gaasterland, T. (1998) Whole-genome phylogenetic analysis. Unpublished manuscript (available on request).

Ragan, M. A. and Lee, A. R. III (1991) Making phylogenetic sense of biochemical and morphological diversity among the protists. In *The Unity of Evolutionary Biology*, (Dudley, T. R., Ed.), Dioscorides Press, Portland OR, pp. 432–441. Proceedings of Fourth International Congress of Systematic and Evolutionary Biology, College Park, MD.

Sagan, L. (1967) On the origin of mitosing cells. *J. Theor. Biol.* 14: 225–274.

Sanderson, M. J., Purvis, A. and Henze, C. (1998) Phylogenetic supertrees: Assembling the trees of life. *Trends Ecol. Evol.* 13: 105–109.

Sankoff, D. and Blanchette, M. (1998) Multiple genome rearrangement and breakpoint phylogeny. *J. Comp. Biol.* 5: 555–570.

Sankoff, D. and El-Mabrouk, N, (2002) Genome rearrangement. in *Current Topics in Computational Biology* (Jiang, T., Smith, T., Xu, Y. and Zhang, M., Eds.), MIT Press, Cambridge MA, pp. 135–155.

Sicheritz-Pontén, T. and Andersson, S. G. E. (2001) A phylogenomic approach to microbial evolution. *Nucleic Acids Res.* 29: 545–552.

Snel, B., Bork, P. and Huynen, M. A. (1999) Genome phylogeny based on gene content. *Nature Genet.* 21: 108–110.

Sogin, M. L. (1991) Early evolution and the origin of eukaryotes. *Curr. Opin. Genet. Dev.* 1: 457–463.

Sokal, R. R. and Sneath, P. H. A. (1963) *The Principles of Numerical Taxonomy*, W.H. Freeman, San Francisco, CA.

Tatusov, R. L., Koonin, E. V. and Lipman, D. J. (1997) A genomic perspective on protein families. *Science* 278: 631–637.

Tekaia, F., Lazcano, A. and Dujon, B. (1999) The genomic tree as revealed from whole proteome comparisons. *Genome Res.* 9: 550–557.

Van de Peer, Y., Ben Ali, A. and Meyer, A. (2000) Microsporidia: Accumulating molecular evidence that a group of amitochondriate and suspectedly primitive eukaryotes are just curious fungi. *Gene* 246: 1–8.

Watson, J. D. and Crick, F. H. C. (1953) Genetical implications of the structure of deoxyribonucleic acid. *Nature* 171: 964–967.

Woese, C. R. (1987) Bacterial evolution. *Microbiol. Rev.* 51: 221–271.

Woese, C. R. (2000) Interpreting the universal phylogenetic tree. *Proc. Natl. Acad. Sci. USA* 97: 8392–8396.

Wolf, Y. I., Rogozin, I. B., Grishin, N. V., Tatusov, R. L. and Koonin, E.V. (2001) Genome trees constructed using five different approaches suggest new major bacterial clades. *BMC Evol. Biol.* 1: 8.

Wolf, Y. I., Rogozin, I. B., Grishin, N. V. and Koonin, E. V. (2002) Genome trees and the Tree of Life. *Trends Genet.* 18: 472–478.

Zuckerkandl, E. and Pauling, L. (1965) Evolutionary divergence and convergence in proteins. In *Evolving Genes and Proteins* (Bryson, B. and Vogel, H. J., Eds.), Academic Press, New York, pp. 97–166.

Chapter 10

Genomics of Microbial Parasites: The Microsporidial Paradigm

Guy Méténier and Christian P. Vivarès

CONTENTS

Abstract

Encephalitozoon cuniculi, a parasite of humans and other mammals, belongs to a large group of amitochondriate unicellular eukaryotes called microsporidia, which conjugate a strict intracellular parasitic lifestyle to the capacity of survival in the environment as resistant spores. Long classified as Protozoa, microsporidia are currently viewed as fungi-related organisms. Illustrating an extreme case of genomic compaction among eukaryotic microbes, the 2.9-Mbp genome of *E. cuniculi* composed of 11 linear chromosomes has been sequenced. We discuss the contribution that the analysis of the *E. cuniculi* genome content has made to our knowledge of the biology of eukaryotic parasites. The energy metabolism of *E.*

0-415-29904-7/04/$0.00+$1.50

cuniculi is greatly reduced and should be supplemented by an import of host ATP, as is the case in some bacterial "energy parasites." Shortening of intergenic and genic sequences appears to guarantee the preservation of most genes required for a minimal eukaryotic organization and reproduction. Microsporidia seem to have retained an organelle (mitosome) resulting from a reductive mitochondrial evolution. As the microsporidian mitosome has been recently identified at the cellular level, the exploration of a novel terminal process of electron transfer and of some ancestral functions related to iron homeostasis should be of great importance to both parasitology and evolutionary biology.

10.1 Introduction

The sequencing of very small bacterial genomes, such as those of mycoplasmas, was useful to evaluate the set of genes that would be sufficient to sustain the life of the simplest unicellular organism (Fraser et al., 1995; Himmelreich et al., 1996). In a review of different minimal genome approaches, Mushegian concluded that the minimal set of protein-coding genes should be close to 300 (Mushegian, 1999). This might be sufficient for growth and replication of a theoretical nanobacterium but probably not for that of a minimal eukaryote-type cell. As recently stressed by Cavalier-Smith (Cavalier-Smith 2002a, b), the development of an endomembrane system and endoskeleton would have been at the basis of the origin of nuclei, mitosis, new cell cycle controls and sexual processes, which obviously involved a significant increase in cellular complexity during the prokaryote–eukaryote transition. In some microsporidian species such as *Encephalitozoon cuniculi*, a rather simple cellular organization is associated with a nuclear genome of only 2.9 Mbp (Biderre et al., 1995). We therefore chose to initiate a genome-sequencing project on *E. cuniculi* both as a representative of eukaryotic intracellular parasites and a prototype of minimal eukaryotes (Vivarès and Méténier, 2000). In a recent review, Keeling and Fast have thoroughly presented several aspects of the biology of microsporidia, with emphasis on the highly reduced character of these parasites at both cellular and molecular levels (Keeling and Fast, 2002). Sequencing of the 11 chromosomes of the *E. cuniculi* genome has been completed, allowing for the prediction of about 2000 protein-coding genes and leading to the hypothesis of the preservation of a cryptic mitochondrion-derived organelle called mitosome (Katinka et al., 2001). In this chapter, we attempt to point out some major features deduced from the genome sequence, especially for energy metabolism and potential mitosomal functions, that are expected to stimulate the interest of researchers in studying microsporidia for a better understanding of the intracellular parasitic lifestyle and of the evolutionary diversification of mitochondria in eukaryotes.

10.2 The Fascinating Microsporidian World

The eukaryotic microorganisms that are named microsporidia and ranked in the phylum Microspora (Sprague, 1977) are obligate intracellular parasites with the capacity of sporulation in order to survive in the environment and to initiate new infections (Vávra and Larsson, 1999; Cali and Tavkorian, 1999). More than 1200 microsporidian species invade most animal groups, including mammals, representing a formidable success for these intracellular parasites. Given that many species remain to be discovered, it seems possible to consider that a microsporidian world is intimately associated with the animal world (Vivarès, 2001). Pathogenicity of microsporidia largely depends on horizontal transmission, but some routes of vertical transmission have been also reported (Dunn and Smith, 2001; Andreadis and Vossbrinck, 2002). In the past two decades, 14 species have been recognized as human pathogens, primarily causing various opportunistic infections in immunocompromised persons (Franzen and Müller, 2001). Gastrointestinal infections are the most frequent, with the incidence of microsporidiosis up to 50% in HIV-infected patients with chronic diarrhea. Treatments with some drugs such as albendazole or a fumagillin derivative are more or less effective (Costa and Weiss, 2000; Molina et al., 2002), but restoration of immunity to microsporidia has been obtained through highly active antiretroviral therapy (HAART) based on the combined use of several anti-HIV compounds (Miao and Gazzard, 2000). Serological data argue for a high frequency of immunocompetent persons having been in contact with microsporidia and involvement of the latter in travelers'diarrhea seems likely (Raynaud et al., 1998; Okhuysen, 2001). In addition to human cross-contamination and possible zoonoses via pets and cattle, risks of microsporidiosis can be directly related to the use of contaminated water and the ingestion of uncooked or undercooked food (Franzen and Müller, 2001; Mota et al., 2000).

The microsporidian invasion strategy depends on spore germination, a complex motile process relying on the activity of a specialized intrasporal apparatus. This apparatus exhibits a long cylindrical and coiled structure called the polar tube, which can be explosively extruded at the spore apex and then used as a channel to introduce the remaining spore contents including the nucleus (sporoplasm) into a target eukaryotic cell (Keohane and Weiss, 1999). In most species, the fully extruded tube can extend over several tens of micrometers (more than 60 times the spore body length). Therefore, the polar tube is highly adapted to mediate the transport of a potentially replicative, wall-lacking parasitic cell toward the interior of a more or less distant host cell. Intracellular development is commonly split into a merogonial phase (merogony), involving divisions of the simplest cells (meronts), and a multistep sporogonial phase (sporogony), leading to highly differentiated spores (Vávra and Larsson, 1999). The switch from merogony to sporogony is morphologically marked by the appearance of an electron-dense cell coat. The corresponding sporonts undergo one or more divisions, producing sporoblasts that evolve into spores. Nuclear division is always of closed type, and dense plaques known as spindle pole bodies play the role of microtubule organizing centers. In contrast, the cell fission mode is variable, being simple or multiple according to species or developmental phase (Cali and Tavkorian, 1999). A multinucleate plasmodium is frequently formed during sporogony, and individual cells are further separated through either successive cleavages or synchronous budding.

Microsporidia lack morphologically recognizable mitochondria and have some prokaryote-like features, such as 70S ribosomes (Ishihara and Hayashi, 1968), 16S and 23S rRNAs (Curgy et al., 1980) and covalent linkage of 5.8S sequence to 23S rRNA 5′ end (Vossbrinck and Woese, 1986; Peyretaillade et al., 2001). An early origin of microsporidia was inferred from phylogenetic trees constructed with small subunit rRNA (Vossbrinck et al., 1987), isoleucyl-tRNA synthetase (Brown and Doolitle, 1995) and two translation elongation factors (Kamaishi et al., 1996a, b), which led to them being considered very ancient eukaryotes or archezoans, as formerly postulated by Cavalier-Smith, who had created the phylum Archezoa to unite amitochondriate protists having putatively diverged before the mitochondrial endosymbiosis (Cavalier-Smith, 1983). However, a radically different conception arose initially from tubulin phylogenies (Edlind et al., 1996; Keeling and Doolittle, 1996), then from the identification of a nuclear gene encoding a mitochondrial-type Hsp70 (mtHSP70; Germot et al., 1997; Hirt et al., 1997; Peyretaillade et al., 1998b), analysis of 23S rRNA sequence features (Peyretaillade et al., 1998a), justification of the erroneous character of a basal placement of microsporidia in molecular trees (Hirt et al., 1999; Van de Peer et al., 2000) and congruence of the data from phylogenies based on several conserved proteins (Brown and Doolitle, 1999; Fast et al., 1999; Hirt et al., 1999; Baldauf et al., 2000; Keeling et al., 2000; Katinka et al., 2001; Keeling and Fast, 2002; Vivarès et al., 2002). To summarize, microsporidia now appear to be fungi-related organisms originating from a mitochondriate ancestor. Both zygomycetes and ascomycetes have been proposed as candidates for being the closest relatives of microsporidia (Cavalier-Smith, 1998; Keeling et al., 2000; Keeling and Fast, 2002). The reductive evolution of microsporidia was thought to have involved either complete loss of mitochondria or genesis of derived organelles. Recently, small vesicular structures surrounded by a double membrane and reacting specifically with an anti-mtHSP70 antibody have been visualized in *Trachipleistophora hominis* (Williams et al., 2002), which strongly supports the second alternative that was also predicted through *E. cuniculi* genome analysis (Katinka et al., 2001; Vivarès et al., 2002).

10.3 Why Study Microsporidian Genomes?

As emerging human pathogens, microsporidia warrant more detailed investigation from medical and biological researchers, because most aspects of their nutrition, metabolism and physiology are poorly known (Weidner et al., 1999; Méténier and Vivarès, 2001). Some species, e.g., those of the *Encephalitozoon* genus, can be cultivated on commercially available mammalian cell lines, but very few attempts at developing a procedure for separating intracellular stages have been made to date (Green et al., 1999). Thus, experimental data have been derived mainly from the nondividing sporal stage. Collecting raw information about all the potential genes of a microsporidian through a genome-sequencing project was therefore attractive to gain insight into metabolic pathways of this parasite as well as to provide an estimate of the number of microsporidia-specific proteins. Even if the fast evolution of some gene sequences limits functional assignation (through rendering homologues undetectable, a phenomenon further complicated by the possible rarity of sequence data from representatives of their closest relatives), the identification of numerous genes related to essential functions can be reasonably expected. Confidence in the statement that several genes commonly found in eukaryotes are missing in microsporidia can be tested through the correspondence between missing genes and either reduced or lacking intracellular structures (70S ribosomes, nondictyosomal Golgi apparatus, no recognized mitochondria and peroxisomes, no basal bodies, etc.). The demonstration of the expression of an mtHSP70 gene in *T. hominis* (Williams et al., 2002) is an illustration of possible discrepancies between molecular and cytological data. Thus, this approach bears two caveats. First, the search for sequence homologies might reveal some genes that were not expected on the basis of cytological considerations. Second, the evolution of some organelles, such as mitochondria, might have led to simplified structures that could not be recognized.

Obviously, although microsporidian species share several structural and developmental features, it cannot be considered that the genome sequence from a single species is sufficiently representative to authorize systematic generalizations for a highly diversified group of parasites. Diversity of nuclear condition, life cycle and host cytoplasm–parasite interface are already evident. The nuclear apparatus is represented by either a single nucleus or a pair of closely apposed identical nuclei forming a diplokaryon. Some life cycles are dependent on only one host whereas others require an alternation of different hosts, producing up to three different spore types (Dunn and Smith, 2001; Andreadis and Vossbrinck, 2002). Three kinds of cycles are generally distinguished: (1) asexual reproduction of monokaryotic (uninucleate) cells in a single host (*Encephalitozoon* type), (2) asexual reproduction of diplokaryotic cells in a single host (*Nosema* type) and (3) cycle with different hosts and involving an alternation between monokaryotic and diplokaryotic stages via meiosis (*Amblyospora* type). The ploidy states have been evaluated microphotometrically in an *Amblyospora* species undergoing meiosis in a mosquito host (Hazard and Brookbank, 1984). A careful reinterpretation of the corresponding data by other authors suggests not only a typical meiosis but also similarities between *Amblyospora* diplokaryotic state and fungal dikaryotic condition, both resulting in functional diploids

(Flegel and Pasharawipas, 1995). Considering that up to seven distinct differentiation programs exist among microsporidia, the same authors have postulated that program selection occurs in G1 phase of the meront cell cycle and depends on host factors such as cell type and hormones. As regards the interface between host cytoplasm and microsporidia, the most frequent cases are either a direct contact during the whole cycle or a thick layer of material secreted by the parasite and enclosing sporogonial stages (sporophorous vesicle; Vávra and Laarson, 1999; Cali and Tavkorian, 1999). Like apicomplexans or *Leishmania*, other microsporidia are adapted to live inside a parasitophorous vacuole, the membrane of which closely adheres to merogonial stages. Some species can induce hypertrophic growth of the host cell, allowing for a considerable accumulation of parasites within a visible cyst-like structure or xenoma. *Nucleospora salmonis*, a salmon parasite, multiplies inside the nucleus instead of the cytoplasm of lymphoblasts (Chilmonczyk et al., 1991). Such diversity allows us to infer the existence of some interspecific variability of the set of microsporidian genes involved in developmental processes and nutrient acquisition.

The comparison of chromosome complements also indicates a great evolutionary diversification among microsporidia. Pulsed-field gel electrophoresis (PFGE) data were first obtained for insect parasites of three different genera (*Nosema, Vairimopha, Vavraia*), showing patterns of 8 to 14 ethidium-stained bands assimilated to species-specific sets of linear chromosomes (Munderloh et al., 1990; Malone and McIvor, 1993). Further molecular karyotyping studies were performed in species infecting insects, fishes or mammals and belonging to the following genera: *Encephalitozoon* (Biderre et al., 1995, 1999) *Glugea* (Biderre et al., 1994; Amigo et al., 2002), *Microgemma* (Amigo et al., 2002), *Nosema* (Kawakami et al., 1994; Street, 1994) and *Spraguea* (Biderre et al., 1994). The number of chromosomal bands currently ranges between 8 and 18, with the smallest sizes at 130 to 140 kbp and the largest ones at more than 2 Mbp. The widest chromosome size ranges are found in *Glugea* and *Vavraia*, two closely related genera. The estimates of haploid genome sizes are mostly between 5 and 15 Mbp, but that of *G. atherinae* attain 19.5 Mbp. In contrast, the three *Encephalitozoon* species harbor a set of small chromosomes (mostly 200–300 kbp) and thus the smallest haploid genome sizes (2.3–2.9 Mbp). No species having only large chromosomes (>1.0 Mbp) has been identified so far. Such species likely exist, when considering the case of *Amblyospora* in which light microscopy has revealed relatively large meiotic chromosomes (Hazard and Brookbank, 1984). Although under- or overestimations cannot be excluded, especially if polymorphic homologous chromosomes are present as in *E. cuniculi* (Biderre et al., 1999), the extent of intergenera genome size variations remains highly significant. In other words, microsporidia do not escape the C-value paradox. It seems unlikely that this only reflects possible changes in gene diversity in relation with more or less complex life cycles. In various eukaryotic cells, variations in nuclear DNA content correlate positively with those of cell size (Cavalier-Smith, 1985). The size of spores is variable in microsporidians, the extreme lengths being less than 1 μm in *Enterocytozoon bieneusi* and ~40 μm in *Bacillidium filiferum* (Vávra and Larsson, 1999). Figure 10.1 shows spores from three species of different genera. It is evident that *Encephalitozoon* species with small genomes also produce small spores (~2 μm long) whereas *G. atherinae* with a larger genome has larger spores (~8 μm long). Assuming that a good correlation exists, we can predict that *Ent. bieneusi*, a major human-infecting microsporidian, has the most reduced genome. Unfortunately, this species is not currently cultivable *in vitro*, preventing determination of karyotype. The possibility that a variable number of copies of genes commonly present at high copy number, such as rDNA, contributes to genome size variation should be easily testable experimentally. Investigations on ploidy state, nongenic repetitive sequences, multigene families and intron frequency in more or less distant microsporidian species are also highly desirable.

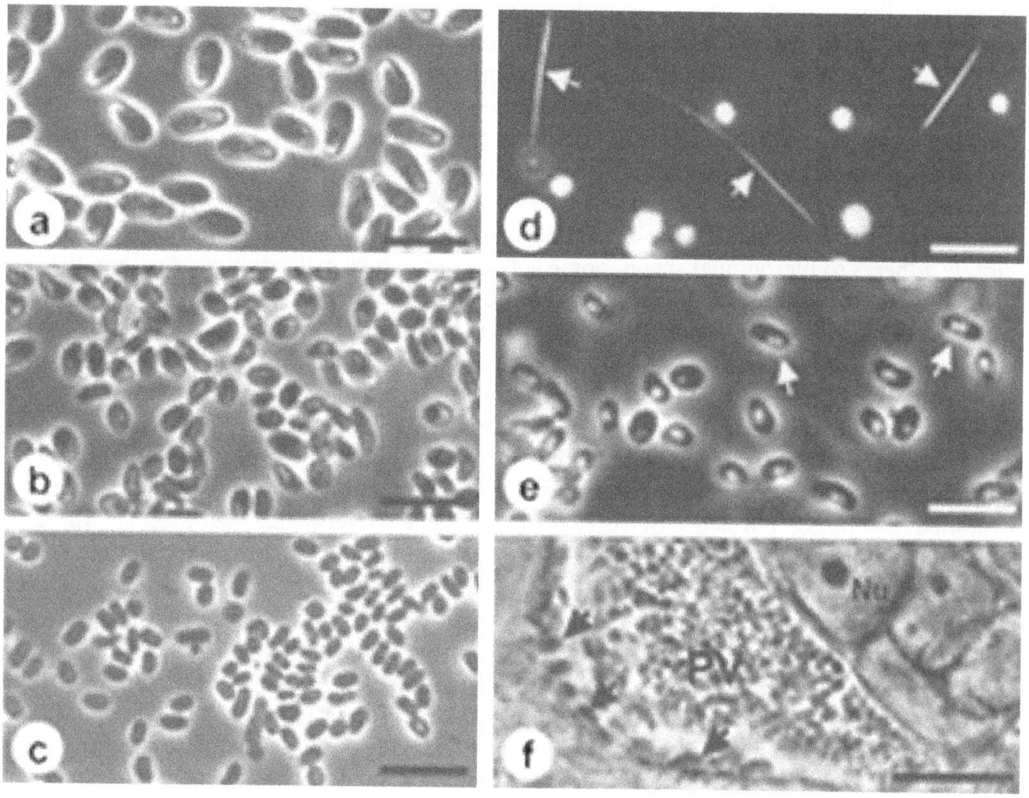

FIGURE 10.1 Purified spores of three microsporidian species: *Glugea atherinae* (a), *Spraguea lophii* (b) and *Encephalitozoon cuniculi* (c). Differences in spore size between these species are evident. DAPI staining reveals a dramatic stretching of the nuclei (arrows) when flowing through extruded polar tubes, as shown in *G. atherinae* (d). Diplokaryotic (binucleate) spores (arrows) always coexist with monokaryotic spores in *S. lophii* (e). A view of a parasitophorous vacuole (PV) containing *E. cuniculi* cells (f). Merogonial stages are located at the periphery of the vacuole (arrows). Nu, host cell nucleus. Bar, 10 μm.

10.4 *E. cuniculi* Genome Content

10.4.1 A Tiny Microsporidian Genome

Encephalitozoon cuniculi harbors a single nucleus for which light and electron microscope studies have failed to identify a nucleolus and determine the number of metaphasic chromosomes. The presence of 11 chromosomes was inferred from a first PFGE analysis of the native sporal DNA from a mouse isolate of *E. cuniculi*, showing a narrow size distribution of DNA bands from 217 kbp (Chromosome I) to 315 kbp (Chromosome XI) and representative of a haploid genome size of ~2.9 Mbp (Biderre et al., 1995). The correspondence between PFGE-separated bands and different linkage groups was further demonstrated through both hybridization experiments with randomly fragmented *E. cuniculi* genomic DNA and mapping of *Mlu*I and *BssH*II restriction fragments by 2D-PFGE techniques (Biderre et al., 1997; Brugère et al., 2000a). Thus, the *E. cuniculi* genome has a prokaryotic-like size, close to that of *Staphylococcus aureus* (2.81 Mbp; Kuroda et al., 2001) and two *Listeria* species (2.94 to 3.01 Mbp; Glaser et al., 2001). This represents less than the largest chromosome (3.29 Mbp) of the malaria parasite *Plasmodium falciparum* for which the 23-Mbp genome sequence has been recently reported (Gardner et al., 2002a). It is clear that nuclear genome miniaturization has been more extensive in *E. cuniculi* than in the smallest autotrophic eukaryote currently represented by the prasinophycean *Ostreococcus tauri* with a haploid genome size of 10.2 Mbp (Courties et al., 1998) or those of the apicomplexan intracellular parasites *Cryptosporidium parvum* and *Theileria parva* with 9 to 10 Mbp (Spano and Crisanti, 2000; Nene et al., 2000) or the fungal parasite *Pneumocystis carinii* with 8.4 Mbp (Cornillot et al., 2002a). As far as we know, only the vestigial nuclei (nucleomorphs) from algal endosymbionts in cryptomonads and chlorarachniophytes have been prone to more severe genomic reduction, with genome sizes as small as ~0.5 Mbp (Gilson et al., 1997; Douglas et al., 2001). The *Encephalitozoon* nucleus is likely diploid, as supported by the ratio between spore DNA content and C value (Biderre et al., 1995), and there is evidence for pairs of homologous chromosomes with length polymorphism in several isolates (Biderre et al., 1999; Brugère et al., 2000b, 2001). For example, in the mouse isolate (GB-M1 isolate, Strain I) that we used in most studies, two Chromosome III homologues can be distinguished by a size difference of ~3 kbp. This results in the electrophoretic comigration of the smallest homologue (IIIb) with Chromosome II and explains the reduced ethidium-staining intensity of Band III by the exclusive presence of the largest homologue (IIIa).

10.4.2 The Relation of One Chromosomal End – One 16S–23S rDNA Unit

After the finding of a striking dispersion of the 16S–23S rDNA transcription units over all the chromosomes (Peyretaillade et al., 1998a), a first conception of the general organization of *E. cuniculi* chromosomes emerged from both construction of *Mlu*I–*BssH*II restriction maps and 16S–23S rDNA hybridizations to either 2D-PFGE separated restriction fragments or combed native DNA molecules from Chromosome XI (Brugère et al., 2000a). Each chromosome is characterized by a variable gene-rich core flanked by two conserved terminal regions at least 15 kbp in size, each including a single 16S–23S rDNA unit. The two distant rDNA units on a given chromosome are divergently oriented, similar to free palindromic rDNA molecules in myxomycetes or in the macronucleus of the ciliate *Tetrahymena*. Interestingly, a similar arrangement has been found in the three nucleomorph chromosomes of the cryptomonad *Guillardia theta* (Douglas et al., 2001). Since these chromosomes are small sized (174 to 196 kbp) and never visible as compact entities during mitosis, Douglas and coworkers have suggested that each chromosome forms a single loop domain and can be packaged inside the tiny nucleus as only 30-nm chromatin fibers. Genes for the core histones (H2A, H2B, H3, H4), proteins of the SMC (structural maintenance of chromosome) family, condensins and other factors involved in chromatin assembly support not only the formation of 30-nm threads but also some higher folding order for *E. cuniculi* chromosomes.

Sequencing of the *E. cuniculi* Chromosome I has provided a good illustration of the near-telomeric location of each rDNA unit (Peyret et al., 2001). Heterogeneous repeats of telomeric DNA can extend over 1.2 kbp, as shown by *Bal*31 digestion kinetics, but the organization of the true terminus is still unknown. As a result of telomerase activity, extensions represented by single-stranded G-rich DNA have been demonstrated in ciliates, yeasts and mammals (Bryan and Cech, 1999). We assume that such extensions also exist in *E. cuniculi*, given the presence of a telomerase gene and the recent identification of a candidate for a telomere-associated protein. Having some similarity to the α-subunit of a well-characterized ciliate protein, a single-stranded telomeric DNA-binding protein was first found in *S. pombe* (Pot1 standing for protection of telomeres) and human cells (Baumann and Cech, 2001). The human homologue of *S. pombe* Pot1 localizes to chromosome ends in interphase nuclei and genes encoding Pot1-like proteins are present in other organisms including *Encephalitozoon* (Baumann et al., 2002). The *E. cuniculi* telomeric repeats should be separated from the end of the 23S rRNA locus by a region of ~9 kbp that was not completely assembled because of numerous short microsatellite sequences. The subtelomeric region upstream of the rDNA unit contains several types of tandem repeats of only 7 to 30 bp in period size. However, the most remarkable feature of Chromosome I is the dyad symmetry offered by two identical ~45-kbp domains extending from each telomere to a transitional region marked by an abrupt decrease in GC content and by an 8-kb duplicated cluster of six genes encoding five known proteins (Peyret et al., 2001). Thus, the organization of this chromosome is clearly of an amphimeric type. The genomes that are called amphimers are characterized by a pair of separated long inverted repeats, e.g., organellar genomes such as the yeast mitochondrial genome, and have been considered as optimally organized to stably retain a duplicated segment (Rayko, 1997).

First obtained through single-specific primer PCR amplifications, the *E. cuniculi* 5S rRNA gene sequence was found to hybridize with Chromosomes V and IX (Peyretaillade et al., 1998a). Genome sequencing has revealed three dispersed gene copies instead of two, the third copy being on Chromosome VII (Katinka et al., 2001). The three 5S rRNA genes are located within chromosome cores and thus fully separated from the 22 subtelomeric 16S–23S rDNA units. This contrasts with the contiguous location commonly observed in prokaryotes and various eukaryotes, including some fungi and the cryptomonad nucleomorph. In the malaria parasite, the two types of rDNA are also distantly located, but only five complete 18S–5.8S–28S rDNA units (two others lack 18S gene) are present and dispersed on several chromosomes. Only two of these occupy a subtelomeric position (Gardner et al., 2002a, b). *Plasmodium falciparum* exhibits exactly the same number of 5S rRNA gene copies as *E. cuniculi* does, but the three genes are tandemly arranged on a single chromosome.

The lack of visible nucleolus and the relation of one chromosome end – one rDNA unit in *E. cuniculi* might signify that the sites of rRNA transcription and processing are not clustered inside a typical nucleolar domain. Genes encoding homologues of conserved nucleolar proteins such as fibrillarin are, however, present. The diplomonad *Giardia lamblia* also apparently lacks nucleoli and the observation of a diffuse immunofluorescence staining of its two nuclei with an antibody to fibrillarin suggested a dispersion of the sites of ribosome assembly (Narcisi et al., 1998). In addition, the genome of this flagellate contains numerous rDNA tandem repeats that are located near the telomeric region of some but not all the chromosomes (Adam et al., 1991; Leblancq and Adam, 1998). Assuming that the *E. cuniculi* telomeres are close to the nuclear envelope, as demonstrated in a number of eukaryotes including *Plasmodium* (Freitas-Junior et al., 2002), it would be interesting to determine whether the subtelomeric rDNA units have a similar location or are fully dispersed throughout the nucleus.

Chromosome size variants of *E. cuniculi* are related to insertions–deletions (indels) that have been mapped to the region upstream of the rDNA (Brugère et al., 2000b, 2001). Genome plasticity in *Giardia* variants is characterized by a common breakpoint also located upstream of rDNA (Upcroft and Upcroft, 1999), but, unlike *E. cuniculi*, most rearrangements in *Giardia* are associated with variations in rDNA copy number (Adam, 1992). Perhaps, the same basic processes occur, but *E. cuniculi* seems to be more deeply engaged in the process of genomic reduction, so that still frequent interchromosomal recombinations involving rDNA are now restricted to isolated units and might thus escape the detection on the only basis of chromosome length polymorphism. The sequence conservation of nucleomorph rDNA units is thought to be maintained by intra- and intermolecular recombination between the terminal repeats, followed by gene conversion (Gilson and Mc Fadden, 1995). This model can be applied to *E. cuniculi* but requires experimental evidence. Unfortunately, the whole genome shotgun strategy did not allow for the complete assembly of subtelomeric sequences in *E. cuniculi* Chromosomes II to XI, hindering useful comparisons to evaluate the role of some repeats in recombination events. Identical DNA stretches are detected within the sequences close to the termini of the final chromosome contigs, confirming interchromosomal conservation of a number of subtelomeric segments. These include minisatellites and identical or truncated copies of potential open reading frames (ORFs) without clear-cut homologues in other organisms. One example is an ORF that we tentatively annotated as having some similarity to glypican-4, a surface proteoglycan of mammals. Two copies (ECU01_0110 and ECU01_1500) are elements of the duplicated terminal domains in Chromosome I, but more or less conserved additional copies are present on other chromosomes. The possibility that they correspond to the members of a multigene family encoding a microsporidia-specific surface protein of importance for pathogenesis should be explored.

10.4.3 A Reduced Set of tRNA Genes

A total of 46 genes for tRNAs (instead of 44, as erroneously indicated in the genome paper published in *Nature*), representative of the families required to interact with all codons, are dispersed among all the *E. cuniculi* chromosomes. This is much less than in yeasts but more than in the nucleomorph of cryptomonad algae. The *G. theta* endosymbiont is thought to import some tRNAs from its host cell, especially $tRNA^{Glu}$ that is not encoded by the nucleomorph genome (Douglas et al., 2001). Such an import cannot be considered for *E. cuniculi*. The *P. falciparum* genome reveals 43 tRNAs with two others possibly located in still nonsequenced regions (Gardner et al., 2002a). Thus, the two different intracellular parasites share a similarly low redundancy of tRNAs as for 5S rRNAs. Curiously, only two tRNA genes are interrupted by a nonspliceosomal intron in *E. cuniculi*, whereas one third of the set of *G. theta* tRNA genes (12 of 37) contains such structures. An intron-containing gene for a $tRNA^{Tyr}$ is common to the two organisms. A supernumerary U residue is located in the D stem of a $tRNA^{Ser}$ gene on *E. cuniculi* Chromosome I, an identical position to the first residue of a very small intron (10 nt) within the same gene of the cryptomonad nucleomorph (Zauner et al., 2000), suggesting that this U residue might be the remnant of an intron (Peyret et al., 2001).

10.4.4 The High Protein-Coding Gene Density of the Chromosome Cores

The *E. cuniculi* genome contains ca. 2000 protein-coding genes, i.e., 4.3-fold the gene number in the genome of the cryptomonad nucleomorph. Compared to the 13.8-Mbp genome of *S. pombe* that was recently claimed to have the smallest number of protein-coding genes yet recorded for a eukaryote (Wood et al., 2002), a simple calculation indicates that *E. cuniculi* is 2.4-fold less rich in genes than is fission yeast (Table 10.1). As *E. cuniculi* is truly a eukaryote but a parasitic one, we think that it would have been better to say that *S. pombe* has the smallest number of genes reported for a free-living eukaryote. With a trend for the G+C content to increase from the two terminal regions toward the central region, the *E. cuniculi* chromosome cores are mainly characterized by a high average density of protein-coding genes (about 1 gene/kbp). As shown in Table 10.1, this density is much less than in *P. falciparum* but close to that in the *G. theta* nucleomorph. Intergenic regions are evidently very short (mean length 129 bp). Overlapping coding sequences are occasionally found, but are considered nonsignificant because of uncertainties in the placement of the initiation codons for genes encoding either hypothetical proteins or defined proteins with poorly conserved N-terminal sequences. In the nucleomorph genome, a more extreme reduction of intergenic regions has entailed the overlapping of 44 genes (Douglas et al., 2001).

Gene compaction in *E. cuniculi,* as well as in the nucleomorph, is related not only to the very low frequency and short length of spliceosomal introns (Table 10.1) but also to a trend toward the shortening of translated regions compared with homologues in other eukaryotes. The mean length of the potential microsporidial proteins is of 359 amino acid

TABLE 10.1 Comparison of Genome Features for Selected Microbial Eukaryotes

			Protein-Coding Genes				
Organism	*Lifestyle*	*Genome Size*	*Chromosome Number*	*Total Number*	*With Introns*	*Gene Density (bp per gene)*	*Mean Protein Length (amino acids)*
Encephalitozoon cuniculi	Intracellular Parasitic	2.9	11	1,997	12	1,025	359
Guillardia theta (nucleomorph)	Endosymbiotic	0.55	3	464	17	977	325
Plasmodium falciparum	Intracellular Parasitic	22.8	14	5,268	2,856	4,338	761[a]
Saccharomyces cerevisiae	Free	13.8	16	5,570–5,651	231	2,088	475
Schizosaccharomyces pombe	Free	13.4	3	4,824	2,226	2,528	470

Note: Some major features of the fully sequenced genomes from *E. cuniculi* [Katinka, M. D. et al. (2001) *Nature* 414: 450–453] and four other unicellular eukaryotic organisms: *G. theta* [Douglas et al. (2001) 1091–1096], *P. falciparum* [Gardner et al. (2002a) 498–511], *S. cerevisiae* [Blandin et al. (2000) 31–36 and Wood et al. (2001) 143–154], and *S. pombe* [Wood et al. (2002) 871–880].

[a] The surprising increased size of *Plasmodium* proteins is partly due to a great proportion of uncharacterized proteins encoded by genes that are more than 4 kbp long [Gardner et al. (2002a)498–511].

(aa) residues. This is close to the value in some bacteria (334 aa in *Rickettsia prowazecki*) or in *G. theta* and less than in budding and fission yeasts. The much higher value in *Plasmodium* (more than 750 aa) is quite surprising (Table 10.1). The comparison of *E. cuniculi* proteins with 350 *S. cerevisiae* homologues indicated a mean size reduction of 14.6% (up to ~50% in extreme cases; Katinka et al., 2001). Deletions are variable in length and position according to the considered protein. The case of a large protein, the cytoplasmic dynein heavy chain (3151 residues in *E. cuniculi*), has been presented in a previous review (Vivarès et al., 2002). The essential AAA- and microtubule-binding domains are preserved, whereas various deletions affect intermediate segments and especially a divergent C-terminal region assigned to a large part of the dynein head. The global shortening represents almost a quarter of the length of the yeast homologue. Another large protein, midasin, a possible nuclear chaperone belonging to the AAA ATPase family, has recently been shown to be conserved in eukaryotes, including *Giardia* and *Encephalitozoon* (Garbarino and Gibbons, 2002). The *E. cuniculi* gene (ECU08_1900) encodes a protein of 2832 aa, which is 42% of the length of *S. cerevisiae* midasin. The N-terminal domain is reduced by 90%, but the six tandem AAA protomers and other essential motifs are retained. As illustrated by the sequences of two dUTPase paralogues and of an adrenodoxin–ferredoxin protein, demonstrated later (see also Figure 10.3 and Figure 10.4A), small proteins do not escape deletion events. A systematic analysis of the compaction of coding sequences will be presented elsewhere (manuscript in preparation).

10.4.5 Gene Duplications

Gene duplications represent an important evolutionary source of new protein functions. The most recent analyses of duplicated genes in fully sequenced genomes generally support an evolution of paralogues under purifying selection, but have led to some discrepancies about the occurrence of an early phase of near-neutrality (Wagner, 2000; Robinson-Rechavi and Laudet, 2001; Kondrashov et al., 2002). The *E. cuniculi* genome obviously contains paralogous genes of ancient origin, forming usual multigene families such as for protein kinases and ABC systems, but large-scale duplications comparable to those assumed to result from tetraploïdization in the *S. cerevisiae* ancestor (Wolfe and Shields, 1997) seem difficult to invoke for the microsporidian species considered here. However, a few surprises arose from examining the distribution of identical or nearly identical gene copies (more than 95% nucleic identity), excluding those of undefined function in extreme chromosomal regions (Méténier et al., 2003).

The largest regional duplications are close to the subtelomeres (upstream of the rDNA unit) and limited to three different chromosomes (Figure 10.2). A block containing a series of six ORFs is associated with each subtelomere of Chromosome I (Peyret et al., 2001), but the distal part of this block (four ORFs) is also present with the same global arrangement near one subtelomere of Chromosome VIII, representing a common segment of more than 8 kbp. Interestingly, the closely spaced but distinct genes for two key enzymes involved in pyrimidine-nucleotide metabolism (dihydrofolate reductase and thymidylate synthase) are retained. Such genes have been recently used for rooting the evolutionary tree of eukaryotes on the basis of the occurrence or absence of a gene fusion event leading to a bifunctional enzyme (Stechmann and Cavalier-Smith, 2002). The *Encephalitozoon* genome is unique in the close linkage of two independent *dhfr* and *ts* genes, thus mimicking a fused *dhfr–ts* gene. Another large block (4.3 kbp) with four embedded ORFs is shared by Chromosomes VIII and X but, in this case, the duplicates (99% nucleic identity) are in opposite orientations, relative to the closest subtelomere (Figure 10.2). A single amino acid change (Ile to Val), related to an A to G transition, is the only difference between the two copies for an HSK1-like protein kinase. These duplicated blocks are assumed to originate from the same type

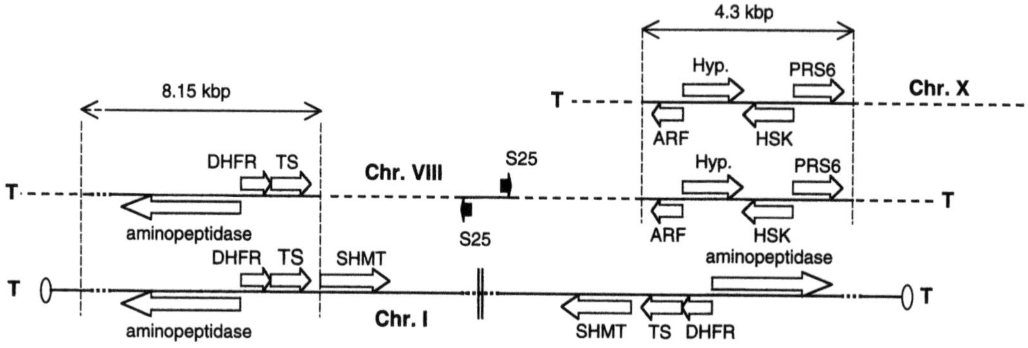

FIGURE 10.2 Schematic view of the largest segmental duplications in the *E. cuniculi* genome. For Chromosome I, only gene clusters encoding aminopeptidase, dihydrofolate reductase (DHFR), thymidylate synthase (TS) and serine hydroxymethyl-transferase (SHMT) are shown to illustrate the two large inverted repeats extending from each telomere (T) to ~4 kbp downstream of the *shmt* gene. A partial copy excluding *shmt* is present on Chromosome VIII. A block comprising four genes is shared by Chromosome VIII (near the other subtelomere) and Chromosome X (ARF: ADP ribosylation factor, Hyp: hypothetical protein, HSK: homologue of yeast hsk1 protein kinase, PRS6: proteasome regulatory subunit 6). A very short and perfect intrachromosomal duplication near the center of Chromosome VIII specifically concerns the gene for ribosomal protein S25 (black arrows).

of rearrangements as those affecting subtelomeric regions over variable distances. Some rare duplications, occupying more internal positions inside the chromosome cores, are suggestive of gene conversion events. They extend over short DNA segments, each including a single protein-coding gene. Characteristic intrachromosomal duplications are represented by two CTP synthase genes on Chromosome XI and two S25 genes on Chromosome VIII. Showing divergent transcriptional orientations, the genes encoding the ribosomal (r-) protein S25 differ by only one neutral nucleotide substitution and are separated by ~2 kbp (Figure 10.2). By contrast, the members of the duplicated gene pair for another r-protein (L3) are located on two different chromosomes. The paucity of duplicated genes can be considered to result from either a low gene conversion rate or a higher deletion rate for these genes in the recent past of *E. cuniculi*. The trend to an extensive size reduction of most genes and intergenic regions seems to favor the second alternative.

Several classical examples can be cited to illustrate functional divergence after an ancient gene duplication. However, in *E. cuniculi* as in other microsporidia, many gene sequences are characterized by an unusually high divergence level, hindering the annotation task and complicating phylogenetic analyses (Vivarès et al., 2002). The divergence is quite apparent by comparing various paralogues. For example, most eukaryotic protein-disulfide isomerases (PDIs) contain two or three thioredoxin-like domains, but the diplomonad *G. lamblia* (*intestinalis*) typically harbors five single-domain PDIs that are reduced in size and are assumed to reflect an ancestral process related to the subdivision of function in protein folding (Mc Arthur et al., 2001). In *E. cuniculi*, two large-sized PDI homologues are different in both the number and location of thioredoxin domains and are difficult to assign to known classes. Secondary domain losses seem likely. Another example is provided by two paralogues of dUTP pyrophosphatase (dUTPase), a crucial enzyme to prevent incorporation of uracil into DNA. The highest similarities are found with dUTPases from vaccinia or chlorella viruses, and the two potential proteins have only 21% identity. The smallest paralogue is typically characterized by a truncated C-terminal region (Figure 10.3). The 15 most C-terminal residues of dUTPase are referred to an arm that is highly flexible (Nord et al., 2001) and needed for enzymatic activity (Nord et al., 2000). This strongly suggests that the

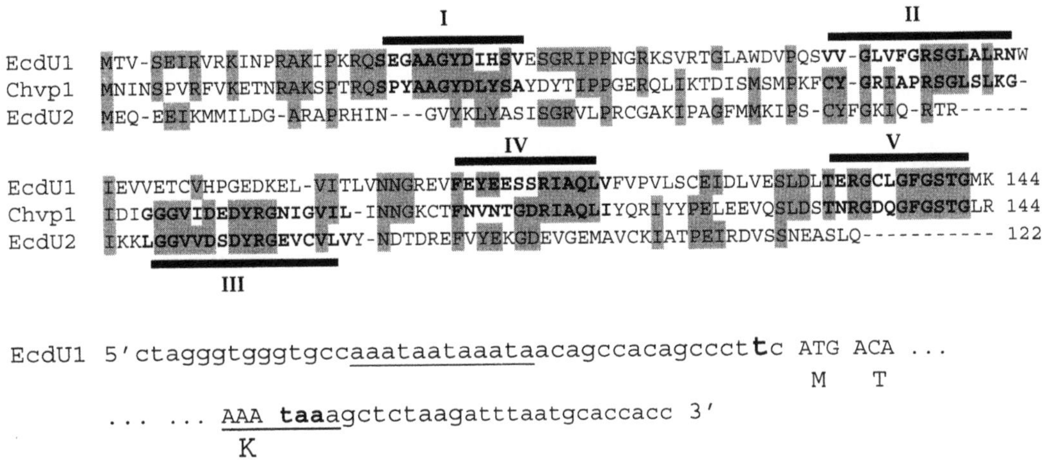

FIGURE 10.3 Two divergent dUTPase genes in *E. cuniculi.* The deduced protein sequences (EcdU1 and EcdU2 for ECU05_0280 and ECU06_0430 genes, respectively) are aligned with *Paramecium bursaria* chlorella virus 1 dUTPase (Chvp1). The best similarity scores are found with viral homologues including chlorella virus and poxviruses. Identical residues at the same position are shaded. All known dUTPases have a set of five conserved sequence motifs (numbered from I to V) that should contain the active site. Clustering of some viral dUTPase sequences with prokaryotic and eukaryotic ones in phylogenetic trees supports horizontal transfer events [see Baldo and McClure (1999) 7710–7721]. More conserved in size, EcdU1 exhibits four typical motifs (in bold). As shown at the bottom of the figure, a promoter sequence including a potential TATA box (underlined) is predicted with the ProScan program to extend just downstream of a very reduced (2 nt) 5′ untranslated region (putative transcription start site in bold). A putative polyadenylation signal overlaps the stop codon. Three motifs are truncated in EcdU2, especially the critical C-terminal motif (V), and no transcription signals are predicted. Although appearing as a degenerated dUTPase, EcdU2 has curiously retained the motif III (in bold), the one that strongly diverges in EcdU1.

shortened version could be inactive for the reaction catalyzed by a dUTPase. An important challenge will be to determine whether this truly reflects gene degradation or this is a prelude to acquiring a novel function.

10.4.6 Preservation of Essential Cellular Processes vs. Metabolic Simplification

A survey of the *E. cuniculi* coding sequences matching proteins of known function (less than 900 ORFs) shows that most can be attributed to the essential machineries for DNA replication and repair, transcription, translation and ribosome biogenesis. It is also obvious that the presence of an endomembrane system, even if the microporidian Golgi apparatus is uneasy to detect at the ultrastructural level, can be correlated with various genes encoding components involved in a typical secretory system (translocon, signal peptidase, COPI and COPII coat complexes, t- and v-SNAREs, glycosyltransferases, adaptin, etc.). A homologue of yeast ORM1 is present and is now known as a member of a newly defined family of ER membrane proteins (ORMDL) thought to participate in protein folding and trafficking in the ER (Hjelmqvist et al., 2002). As supported by the recent finding of genes for Golgi-related syntaxin in both *Giardia* (adictyosomal protist), *Trypanosoma* and algae, the eukaryotic endomembrane system is of ancient origin (Dacks and Doolittle, 2002). It is therefore not surprising that microsporidia have also preserved some important proteins for vesicular transport. Likewise, as expected for a eukaryotic cell, a significant number of homologues can account for basic chromatin organization, nuclear import and export, cell cycle control and mitosis. Genes for the subunits of the proteasome, T-chaperonin complex and V_0V_1-

A. Ferredoxin

```
E. cuniculi      1 MDMFSAPDRIPEQIRIFFKTMK-QVVPAKAVCGSTVLDVAHKNGVDLEGACEGNLACSTCHVILEEPLY
R. prowazekii    1           MLRKIKVTFIINDEEERTVEAPIGLSILEIAHSNDLDLEGACEGSLACATCHVMLEEEFY
                             :  :*:: *   . :   ..:*  * ::*::**.*.:********.***:****:***  :*

E. cuniculi        RKLGEPSDKEYDLIDQAFGATGTSRLGCQLRVDKSFENAVFTVPRATKNMAVDGFKPKPH 128
R. prowazekii      NKLKKPTEAEEDMLDLAFGLTDTSRLGCQIILTEELDGIKVRLPSATRNIKL         112
                   .** :*:: * *::* *** *.*******: : :.::.  . :* :*:*: :

E. cuniculi      1 MDMFSAPDRIPEQIRIFFKTMKQVVPAKAVCGSTVLDVAHKNGVDLEGACEGNLACSTCHVILEEPLY
G. lamblia       1 MSLLSSIRRF-ITFRVVQQGVEHTVSG-AV-GQSLLDAIKAAHIPIQDACEGHLXCGTCGVYLDKKTY
                   *.::*:  *:_  :*:. : :::.*.. ** *.::**. :   : ::.****:* *.** * *::  *

E. cuniculi        RKLGEPSDKEYDLIDQAFGATGTSRLGCQLRVDKSFENAVFTVPRATKNMA--VDGFK---PKPH    128
G. lamblia         KRIPRATKEEAVLLDQVPNPKPTSRLSCAVKLSSMLEGATVRIPSFNKNVLSESDILASEEKKRHGQH 133
                   ::: ..:.:*  *:**. ... ****.* ::::.. :*.*.. :*  .**:     * :     * *
```

C-terminal region (type II-mtFd) **Pro-His motif**

```
E. cuniculi        ---GEPSDKEYDLIDQAFGATGTSRLGCQLRVDKSFENAVFTVPRATKNMAVDGFKPKPH 128
A. thaliana        ---EEPTDEENDMLDLAFGLTATSRLGCQVIAKPELDGVRLAIPSATRNFAVDGFVPKPH 197
D. melanog.(II)    ---KEAEEQEDDLLDMAPFLRENSRLGCQILLDKSMEGMELELPKATRNFYVDGHKPKPH 172
P. falciparum      ---PEPLDNEIDMLELAPCITETSRLGCQIKLSKELDGMKIQLPPMTRNFYVDGHVPTPH 139
                       *. ::* *::: *      .******:  . .::.  : :*  *:*: ***. *.**
```

B. Tim17/22-like protein (ECU07_0240)

```
E. cuniculi      1 MNTKCLREKIEKIKPHLLKTASDT--LQGYVFGCMIGVFSS--SEKPS-LRHVHES----  51
S. cerev.Tim17   1 MSADHSRDPCPIVILNDFGGAFAMGAIGGVVWHGIKGFRNSPLGERGSGAMSAIKARAPV  60
                   *.:.  *:     :  : :  *      : * *:  : *. .*  .*: *     . ::

E. cuniculi        -GKSFAKVSMIYSTTESALQLYGSKNAPLNSLISGAVAGG-LGVND--R-SKKSILTGAA 106
S. cerev.Tim17     LGGNFGVWGGLFSTFDCAVKAVRKREDPWNAIIAGFFTGGALAVRGGWRHTRNSSITCAC 120
                    * .*.  . ::** :.*::   .:: * *::*:* .:** *.*..  * :::* :* *.

E. cuniculi        SFGLYTGMSNIFS--VQGQK                124
S. cerev.Tim17     LLGVIEGVGLMFQRYAAWQAKPMAPPLPEAPSSQPLQA 158
                    :*:  *:. :*.   .   *
```

FIGURE 10.4 (A) Comparison of *E. cuniculi* ferredoxin (Fd) with alpha-proteobacterial (*Rickettsia prowazekii*) and diplomonad (*Giardia lamblia*) homologues. The 10 first amino acid residues (underlined) might represent a putative presequence. In *Giardia*, an intron has been recently discovered [Nixon et al. (2002) 3359–3361], occupying a position (_) just downstream of the Fd gene region encoding an N-terminal extension with the same length. Similarity with rickettsial Fd is especially high in a domain including three conserved cysteine residues (_) for possible binding to [2Fe2S] cluster. The *E. cuniculi* Fd clearly diverges from both bacterial and diplomonad homologues in the most C-terminal residues (in boldface characters). As shown by Seeber, variability of the extreme C-terminal region among bacterial Fds and mitochondrial(mt)-Fds of eukaryotes allows for the distinction of two types, leading to consider that *E. cuniculi* Fd is of Type II on the basis of a common Pro–His motif [Seeber (2002a) 545–547]. A rather puzzling distribution is thus observed. Like prokaryotes but unlike *Encephalitozoon*, various fungi harbor only the Type I (heterogeneous C-terminus). Sequence alignment in the bottom part of Figure 4A illustrates that the Type II (with conserved Pro–His motif, possibly required to mediate self-assembly) is shared by some Fd isoforms in several protists and animals (e.g., in *Drosophila melanogaster*) but unique in all plant and apicomplexan mtFds (e.g., in *Arabidopsis thalania* and *Plasmodium falciparum*). (B) An *E. cuniculi* sequence candidate for a Tim-like subunit that might associated with the mitosomal inner membrane. Having a weak similarity to both Tim22 and Tim17, the sequence is closer to Tim17 as judged by protein length and domain prediction. Only the alignment with *S. cerevisiae* Tim17 is shown. The C-terminal part is reduced and no tryptophan residue (in bold) is found. Three transmembrane regions (19 to 20 residues) predicted with TmPred program are underlined.

ATPase complex argue for absolute requirements to control protein quality and to acidify an endosomal or vacuolar compartment. Whether the latter ATPase also serve to generate a proton motive force across the plasma membrane is an open question. Endocytosis is assumed to occur, but no clathrin homologue has been found. Genes for conserved cytoskeletal proteins such as actin, α -, β- γ-tubulins, major motor proteins (myosins, kinesins, dynein) and some other actin- and microtubule-associated proteins confirm the involvement of a yet ill-characterized microsporidian cytoskeleton in division processes and intracellular transport.

Corresponding to the strict host dependency of microsporidia for nutrient acquisition, the set of enzymes for the biosynthesis of usual aminoacids is extremely limited. Nucleotide metabolism is restricted to some usual interconversions, but dTMP can be synthesized from thymidine. Trehalose should be the only putative storage sugar and should also play a major role in the protection against various injuries, including osmotic stress, as is the case in yeasts (Arguelles, 2000). Some large gene deficits are unambiguously related to the lack of mitochondria. No components for Krebs cycle, fatty acid β-oxidation, respiratory chain, F_0F_1-ATPase complex and mitochondrial ribosomes are predicted. Amitochondriate eukaryotes, such as *Trichomonas, Giardia and Entamoeba* species, also lack oxidative phosphorylation linked to aerobic electron transport chain and thus generate ATP through substrate-level phosphorylation. In most of these organisms, various enzymes required for energy metabolism have been well characterized through biochemical and molecular studies, revealing pathways from either glucose or aspartate to the synthesis of pyruvate and different fates of the pyruvate (see, e.g., Adam, 2001). Major organic end-products can be acetate, ethanol or alanine. Genes for glucose transporters and for all the glycolytic enzymes ensuring the conversion of glucose to pyruvate were identified. As in the aforementioned amitochondriate protists, the production of fructose-1,6-bisphosphate is only catalyzed by a pyrophosphate(PPi)-dependent phosphofructokinase that is not posttranslationally regulated, in contrast to the ATP-dependent enzyme. As regards the final step leading from phosphoenolpyruvate to pyruvate, the energy advantage relying on the possession of two kinds of enzymes (ATP-dependent pyruvate kinase and PPi-dependent pyruvate dikinase; Mertens et al., 1992; Philipps and Li, 1995; Reeves et al., 1974) cannot be applied to the microsporidian species. The potential lack of PPi-dependent pyruvate dikinase seems to correlate with that of adenylate kinase for a possibly coupled reaction needed for the production of two ATP molecules. Thus, substrate-level phosphorylations during glycolysis should be mediated by only phosphoglycerate kinase and ATP-dependent pyruvate kinase.

Terminal reactions leading to the production of compounds such as acetyl-CoA, acetate, ethanol or lactate are still unclear. We did not identify homologues for alcohol- and lactate dehydrogenases as well as for pyruvate-ferredoxin oxidoreductase, an enzyme characteristic of several amitochondriate protists. A surprising fact was that the only genes assigned to a function in pyruvate catabolism were those encoding the E1 component of the mitochondrial pyruvate dehydrogenase (PDH) complex. It is also noteworthy that a potential key enzyme of *E. cuniculi* is homologous to fungal acetyl-CoA synthetases including yeast ACS1, belonging to the ubiquitous AMP-forming type and thus mediating the activation of acetate to acetyl-CoA, which contrasts with the *Giardia* ADP-forming type that allows for ATP production through the conversion of acetyl-CoA to acetate (Sánchez et al., 2000). This suggests that as a prerequisite to the formation of acetyl-CoA, *E. cuniculi* should be able to use exogenous acetate or to generate acetate. Under the second hypothesis, we have assumed that pyruvate decarboxylation catalyzed by E1-PDH is coordinated with an oxidation process involving ferredoxin and leading to acetate (Katinka et al., 2001). Trichomonads as well as chytrid fungi and the anaerobic ciliate *Nyctotherus* can eliminate protons in the form of molecular hydrogen via a Fe-hydrogenase located in the organelles called hydrogenosomes (Martin and Müller, 1998). However, two nonhydrogenosomal parasites (*Entamoeba* and *Spironucleus*) also contain a Fe-hydrogenase, although they do not apparently make hydrogen (Horner et al., 2000). Homologues of the human gene for nuclear protein NARF (nuclear prelamin A recognition factor), with similarity to the C-terminal domain of *Clostridium* Fe-hydrogenase but with only one motif for binding to a [4Fe4S] cluster, have been recently identified in the genomes of various eukaryotes, including that of *E. cuniculi* (ECU05_0970; Horner et al., 2002). The new NARF-like proteins do not seem to be involved in energy metabolism, but their ubiquity supports a crucial function (nuclear organization maintenance, sensing of oxidative stress, etc.). We do not know the terminal acceptor of protons and electrons in *E. cuniculi*, but it is clear that the parasite is potentially equipped with major enzymes for redox regulation and protection against reactive oxygen species (Mn-superoxide dismutase, glutathione peroxidase, thioredoxin and glutaredoxin systems). Discussing the importance of these systems is beyond the scope of this article, but it can be recalled that their evolutionary history is especially complex. For example, two types of thioredoxin reductase (L and H) are distributed differentially among both free-living and parasitic eukaryotes. The L-type (~35 kDa) is found in most prokaryotes, fungi, plants and the parasites *Entamoeba, Giardia and Spironucleus*, whereas the H-type (~55 kDa) is present in animals and curiously also in apicomplexan parasites (horizontal gene transfer?; Hirt et al., 2002). In *E. cuniculi*, an ORF (ECU02_0940) codes for a slightly shortened L-type thioredoxin reductase (309 aa). A much shorter ORF (ECU01_0680, 156 aa) has only weak similarity with the N-terminal domain of thioredoxin reductases and lacks redox-active cysteines. No H-type homologue has been identified. Thus, unlike api-complexans, *E. cuniculi* does not make exception in the conservation of the L-type.

In addition to a set of poorly diversified ABC systems lacking homologues of known multidrug pumps (Cornillot et al., 2002b), a limited complement of membrane transporters for aminoacids, sugars, nucleotides, water and inorganic ions has been inferred from the genome sequence. Four ADP/ATP exchanger paralogues drew our attention because of their possible role in the import of host ATP, as judged by strong similarities with the sequences of both intracellular parasitic bacteria (*Rickettsia, Chlamydia*) and of chloroplasts but not of mitochondria (Katinka et al., 2001; Vivarès et al., 2002). It is tempting to hypothesize that a common scenario involving a late horizontal gene transfer from a parasitic α-proteobacterium occurred in both plant and microsporidian ancestors. In the case of plastids, the novel ADP/ATP translocase would have been targeted to the protochloroplast in a mitochondria-bearing plant ancestor, providing a great advantage for increasing the biosynthetic capacities of plastids through the import of cytosolic ATP (Winkler and Neuhaus, 1999). Perhaps the divergence of the microsporidian lineage has been marked by a similar transfer but in the absence of protochloroplast, the ADP/ATP translocase became available to the plasma membrane and was then used to import ATP from host cells. One can also speculate that this event rendered the mitochondrial machinery for ATP production dispensable and therefore initiated a regressive evolution of mitochondria in these parasites.

10.5 Microsporidial Genes Encoding Mitochondrial-Type Proteins

The *E. cuniculi* genome sequence did not provide evidence of components of the Krebs cycle as well as of respiratory chain and oxidative phosphorylation, which is consistent with the lack of intracellular structures referred to as mitochondria and the failure to detect malate and succinate dehydrogenase activities in *N. grylli* spores (Dolgikh et al., 1997; Weidner et al., 1999). As inferred from studies of an mtHsp70 gene in three different species, microsporidia might have evolved from a mitochondrion-bearing ancestor (Germot et al., 1997; Hirt et al., 1997; Peyretaillade et al., 1998b). The hypothesis that the corresponding chaperone is targeted to a cryptic compartment was more difficult to accept on only the basis of a rather short N-terminal extension. However, other *E. cuniculi* genes encode additional mitochondrial-type proteins, six of these clustering unambiguously with both mitochondrial and alpha-proteobacterial sequences in phylogenetic trees (Katinka et al., 2001; Vivarès et al., 2002).

10.5.1 Pyruvate Fate and NADH Reoxidation: Two Major Uncertainties

Interesting examples of mitochondria-related genes are offered by those coding for the two subunits of pyruvate dehydrogenase (PDH) E1 component, which have been identified in both *E. cuniculi* (Katinka et al., 2001) and the insect microsporidian *N. locustae* (Fast and Keeling, 2001). Separated and combined phylogenies of these sequences strongly support a mitochondrial origin (Fast and Keeling, 2001; Keeling and Fast, 2002). As for other known enzymes of the pyruvate catabolism, we failed to detect homologues for the E2 (dihydrolipoyl dehydrogenase) and E3 (dihydrolipoyl transacetylase) components of the PDH complex. This leads us to wonder whether the potential function of PDH-E1 in the pyruvate decarboxylation step producing hydroxethyl-thiamine pyrophosphate (HETPP) represents an essential part of core microsporidian metabolism. A significant glycerol-3-P dehydrogenase (GPDH) activity has been detected in *N. grylli*, justifying the proposition of a role in the reoxidation of the NADH produced during glycolysis (Weidner et al., 1999). Accordingly, an *E. cuniculi* gene encodes cytosolic GPDH, an NAD-dependent enzyme that catalyzes the conversion of dihydroxyacetone-3-P into glycerol-3-P coupled with NADH oxidation. More surprising is a gene coding for mitochondrial GPDH (mtGPDH) that mediates the opposite reaction and uses FAD as cofactor. The FAD-binding domain is present, and a significant N-terminal extension appears as a putative mitochondrial targeting presequence. Thus, *E. cuniculi* might have preserved a glycerol-3-P shuttle, one way used to transfer reducing equivalents from the cytosol toward mitochondrion against an NADH gradient in mitochondriate eukaryotes. That mtGPDH is located in the external face of the inner mitochondrial membrane encouraged us to hypothesize that the preservation of glycerol-3-P shuttle in microsporidia should depend on the presence of an organelle maintaining a double membrane that we called mitosome, a term originally proposed to designate a mitochondrial remnant in *Entamoeba histolytica* (Tovar et al., 1999). Obviously, this does not explain the possible role of the glycerol-3P shuttle in an organism for which there is no evidence of an electron transport chain. Like for HETPP, the fate of the electron–proton pair in the $FADH_2$ associated with mtGPDH is open to speculations.

In several anaerobic and microaerotolerant protozoa, such as trichomonads, *Giardia* and *Entamoeba*, oxidative decarboxylation of the pyruvate is ensured by a unique enzyme pyruvate:ferredoxin oxidoreductase (PFOR), an iron–sulfur protein targeted to hydrogenosomes, when present, and of which the evolutionary origin is highly debated (Upcroft and Upcroft, 1999; Horner et al., 1999; Rotte et al., 2001). Because two *E. cuniculi* genes encode a [2Fe2S]ferredoxin (or adrenodoxin) and a ferredoxin:NAD(P)H oxidoreductase, we tentatively suggested that E1-mediated pyruvate decarboxylation could be followed by a dehydrogenation step coupled with ferredoxin reduction, mimicking the activity of PFOR (Katinka et al., 2001). This hypothesis has also been considered by Keeling and Fast, who have assumed that a new enzyme (HETPP:ferredoxin oxidoreductase) serves as an intermediate protein, specifically mediating the electron transfer between HETPP and ferredoxin (Keeling and Fast, 2002). The *E. cuniculi* ferredoxin gene encodes a polypeptide of only 128 aa, which is clearly less than its *S. cerevisiae* homologue (172 aa) and closer to rickettsial ferredoxins (112 aa), with which it also shares the highest level of sequence identity (Figure 10.4A, upper part). The size difference is partly related to the lost or reduced mitochondrial-type presequence. Analyses of several hydrogenosomal matrix proteins in the flagellate *Trichomonas vaginalis* have shown that their presequences are similar to those of mitochondrial counterparts, but they are frequently shorter (Dyall and Johnson, 2000). For example, the experimentally determined cleavages sites of both PFOR A and B are between the fifth and sixth residues. As explained in Figure 10.4A, the *E. cuniculi* ferredoxin would be related to the nonfungal ferredoxin type II (Seeber, 2002a).

10.5.2 A Putative Organellar Protein Import System

The protein import machinery of mitochondria harbors various proteins that mediate steps of recognition, translocation, membrane insertion, processing and chaperoning. Receptor proteins of the TOM complex are associated with the cytosolic face of mitochondria and recognize cargo preproteins that pass through the general import pore (GIP) of the outer membrane (OM) to be transported via distinct complexes into the matrix, inner membrane (IM) or intermembrane space (Rehling et al., 2001). An important receptor predicted to be present in *E. cuniculi* is similar to mitochondrial Tom70, a peripheral receptor that preferentially interacts with preproteins utilizing cytosolic chaperones and having internal targeting signals, such as the IM-located ADP/ATP carrier (Wiedemann et al., 2001). The microsporidial Tom70 sequence comprises a reduced but significant number of tetratricopeptide repeat motifs (five instead of seven in yeasts) that are required for protein–protein interactions, including interactions with cytoskeletal elements such as microtubules. A potential *E. cuniculi* protein (ECU07_0240) displays some sequence similarity to both Tim22 and Tim17 proteins of the mitochondrial IM. Two TIM complexes, called TIM23 and TIM22, are involved in the import of different proteins. [For a recent review, see Jensen and Dunn (2002).] The TIM23 complex is specialized for the import of most matrix proteins, whereas the TIM22 complex mediates the insertion of polytopic proteins into the IM. Because Tim23, Tim22 and Tim17 are homologous proteins that are not all located in the same translocon, it was important to reexamine the *E. cuniculi* sequence through multiple alignment and domain searches. A better fit is found with yeast Tim17 (Figure 10.4B) and a Tim17 Pfam domain is predicted. Tim17 is known to be associated with Tim23 in a specific channel of the mitochondrial IM. With only 124 aa and three predicted transmembrane domains, the *E. cuniculi* protein might thus be a shortened version of Tim17. Given that at least two of the four essential components of the TIM23 complex (Tim23, Tim17, Tim44 and mtHSP70) could exist in *E. cuniculi*, we assume that a TIM-like translocase permits the transfer of proteins into a cryptic organelle (Figure 10.5).

If necessary, an N-terminal presequence should be cleaved by a matrix processing peptidase (MPP), as suggested by a number of *E. cuniculi* proteins sharing an arginine residue at the –2 position with respect to a predicted cleavage site (Katinka et al., 2001). Mitochondrial processing peptidases are heterodimeric and can be of two types, one an integral component of the cytochrome bc_1 complex and the other (MPP) released in the matrix (Gakh et al., 2002). Possibly, during evolution, the two MPP subunits separated from the bc_1 complex through gene duplication events. The alpha- and beta-MPP subunits show more or less conserved domains, including a zinc-binding motif, more characteristic of beta-MPP having catalytic activity, and a glycine-rich loop for alpha-MPP involved in substrate recognition. The strongest similarities with putative *E. cuniculi* proteins were found only for a larger putative zinc-dependent protease homologous to human insulin-degrading enzyme and some yeast proteases of the same family. Thus, the preservation of alpha- and beta-MPP genes in *E. cuniculi* seems doubtful. This does not exclude the possibility that a divergent peptidase might play a role similar to MPP. A highly regressive evolution of mitochondria associated with the loss of the numerous proteins responsible for a large part of the oxidative catabolism as that of mitochondrial r-proteins and various enzymes required for replication and transcription of an organellar genome might also have been related to some simplification of the protein import machinery.

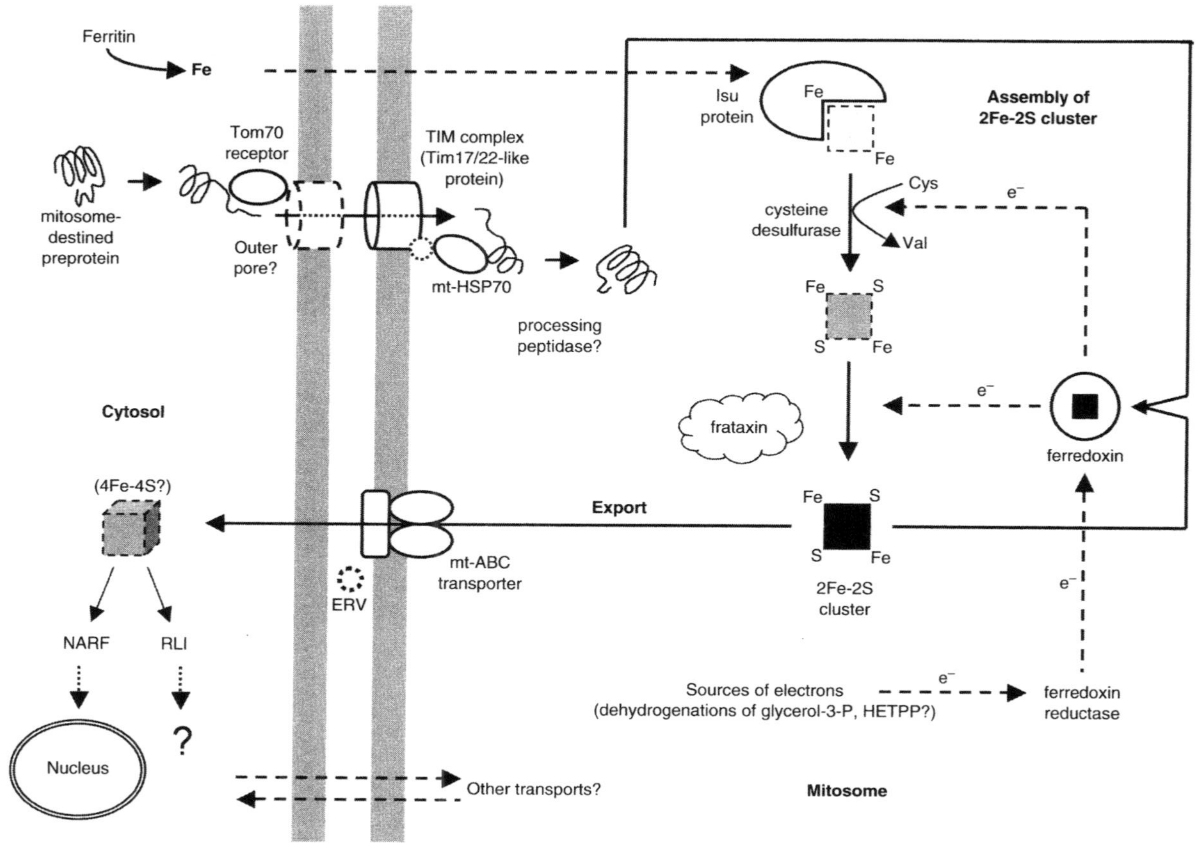

FIGURE 10.5 Schematic representation of some major steps of Fe–S cluster assembly assumed to occur inside the microsporidian mitosome, on the basis of the finding of *E. cuniculi* coding sequences for more or less divergent homologues of mitochondrial proteins involved in this process. After its release from cytosolic ferritin, iron should enter the mitosome (iron transport?). Protein import is predicted to depend on a recognition step mediated by a Tom70-like receptor, followed by passage through an undefined outer pore and at least one channel of the inner membrane having a protein similar to Tim17. Ferredoxin maturation would be completed in the organellar matrix and the ferredoxin – ferredoxin reductase system is thought to play a central role in some specific oxidoreductions. A mitochondrial-type ABC transporter and perhaps an ERV/ALR homologue could mediate Fe–S cluster export to the cytosol. Two potential *E. cuniculi* proteins have a [4Fe4S] binding domain — a nuclear protein (NARF) and a soluble ABC protein (RNase L inhibitor or RLI) — but their functions also need investigations.

10.5.3 Iron–Sulfur Cluster Assembly: A Ubiquitous Function for the E. Cuniculi Mitosome

The most significant set of *E. cuniculi* genes that we considered to support the mitosome hypothesis was represented by the characteristic assembly of Fe–S proteins in mitochondria, as illustrated in Figure 10.5. The earliest data about the maturation of such proteins was derived from studies on bacterial *nif* (nitrogen fixation) and *isc* (iron–sulfur cluster assembly) operons, justifying designations such as NifU-like or Isu proteins for some eukaryotic homologues. Through comparisons with a list of 11 *S. cerevisiae* proteins involved in Fe–S cluster biosynthesis (Lill and Kispal, 2000), it was obvious that the *E. cuniculi* genome sequence predicts homologues of several of these proteins. In addition to [2Fe2S]ferredoxin, ferredoxin:NAD(P)H oxidoreductase and mtHSP70, specific candidates were a NifS-like protein, a NifU-like protein and a mtABC transporter similar to yeast Nfs1, Isu1/Isu2 and Atm1 proteins, respectively, and are all located in mitochondria. The NifS/IscS bacterial proteins are pyridoxal-5′-phosphate (PLP)-dependent cysteine desulfurases, which act as donors of elemental sulfur from cysteine. The *E. cuniculi* homologue contains the critical residues involved in cysteine desulfurase activity, including a conserved histidine for substrate deprotonation as well as essential residues for PLP- and substrate-binding sites. The C-terminal region presents a consensus 20-aa sequence signature that differentiates proteobacteria and eukaryotes from other organisms, as revealed by a phylogenetic analysis of *Trichomonas* and *Giardia* Isc homologues (Tachezy et al., 2001). This analysis supports both a conserved pathway for Fe–S cluster assembly in eukaryotes and a common origin of mitochondria and hydrogenosomes. Similar to all IscU proteins in nondiazotrophic bacteria and eukaryotes, the microsporidial Nifu-like sequence shows homology to only the N-terminal region of the *Azotobacter* NifU protein. The function of IscU should consist of binding iron and providing a scaffold for the assembly of a transient Fe–S cluster (Lill and Kispall, 2000; Muhlenoff and Lill, 2000; Seeber, 2002b). In contrast to *S. cerevisiae* in which the two IscU homologues (Isu1 and Isu2) have been demonstrated to be essential for viability, *E. cuniculi* can be thought to require only one IscU. Because IscU proteins do not possess the central Fe–S cluster of NifU, the [2Fe2S]ferredoxin – ferredoxin:NAD(P)H oxidoreductase system is of great importance as a source of electrons for reducing sulfan sulfur and iron within a nascent Fe–S cluster and possibly also for protecting the cluster from oxidation (Lange et al., 2000).

As indicated by the accumulation of iron within mitochondria of yeast mutants in mtHSP70 gene (Knight et al., 1998), the chaperone mtHSP70 is required for Fe–S cluster assembly, but its precise role in this process is not yet defined. In contrast, the mitochondrial yeast ABC transporter (Atm1) has been proven to be specifically involved in the maturation of cytosolic Fe–S proteins (Kispal et al., 1999). Through a comparative study of the 13 ABC transporter genes in the *E. cuniculi* genome, we have ascertained that 6 genes (including a duplicated gene on Chromosome I) can be assigned to the HMT family comprising all known mitochondrial ABC transporters (Cornillot et al., 2002b). The best candidate to an *Atm1* orthologue is present on Chromosome XI (ECU11_1200). Therefore, the corresponding ABC transporter might be located in the IM of a putative mitosome and, like the Atm1 protein, could mediate the export of stabilized Fe clusters toward the cytosol. Another potential *E. cuniculi* protein is of great interest as it is similar to the yeast Erv1/Erv2 and human augmenter of liver regeneration (ALR) that are characterized by a C-terminal sulfhydryloxidase domain. The Erv1/ALR proteins are now recognized as being primarily required for maturation of cytosolic Fe–S proteins and their location in the mitochondrial intermembrane space is suggestive of a crucial step subsequent to that mediated by Atm1 transporter (Lange et al., 2001). Mitochondrial iron accumulation is also known to occur in Δyfh1 yeast mutants resulting from the deletion of a gene encoding a matrix protein called frataxin (Foury and Cazzalini, 1997). The name of this protein is derived from Friedreich's ataxia, a prevalent human neurodegenerative disease caused by a defective frataxin gene and marked by an extensive iron deposition within heart myofibrils. The analysis of the Chromosome I sequence has revealed the presence of a frataxin gene in *E. cuniculi* (Peyret et al., 2000). Frataxin should participate in the regulation of the import and export of mitochondrial iron (Becker and Richardson, 2001). Recently, a significant reduction of the incorporation of Fe–S clusters into apoferredoxin within isolated mitochondria of Δyfh1 yeast mutants has been demonstrated, indicating that frataxin stimulates Fe–S cluster assembly (Duby et al., 2001). In summary, the amitochondriate *E. cuniculi* seems to have preserved the majority of the currently known components of a mitochondrial apparatus devoted to Fe–S cluster assembly, an essential function common to all organisms investigated so far. This made us more confident in proposing that these proteins should reside in an organelle that could not be identified in microsporidia on the basis of a full conservation of the usual morphology of mitochondria.

10.5.4 First Experimental Evidence of the Microsporidial Mitosome and Related Questions

The mitosome hypothesis has been tested experimentally through Western blotting and immunocytochemical analyses of the expression of mtHSP70 in the human parasite *T. hominis* (Williams et al., 2002). A polyclonal antibody raised against the mtHsp70 fusion protein reacts specifically with a 60-kDa protein band in total extracts from purified *T. hominis* spores as well as from infected host cells. Immunofluorescence data provide clear evidence of the localization of the mtHSP70 protein within numerous small structures distributed throughout the cytoplasm of meronts. Finally, electron microscopy reveals that the anti-mtHSP70 labeling is confined to the matrix of ovoid structures of uniform size (90 nm × 50 nm) that are surrounded by two membranes but lack cristae (see also Chapter 13). This represents the first visualization of mitochondria-derived organelles retained by a microsporidian and therefore encourages investigations into the location of candidate proteins for these novel organelles.

The need of [2Fe2S]ferredoxin for an electron transport associated with a terminal step of carbohydrate catabolism might be the main justification for Fe–S cluster assembly in the microsporidial mitosome. No additional *E. cuniculi* protein containing a [2Fe2S] cluster is predicted. As Atm1 and Erv1/ARL proteins are related to export processes, the question arises whether *E. cuniculi* might synthesize other Fe–S proteins. Besides the NARF-like protein possibly destined to the nucleus, the only candidate so far inferred from the genome sequence is homologous to RNase L inhibitor (RLI), a nontransporter ABC protein that harbors an N-terminal [4Fe4S]-binding domain (Cornillot et al., 2002b). In humans, RLI downregulates the 2′-5′ oligoadenylate synthetase/RNase L system, an interferon-inducible RNA degradation pathway that participates to the defense against viruses including HIV (Martinand et al., 1999). It has also been shown that RNase L and RLI are partly localized in the mitochondria of human H9 cells and regulate the stability of mitochondrial mRNAs (Le Roy et al., 2001). However, the 2 ′-5′ oligoadenylate synthetase/RNase L system has so far been found only in vertebrates, whereas RLI homologues are present in many prokaryotes and eukaryotes. Even the miniature genome of a cryptomonad endosymbiont has retained the *RLI* gene (Douglas et al., 2001). It seems unlikely that the basic function of RLI proteins is related to mRNA degradation control. The conservation of a [4Fe4S]-binding domain suggests a role in a ubiquitous electron transfer process that would not be dispensable in microsporidia or other organisms.

Mitochondria might facultatively function under anaerobic conditions in various eukaryotes by using electron acceptors different from oxygen and of either external (nitrate, nitrite) or internal (fumarate) origin, raising the issue of the possible diversity of evolutionary descent of these organelles. [For a recent review, see Tielens et al. (2002).] Microsporidia, diplomonads and *Entamoeba*, collectively viewed as of Type I amitochondriate eukaryotes (Martin and Müller, 1998), seem to have lost the capacity of ATP synthesis inside possible remnants of mitochondria. However, this does not imply that these so-called mitosomes had the same history. Whereas mt-HSP70 is present in all these organisms (Arisue et al., 2002), another mitochondrial chaperone (Cpn60 or Hsp60) has been preserved in *Entamoeba* (Mai et al., 1999; Tovar et al., 1999) and *Giardia* (Clark and Roger, 1995) but not in *Encephalitozoon*. This suggests that the adaptation of various parasitic eukaryotes to more or less anaerobic niches has involved differential losses of the components from a common ancestral system at the origin of both mitochondria, hydrogenosomes and mitosomes (see also Chapter 13).

10.6 Conclusions

The compaction of the whole genome and individual genes in *E. cuniculi* leads us to conclude that microsporidia tend to do the maximum with the minimum in terms of basic functions for cell reproduction and for protection against environmental stresses. There is evidently a widespread loss of biosynthetic pathways for the simplest precursors of informational macromolecules and storage polysaccharides as well as for most fatty acids and coenzymes. The auxotrophic character is exacerbated by the lack of oxidative phosphorylation. Microsporidia thus appear to be energy parasites and are predicted to use an ADP/ATP exchanger for importing ATP from host cell. However, these organisms seem to have retained a mitochondrion-derived organelle whose molecular composition and functions deserve further investigation. As indicated by the presence of genes encoding ferritin, ferredoxin and a set of proteins involved in Fe–S assembly, the life of these parasites should be highly dependent on iron metabolism and homeostasis, as in several pathogenic fungi (Howard, 1999). Various mechanisms of iron acquisition exist among these fungi, but iron storage through ferritin-like compounds is known in only the zygomycetes. Thus, the *E. cuniculi* ferritin gene might be an indicator of a closer evolutionary relationship between microsporidia and zygomycetes.

Finally, it should not be forgotten that the *E. cuniculi* genome contains about 1100 orphan genes. Even if this number is less high than in *Plasmodium* (3200), it is obvious that insights into new functions related to parasitism will depend on the development and applicability of powerful postgenomic technologies. Waiting for experimental approaches, we believe that comparative genomics of microsporidia can help evaluate the extent of possible intraphylum variations in genes for alternative life cycle pathways and host–parasite interactions. The initial genome survey sequences from *Vittaforma corneae* (Nosematidae) have revealed some interesting data, especially as regards possible non-LTR retrotransposons (Mittleider et al., 2002). There is no doubt that an inventory of the differences in gene content between two species that differ in host specificity, nuclear configuration and life cycle will be important to orient functional studies. With the help of the Genoscope-CNS center (Evry, France) in sequencing at least one large-sized chromosome from the xenoma-forming parasite *Glugea atherinae,* we also expect to obtain useful data. Microsporidia are not easily tractable experimentally, but we are convinced that studying usual model organisms or the most dangerous human pathogens is insufficient for unraveling all the secrets of the evolution of eukaryotic genomes and organelles.

Acknowledgments

The authors are grateful to Drs. Robert P. Hirt and David S. Horner for valuable comments and help in improving our English on a draft of this manuscript.

References

Adam, R. D. (1992) Chromosome-size variation in *Giardia lamblia*: The role of rDNA repeats. *Nucleic Acids Res.* 20: 3057–3061.

Adam, R. D. (2001) Biology of *Giardia lamblia. Clin. Microbiol. Rev.* 14: 447–475.

Adam, R. D., Nash, T. E. and Wellems, T. E. (1991) Telomeric location of *Giardia* rDNA genes. *Mol. Cell Biol.* 11: 3326–3330.

Amigo, J. M., Gracia P. M., Salvado H. and Vivarès, C. P. (2002) Pulsed field gel electrophoresis of three microsporidian parasites of fish. *Acta Protozool.* 41: 11–16.

Andreadis, T. G. and Vossbrinck, C. R. (2002) Life cycle, ultrastructure and molecular phylogeny of *Hyalinocysta chapmani* (Microsporidia: Thelohaniidae), a parasite of *Culiseta melanura* (Diptera: Culicidae) and *Orthocyclops modestus* (Copepoda: Cyclopidae). *J. Euk. Microbiol.* 49: 350–364.

Arguelles, J. C. (2000) Physiological roles of trehalose in bacteria and yeasts: A comparative analysis. *Arch. Microbiol.* 174: 217–224.

Arisue, N., Sanchez, L. B., Weiss, L. M., Muller, M. and Hashimoto, T. (2002) Mitochondrial-type Hsp70 genes of the amitochondriate protists, *Giardia intestinalis, Entamoeba histolytica* and two microsporidians. *Parasitol. Int.* 51: 9–16.

Baldauf, S. L., Roger, A. J., Wenk-Siefert, I., Doolittle, W. F. (2000) A kingdom-level phylogeny of eukaryotes based on combined protein data. *Science* 290: 972–977.

Baldo, A. M. and McClure, M. A. (1999) Evolution and horizontal transfer of dUTPase-encoding genes in viruses and their hosts. *J. Virol.* 73: 7710–7721.

Baumann, P. and Cech, T. R. (2001) Pot1, the putative telomere end-binding protein in fission yeast and humans. *Science* 292: 1171–1175.

Baumann, P., Podell, E. and Cech, T. R. (2002) Human pot1 (protection of telomeres) protein: Cytolocalization, gene structure, and alternative splicing. *Mol. Cell. Biol.* 22: 8079–8087.

Becker, E. and Richardson, D. R. (2001) Frataxin: Its role in iron metabolism and the pathogenesis of Friedreich's ataxia. *Int. J. Biochem. Cell Biol.* 33: 1–10.

Biderre, C., Duffieux, F., Peyretaillade, E., Glaser, P., Peyret, P., Danchin, A., Pagès, M., Méténier, G. and Vivarès, C. P. (1997) Mapping of repetitive and non-repetitive DNA probes to chromosomes of the microsporidian *Encephalitozoon cuniculi. Gene* 191: 39–45.

Biderre, C., Mathis, A., Deplazes, P., Weber, R., Méténier, G., Vivarès, C. P. (1999) Molecular karyotype diversity in the microsporidian *Encephalitozoon cuniculi. Parasitology* 118: 439–445.

Biderre, C., Pages, M., Méténier, G., Canning, E. U., Vivarès, C. P. (1995) Evidence for the smallest nuclear genome (2.9 Mb) in the microsporidium *Encephalitozoon cuniculi. Mol. Biochem. Parasitol.* 74: 229–231.

Biderre, C., Pages, M., Méténier, G., David, D., Bata, J., Prensier, G., Vivarès, C. P. (1994) On small genomes in eukaryotic organisms: Molecular karyotypes of two microsporidian species (Protozoa) parasites of vertebrates. *C. R. Acad. Sci. III* 317: 399–404.

Blandin, G., Durrens, P., Tekaia, F., Aigle, M., Bolotin-Fukuhara, M., Bon, E., Casarégola, S., de Montigny, J., Gaillardin, C., Lépingle, A et al. (2000) Genomic exploration of the hemiascomycetous yeasts: 4. The genome of *Saccharomyces cerevisiae* revisited. *FEBS Lett.* 487: 31–36.

Brown, J. R., Doolittle, W. F. (1995) Root of the universal tree of life based on ancient aminoacyl-tRNA synthetase gene duplications. *Proc. Natl. Acad. Sci. USA* 92: 2441–2445.

Brown, J. R. and Doolittle, W. F. (1999) Gene descent, duplication, and horizontal transfer in the evolution of glutamyl- and glutaminyl-tRNA synthetases. *J. Mol. Evol.* 49: 485–495.

Brugère, J. F., Cornillot, E., Bourbon, T., Méténier, G. and Vivarès, C. P. (2001) Inter-strain variability of insertion/deletion events in the *Encephalitozoon cuniculi* genome: A comparative KARD-PFGE analysis. *J. Euk. Microbiol.* Suppl: 50S–55S.

Brugère, J. F., Cornillot, E., Méténier, G., Bensimon, A. and Vivarès, C. P. (2000a) *Encephalitozoon cuniculi* (Microspora) genome: Physical map and evidence for telomere-associated rDNA units on all chromosomes. *Nucleic Acids Res.* 28: 2026–2033.

Brugère, J. F., Cornillot, E., Méténier, G. and Vivarès, C. P. (2000b) Occurence of subtelomeric rearrangements in the genome of the microsporidian parasite *Encephalitozoon cuniculi*, as revealed by a new fingerprinting procedure based on two-dimensional pulsed field gel electrophoresis. *Electrophoresis* 21: 2576–2581.

Bryan, T. M. and Cech, T. R. (1999) Telomerase and the maintenance of chromosome ends. *Curr. Opin. Cell Biol.* 11: 318-324.

Cali, A. and, Tavkorian, P. M. (1999) Developmental morphology and life cycles of the microsporidia. In *The Microsporidia and Microsporidiosis* (Wittner, M. and Weiss, L. M., Eds.) ASM Press, Washington, pp. 85–128.

Cavalier-Smith, T. (1983) A 6-kingdom classification and a unified phylogeny. In *Endocytobiology II: Intracellular Space as Oligogenetic* (Schenk, H. E. A. and Schwemmler, W. S., Eds.), Walter deGruyter, Berlin, pp. 1027–1034.

Cavalier-Smith, T. (1985) *The Evolution of Genome Size*, John Wiley & Sons, Chichester.

Cavalier-Smith, T. (1998) A revised six-kingdom system of life. *Biol. Rev. Camb. Philos. Soc.* 73: 203–266.

Cavalier-Smith, T. (2002a) Origins of the machinery of recombination and sex. *Heredity* 88: 125–141.

Cavalier-Smith, T. (2002b) The phagotrophic origin of eukaryotes and phylogenetic classification of Protozoa. *Int. J. Syst. Evol. Microbiol.* 52: 297–354.

Chilmonczyk, S., Cox, W. T. and Hedrick, R. P. (1991) *Enterocytozoon salmonis* n. sp.: An intranuclear microsporidium from salmonid fish. *J. Protozool.* 38: 264–269.

Clark, C. G. and Roger, A. J. (1995) Direct evidence for secondary loss of mitochondria in *Entamoeba histolytica. Proc. Natl. Acad. Sci. USA* 92: 6518–6521.

Cornillot, E., Keller, B., Cushion, M. T., Méténier, G. and Vivarès, C. P. (2002a) Fine analysis of the *Pneumocystis carinii* f. sp. *carinii* genome by two-dimensional pulsed-field gel electrophoresis. *Gene* 293: 87–95.

Cornillot, E., Méténier, G., Vivarès, C. P. and Dassa, E, (2002b) Comparative analysis of sequences encoding ABC systems in the genome of the microsporidian *Encephalitozoon cuniculi. FEMS Microbiol. Lett.* 210: 39–47.

Costa, S. F. and Weiss, L. M. (2000) Drug treatment of microsporidiosis. *Drug Resist. Update* 3: 384–399.

Courties, C., Perasso, R., Chrétiennot-Dinet, M. J., Gouy, M., Guillou, L. and Trousselier, M. (1998) Phylogenetic analysis and genome size of *Ostreococcus tauri* (Chlorophyta, Prasinophyceae). *J. Phycol.* 34: 844–849.

Curgy, J. J., Vávra, J. and Vivarès, C. P. (1980) Presence of ribosomal RNAs with prokaryotic properties in Microsporidia. *Biol. Cell* 38: 49–52.

Dacks, J. B. and Doolittle, W. F. (2002) Novel syntaxin gene sequences from *Giardia*, *Trypanosoma* and algae: Implications for the ancient evolution of the eukaryotic endomembrane system. *J. Cell Sci.* 115:1635–1642.

Dolgikh, V. V., Sokolova, J. J. and Issi, I. V. (1997) Activities of enzymes of carbohydrate and energy metabolism of the spores of the microsporidian, *Nosema grylli. J. Euk. Microbiol.* 44: 246–249.

Douglas, S., Zauner, S., Fraunholz, M., Beaton, M., Penny, S., Deng, L. T., Wu, X., Reith, M., Cavalier-Smith, T. and Maier, U. G. (2001) The highly reduced genome of an enslaved algal nucleus. *Nature* 410: 1091–1096.

Duby, G., Foury, F., Ramazzotti, A., Herrmann, J. and Lutz, T. (2002) A non-essential function for yeast frataxin in iron-sulfur cluster assembly. *Hum. Mol. Genet.* 11: 2635–2643.

Dunn, A. M. and Smith, J. E. (2001) Microsporidian life cycles and diversity: The relationship between virulence and transmission. *Microb. Infect.* 3: 381–388.

Dyall, S. D. and Johnson, P. J. (2000) Origins of hydrogenosomes and mitochondria: Evolution and organelle biogenesis. *Curr. Opin. Microbiol.* 3: 404–411.

Edlind, T. D., Li, J., Visvesvara, G. S., Vodkin, M. H., McLaughlin, G. L. and Katiyar, S. K. (1996) Phylogenetic analysis of beta-tubulin sequences from amitochondrial protozoa. *Mol. Phylogenet. Evol.* 5: 359–367.

Fast, N. M. and Keeling, P. J. (2001) Alpha and beta subunits of pyruvate dehydrogenase E1 from the microsporidian *Nosema locustae*: Mitochondrion-derived carbon metabolism in microsporidia. *Mol. Biochem. Parasitol.* 117: 201–209.

Fast, N. M., Logsdon, J. M., Jr. and Doolittle, W. F. (1999) Phylogenetic analysis of the TATA box binding protein (TBP) gene from *Nosema locustae*: Evidence for a microsporidia-fungi relationship and spliceosomal intron loss. *Mol. Biol. Evol.* 16: 1415–1419.

Flegel, T. W. and Pasharawipas, T. (1995) A proposal for typical eukaryotic meiosis in microsporidians. *Can. J. Microbiol.* 41: 1–11.

Foury, F. and Cazzalini, O. (1997) Deletion of the yeast homologue of the human gene associated with Friedreich's ataxia elicits iron accumulation in mitochondria. *FEBS Lett.* 411: 373–377.

Franzen, C. and Müller, A. (2001) Microsporidiosis: Human diseases and diagnosis. *Microb. Infect.* 3: 389–400.

Fraser, C. M., Gocayne, J. D., White, O., Adams, M. D., Clayton, R. A., Fleischmann, R. D., Bult, C. J., Kerlavage, A. R., Sutton, G., Kelley, J. M. et al. (1995) The minimal gene complement of *Mycoplasma genitalium*. *Science* 270: 397–403.

Freitas-Junior, L. H., Bottius, E., Pirrit, L. A., Deitsch, K. W., Scheidig, C., Guinet, F., Nehrbass, U., Wellems, T. E. and Scherf, A. (2000) Frequent ectopic recombination of virulence factor genes in telomeric chromosome clusters of *P. falciparum*. *Nature* 407: 1018–1022.

Gakh, O., Cavadini. P. and Isaya, G. (2002) Mitochondrial processing peptidases. *Biochim. Biophys. Acta* 1592: 63–77.

Garbarino, J. E. and Gibbons, I. R. (2002) Expression and genomic analysis of midasin, a novel and highly conserved AAA protein distantly related to dynein. *BMC Genomics* 3: 18.

Gardner, M. J., Hall, N., Fung, E., White, O., Berrriman, M., Hyman, R. W., Carlton, J. M., Pain, A., Nalson, K. E., Bolwman, S. et al. (2002a) Genome sequence of the human malaria parasite *Plasmodium falciparum*. *Nature* 419: 498–511.

Gardner, M. J., Shallom, S. J., Carlton, J. M., Salzberg, S. L., Nene, V., Shoaibi, A., Ciecko, A., Lynn, J., Rizzo, M., Weaver, B. et al. (2002b) Sequence of *Plasmodium falciparum* chromosomes 2, 10, 11 and 14. *Nature* 419: 531–534.

Germot, A., Philippe, H. and Le Guyader, H. (1997) Evidence for loss of mitochondria in Microsporidia from a mitochondrial-type HSP70 in *Nosema locustae*. *Mol. Biochem. Parasitol.* 87: 159–168.

Gilson, P. and McFadden, G. I. (1995) The chlorarachniophyte: A cell with two different nuclei and two different telomeres. *Chromosoma* 103: 635–641.

Gilson, P. R., Maier, U. G. and McFadden, G. I. (1997) Size isn't everything: Lessons in genetic miniaturisation from nucleomorphs. *Curr. Opin. Genet. Dev.* 7: 800–886.

Glaser, P., Frangeul, L., Buchrieser, C., Rusniok, C., Amend, A., Baquero, F., Berche, P., Bloecker, H., Brandt, P., Chakraborty, T. et al. (2001) Comparative genomics of *Listeria* species. *Science* 294: 849–852.

Green, L. C., Didier, P. J. and Didier, E. S. (1999) Fractionation of sporogonial stages of the microsporidian *Encephalitozoon cuniculi* by Percoll gradients. *J. Euk. Microbiol.* 46: 434–438.

Hausmann, S., Vivarès, C. P. and Shuman, S. (2002) Characterization of the mRNA capping apparatus of the microsporidian parasite *Encephalitozoon cuniculi*. *J. Biol. Chem.* 277: 96–103.

Hazard, E. I. and Brookbank, J. W. (1984) Karyogamy and meiosis in an *Amblyospora* sp. (Microspora) in the mosquito *Culex salinarius*. *J. Invertebr. Pathol.* 44: 3–11.

Himmelreich, R., Hilbert, H., Plagens, H., Pirkl, E., Li, B. C. and Herrmann, R. (1996) Complete sequence analysis of the genome of the bacterium *Mycoplasma pneumoniae. Nucleic Acids Res.* 24: 4420–4449.

Hirt, R. P., Healy, B., Vossbrinck, C. R., Canning, E. U. and Embley, T. M. (1997) A mitochondrial Hsp70 orthologue in *Vairimorpha necatrix*: Molecular evidence that microsporidia once contained mitochondria. *Curr. Biol.* 7: 995–958.

Hirt, R. P., Logsdon, J. M., Jr., Healy, B., Dorey, M. W., Doolittle, W. F. and Embley, T.M. (1999) Microsporidia are related to Fungi: Evidence from the largest subunit of RNA polymerase II and other proteins. *Proc. Natl. Acad. Sci. USA* 96: 580–585.

Hirt, R. P., Müller, S., Embley, T. M. and Coombs, G. H. (2002) The diversity and evolution of thioredoxin reductase: New perspectives. *Trends Parasitol.* 18: 302–308.

Hjelmqvist, L., Tuson, M., Marfany, G., Herrero, E., Balcells, S. and Gonzalez-Duarte, R. (2002) ORMDL proteins are a conserved new family of endoplasmic reticulum membrane proteins. *Genome Biol.* 3: RESEARCH0027.

Horner, D. S., Foster, P. G. and Embley, T. M. (2000) Iron hydrogenases and the evolution of anaerobic eukaryotes. *Mol. Biol. Evol.* 17: 1695–1709.

Horner, D. S., Heil, B., Happe, T., Embley, T. M. (2002) Iron hydrogenases: Ancient enzymes in modern eukaryotes. *Trends Biochem. Sci.* 27: 148–153.

Horner, D. S., Hirt, R. P. and Embley, T. M. (1999) A single eubacterial origin of eukaryotic pyruvate:ferredoxin oxidoreductase genes: Implications for the evolution of anaerobic eukaryotes. *Mol. Biol. Evol.* 16: 1280–1291.

Howard, D. H. (1999) Acquisition, transport and storage of iron by pathogenic fungi. *Clin. Microb. Rev.* 12: 394–404.

Ishihara, R. and Hayashi, Y. (1968) Some properties of ribosomes from the sporoplasm of *Nosema bombycis. J. Invertebr. Pathol.* 11: 243–248.

Jensen, R. and Dunn, C. (2002) Protein import into and across the mitochondrial inner membrane: Role of the TIM23 and TIM22 translocons. *Biochim. Biophys. Acta* 1592: 25–34.

Kamaishi, T., Hashimoto, T., Nakamura, Y., Masuda, Y., Nakamura, F., Okamoto, K., Shimizu, M. and Hasegawa, M. (1996a) Complete nucleotide sequences of the genes encoding translation elongation factors 1 alpha and 2 from a microsporidian parasite, *Glugea plecoglossi*: Implications for the deepest branching of eukaryotes. *J. Biochem. (Tokyo)* 120: 1095–1103.

Kamaishi, T., Hashimoto, T., Nakamura, Y., Nakamura, F., Murata, S., Okada, N., Okamoto, K., Shimizu, M. and Hasegawa, M. (1996b) Protein phylogeny of translation elongation factor EF-1 alpha suggests microsporidians are extremely ancient eukaryotes. *J. Mol. Evol.* 42: 257–263.

Katinka, M. D., Duprat, S., Cornillot, E., Méténier, G., Thomarat, F., Prensier, G., Barbe, V., Peyretaillade, E., Brottier, P., Wincker, P. et al. (2001) Genome sequence and gene compaction of the eukaryote parasite *Encephalitozoon cuniculi. Nature* 414: 450–453.

Kawakami, Y., Inoue, T., Ito, K., Kimatizu, K., Hawana, C., Ando, T., Iwano, H. and Ishihara, R. (1994) Identification of a chromosome harboring the small unit ribosomal gene of *Nosema bombycis. J. Invertebr. Pathol.* 64: 147–148.

Keeling, P.J. and Doolittle, W.F. (1996) Alpha-tubulin from early diverging eukaryotic lineages and the evolution of the tubulin family. *Mol. biol. Evol.* 13: 1297–1305.

Keeling, P. J. and Fast, N. M. (2002) Microsporidia: Biology and evolution of highly reduced intracellular parasites. *Annu. Rev. Microbiol.* 56: 93–116.

Keeling, P. J., Luker, M. A. and Palmer, J. D. (2000) Evidence from beta-tubulin phylogeny that microsporidia evolved from within the fungi. *Mol. Biol. Evol.* 17: 23–31.

Keohane, E. M. and Weiss, L. M. (1999) The structure, function, and composition of the microsporidian polar tube. In *The Microsporidia and Microsporidiosis* (Wittner, M. and Weiss L.M., Eds.), ASM Press, Washington, pp. 196–224.

Kispal, G., Csere, P., Prohl, C. and Lill, R. (1999) The mitochondrial proteins Atm1p and Nfs1p are essential for biogenesis of cytosolic Fe/S proteins. *EMBO J.* 18: 3981–3989.

Knight, S. A. B., Sepuri, N. B., Pain, D. and Dancis, A. (1998) Mt-Hsp70 homolog, Ssc2p, required for maturation of yeast frataxin and mitochondrial iron homeostasis. *J. Biol. Chem.* 273: 18389–18393.

Kondrashov, F. A., Rogozin, I. B., Wolf, Y. I. and Koonin, E. V. (2002) Selection in the evolution of gene duplications. *Genome Biol.* 3: RESEARCH0008.

Kuroda, M., Ohta, T., Uchiyama, I., Baba, T., Yuzawa, H., Kobayashi, I., Cui, L., Oguchi, A., Hirakawa, H., Kuhara, S. et al. (2001) Whole genome sequencing of meticillin-resistant *Staphylococcus aureus*. *Lancet* 357: 1225–1240.

Lange, H., Kaut, A., Kispal, G. and Lill, R. (2000) A mitochondrial ferredoxin is essential for biogenesis of cellular iron-sulfur proteins. *Proc. Natl. Acad. Sci. USA* 97: 1050–1055.

Lange, H., Lisowsky, T., Gerber, J., Muhlenhoff, U., Kispal, G. and Lill, R. (2001) An essential function of the mitochondrial sulfhydryl oxidase Erv1p/ALR in the maturation of cytosolic Fe/S proteins. *EMBO Rep.* 2: 715–720.

Le Blancq, S. M. and Adam, R. D. (1998) Structural basis of karyotype heterogeneity in *Giardia lamblia*. *Mol. Biochem. Parasitol.* 97: 199–208.

Le Roy, F., Bisbal, C., Silhol, M., Martinand, C., Lebleu, B. and Salehzada, T. (2001) The 2-5A/RNase L/RNase L inhibitor (RNI) pathway regulates mitochondrial mRNAs stability in interferon-treated H9 cells. *J. Biol. Chem.* 276: 48473–48482.

Lill, R. and Kispal, G. (2000) Maturation of cellular Fe-S proteins: An essential function of mitochondria. *Trends Biochem. Sci.* 25: 352–356.

Mai, Z., Ghosh, S., Frisardi, M., Rosenthal, B., Rogers, R. and Samuelson, J. (1999) Hsp60 is targeted to a cryptic mitochondrion-derived organelle ("crypton") in the microaerophilic protozoan parasite *Entamoeba histolytica*. *Mol. Cell Biol.* 19: 2198–2205.

Malone, L. A. and McIvor, C. A. (1993) Pulsed field gel electrophoresis of DNA from four microsporidian isolates. *J. Invertebr. Pathol.* 61: 203–205.

Martin, W. and Müller, M. (1998) The hydrogen hypothesis for the first eukaryote. *Nature* 392: 37–41.

Martinand, C., Montavon, C., Salehzada, T., Silhol, M., Lebleu, B. and Bisbal, C. (1999) RNase L Inhibitor is induced during Human Immunodeficiency Virus type 1 infection and down regulates the 2-5A/RNase L pathway in human T cells. *J. Virol.* 73: 290–296.

McArthur, A. G., Knodler, L. A., Silberman, J. D., Davids, B. J., Gillin, F. D. and Sogin, M. L. (2001) The evolutionary origins of eukaryotic protein disulfide isomerase domains: New evidence from the amitochondriate protist *Giardia lamblia*. *Mol. Biol. Evol.* 18: 1455–1463.

Mertens, E. and Müller, M. (1990) Glucokinase and fructokinase of *Trichomonas vaginalis* and *Tritrichomonas foetus*. *J. Protozool.* 37: 384–388.

Méténier, G., Prensier, G., Delbac, F. and Vivarès, C. P. (2003) Analysis of duplicated genes in the genome of the microsporidian *Encephalitozoon cuniculi*. *J. Euk. Microbiol.*50: 31A. (Abstracts, GPLF 40th Annual Meeting 2002, La Rochelle).

Méténier, G. and Vivarès, C. P. (2001) Molecular characteristics and physiology of microsporidia. *Microb. Infect.* 3: 407–415.

Miao, Y. M. and Gazzard, B. G. (2000) Management of protozoal diarrhoea in HIV disease. *HIV Med.* 1: 194–199.

Mittleider, D., Green, L. C., Mann, V. H., Michael, S. F., Didier, E. S. and Brindley, P. J. (2002) Sequence survey of the genome of the opportunistic microsporidian pathogen, *Vittaforma corneae*. *J. Euk. Microbiol.* 49: 393–401.

Molina, J. M., Tourneur, M., Sarfati, C., Chevret, S., de Gouvello, A., Gobert, J. G., Balkan, S. and Derouin, F. (2002) Fumagillin treatment of intestinal microsporidiosis. *N. Engl. J. Med.* 346: 1963–1969.

Mota, P., Rauch, C. A. and Edberg, S. C. (2000) Microsporidia and Cyclospora: Epidemiology and assessment of risk from the environment. *Crit. Rev. Microbiol.* 26: 69–90.

Muhlenhoff, U. and Lill, R. (2000) Biogenesis of iron-sulfur proteins in eukaryotes: A novel task of mitochondria that is inherited from bacteria. *Biochim. Biophys. Acta* 1459: 370–382.

Munderloh, U. G., Kurti, T. J. and Ross, S. E. (1990) Electrophoretic characterization of chromosomal DNA from two microsporidia. *J. Invertebr. Pathol.* 56: 243–248.

Mushegian, A. (1999) The minimal genome concept. *Curr. Opin. Genet. Dev.* 9: 709–714.

Narcisi, E. M., Glover, C. V. and Fechheimer, M. (1998) Fibrillarin, a conserved pre-ribosomal RNA processing protein of *Giardia*. *J. Euk. Microbiol.* 45: 105–111.

Nene, V., Bishop, R., Morzaria, S., Gardner, M. J., Sugimoto, C., Ole-MoiYoi, O. K., Fraser, C. M. and Irvin, A. (2000) *Theileria parva* genomics reveals an atypical apicomplexan genome. *Int. J. Parasitol.* 30: 465–474.

Nixon, J. E., Wang, A., Morrison, H. G., McArthur, A. G., Sogin, M. L., Loftus, B. J. and Samuelson, J. (2002) A spliceosomal intron in *Giardia lamblia*. *Proc. Natl. Acad. Sci. USA* 99: 3359–3361.

Nord, J., Kiefer, M., Adolph, H. W., Zeppezauer, M. M. and Nyman, P. O. (2000) Transient kinetics of ligand binding and role of the C-terminus in the dUTPase the microaerophilic protozoan parasite *Entamoeba histolytica*. *Mol. Cell Biol.* 19: 2198–2205.

Nord, J., Nyman, P., Larsson, G. and Drakenberg, T. (2001) The C-terminus of dUTPase: Observation on flexibility using NMR. *FEBS Lett.* 492: 228–232.

Okhuysen, P. C. (2001) Traveler's diarrhea due to intestinal protozoa. *Clin. Infect. Dis.* 33: 110–114.

Peyret, P., Katinka, M. D., Duprat, S., Duffieux, F., Barbe, V., Barbazanges, M., Weissenbach, J., Saurin, W. and Vivares, C. P. (2001) Sequence and analysis of chromosome I of the amitochondriate intracellular parasite *Encephalitozoon cuniculi* (Microspora). *Genome Res.* 11: 198–207.

Peyretaillade, E., Biderre, C., Peyret, P., Duffieux, F., Méténier, G., Gouy, M., Michot, B. and Vivarès, C. P. (1998a) Microsporidian *Encephalitozoon cuniculi*, a unicellular eukaryote with an unusual chromosomal dispersion of ribosomal genes and a LSU rRNA reduced to the universal core. *Nucleic Acids Res.* 26: 3513–3520.

Peyretaillade, E., Broussolle, V., Peyret, P., Méténier, G., Gouy, M. and Vivarès, C. P. (1998b) Microsporidia, amitochondrial protists, possess a 70-kDa heat shock protein gene of mitochondrial evolutionary origin. *Mol. Biol. Evol.* 15: 683–689.

Peyretaillade, E., Peyret, P., Méténier, G., Vivarès, C. P. and Prensier, G. (2001) The identification of rRNA maturation sites in the microsporidian *Encephalitozoon cuniculi* argues against the full excision of presumed ITS1 sequence. *J. Euk. Microbiol.* Suppl: 60S–62S.

Philipps, N. F. and Li, Z. (1995) Kinetic mechanism of pyrophosphate-dependent phosphofructokinase from *Giardia lamblia*. *Mol. Biochem. Parasitol.* 73: 43–51.

Rayko, E. (1997) Organization, generation and replication of amphimeric genomes: A review. *Gene* 199: 1–18.

Raynaud, L., Delbac, F., Broussolle, V., Rabodonirina, M., Girault, V., Wallon, M., Cozon, G., Vivares, C. P. and Peyron, F. (1998) Identification of *Encephalitozoon intestinalis* in travelers with chronic diarrhea by specific PCR amplification. *J. Clin. Microbiol.* 36: 37–40.

Reeves, R. E., South, D. J., Blytt, H. J. and Warren, L. G. (1974) Pyrophosphate:D-fructose 6-phosphate 1-phosphotransferase: A new enzyme with the glycolytic function of 6-phosphofructokinase. *J. Biol. Chem.* 249:7737–7741.

Rehling, P., Wiedemann, N., Pfanner, N. and Truscott, K. N. (2001) The mitochondrial import machinery for preproteins. *Crit. Rev. Biochem. Mol. Biol.* 36: 291–336.

Robinson-Rechavi, M. and Laudet, V. (2001) Evolutionary rates of duplicate genes in fish and mammals. *Mol. Biol. Evol.* 18: 681–683.

Rotte, C., Stejskal, F., Zhu, G., Keithly, J. S. and Martin, W. (2001) Pyruvate : NADP+ oxidoreductase from the mitochondrion of *Euglena gracilis* and from the apicomplexan *Cryptosporidium parvum*: A biochemical relic linking pyruvate metabolism in mitochondriate and amitochondriate protists. *Mol. Biol. Evol.* 18: 710–720.

Sánchez, L. B., Galperin, M. Y. and Müller, M. (2000) Acetyl-CoA synthetase from the amitochondriate eukaryote *Giardia lamblia* belongs to the newly recognized superfamily of acyl-CoA synthetases (nucleoside diphosphate-forming). *J. Biol. Chem.* 275: 5794–5803.

Seeber, F. (2002a) Eukaryotic genome contain a [2Fe-2S] ferredoxin isoform with a conserved C-terminal sequence motif. *Trends Biochem. Sci.* 27: 545–547.

Seeber, F. (2002b) Biogenesis of iron-sulfur clusters in amitochondriate and apicomplexan protists. *Int. J. Parasitol.* 32: 1207–1217.

Spano F. and Crisanti, A. (2000) *Cryptosporidium parvum*: The many secrets of a small genome. *Int. J. Parasitol.* 30: 553–565.

Sprague, V. (1977) Systematics of the Microsporidia. In *Comparative Pathobiology*, Vol. 2 (Bulla, L. A., Jr. and Cheng, T. C., Eds.), Plenum Press, New York, pp. 1–510.

Stechmann, A., Cavalier-Smith, T. (2002) Rooting the eukaryote tree by using a derived gene fusion. *Science* 297: 89–91.

Streett, D. A. (1994) Analysis of *Nosema locustae* (Microsporidia, Nosematidae) chromosomal DNA with pulsed field gel electrophoresis. *J. Invertebr. Pathol.* 63: 301–303.

Tachezy, J., Sanchez, L. B. and Müller, M. (2001) Mitochondrial type iron-sulfur cluster assembly in the amitochondriate eukaryotes *Trichomonas vaginalis* and *Giardia intestinalis*, as indicated by the phylogeny of IscS. *Mol. Biol. Evol.* 18: 1919–1928.

Tielens, A. G., Rotte, C., van Hellemond, J. J. and Martin, W. (2002) Mitochondria as we don't know them. *Trends Biochem. Sci.* 27: 564–572.

Tovar, J., Fischer, A. and Clark, C. G. (1999) The mitosome, a novel organelle related to mitochondria in the amitochondrial parasite *Entamoeba histolytica*. *Mol. Microbiol.* 32: 1013–1021.

Upcroft, P. and Upcroft, J. A. (1999) Organization and structure of the *Giardia* genome. *Protist* 150:17–23.

Van de Peer, Y., Ben Ali, A. and Meyer, A. (2000) Microsporidia: Accumulating molecular evidence that a group of amitochondriate and suspectedly primitive eukaryotes are just curious fungi. *Gene* 246: 1–8.

Vávra, J. and Larsson, J. I. R. (1999) Structure of the microsporidia. In *The Microsporidia and Microsporidiosis* (Wittner, M. and Weiss, L. M., Eds.), ASM Press, Washington, pp. 7–84.

Vivarès, C. P. (2001) Introduction: The microsporidial world, a paragon for analyzing intracellular parasitism. *Microb. Infect.* 3: 371–372.

Vivarès, C. P., Gouy, M., Thomarat, F. and Méténier, G. (2002) Functional and evolutionary analysis of a eukaryotic parasitic genome. *Curr. Opin. Microbiol.* 5: 499–504.

Vivarès, C. P. and Méténier, G. (2000) Towards the minimal eukaryotic parasitic genome. *Curr. Opin. Microbiol.* 3: 463–467.

Vivarès, C. P. and Méténier, G. (2001) The microsporidian *Encephalitozoon*. *Bioessays* 23: 194–202.

Vossbrinck, C. R., Maddox, J. V., Friedman, S., Debrunner-Vossbrinck, B. A. and Woese, C. R. (1987) Ribosomal RNA sequence suggests microsporidia are extremely ancient eukaryotes. *Nature* 326: 411–414.

Vossbrinck, C. R. and Woese, C. R. (1986) Eukaryotic ribosomes that lack a 5.8S RNA. *Nature* 320: 287–288.

Wagner, A. (2000) Decoupled evolution of coding region and mRNA expression patterns after gene duplication: Implications for the neutralist-selectionist debate. *Proc. Natl. Acad. Sci. USA* 97: 6579–6584.

Weidner, E., Findley, A. M., Dolgikh, V. and Sokolova, J. (1999) Microsporidian biochemistry and physiology. In *The Microsporidia and Microsporidiosis* (Wittner, M. and Weiss, L. M., Eds.), ASM Press, Washington, pp. 172–195.

Wiedemann, N., Pfanner, N. and Ryan, M. T. (2001) The three modules of ADP/ATP carrier cooperate in receptor recruitment and translocation into mitochondria. *EMBO J.* 20: 951–260.

Williams, B. A. P., Hirt, R. P., Lucocq, J. M. and Embley, T. M. (2002) A mitochondrial remnant in the microsporidian *Trachipleistophora hominis*. *Nature* 418: 865–869.

Winkler, H. H., Neuhaus, H. E. (1999) Non-mitochondrial ATP transport. *Trends Biochem. Sci.* 24: 64–68.

Wolfe, K. H. and Shields, D. C. (1997) Molecular evidence for an ancient duplication of the entire yeast genome. *Nature* 387: 708–713.

Wood, V., Rutherford, K. M., Ivens, A., Rajandream, M. A. and Barrell, B. (2001) A re-annotation of the *Saccharomyces* genome. *Comp. Funct. Genom.* 2: 143–154.

Wood, V., Gwilliam, R., Rajandream, M. A., Lyne, M., Lyne, R., Stewart. A., Sgouros, J., Peat, N., Hayles, J., Baker S et al. (2002) The genome sequence of *Schizosaccharomyces pombe.* *Nature* 415: 871–880.

Zauner, S., Fraunholz, M., Wastl, J., Penny, S., Beaton, M., Cavalier-Smith, T., Maier, U. G. and Douglas, S. (2000) Chloroplast protein and centrosomal genes, a tRNA intron, and odd telomeres in an unusually compact eukaryotic genome, the cryptomonad nucleomorph. *Proc. Natl. Acad. Sci. USA* 97: 200–205.

Chapter 11

Evolutionary Contribution of Plastid Genes to Plant Nuclear Genomes and Its Effects on Composition of the Proteomes of All Cellular Compartments

Dario Leister and Anja Schneider

CONTENTS

Abstract

The acquisition of a cyanobacterial endosymbiont by an ancestral host cell resulted in the establishment of chloroplasts. The evolution of the new organelle occurred mainly via gene loss, together with a massive transfer of genes to the nucleus. Genome-wide comparative analyses have substantially improved our understanding of the extent and consequences of this large-scale gene transfer. Genes from the plastid ancestor have been a rich source of genetic material for the evolution of new cellular functions, as is documented in this chapter.

0-415-29904-7/04/$0.00+$1.50

A novel observation is that the acquisition of cyanobacterial genes by plants led not only to the successful establishment of a new organelle but also to substantial changes in the composition of the proteomes of other compartments in the plant cell.

11.1 Introduction

Plastids of plant cells contain their own genome, the plastome. Although many plastome copies with the same DNA information are contained in each cell of a plant, the organelles themselves can widely vary in morphology and function. Proplastids can develop into green chloroplasts, red or yellow chromoplasts or into other variants specialized for the storage of starch, lipids or proteins. Chloroplasts have a high rate of transcription and translation, allowing the expression of large amounts of the enzyme ribulose bisphosphate carboxylase (Rubisco) and enabling the rapid renewal of electron transfer components, two features crucial for efficient photosynthetic CO_2 fixation. Besides photosynthesis, chloroplasts carry out other essential functions, such as the synthesis of amino acids, fatty acids and lipids, plant hormones, nucleotides, vitamins and secondary metabolites.

It is now widely accepted that current plastids are the endosymbiotic remnants of a free-living cyanobacterial progenitor, and plastids have, over evolutionary time, lost the vast majority of their original gene complement. Depending on the organism considered, contemporary plastomes contain only 60 to 200 open reading frames (ORFs), whereas most of their original genetic information has been transferred to the nucleus. Nucleus-encoded plastid proteins are synthesized as precursors containing N-terminal presequences (cTPs, chloroplast transit peptides), which target the proteins to the chloroplast. After import, cTPs are cleaved by the stromal processing peptidase, releasing the mature proteins into the stroma for distribution to other subcompartments of the organelle if necessary.

The precise lineage of cyanobacteria that gave rise to plastids is still unknown, as it is difficult to reconstruct early events in plastid evolution. Nevertheless, with the increasing number of sequenced genomes available, including those of plastids, cyanobacteria and of the flowering plant species *Arabidopsis thaliana*, genome-wide comparisons have become feasible. Such analyses allow us to address questions concerning the number and functions of the plastid genes transferred to the nucleus, the phylogeny of contemporary plastomes and the reasons why plastids have retained a genome. This chapter introduces the current view of the mechanism of plastid-to-nucleus gene transfer, its consequences for the regulation of plastid function in terms of intercompartmental signaling and the extent to which this large-scale horizontal gene transfer has affected the evolution of the plant lineage in comparison to that of animal phyla.

11.2 How Many Nuclear Genes of Plants Are of Cyanobacterial Origin?

Genomes of contemporary cyanobacteria encode several thousand proteins, ranging from less than 2000 ORFs in *Prochlorococcus* to more than 7000 ORFs in *Nostoc* (http://www.jgi.doe.gov/JGI_microbial/html/). The first estimate of the number of cyanobacterial genes in a plant genome (Abdallah et al., 2000) employed comparative BLAST analysis of the then partially sequenced genome of *A. thaliana* and the fully sequenced genome of the cyanobacterium *Synechocystis* (Kaneko et al. 1996a, b), using the genome of the yeast *Saccharomyces cerevisiae* (Goffeau et al., 1996) as a reference. Between 1400 and 1500 proteins of cyanobacterial origin were predicted to be encoded in the *Arabidopsis* nuclear genome, about half of which were targeted to the chloroplast as indicated by the presence of cTPs (Abdallah et al., 2000). These authors also predicted the existence of more than 1000 chloroplast proteins of noncyanobacterial descent. Other studies, based on phylogenetic analysis of the entire *A. thaliana* genome or of samples of it, concluded that 400 to

2200 *Arabidopsis* genes were of cyanobacterial origin (The *Arabidopsis* Genome Initiative, 2000; Rujan and Martin, 2001).

The most thorough analysis of the cyanobacterial heritage in the *Arabidopsis* nuclear genome so far undertaken compared it with the genomes of 3 cyanobacterial species, 16 other reference prokaryotes and yeast (Martin et al., 2002). Of the 24,990 *Arabidopsis* proteins considered, a subset of 9368 was sufficiently conserved for primary sequence comparison. Among these, 866 had homologues only in cyanobacteria, and an additional set of 834 proteins were grouped together with cyanobacterial proteins in phylogenetic trees. Assuming that *Arabidopsis* proteins that originated from cyanobacteria do not preferentially belong to the set of the 9368 most conserved ones, it can be calculated that 18% [(866 + 834)/9368] of the entire *Arabidopsis* genome, or 4500 genes, is derived from cyanobacterial progenitor genes (Table 11.1).

How many genes did the ancient endosymbiont possess, and how many of these were transferred to the nucleus before their number was modified by molecular events in the nuclear genome itself? Unfortunately, it is not known how many genes were lost during plastid evolution, or how many — and which — genes the cyanobacterial ancestor brought with it. Furthermore, the estimate of 4500 genes of cyanobacterial descent does not imply that the ancestor of plastids actually donated this number of genes to the nucleus; possibly far fewer were transferred, which were subsequently amplified in the nucleus by duplications of single genes or entire chromosomal segments (The *Arabidopsis* Genome Initiative, 2000).

TABLE 11.1 Prediction of the Intracellular Localization of *Arabidopsis* Proteins Encoded by Cyanobacterium-Derived Genes

Targeting Signal	*Number of Genes Estimated by TargetP Analysis of 1700 Protein-Coding Genes*	*Number in the Entire Genome (× 25678/9368)*	*Final Estimate (after Correction for Accuracy of TargetP)*
cTP	571	1565	1270
mTP	140	384	421
SP	302	828	872
None	687	1883	1727
Total	1700	4660	4290

Note: Of 9368 *Arabidopsis* proteins sufficiently similar to be recognized in different taxa, 1700 were presumed to stem from cyanobacteria [Martin et al. (2002) 12246–12251]. Their subcellular localization was predicted by TargetP [Emanuelsson et al. (2000)1005-1016] (Column 2), and extrapolation to the entire genome (25,678 genes) was performed by multiplying by a factor of 2.74 (25678/9368; Column 3). The accuracy of the TargetP program was tested with respect to both the specificity and the sensitivity of the program [Emanuelsson et al. (2000) 1005–1016 and Leister (2003) 47–56] (Column 4). The genome-wide extrapolation results in 4660 genes of cyanobacterial origin [see Column 3; 4500 genes estimated by Martin, W. et al. (2002) *Proc. Natl. Acad. Sci. USA* 99: 12246–12251, who considered only 24,990 of the 25,678 *Arabidopsis* genes]. After the second extrapolation step (Column 4), the total number of cyanobacterial proteins comes to ca. 4300. The decrease is due to the slightly different composition of the data set (with respect to the frequencies of cTPs, mTPs, SPs = signal peptide for secretory pathway, or no N-terminal presequence) used to calculate the accuracy of TargetP prediction with respect to both the specificity and the sensitivity of the program [Emanuelsson et al. (2000) 1005–1016 and Leister (2003) 47–56] and the data set representing all protein-coding genes of *Arabidopsis*.

11.3 Reconstruction of More Recent Instances of Plastid-to-Nucleus Gene Transfers

The comparative analysis of sequenced plastid genomes from different species allows us to reconstruct the phylogeny of plastomes. In one of the first studies, Martin et al. (1998) compared the plastomes of a glaucocystophyte, a rhodophyte, a diatom, a euglenophyte and five land plants. In total, 210 different protein-coding genes were detected, of which 45 were common to all these species and to the cyanobacterium *Synechocystis*. A phylogenetic tree of the nine plastomes based on the 11,039 amino acid positions of the 45 common proteins allowed the authors to discern the pattern of gene losses from chloroplast genomes, revealing that independent parallel losses in multiple lineages outnumbered unique losses. Moreover, for 44 different plastid-encoded proteins, functional nuclear genes of chloroplast origin were identified.

This type of comparative analysis has recently been extended to additional plant species whose plastome sequences are known. Lemieux et al. (2000) have compared the plastome sequence of the flagellate *Mesostigma* with plastomes of three land plants, three chlorophyte green algae, as well as the red alga *Porphyra purpurea* and *Synechocystis* as outgroups. They concluded that *Mesostigma* represents a lineage that emerged before the divergence of the Streptophyta (land plants and their closest green algal relatives, the charophytes) and Chlorophyta (green algae other than charophytes).

The most comprehensive phylogenetic analysis of chloroplast genomes considered 15 sequenced plastomes for a total of 274 different protein-coding genes, pinpointing 117 nucleus-encoded proteins that are also still encoded in at least one chloroplast genome (Martin et al., 2002). A phylogenetic tree of the 15 chloroplast genomes based on 8303 amino acid positions in 41 proteins provided support for independent secondary endosymbiotic events for Euglena, Guillardia and Odontella. In contrast to Lemieux et al. (2000), these authors concluded that *Mesostigma* branched off basal to land plants but later than the chlorophyte algae *Chlorella* and *Nephroselmis* (Figure 11.1).

11.4 Functions of Cyanobacterium-Derived Proteins in the Plant Cell

11.4.1 Chloroplasts: Cyanobacterial Inheritance vs. Eukaryotic Invention

Contemporary plastids resemble their cyanobacterial relatives in many respects: the existence of thylakoid membranes (Vothknecht and Westhoff, 2001); 70S-type ribosomes (Yamaguchi and Subramanian, 2000; Yamaguchi et al., 2000); cell division proteins (Osteryoung and McAndrew, 2001); light-dependent chlorophyll biosynthesis (Suzuki and Bauer, 1995); and the Sec-, Tat- and SRP-types of protein targeting to thylakoids (Robinson et al., 2001). However, no more than half of the plant proteins that originated from cyanobacteria are targeted back to the plastid (Abdallah et al., 2000; Pesaresi et al., 2001; Martin et al., 2002). Furthermore, a substantial fraction of proteins for which no cyanobacterial origin has been established is destined for roles in plastids (Abdallah et al., 2000; Leister, 2003). Even in the case of well-known elements of the cyanobacterial heritage of the chloroplast, a closer look allows us to identify eukaryotic additions, such as novel (i.e., plant-specific) photosynthetic (Scheller et al., 2001) or ribosomal (Yamaguchi and Subramanian, 2000; Yamaguchi et al., 2000) proteins, as well as novel domains in otherwise conserved proteins (see later). Well-studied plastid functions that appear to be completely new include the machinery responsible for importing proteins across the plastid envelope (Jarvis and Soll, 2001; see also Chapter 12), the spontaneous targeting pathway for thylakoid membrane proteins (Robinson et al., 2001) and the light-harvesting antenna complexes (LHCs) of the photosystems (Montane and Kloppstech, 2000).

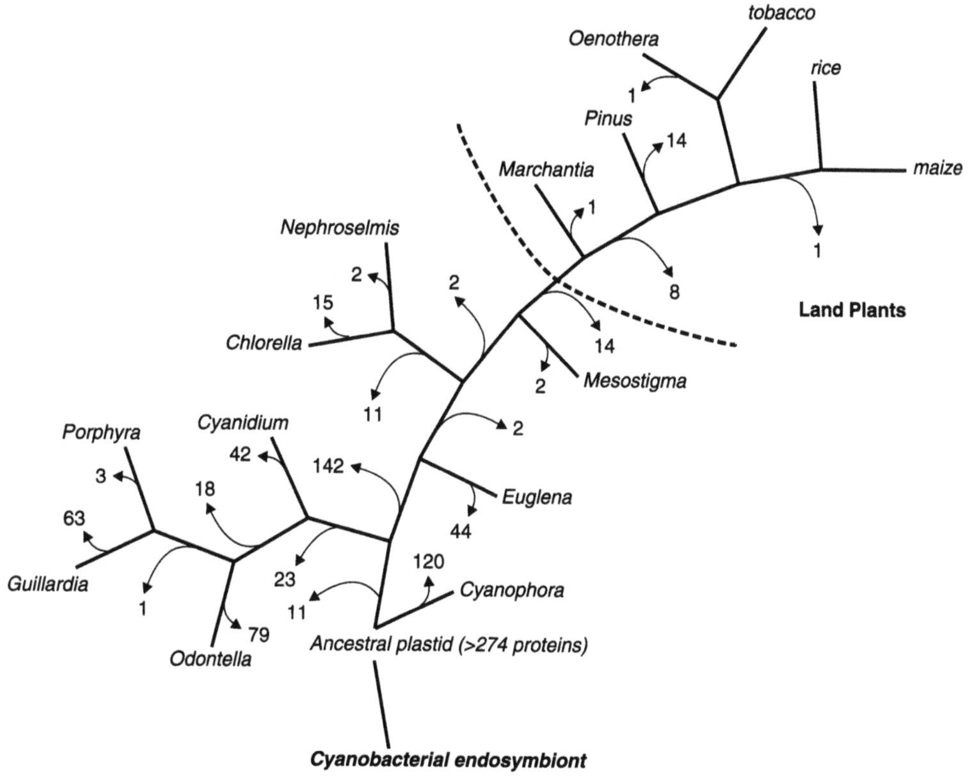

FIGURE 11.1 Phylogeny of plastid genomes. A simplified scheme based on the topology provided by Martin, W. et al. (2002) *Proc. Natl. Acad. Sci. USA* 99: 12246–12251 is shown. Arrows point to the number of genes lost from chloroplast genomes (and might have been transferred to the nucleus) during evolution.

11.4.2 Photosystem I and Plastid Ribosomes

This collection of cyanobacterial and eukaryotic proteins operating in plastids has been studied in detail. In plants, the photosystem I (PSI) complex is a mosaic of plastid- and nucleus-encoded protein subunits; of the latter, three (PSI-D, -E and -H) are each encoded by two functional gene copies in *A. thaliana*. Cross-species comparison of PSI raises the question why the subunits D, E and F in higher plants contain N-terminal extensions not present in their cyanobacterial counterparts. In the case of PSI-F, the extra 18 residues in the eukaryotic protein form an amphipathic helix located on the lumenal side of the thylakoid membrane, which allows more efficient oxidation of plastocyanin, a characteristic of eukaryotic PSI (Hippler et al., 1999). A further question concerns the function of PSI-G, -H, -N and -O, which are absent in cyanobacteria. Genetic knockout of PSI-G (Varotto et al., 2002; Jensen et al., 2002), -H (Naver et al., 1999), -N (Haldrup et al., 1999) or -O (unpublished results of our laboratory) individually does not affect plant viability, but leads to subtle alterations in photosynthetic electron flow, suggesting that plant-specific PSI subunits rather than being essential play accessory or regulatory roles in PSI function.

The 70S-type ribosome of *E. coli* consists of 54 different proteins. Twenty-one (RPS1 to RPS21) make up the small 30S subunit and thirty-three (RPL1 to RPL6, RPL9 to RPL25 and RPL27 to RPL36) are located in the large 50S subunit (Wittmann, 1982). The complete set of 70S-type chloroplast ribosomal proteins in higher plants has been identified recently

(Yamaguchi and Subramanian, 2000; Yamaguchi et al., 2000). Based on sequence similarity to their bacterial orthologues, these plastid ribosomal proteins (PRPs) are designated PRPL1 to PRPL36 (for the large subunit) and PRPS1 to PRPS21 (for the small subunit). The 50S subunit of chloroplast ribosomes from spinach contains 33 proteins, of which the plastid-specific ribosomal protein 5 (PSRP5) and PSRP6 are specific to the plastid, having no cyanobacterial counterparts. Twenty-five of the thirty-three proteins are nucleus encoded and are imported into the chloroplast via a cTP (Yamaguchi and Subramanian, 2000). In the same plant, the 30S subunit contains 25 proteins, of which four (PSRP1 to PSRP4) are specific to chloroplasts. Thirteen of the 25 are encoded in the nucleus and twelve in the plastome (Yamaguchi et al., 2000). Plastid-specific ribosomal proteins might have evolved to facilitate plant-specific ribosomal functions, such as association with thylakoid membranes and light-dependent regulation of translation (Yamaguchi et al., 2000; Yamaguchi and Subramanian, 2000).

11.4.3 The Calvin Cycle and Glycolysis

Studies of the evolutionary origin of proteins involved in plant metabolism are available for the enzymes of the Calvin cycle (located in the plastid) and glycolysis. The Calvin cycle, or reductive pentose phosphate pathway (Calvin, 1956), is the only eukaryotic CO_2 fixation pathway known. CO_2 fixed as triosephosphate is exported from chloroplasts to the cytosol by the phosphate translocator, primarily as glyceraldehyde-3-phosphate, in exchange for phosphate (Flügge and Heldt, 1991). In the cytosol, glyceraldehyde-3-phosphate can either enter the glycolytic pathway for ATP synthesis in mitochondria or the gluconeogenetic pathway for the synthesis of sucrose. Several, but not all, Calvin cycle enzymes are derived from cyanobacteria. Martin and Schnarrenberger (1997) have thoroughly investigated the 11 proteins that catalyze the reactions of the Calvin cycle in higher-plant chloroplasts. In spinach, five isoforms that operate in the glycolytic or gluconeogenetic pathway are known; however, this number can vary, depending on the tissue or species (Plaxton, 1996). All Calvin cycle enzymes are encoded in the nucleus, with the exception of the large subunit of Rubisco (RbcL), which is encoded by the plastid gene *rbcL*. The gene encoding the small subunit of Rubisco (RbcS) must have been transferred to the nucleus early in chlorophyte evolution, as deduced from comparative phylogenetic analyses (Delwiche and Palmer, 1996).

Higher plants possess chloroplast and cytosolic isoforms of phosphoglycerate kinase (PGK), both of cyanobacterial origin. The two enzymes arose through the duplication of a precursor gene, and the chloroplast enzyme has acquired a cTP (Martin and Schnarrenberger, 1997; Brinkmann and Martin, 1996). The cyanobacterial PGK genes replaced a preexisting nuclear gene, which might have been of mitochondrial descent, coding for cytosolic PGK (Martin and Schnarrenberger, 1997). For the chloroplast and cytosolic isoforms of glyceraldehyde-3-phosphate dehydrogenase (GAPDH), the phylogenetic situation is different. [For a recent review, see Figge et al. (1999) and (2001).] The origin of the gene for the cytosolic GAPDH has been a matter of much debate, but the weight of evidence suggests that it derives from the mitochondria (Henze et al., 1995). The nucleus-encoded gene for the Calvin-cycle GAPDH has a different phylogeny. Its product is more similar to cyanobacterial homologues than it is to the enzyme from any other eubacterial or eukaryotic source, indicating that the protein is now reimported into the organelle that donated the coding sequence (Martin et al., 1993). Also of cyanobacterial origin are the genes that code for the chloroplast enzymes transketolase (TKL), ribulose-5-phosphate 3-epimerase (RPE) and phosphoribulokinase (PRK; Martin and Schnarrenberger, 1997). Where cytosolic isoforms exist, e.g., the cytosolic TKL of the dehydratable angiosperm *Craterostigma* (Bernacchia et al., 1995), these are also of cyanobacterial origin.

The enzymes discussed previously are of cyanobacterial origin, but must now be rerouted to the plastid organelle or have replaced cytosolic proteins originally supplied by the host genome. In contrast, the chloroplast enzymes triosephosphate isomerase (TPI) and fructose-1,6-bisphophatase (FBP) in higher plants, as well as their cytosolic isoforms, are of mitochondrial origin (Martin and Schnarrenberger, 1997). For the remaining three enzymes of the Calvin cycle, fructose-1,6-bisphosphate aldolase (FBA), sedoheptulose-1,7-bisphosphatase (SBP) and ribose-5-phosphate isomerase (RPI), no clear origin has yet been defined. However, it can be concluded that most Calvin-cycle enzymes originated from cyanobacteria, with a few exceptions that represent proteobacterial proteins encoded by genes descended from mitochondria.

Enzymes of the glycolytic pathway present in the cytosol of higher plants are mostly of eubacterial (mitochondrial or cyanobacterial) origin (Martin and Herrmann, 1998). Dependent on the nature of the presumptive archaebacterial host (Doolittle, 1998), which incorporated the mitochondrial endosymbiont, several scenarios are possible. If the archaebacterium-like form had its own repertoire of glycolytic and gluconeogenetic enzymes, the archaebacterial enzymes must have been replaced by the products of genes donated to the nucleus by the eubacterial symbionts, chloroplasts and mitochondria. However, at the present time it is uncertain whether the presumptive archaebacterial host was a heterotrophic organism with a glycolytic pathway: the hydrogen hypothesis of Martin and Müller (1998) proposes that the host that incorporated the mitochondrial endosymbiont possessed a hydrogen-dependent metabolism and was a strict anaerobic autotroph.

11.4.4 Massive Rerouting of Cyanobacterial Proteins

Martin et al. (2002) have shown that of ca. 4500 proteins predicted to be of cyanobacterial origin, only ca. 1300 code for chloroplast-targeted proteins. This implies that a massive redistribution of cyanobacterium-derived proteins to other cellular compartments has occurred (Figure 11.2). Thus, gene origin and protein compartmentation do not correspond, or, in other words, most plant proteins that derive from cyanobacteria do not appear to be rerouted to their original compartment, the plastid. The genes of cyanobacterial origin in *Arabidopsis* cover all functional categories, including many not represented in contemporary cyanobacteria, such as disease resistance and intracellular protein routing (Martin et al., 2002). This allows us to conclude that genes inherited from the ancestor of plastids have been a rich source of material for the evolution of novel functions, and, furthermore, that the acquisition of these bacterial genes led not only to the successful establishment of a new organelle but also to a substantial change in the composition of the proteomes of other cell compartments. This massive redistribution of processes between different compartments is at variance with the view that such relocation events are rare (Zerges, 2002); future studies will clarify whether complex processes involving many proteins have indeed been transferred as a whole from the endosymbiont to other compartments.

11.5 Why Have Plastids Retained a Genome?

The transfer of genes from plastids and mitochondria to the nucleus has several advantages, e.g., a lower mutation rate and a Mendelian type of inheritance (Martin and Herrmann, 1998; Blanchard and Lynch, 2000). Foreign DNA insertions occur more often into the nuclear genome than into organellar genomes, simply because of the difference in their sizes (Doolittle, 1998), and, last but not least, gene losses could be positively selected for in organellar genomes, because smaller genomes mean higher intraorganellar competitiveness (Cavalier-Smith, 1987).

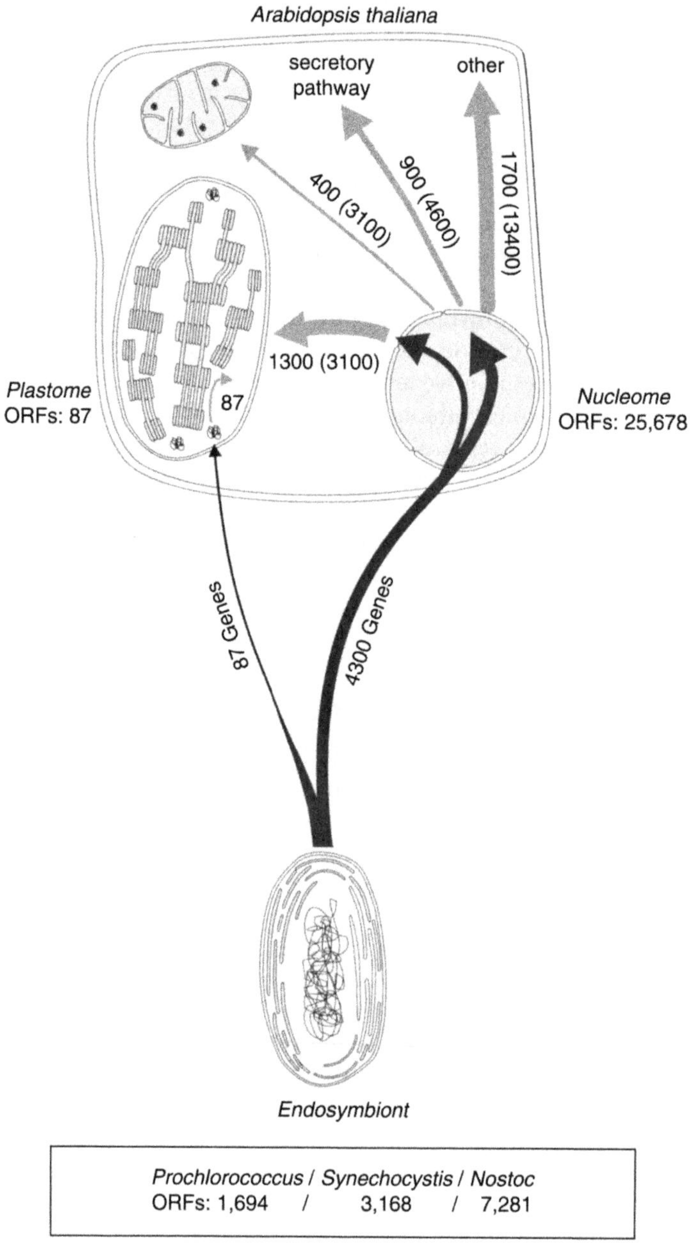

FIGURE 11.2 Intracellular targeting of the proteins of cyanobacterial descent in *Arabidopsis*. Chloroplasts bounded by double membranes, such as those of *Arabidopsis*, originated from a cyanobacterial-like endosymbiont. Because the identity of the endosymbiont is unknown, the three cyanobacterial species whose genome sequences are known (*Synechocystis* sp. PCC6803, *Prochlorococcus marinus* and *Nostoc punctiforme*) were used to reconstruct protein phylogenies [Martin et al. (2002) 12246–12251]. Estimation of the number and intracellular localization of cyanobacterium-derived gene products is illustrated in Table 11.1, and the number of nuclear genes of cyanobacterial origin comes to ca. 4300. Black lines trace the origin of the 87 plastome and the 4300 nuclear genes of cyanobacterial origin. Black numbers report the number of products (grey arrows) of nuclear genes of cyanobacterial origin. Numbers in parentheses indicate the total number of predicted proteins targeted to the respective compartment as in Leister (2003) 47–56.

A small subset of 60 to 200 protein-coding genes is retained within chloroplasts (Stoebe et al., 1998). Of these, ca. 40 are present in all chloroplast genomes analyzed so far (Martin et al., 2002). Only parasitic protists lacking photosynthesis have smaller plastomes, comprising ca. 35 kb of DNA with 30 protein-coding genes (Wilson and Williamson, 1997). One possible reason for the retention of genes in plastid DNA is that although these genes can be also expressed in the nucleus, the resulting proteins are too hydrophobic to be imported across the double membrane surrounding the plastids (Palmer, 1997). An alternative explanation is that idiosyncratic features, such as variant codes, editing or other complex processing patterns, might prevent nuclear expression of some organellar genes (Doolittle, 1998). One example is RbcL, the large subunit of Rubisco, which is highly hydrophobic and is always encoded in the plastome (Martin et al., 1998, 2002). By inactivating the plastome copy of *rbcL* and introducing a corresponding construct into the nuclear genome by *Agrobacterium*-mediated gene transfer, the plastid *rbcL* gene of tobacco has been relocated to the nuclear genome (Kanevski and Maliga, 1994). The chloroplast-targeted RbcL protein was able to restore autotrophic growth, albeit with severely reduced levels of Rubisco activity (3 to 10% of wild type), indicating that neither hydrophobicity nor codon usage can explain why the *rbcL* gene has been retained in the plastid genome.

A further hypothesis explains the presence of genes in plastids as an intermediate state in a long-lasting process, which will eventually lead to the complete transfer of organelle genes to the nucleus (Herrmann, 1997). This idea is based on the observation that mitochondria, which derive from a more ancient endosymbiotic event than plastids do, contain significantly fewer genes than plastids (5 to 60; Gray et al., 1998). However, the common set of ca. 40 genes found in all chloroplast genomes sequenced so far strongly argues that relocation is not simply a stochastic process by which all organellar genes will ultimately be transferred to the nucleus. It appears more likely that selection, rather than chance, accounts for the independent retention of this set of genes in the plastome DNA of many different lineages.

An intriguing hypothesis was proposed by Allen (1993) and later revisited by Race et al. (1999). This argues that the efficient regulation of some chloroplast processes requires the expression of specific proteins from organellar genes. Thus, the expression of plastid genes encoding proteins with key roles in electron transport and energy coupling is, in fact, thought to be tightly regulated via redox potentials generated by the electron transfers. The basic postulate is that the photosynthetic electron transport could become harmful for the organelle if a structural component of the electron transport chain is not available in sufficient quantity, and electrons are taken from or transferred to inappropriate donors or acceptors. This, in turn, could lead to damage to the photosynthetic membrane and the rest of the cell. Thus, organelle genomes have persisted to ensure tight control of electron flows, allowing their rapid regulation in response to changes in the redox state (Allen, 1993). In the interests of efficient redox control, the plastid genome should therefore primarily encode and rapidly express structural proteins of the photosynthetic membrane, and, in addition, of the ribosomal machinery. Indeed, the two main functional categories of genes in extant plastid genomes code for proteins of the photosynthetic membrane or products involved in gene expression (Race et al., 1999): of the 49 proteins common to nine functional chloroplast genomes (Martin et al., 1998), 24 are constituents of the photosynthetic membrane, 16 are ribosomal proteins, 3 are subunits of the RNA polymerase, 1 is the large subunit of Rubisco and the remaining 2 have unknown functions. A closer inspection of the composition of protein complexes within the thylakoid membrane reveals that proteins whose genes are more resistant to nuclear transfer are located at or close to the functional physical core of the photosynthetic reaction centers. These proteins serve to maintain the redox balance in the photosynthetic membrane. Even the existence of a plastid gene coding for a subunit of Rubisco makes sense with respect to redox balancing, because if electrons from the thylakoid membrane cannot be transferred rapidly enough to CO_2, photooxidation of the membrane and cell death result (Allen, 1993). The redox control hypothesis of Allen (1993), however, is somewhat challenged by the finding that expression and thylakoid membrane targeting of a plastid-encoded protein is not necessarily faster than that of a nucleus-encoded counterpart. In *Chlamydomonas*, expression of light-harvesting proteins (LHCPs) from the nuclear genome is as rapidly and precisely regulated as that of the corresponding plastome-encoded proteins (Grossman et al., 1995).

A further hypothesis was recently formulated by Zerges (2002). The central and most hydrophobic subunits of thylakoid membrane complexes of the photosynthetic apparatus are encoded by plastome genes (see above). Thus, hydrophobicity alone might hamper the expression and targeting of such proteins from nuclear genes. However, other hydrophobic proteins such as RbcL (see above) or LHCPs can enter the chloroplast. In this context, the idea was developed that genes are retained in the chloroplast because the assembly of photosystems has to take place on the spot, possibly in a cotranslational manner and with the help of plastid mRNAs as assembly factors (reviewed by Zerges, 2002).

11.6 Consequences of Plastid-to-Nucleus Gene Transfer for Transport, Signaling and Development

11.6.1 Transport

Communication between plastids and the surrounding cytosol occurs across the plastid envelope, a double membrane. The inner membrane, and to a lesser extent the outer, contains a variety of transporters that mediate the exchange of metabolites between the compartments (Flügge, 1998; Soll et al., 2000; see also Chapter 12). The need for metabolite exchange must have arisen during the course of organelle evolution. What contribution did the endosymbiont make to these metabolite exchanges? In the case of membrane proteins, appropriate databases can be consulted (Schwacke et al., 2002) to investigate which plastid membrane proteins have homologues in cyanobacteria. We have found that of the 6475 tentative membrane proteins encoded in the *Arabidopsis* genome, 8% have a homologue in *Synechocystis* (unpublished results of our laboratory). When the *Arabidopsis* membrane proteins are analyzed for their intracellular location based on their N-terminal presequences, 658 contain cTPs, as independently established by three prediction programs. Of these 658 proteins, 16.5% have a *Synechocystis* homologue, whereas only 9% of mitochondrial and 1.3% of membrane proteins entering the secretory pathway have a homologue in *Synechocystis*. The 658 membrane proteins with chloroplast targeting sequences can be further divided into two subgroups: those with one to three transmembrane domains (e.g., protein kinases) and those with four or more transmembrane domains (e.g., metabolite transporters). Strikingly, of the first set, only 8% have cyanobacterial homologues, whereas 25% of the second group have a cyanobacterial pendant (unpublished results of our laboratory). Although these data need further investigation, they demonstrate that certain classes of membrane proteins are more likely to be derived from the cyanobacterial ancestor than others.

11.6.2 Plastid-to-Nucleus Signaling

The coordination between nucleus and chloroplasts necessarily involves the exchange of information between the two compartments. Most information exchange should flow from the nucleus to the chloroplast; but signals from the chloroplast that influence the nucleus and the rest of the cell have also been described (Oelmüller, 1989; Susek and Chory, 1992). Intermediates of chlorophyll biosynthesis, such as the porphyrin Mg-protoporphyrin IX (Strand et al., 2003), function as plastid signals, regulating the expression of nuclear genes active in both photosynthesis and nonphotosynthetic processes (Kropat et al., 1997; Muramoto et al., 1999; Davis et al., 1999; La Rocca et al., 2001; Meskauskiene et al., 2001).

Recently, investigations of the genomes-uncoupled 5 (*gun5*) and long after far red 6 (*laf6*) mutants of *A. thaliana* (Mochizuki et al., 2001; Moller et al., 2001) have provided a deeper insight into the molecular mechanism of porphyrin-based signaling. The *GUN5* gene encodes the ChlH subunit of the plastid Mg^{2+} chelatase and the *LAF6* gene codes for the ATP-binding cassette (ABC) protein AtABC1; both genes are thought to originate from cyanobacteria. In *gun5* mutants, the repression of nuclear photosynthetic genes on photooxidation fails to function, and it has been suggested that ChlH has both signaling and catalytic functions in the process. It is plausible that the plastid signal could reflect porphyrin concentrations, changes in which might be expected to occur in damaged chloroplasts. In this model ChlH senses the porphyrin concentration, integrating the information into a plastid signal, which is then transmitted to the nucleus.

In addition, the AtABC1 protein is thought to be involved, albeit indirectly, in the biosynthesis of chlorophyll from protoporphyrinogen IX (Moller et al., 2001). Protoporphyrinogen IX is exported from the stroma to the envelope, where it is oxidized to protoporphyrin IX. AtABC1 mediates the reimport of protoporphyrin IX into the stroma, where chlorophyll biosynthesis proceeds. In the *laf6* mutant, which lacks AtABC1, protoporphyrin IX accumulates in the envelope and leaks out into the cytosol and is therefore not accessible to downstream enzymes of the chlorophyll biosynthetic pathway. The *laf6* mutant was originally isolated in a screen for *Arabidopsis* mutants with defects in developmental responses to light (photomorphogenesis; Moller et al., 2001). This demonstrates that (1) the influence of the chloroplast extends far beyond the regulation of nuclear genes encoding products for its direct use and (2) a chloroplast protein of cyanobacterial origin is involved in the transduction of photomorphogenetic signals.

Interactions between leaf morphogenesis and plastid signal transduction pathways have been addressed in numerous studies. Pigment mutants in which leaf morphogenesis is affected include the mutants *cla1* (Mandel et al., 1996; Estevez et al., 2000), *cue1* (Li et al., 1995; Streatfield et al., 1999) and *immutans* (Wu et al., 1999, Carol et al., 1999) from *Arabidopsis*; *pac1* (Reiter et al., 1994) and *dag* (Chatterjee et al., 1996) from *Antirrhinum*; and *ghost* (Josse et al., 2000) and *dcl* (Keddie et al., 1996) from tomato. The genes affected in these mutants code for plastid proteins that are required for chloroplast biogenesis. Because the mutations also affect cell differentiation, it was suggested that the developmental or metabolic status of the plastid plays a role in regulating cell differentiation and leaf morphogenesis (Rodermel, 2001). The functions of the encoded proteins are as various as their evolutionary origin might be.

11.6.3 Development

A very important contribution of the cyanobacterial progenitor of chloroplasts to developmental processes in plants was the transfer to the nucleus of genes for members of the two-component receptor family, which are related to the bacterial two-component histidine kinase receptors; no similar transfer occurred in the animal kingdom (Meyerowitz, 2002). The proteins of this family are used in plants for at least four different functions (Urao et al., 2000; Hwang et al., 2002): (1) perception of red and far-red light, (2) recognition and responses to the plant hormones ethylene and (3) cytokinin and (4) response to changes in osmotic pressure. The cyanobacterial phytochrome is a light-regulated histidine kinase (Hughes et al., 1997; Lamparter et al., 1997; Yeh et al., 1997) that mediates red–far-red reversible phosphorylation of a small response regulator, Rcp1 (response regulator for cyanobacterial phytochrome). This implies that protein phosphorylation–dephosphorylation is the initial step in light signal transduction by phytochrome (Yeh et al., 1997). In *Arabi-*

dopsis, five phytochrome photoreceptors exist: PHYA, PHYB, PHYC, PHYD and PHYE (Sharrock and Quail, 1989; Clack et al., 1994); all are soluble proteins with structural features similar to those of sensor histidine kinases, with an N-terminal sensor and a C-terminal histidine protein kinase domain. Analysis of *phyb* mutants indicates that the histidine-kinase-related domain is important for PHYB signaling, but removal of this domain does not eliminate PHYB activity (Krall and Reed, 2000). In addition to their direct interactions with transcription factors (Smith, 2000), plant phytochromes might express Ser/Thr, not His, kinase activity in response to light (McMichael and Lagarias, 1990). Oat phytochrome can be autophosphorylated on Ser/Thr residues in a light-dependent manner (Yeh and Lagarias, 1998), indicating that plant phytochromes have diverged from an ancestral histidine protein kinase to become Ser/Thr kinases with a new activity.

Moreover, five genes coding for ethylene receptors exist in the *Arabidopsis* genome. Loss-of-function mutations in these genes lead to a constitutive ethylene response (Hua and Meyerowitz, 1998). In the absence of ethylene, the receptors repress the response to the hormone, which includes effects on seedling growth and on senescence. When bound to ethylene, the receptors are inactivated and the ethylene response ensues. All five gene products resemble bacterial two-component receptors, and at least one of the encoded proteins, ETR1, has been shown *in vitro* to be a histidine kinase (Gamble et al., 1998). In databases, the only domains related to the N-terminal ethylene-binding domain of the ethylene receptors are other putative plant ethylene receptors and proteins encoded in cyanobacterial genomes (Rodriguez et al., 1999). The ethylene receptors function, at least in part, by interacting with a plant Raf (mitogen-activated protein kinase kinase kinase) homologue, a typical eukaryotic signal transduction protein (Clark et al., 1998). Thus, these receptors appear to be a legacy from the cyanobacterium, and have since evolved to form the upstream portion of a typically eukaryotic signal transduction pathway.

A third type of histidine protein kinase receptors can sense the plant hormone cytokinin. Cytokinins are perceived by multiple histidine protein kinases at the plasma membrane. On reception of the cytokinin signal, histidine protein kinases initiate a signaling cascade via a phosphorelay, which results in the translocation to the nucleus of histidine-containing phosphotransferase proteins (Hwang and Sheen, 2001). The fourth type of receptor histidine protein kinases, AtHK1, has been implicated in osmosensing, but the elucidation of its precise function requires further investigation (Urao et al., 1999).

11.7 Conclusions

Plastid-to-nucleus gene transfer played an important role in the early evolution of plants. This transfer of genes from plastids (and mitochondria) to the nucleus is still active, as is evident from the presence of inactive copies of plastid genes (17 insertions totalling 11 kb; The *Arabidopsis* Genome Initiative, 2000), as well as a 620-kb segment comprising more than 75% of the mitochondrial genome, in the nuclear genome of *Arabidopsis* (Stupar et al., 2001). In the rice Chromosome 10, a contiguous 33-kb segment of chloroplast DNA was found (Yuan et al., 2002), indicating that organelle-to-nucleus transfers occur as bulk DNA integration (Henze and Martin, 2001). In the case of the ca. 40 plastid genes that are still present in all contemporary chloroplasts, positive selection for transcription and translation in the organelle seems to account for their failure to be successfully incorporated into the nuclear genome (see above).

The genetic systems of plastids and nuclei are very different. Genome and gene organization in plastids is, generally speaking, prokaryotic. When a plastid gene is transferred to the nucleus, it moves from a compact genetic apparatus organized into operons and almost without introns into a much larger setting in which genes are organized in introns and exons

(Herrmann, 1997). Before a plastid gene coding for a product required for plastid fitness can be lost from the plastid genome, it must first be successfully established in the nucleus; that is, it must acquire both appropriate expression and targeting signals. Apparently more than 1000 genes of cyanobacterial origin in the *Arabidopsis* genome that code for chloroplast proteins have undergone this metamorphosis. However, an even larger number of chloroplast proteins do not derive from the endosymbiont and have arisen either by the replacement of protochloroplast functions by host gene products or the establishment of novel functions in the organelle during transformation of the protochloroplast into an organelle.

It seems plausible that the successful achievement of stable expression, rather than the acquisition of proper targeting signals, is rate limiting for the integration of chloroplast genes into the nuclear chromosomes (Martin and Herrmann, 1998). On this assumption, the first step in endosymbiotic gene transfer would be chromosomal integration and nuclear expression; the acquisition of a viable organelle-targeting signal by mechanisms such as duplication, recombination (Kadowaki et al., 1996) or exon shuffling (Long et al., 1996) probably occurred later. The fate of a transferred cyanobacterial gene that is stably expressed but lacks a targeting signal depends on its usefulness for the plant cell. Genes that are no longer required for the functioning of the chloroplast organelle and do not contribute useful extra-plastid functions are likely to be lost from both genomes. In this context mitochondrial, or chloroplast genes, or both, that have replaced their archaebacterial orthologues in plants (e.g., the glycolytic pathway, see above) are known, indicating that eubacterial enzymes were in these cases more active than their archaebacterial pendants. When a gene is necessary for chloroplast function, it will establish itself in the nuclear genome after acquiring a chloroplast-targeting signal; if its product also serves the plant cell outside chloroplasts, it can exist in several versions, one with a chloroplast-targeting signal and another without. Additional well-documented examples for postendosymbiotic replacement and metamorphoses of genes of endosymbiotic origin come from secondary red algae (cryptomonads, heterokonts, dinoflagellates). Here a duplicated isoform (called GapC-I) of the host cell NAD-specific GapC enzyme of cytosolic glycolysis, which is of mitochondrial origin (Henze et al., 1995), was retargeted to the chloroplast and transformed into an anabolic Calvin-cycle enzyme displaying NADPH activity (Liaud et al., 1997; Fagan et al., 1998). Moreover, in heterokonts (diatoms and oomycetes), an additional glycolytic GapC-III isoform is targeted to the mitochondria, which is encoded and expressed as a TPI–GAPDH fusion protein forming a bifunctional tetrameric mitochondrial enzyme complex of 250 kDa (Liaud et al., 2000).

The large fraction of cyanobacterial genes in the nuclear genome that code for proteins not redirected to plastids indicates that most genes provided by the cyanobacterial endosymbiont have acquired a role outside chloroplasts. This is a strong support for the proposal of Martin and Herrmann (1998) that successful expression of transferred protochloroplast genes precedes the acquisition of sequences that target the gene product to the chloroplast or other compartments. In this sense, plastid-to-nucleus gene transfer not only reveals the redirection of preexisting gene products to the chloroplast but also uncovers a fascinating facet of eukaryotic evolution, i.e., that any endogenous or introduced gene can be tested and selected for its usefulness by redirecting it to the different compartments of the cell. In particular, during early chloroplast evolution before the machinery for the import of cytoplasmic proteins was established, genes of cyanobacterial descent could fix only when they increased the overall fitness of the cell, contributing significantly to the redirection of cyanobacterial gene products to other compartments. The consequence is that plant cells changed dramatically after the uptake of the cyanobacterial endosymbiont by (1) shaping a new organelle that in many, if not most, functions differs from its cyanobacterial progenitor, and (2) recruiting prokaryotic proteins to perform eukaryotic functions outside of chloroplasts.

Acknowledgments

We thank Francesco Salamini and Paul Hardy for their critical reading of the manuscript. This work was supported by the European Community's Human Potential Program (contract no. HPRN-CT-2002-00248 [PSI-CO]), and by a Heisenberg Fellowship provided by the Deutsche Forschungsgemeinschaft (DL 1265-8).

References

Abdallah, F., Salamini, F. and Leister, D. (2000) A prediction of the size and evolutionary origin of the proteome of chloroplasts of *Arabidopsis*. *Trends Plant Sci.* 5: 141–142.

Allen, J. F. (1993) Control of gene expression by redox potential and the requirement for chloroplast and mitochondrial genomes. *J. Theor. Biol.* 165: 609–631.

Bernacchia, G., Schwall, G., Lottspeich, F., Salamini, F. and Bartels, D. (1995) The transketolase gene family of the resurrection plant *Craterostigma plantagineum*: Differential expression during the rehydration phase. *EMBO J.* 14: 610–618.

Blanchard, J. L. and Lynch, M. (2000) Organellar genes: Why do they end up in the nucleus? *Trends Genet.* 16: 315–320.

Brinkmann, H. and Martin, W. (1996) Higher-plant chloroplast and cytosolic 3-phosphoglycerate kinases: A case of endosymbiotic gene replacement. *Plant Mol. Biol.* 30: 65–75.

Calvin, M. (1956) The photosynthetic carbon cycle. *J. Chem. Soc.* 78: 1895–1915.

Carol, P., Stevenson, D., Bisanz, C., Breitenbach, J., Sandmann, G., Mache, R., Coupland, G. and Kuntz, M. (1999) Mutations in the *Arabidopsis* gene *IMMUTANS* cause a variegated phenotype by inactivating a chloroplast terminal oxidase associated with phytoene desaturation. *Plant Cell* 11: 57–68.

Cavalier-Smith, T. (1987) The simultaneous symbiotic origin of mitochondria, chloroplasts, and microbodies. *Ann. N. Y. Acad. Sci.* 503: 55–71.

Chatterjee, M., Sparvoli, S., Edmunds, C., Garosi, P., Findlay, K. and Martin, C (1996) DAG, a gene required for chloroplast differentiation and palisade development in *Antirrhinum majus*. *EMBO J.* 15: 4194–4207.

Clack, T., Mathews, S. and Sharrock, R. A. (1994) The phytochrome apoprotein family in *Arabidopsis* is encoded by five genes: The sequences and expression of *PHYD* and *PHYE*. *Plant Mol. Biol.* 25: 413–427.

Clark, K. L., Larsen, P. B., Wang, X. and Chang, C. (1998) Association of the *Arabidopsis* CTR1 Raf-like kinase with the ETR1 and ERS ethylene receptors. *Proc. Natl. Acad. Sci. USA* 95: 5401–5406.

Davis, S. J., Kurepa, J. and Vierstra, R. D. (1999) The *Arabidopsis thaliana HY1* locus, required for phytochrome-chromophore biosynthesis, encodes a protein related to heme oxygenases. *Proc. Natl. Acad. Sci. USA* 96: 6541–6546.

Delwiche, C. F. and Palmer, J. D. (1996) Rampant horizontal transfer and duplication of Rubisco genes in eubacteria and plastids. *Mol. Biol. Evol.* 13: 873–882.

Doolittle, W. F. (1998) You are what you eat: A gene transfer ratchet could account for bacterial genes in eukaryotic nuclear genomes. *Trends Genet.* 14: 307–311.

Emanuelsson, O., Nielsen, H., Brunak, S. and von Heijne, G. (2000) Predicting subcellular localization of proteins based on their N-terminal amino acid sequence. *J. Mol. Biol.* 300: 1005–1016.

Estevez, J. M., Cantero, A., Romero, C., Kawaide, H., Jimenez, L. F., Kuzuyama, T., Seto, H., Kamiya, Y. and Leon, P. (2000) Analysis of the expression of *CLA1*, a gene that encodes the 1- deoxyxylulose 5-phosphate synthase of the 2-C-methyl-D-erythritol-4- phosphate pathway in *Arabidopsis*. *Plant Physiol.* 124: 95–104.

Fagan, T., Hastings, J. W. and Morse, D. (1998) The phylogeny of glyceraldehyde-3-phosphate dehydrogenase indicates lateral gene transfer from cryptomonads to dinoflagellates. *J. Mol. Evol.* 47: 633–639.

Figge R. M., Schubert, M., Brinkmann, H. and Cerff, R. (1999) Glyceraldehyde-3-phosphate dehydrogenase gene diversity in eubacteria and eukaryotes: Evidence for intra- and inter-kingdom gene transfer. *Mol. Biol. Evol.* 16: 429–440.

Figge, R. M. and Cerff, R. (2001) GAPDH gene diversity in spirochetes: A paradigm for genetic promiscuity. *Mol. Biol. Evol.* 18: 2240–2249.

Flügge, U. I. (1998) Metabolite transporters in plastids. *Curr. Opin. Plant Biol.* 1: 201–206.

Flügge, U. I. and Heldt, H. W. (1991) Metabolite translocators of the chloroplast envelope. *Annu. Rev. Plant Physiol. Plant Mol. Biol.* 24: 129–144.

Gamble, R. L., Coonfield, M. L. and Schaller, G. E. (1998) Histidine kinase activity of the ETR1 ethylene receptor from *Arabidopsis*. *Proc. Natl. Acad. Sci. USA* 95: 7825–7829.

Goffeau, A., Barrell, B. G., Bussey, H., Davis, R. W., Dujon, B., Feldmann, H., Galibert, F., Hoheisel, J. D., Jacq, C., Johnston, M., Louis, E. J., Mewes, H. W., Murakami, Y., Philippsen, P., Tettelin, H. and Oliver, S. G. (1996) Life with 6000 genes. *Science* 274: 546–567.

Gray, M. W., Lang, B. F., Cedergren, R., Golding, G. B., Lemieux, C., Sankoff, D., Turmel, M., Brossard, N., Delage, E., Littlejohn, T. G., Plante, I., Rioux, P., Saint-Louis, D., Zhu, Y. and Burger, G. (1998) Genome structure and gene content in protist mitochondrial DNAs. *Nucleic Acids Res.* 26: 865–878.

Grossman, A. R., Bhaya, D., Apt, K. E. and Kehoe, D. M. (1995) Light-harvesting complexes in oxygenic photosynthesis: Diversity, control, and evolution. *Annu. Rev. Genet.* 29: 231–288.

Haldrup, A., Naver, H. and Scheller, H. V. (1999) The interaction between plastocyanin and photosystem I is inefficient in transgenic *Arabidopsis* plants lacking the PSI-N subunit of photosystem I. *Plant J.* 17: 689–698.

Henze, K., Badr, A., Wettern, M., Cerff, R. and Martin, W. (1995) A nuclear gene of eubacterial origin in *Euglena gracilis* reflects cryptic endosymbioses during protist evolution. *Proc. Natl. Acad. Sci. USA* 92: 9122–9126.

Henze, K. and Martin, W. (2001) How do mitochondrial genes get into the nucleus? *Trends Genet.* 17: 383–387.

Herrmann, R. G. (1997) *Eukaryotism: Towards a New Interpretion*, Springer Verlag, Heidelberg.

Hippler, M., Drepper, F., Rochaix, J. D. and Muhlenhoff, U. (1999) Insertion of the N-terminal part of PsaF from *Chlamydomonas reinhardtii* into photosystem I from *Synechococcus elongatus* enables efficient binding of algal plastocyanin and cytochrome c_6. *J. Biol. Chem.* 274: 4180–4188.

Hua, J. and Meyerowitz, E. M. (1998) Ethylene responses are negatively regulated by a receptor gene family in *Arabidopsis thaliana*. *Cell* 94: 261–271.

Hughes, J., Lamparter, T., Mittmann, F., Hartmann, E., Gartner, W., Wilde, A. and Börner, T. (1997) A prokaryotic phytochrome. *Nature* 386: 663.

Hwang, I., Chen, H. C. and Sheen, J. (2002) Two-component signal transduction pathways in *Arabidopsis*. *Plant Physiol.* 129: 500–515.

Hwang, I. and Sheen, J. (2001) Two-component circuitry in *Arabidopsis* cytokinin signal transduction. *Nature* 413: 383–389.

Jarvis, P. and Soll, J. (2001) Toc, Tic, and chloroplast protein import. *Biochim. Biophys. Acta* 1541: 64–79.

Jensen, P. E., Rosgaard, L., Knoetzel, J. and Scheller, H. V. (2002) Photosystem I activity is increased in the absence of the PSI-G subunit. *J. Biol. Chem.* 277: 2798–2803.

Josse, E. M., Simkin, A. J., Gaffe, J., Laboure, A. M., Kuntz, M. and Carol, P. (2000) A plastid terminal oxidase associated with carotenoid desaturation during chromoplast differentiation. *Plant Physiol.* 123: 1427–1436.

Kadowaki, K., Kubo, N., Ozawa, K. and Hirai, A. (1996) Targeting presequence acquisition after mitochondrial gene transfer to the nucleus occurs by duplication of existing targeting signals. *EMBO J.* 15: 6652–6661.

Kaneko, T., Sato, S., Kotani, H., Tanaka, A., Asamizu, E., Nakamura, Y., Miyajima, N., Hirosawa, M., Sugiura, M., Sasamoto, S. et al. (1996a) Sequence analysis of the genome of the unicellular cyanobacterium *Synechocystis* sp. strain PCC6803. II. Sequence determination of the entire genome and assignment of potential protein-coding regions. *DNA Res.* 3: 109–136.

Kaneko, T., Sato, S., Kotani, H., Tanaka, A., Asamizu, E., Nakamura, Y., Miyajima, N., Hirosawa, M., Sugiura, M., Sasamoto, S. et al. (1996b) Sequence analysis of the genome of the unicellular cyanobacterium *Synechocystis* sp. strain PCC6803. II. Sequence determination of the entire genome and assignment of potential protein-coding regions (supplement). *DNA Res.* 3: 185–209.

Kanevski, I. and Maliga, P. (1994) Relocation of the plastid *rbcL* gene to the nucleus yields functional ribulose-1,5-bisphosphate carboxylase in tobacco chloroplasts. *Proc. Natl. Acad. Sci. USA* 91: 1969–1973.

Keddie, J. S., Carroll, B., Jones, J. D. and Gruissem, W. (1996) The *DCL* gene of tomato is required for chloroplast development and palisade cell morphogenesis in leaves. *EMBO J.* 15: 4208–4217.

Krall, L. and Reed, J. W. (2000) The histidine kinase-related domain participates in phytochrome B function but is dispensable. *Proc. Natl. Acad. Sci. USA* 97: 8169–8174.

Kropat, J., Oster, U., Rudiger, W. and Beck, C. F. (1997) Chlorophyll precursors are signals of chloroplast origin involved in light induction of nuclear heat-shock genes. *Proc. Natl. Acad. Sci. USA* 94: 14168–14172.

Lamparter, T., Mittmann, F., Gartner, W., Börner, T., Hartmann, E. and Hughes, J. (1997) Characterization of recombinant phytochrome from the cyanobacterium *Synechocystis*. *Proc. Natl. Acad. Sci. USA* 94: 11792–11797.

La Rocca, N., Rascio, N., Oster, U. and Rüdiger, W. (2001) Amitrole treatment of etiolated barley seedlings leads to deregulation of tetrapyrrole synthesis and to reduced expression of *Lhc* and *RbcS* genes. *Planta* 213: 101–108.

Leister, D. (2003) Chloroplast research in the genomic age. *Trends Genet.* 19: 47–56.

Lemieux, C., Otis, C. and Turmel, M. (2000) Ancestral chloroplast genome in *Mesostigma viride* reveals an early branch of green plant evolution. *Nature* 403: 649–652.

Li, H., Culligan, K., Dixon, R. A. and Chory, J. (1995) *CUE1*: A mesophyll cell-specific positive regulator of light-controlled gene expression in *Arabidopsis*. *Plant Cell* 7: 1599–1610.

Liaud, M. -F., Brandt, U., Scherzinger, M. M. and Cerff, R. (1997) Evolutionary origin of cryptomonad microalgae: Two novel chloroplast/cytosol specific GAPDH genes as potential markers of ancestral endosymbiont and host cell components. *J. Mol. Evol.* 44: S28–S37.

Liaud M.-F., Lichtlé, C., Apt, K., Martin, W. and Cerff, R. (2000) Compartment-specific isoforms of TPI and GAPDH are imported into diatom mitochondria as a fusion protein: Evidence in favour of a mitochondrial origin of the eukaryotic glycolytic pathway. *Mol. Biol. Evol.* 17: 213–223.

Long, M., de Souza, S. J., Rosenberg, C. and Gilbert, W. (1996) Exon shuffling and the origin of the mitochondrial targeting function in plant cytochrome c_1 precursor. *Proc. Natl. Acad. Sci. USA* 93: 7727–7731.

Mandel, M. A., Feldmann, K. A., Herrera-Estrella, L., Rocha-Sosa, M. and Leon, P. (1996) *CLA1*, a novel gene required for chloroplast development, is highly conserved in evolution. *Plant J.* 9: 649–658.

Martin, W., Brinkmann, H., Savona, C. and Cerff, R. (1993) Evidence for a chimaeric nature of nuclear genomes: Eubacterial origin of eukaryotic glyceraldehyde-3-phosphate dehydrogenases. *Proc. Natl. Acad. Sci. USA* 90: 8692–8696.

Martin, W. and Herrmann, R. G. (1998) Gene transfer from organelles to the nucleus: How much, what happens, and why? *Plant Physiol.* 118: 9–17.

Martin, W. and Müller, M. (1998) The hydrogen hypothesis for the first eukaryote. *Nature* 392: 37–41.
Martin, W., Rujan, T., Richly, E., Hansen, A., Cornelsen, S., Lins, T., Leister, D., Stoebe, B., Hasegawa, M. and Penny, D. (2002) Evolutionary analysis of *Arabidopsis*, cyanobacterial, and chloroplast genomes reveals plastid phylogeny and thousands of cyanobacterial genes in the nucleus. *Proc. Natl. Acad. Sci. USA* 99: 12246–12251.
Martin, W. and Schnarrenberger, C. (1997) The evolution of the Calvin cycle from prokaryotic to eukaryotic chromosomes: A case study of functional redundancy in ancient pathways through endosymbiosis. *Curr. Genet.* 32: 1–18.
Martin, W., Stoebe, B., Goremykin, V., Hapsmann, S., Hasegawa, M. and Kowallik, K. V. (1998) Gene transfer to the nucleus and the evolution of chloroplasts. *Nature* 393: 162–165.
McMichael, R. W., Jr. and Lagarias, J. C. (1990) Phosphopeptide mapping of *Avena* phytochrome phosphorylated by protein kinases in vitro. *Biochemistry* 29: 3872–3878.
Meskauskiene, R., Nater, M., Goslings, D., Kessler, F., op den Camp, R. and Apel, K. (2001) FLU: A negative regulator of chlorophyll biosynthesis in *Arabidopsis thaliana*. *Proc. Natl. Acad. Sci. USA* 98: 12826–12831.
Meyerowitz, E. M. (2002) Plants compared to animals: The broadest comparative study of development. *Science* 295: 1482–1485.
Mochizuki, N., Brusslan, J. A., Larkin, R., Nagatani, A. and Chory, J. (2001) *Arabidopsis genomes uncoupled 5* (*GUN5*) mutant reveals the involvement of Mg-chelatase H subunit in plastid-to-nucleus signal transduction. *Proc. Natl. Acad. Sci. USA* 98: 2053–2058.
Moller, S. G., Kunkel, T. and Chua, N. H. (2001) A plastidic ABC protein involved in intercompartmental communication of light signaling. *Genes Dev.* 15: 90–103.
Montane, M. H. and Kloppstech, K. (2000) The family of light-harvesting-related proteins (LHCs, ELIPs, HLIPs): Was the harvesting of light their primary function? *Gene* 258: 1–8.
Muramoto, T., Kohchi, T., Yokota, A., Hwang, I. and Goodman, H. M. (1999) The *Arabidopsis* photomorphogenic mutant *hy1* is deficient in phytochrome chromophore biosynthesis as a result of a mutation in a plastid heme oxygenase. *Plant Cell* 11: 335–348.
Naver, H., Haldrup, A. and Scheller, H. V. (1999) Cosuppression of photosystem I subunit PSI-H in *Arabidopsis thaliana*. Efficient electron transfer and stability of photosystem I is dependent upon the PSI-H subunit. *J. Biol. Chem.* 274: 10784–10789.
Oelmüller, R. (1989) Photooxidative destruction of chloroplasts and its effect on nuclear gene expression and extraplastidic enzyme levels. Photochem. Photobiol. 49: 229–239.
Osteryoung, K. W. and McAndrew, R. S. (2001) The plastid division machine. *Annu. Rev. Plant Physiol. Plant Mol. Biol.* 52: 315–333.
Palmer, J. D. (1997) Organelle genomes: Going, going, gone! *Science* 275: 790–791.
Pesaresi, P., Varotto, C., Richly, E., Kurth, J., Salamini, F. and Leister, D. (2001) Functional genomics of *Arabidopsis* photosynthesis. *Plant Physiol. Biochem.* 39: 285–294.
Plaxton, W. C. (1996) The organization and regulation of plant glycolysis. *Annu. Rev. Plant Physiol. Plant Mol. Biol.* 47: 185–214.
Race, H. L., Herrmann, R. G. and Martin, W. (1999) Why have organelles retained genomes? *Trends Genet.* 15: 364–370.
Reiter, R. S., Coomber, S. A., Bourett, T. M., Bartley, G. E. and Scolnik, P. A. (1994) Control of leaf and chloroplast development by the *Arabidopsis* gene *pale cress*. *Plant Cell* 6: 1253–1264.
Robinson, C., Thompson, S. J. and Woolhead, C. (2001) Multiple pathways used for the targeting of thylakoid proteins in chloroplasts. *Traffic* 2: 245–251.
Rodermel, S. (2001) Pathways of plastid-to-nucleus signaling. *Trends Plant Sci.* 6: 471–478.
Rodriguez, F. I., Esch, J. J., Hall, A. E., Binder, B. M., Schaller, G. E. and Bleecker, A. B. (1999) A copper cofactor for the ethylene receptor ETR1 from *Arabidopsis*. *Science* 283: 996–998.
Rujan, T. and Martin W. (2001) How many genes in *Arabidopsis* come from cyanobacteria? An estimate from 386 protein phylogenies. *Trends Genet.* 17: 113–120.

Scheller, H. V., Jensen, P. E., Haldrup, A., Lunde, C. and Knoetzel, J. (2001) Role of subunits in eukaryotic photosystem I. *Biochim. Biophys. Acta* 1507: 41–60.

Schwacke, R., Schneider, A., van der Graaff, E., Fischer, K., Desimone, M., Catoni, E., Frommer, W., Flügge, U. I. and Kuntze R. (2002) ARAMEMNON: A new *Arabidopsis thaliana* membrane protein database. *Plant Physiol.* 131: 16–26.

Sharrock, R. A. and Quail, P. H. (1989) Novel phytochrome sequences in *Arabidopsis thaliana*: Structure, evolution, and differential expression of a plant regulatory photoreceptor family. *Genes Dev.* 3: 1745–1757.

Smith, H. (2000) Phytochromes and light signal perception by plants: An emerging synthesis. *Nature* 407: 585–591.

Soll, J., Bolter, B., Wagner, R. and Hinnah, S. C. (2000)...response : The chloroplast outer envelope: a molecular sieve? *Trends Plant Sci.* 5: 137–138.

Stoebe, B., Martin, W. and Kowallik K. V. (1998) Distribution and nomenclature of protein-coding genes in 12 sequenced chloroplast genomes. *Plant Mol. Biol. Rep.* 16: 243–255.

Strand, A., Asami, T., Alonso, J., Ecker, J. R. and Chory, J. (2003) Chloroplast to nucleus communication triggered by Mg-protoporphyrinIX accumulation. *Nature* 421: 79–83.

Streatfield, S. J., Weber, A., Kinsman, E. A., Hausler, R. E., Li, J., Post-Beittenmiller, D., Kaiser, W. M., Pyke, K. A., Flugge, U. I. and Chory J. (1999) The phosphoenolpyruvate/phosphate translocator is required for phenolic metabolism, palisade cell development, and plastid-dependent nuclear gene expression. *Plant Cell* 11: 1609–1622.

Stupar, R. M., Lilly, J. W., Town, C. D., Cheng, Z., Kaul, S., Buell, C. R. and Jiang, J. (2001) Complex mtDNA constitutes an approximate 620-kb insertion on *Arabidopsis thaliana* chromosome 2: Implication of potential sequencing errors caused by large-unit repeats. *Proc. Natl. Acad. Sci. USA* 98: 5099–5103.

Susek, R. E. and Chory, J. (1992) A tale of two genomes: Role of a chloroplast signal in coordinating nuclear and plastid genome expression. *Aust. J. Plant Physiol.* 19: 387–399.

Suzuki, J. Y. and Bauer, C. E. (1995) A prokaryotic origin for light-dependent chlorophyll biosynthesis of plants. *Proc. Natl. Acad. Sci. USA* 92: 3749–3753.

The *Arabidopsis* Genome Initiative (2000) Analysis of the genome sequence of the flowering plant *Arabidopsis thaliana. Nature* 408: 796–815.

Urao, T., Yakubov, B., Satoh, R., Yamaguchi-Shinozaki, K., Seki, M., Hirayama, T. and Shinozaki, K. (1999) A transmembrane hybrid-type histidine kinase in *Arabidopsis* functions as an osmosensor. *Plant Cell* 11: 1743–1754.

Urao, T., Yamaguchi-Shinozaki, K. and Shinozaki, K. (2000) Two-component systems in plant signal transduction. *Trends Plant Sci.* 5: 67–74.

Varotto, C., Pesaresi, P., Jahns, P., Lessnick, A., Tizzano, M., Schiavon, F., Salamini, F. and Leister, D. (2002) Single and double knockouts of the genes for photosystem I subunits G, K, and H of *Arabidopsis*: Effects on photosystem I composition, photosynthetic electron flow, and state transitions. *Plant Physiol.* 129: 616–624.

Vothknecht, U. C. and Westhoff, P. (2001) Biogenesis and origin of thylakoid membranes. *Biochim. Biophys. Acta* 1541: 91–101.

Wilde, A., Fiedler, B. and Börner, T. (2002) The cyanobacterial phytochrome Cph2 inhibits phototaxis towards blue light. *Mol. Microbiol.* 44: 981–988.

Wilson, R. J. and Williamson, D. H. (1997) Extrachromosomal DNA in the Apicomplexa. Microbiol. *Mol. Biol. Rev.* 61: 1–16.

Wittmann, H. G. (1982) Components of bacterial ribosomes. *Annu. Rev. Biochem.* 51: 155–183.

Wu, D., Wright, D. A., Wetzel, C., Voytas, D. F. and Rodermel, S. (1999) The *IMMUTANS* variegation locus of *Arabidopsis* defines a mitochondrial alternative oxidase homolog that functions during early chloroplast biogenesis. *Plant Cell* 11: 43–55.

Yamaguchi, K. and Subramanian, A. R. (2000) The plastid ribosomal proteins: Identification of all the proteins in the 50 S subunit of an organelle ribosome (chloroplast). *J. Biol. Chem.* 275: 28466–28482.

Yamaguchi, K., von Knoblauch, K. and Subramanian, A. R. (2000) The plastid ribosomal proteins. Identification of all the proteins in the 30 S subunit of an organelle ribosome (chloroplast) *J. Biol. Chem.* 275: 28455–28465.
Yeh, K. C. and Lagarias, J. C. (1998) Eukaryotic phytochromes: Light-regulated serine/threonine protein kinases with histidine kinase ancestry. *Proc. Natl. Acad. Sci. USA* 95: 13976–13981.
Yeh, K. C., Wu, S. H., Murphy, J. T. and Lagarias, J. C. (1997) A cyanobacterial phytochrome two-component light sensory system. *Science* 277: 1505–1508.
Yuan, Q., Hill, J., Hsiao, J., Moffat, K., Ouyang, S., Cheng, Z., Jiang, J. and Buell, C. R. (2002) Genome sequencing of a 239-kb region of rice chromosome 10L reveals a high frequency of gene duplication and a large chloroplast DNA insertion. *Mol. Genet. Genom.* 267: 713–720.
Zerges, W. (2002) Does complexity constrain organelle evolution? *Trends Plant Sci.* 7: 175–182.

Section III

Evolutionary Cell Biology and Epigenetics

Chapter 12

Protein Translocation Machinery in Chloroplasts and Mitochondria: Structure, Function and Evolution

Hrvoje Fulgosi, Jürgen Soll and Masami Inaba-Sulpice

CONTENTS

Abstract

Mitochondria and chloroplasts originated from free-living bacteria as a result of endosymbiosis. After the endosymbiotic event, most of the prokaryotic genes were lost or transferred to the host nucleus (see Chapter 11). The nuclear-encoded organellar proteins are synthesized as precursors with an amino-terminal cleavable presequence, which carries targeting and sorting information. The outer and inner membranes of mitochondria contain the TOM and TIM complex, respectively, which are responsible for selective protein translocation across the membranes. The outer and inner membranes of chloroplasts contain functional equivalents, the TOC and TIC complexes, respectively. The main subunits of these complexes have been identified and characterized *in vitro* and *in vivo*. Some subunits of the TOC and

0-415-29904-7/04/$0.00+$1.50

TIC complexes have homologues in cyanobacteria, suggesting that the chloroplast protein import machinery might have been built around a prokaryotic core. In contrast, so far there is no report on bacterial proteins that share significant homology to the subunits of the TOM and TIM complexes. Translocation into or across the thylakoid membranes of chloroplasts is catalyzed by the Sec, the signal recognition particle (SRP), the pH-dependent or Tat and the YidC/Oxa1p/Alb3 pathways. Translocation of proteins into or across the inner membrane of mitochondria also involves SRP components and Oxa1p. Phylogenetic evidence suggests that these intraorganellar pathways are derived from preexisting prokaryotic systems for protein export.

12.1 Introduction

Evidence has been presented that mitochondria and chloroplasts originate from free-living α-proteobacteria (Andersson et al., 1998) and cyanobacteria (McFadden, 1999), respectively, as a result of endosymbiosis. They still possess their own genome and perpetuate themselves by division. However, the organellar genome encodes only a minor fraction of their proteins (Unseld et al., 1997; Sato et al., 1999). Most of the endosymbiont genes have been transferred to the nucleus of the host (Martin et al., 1998; Karlberg et al., 2000). This led to the transformation of the endosymbiont into the semiautonomous organelle that is subject to regulation by the nucleus. The establishment of the organelles necessitated the evolution of a machinery to transport the nuclear-encoded, cytoplasmically synthesized proteins into the organelles.

Mitochondria and chloroplasts have complicated membrane structures. They are surrounded by an outer and an inner membrane, which define two spaces, the intermembrane space and the inner space. The latter is termed the matrix for mitochondria and the stroma for chloroplasts. Chloroplasts contain an internal closed membrane system: the thylakoid membrane, which defines an additional space, the lumen. During the development of functional organelles, their proteins, whether nuclear-encoded or organellar-encoded, have to be sorted to their final destination.

Several systems to import and sort proteins to mitochondrial or chloroplast subcompartments have evolved in these organelles. Most of the nuclear-encoded proteins destined for organelles are synthesized as precursors with an amino-terminal cleavable extension: the targeting sequence (also called a presequence for mitochondrial precursor proteins and a transit peptide or transit sequence for chloroplast precursor proteins), which carries targeting and sorting information. Internal and carboxyl-terminal targeting signals have also been described (Diekert et al., 1999; Mitoma and Ito, 1992). These targeting signals are recognized by receptors on the outer membranes of the organelles. These receptors are part of the TOM or TOC (for translocase of outer mitochondrial or chloroplast membrane, respectively) complexes. The TOM and TOC complexes mediate the transport of the precursors across the outer membrane, and the TIM and TIC (for translocase of inner mitochondrial or chloroplast membrane, respectively) complexes mediate the transport across or into the inner membrane. Once inside the organelle, imported proteins can be sorted to their final destinations via different pathways. Translocation of proteins into or across the thylakoid membranes of chloroplasts is mediated by machinery similar to that used by proteins of bacterial plasma membranes. In both these systems, Sec, a Tat/ΔpH-dependent, and signal recognition particle (SRP) pathways are involved. Translocation of proteins into or across the inner membrane of mitochondria also involves a SRP pathway. A gene homologous to SecY has been found in the mitochondrial genome of *Reclinomonas Americana* (Lang et al., 1996). However, no Sec pathway has been characterized in mitochondria to date. A conserved pathway has recently been identified in mitochondria, chloroplasts and eubacteria. This pathway involves the Oxa1 protein for translocation of proteins (Luirink et al., 2001). We summarize the current information on the structure and function of the protein translocases and discuss how they have been established during the course of evolution.

12.2 TOM and TIM Translocases of Mitochondria

Translocation of precursor proteins across or into the mitochondrial membranes is mediated by the TOM and TIM complexes. [For reviews see Bauer, M. F. et al. (2000), Koehler (2000) and Krimmer et al. (2001).] Components of these translocases are systematically named Tom proteins and Tim proteins, with a number referring to the molecular mass of the protein (Pfanner et al., 1996). Several Tom and Tim proteins have been identified and subsequently characterized by biochemical and molecular approaches (Kübrich et al., 1995). Most studies have been carried out with mitochondria from the yeast *Saccharomyces cerevisiae*. However, the identification of several Tom and Tim proteins in plant (Werhahn et al., 2001) and mammalian genomes (Saeki et al., 2000) indicates that the mitochondrial protein import systems are basically conserved among eukaryotic organisms.

12.2.1 TOM Translocase

The mitochondrial outer membrane contains one translocase, the TOM complex. The TOM complex consists of the receptors Tom20 (Söllner et al., 1989), Tom22 (Mayer et al., 1995), Tom37 (Gratzer et al., 1995) and Tom70 (Hines et al., 1990), and a protein-conducting channel comprising Tom40 (Vestweber et al., 1989; Hill et al., 1998), Tom7 (Moczko et al., 1992; Söllner et al., 1992; Alconada et al., 1995), Tom6 (Kassenbrock et al., 1993; Alconada et al., 1995) and Tom5 (Moczko et al., 1992; Söllner et al., 1992; Alconada et al., 1995). The receptors contain tetratricopeptide repeat (TPR) motifs, which are generally thought to mediate dynamic protein–protein interactions (Blatch and Lassle, 1999). Tom20, Tom37 and Tom70 are loosely attached to the core complex formed by the other Tom proteins, namely, Tom22 plus Tom40, Tom7, Tom6 and Tom5 (Dekker et al., 1998). Tom40 is the major constituent of the channel (Hill et al., 1998; Stan et al., 2000; Ahting et al., 2001), whereas Tom7 and Tom6 are suggested to modulate the stability of the TOM complex (Alconada et al., 1995; Hönlinger et al., 1996). Tom5 accepts precursors from the receptors and mediates their insertion into the channel (Dietmeier et al., 1997). It is proposed that Tom20 and Tom22 also play a role in the structural organization of the TOM complex (Model et al., 2002).

Precursors that carry a cleavable presequence, which are mainly hydrophilic proteins, are recognized by the receptor Tom20 in cooperation with Tom22 (Mayer et al., 1995; Brix et al., 1997). Precursors that carry internal targeting signals, typically members of the inner membrane ADP/ATP carrier family (Ryan et al., 1999), are recognized by the receptor Tom70 in cooperation with Tom37 (Gratzer et al., 1995; Brix et al., 1997). Both types of precursor merge at Tom5 and are subsequently transferred to the channel, through which the translocating precursor passes (Dietmeier et al., 1997). The mechanism that drives this protein import is not clear. After passage through the channel, precursors are sorted either directly into the outer membrane, the intermembrane space or the TIM translocases.

12.2.2 TIM Translocases (TIM23 and TIM22)

The mitochondrial inner membrane contains two translocases, TIM23 and TIM22. In general, the TIM23 translocase mediates protein translocation into the matrix, and the TIM22 translocase mediates insertion into the inner membrane (Rassow et al., 1999; Bauer,

M. F. et al., 2000). It has been proposed that the TIM23 translocase might form a contact site with the TOM translocase during translocation (Reichert et al., 2002).

12.2.2.1 TIM23 Translocase

Precursors that carry a cleavable presequence are imported into the matrix by the TIM23 translocase. This translocase contains Tim17 (Ryan et al., 1994; Maarse et al., 1994), Tim23 (Dekker et al., 1993; Emtage and Jensen, 1993) and Tim44 (Maarse et al., 1992; Scherer et al., 1992). Tim17 and Tim23 are components of the protein-conducting channel across the inner membrane (Dekker et al., 1997). Tim23 consists of an amino-terminal hydrophilic domain followed by four membrane-spanning α helices that are connected by short loops. The amino-terminal domain is exposed to the inner membrane space. When reconstituted, Tim23 can form a cation-selective, voltage-sensitive channel that is activated by a membrane potential and a mitochondrial presequence (Truscott et al., 2001). Tim17 shares amino acid sequence similarity with the carboxy-terminal domain of Tim23 (25% identity) and interacts with this domain of Tim23 (Ryan et al., 1998).

The protein translocation through the Tim17/Tim23 channel is catalyzed by the motor complex consisting of Tim44, a mitochondrial heat-shock protein mtHsp70 (Ssc1) and its cochaperone mGrpE (Mge1p). Tim40 is stably associated with the inner membrane but mainly exposed at the matrix side and serves as an adapter protein that recruits, targets the soluble mtHsp70 to the import sites of the inner membrane. mtHsp70 is the ATPase that drives the membrane transport of precursors into the matrix. mGrpE acts as a nucleotides exchange factor for mtHsp70. This translocation is dependent on a membrane potential and generally requires ATP hydrolysis by mtHsp70. The mechanism involved in the translocation process is not fully understood, but comprehensive models for the mechanism are described in several reviews (Rassow et al., 1999; Bauer, M. F. et al., 2000; Strub et al., 2000).

Additional components of the TIM23 translocase have been identified, but their role in the protein import process is not clear (Tokatlidis et al., 1996; Kanamori et al., 1997). One such component, Tim11, has been identified because of its closed association with the Tim17/Tim23 complex (Tokatlidis et al., 1996). Tim11 interacts with the sorting signal of a cytochrome *b*2 (Tokatlidis et al., 1996) and is required for insertion of the mitochondrial NADH-cytochrome reductase *b*5 (Haucke et al., 1997).

12.2.2.2 TIM22 Translocase

Many mitochondrial inner membrane proteins lack a cleavable presequence but instead contain internal targeting signals. Insertion of these proteins into the inner membrane is mediated by the TIM22 translocase. This translocase consists of the integral membrane proteins Tim18 (Koehler et al., 2000), Tim22 (Sirrenberg et al., 1996) and Tim54 (Kerscher et al., 1997), and a group of related proteins, Tim8 (Koehler et al., 1999), Tim9 (Koehler et al., 1998; Adam et al., 1999), Tim10 (Sirrenberg et al., 1998), Tim12 (Sirrenberg et al., 1998) and Tim13 (Koechler et al., 1998) as cooperating proteins in the intermembrane space. Tim22 was identified based on its amino acid sequence similarity to Tim23 and Tim17 (Sirrenberg et al., 1996). Tim22 can form a voltage-activated channel when reconstituted (Kovermann et al., 2002). Tim54 is tightly associated with Tim22, and the Tim54/Tim22 complex catalyzes the protein insertion into the inner membrane in a membrane potential-dependent manner (Kerscher et al., 1997). Tim18 was identified based on its interaction with Tim54 and is suggested to function in the assembly and stabilization of the TIM22 complex but does not directly participate in protein insertion into the inner membrane (Kerscher et al., 2000; Koehler et al., 2000). The small Tim proteins (Tim8, Tim9, Tim10, Tim12 and Tim13) share amino acid sequence similarities with each other (25% identity). Tim9 and Tim10 form a complex in the intermembrane space (Koehler et al., 1998; Adam et al., 1999). The Tim9/Tim10 complex consists of three Tim9 and three Tim10 and binds to the ADP/ATP carrier proteins (Curran et al., 2002). Tim12 also interacts with the carrier proteins but is peripherally bound to the outer surface of the inner membrane (Sirrenberg et al., 1998). Tim8 and Tim13 form a complex in the intermembrane space, and the Tim8/Tim13 complex most likely works in parallel with the Tim9/Tim10 complex (Leuenberger et al., 1999; Paschen et al., 2000).

12.2.3 Origin of TOM and TIM Translocases

The endosymbiotic origin of mitochondria raises the question of whether the TOM and TIM translocases have been derived from preexisting bacterial systems. So far there is no report on bacterial proteins that share significant sequence similarity to Tom and Tim proteins. However, Gabriel et al. (2001) proposed that Tom40 might originate from a bacterial protein, because it is predicted to be composed mainly of β strands, similar to the typical structure of outer membrane porins of Gram-negative bacteria (Stan et al., 2000). The same argument was made for Toc75, the major channel-forming protein of the TOC translocase of chloroplasts (see Section 12.3.3).

Tim22 shares a sequence similarity to Tim17 and Tim23. This suggests that the TIM23 and TIM22 translocases might have evolved by gene-duplication events. A protein that shows a significant similarity to Tim17, Tim22 and Tim23 was also found in the outer envelope membrane of chloroplasts. This protein, OEP16, is present in homodimers and forms a channel that is selective for amino acids (Pohlmeyer et al., 1997). OEP16 has a size similar to that of Tim proteins, and the similarity extends over the entire sequences. Tim17, Tim22, Tim23 and OEP16 contain a conserved sequence motif of ca. 50 residues in their membrane-spanning domains. This region also shows a similarity to a part of LivH, an amino acid permease of the plasma membrane of *Escherichia coli* (Adams et al., 1990). LivH functions in the uptake of branched-chain amino acids in cooperation with other proteins. The similarities in the sequences and functions of the Tim proteins, OEP16 and LivH suggest that the Tim proteins and OEP16 might have been derived from a prokaryotic amino acid transporter (Rassow et al., 1999).

12.3 TOC and TIC Translocases of Chloroplasts

Translocation of proteins across the envelope membranes of chloroplasts is mediated by the TOC and TIC complexes (Jarvis and Soll, 2001). Most chloroplast proteins, except the outer envelope proteins (Schleiff and Klösgen, 2001), are synthesized in the cytosol as precursors that carry an amino-terminal cleavable extension, called the transit peptide. The transit peptide can be phosphorylated in the cytosol by specific protein kinases (Waegemann et al., 1996), and then bound by a so-called guidance complex consisting of a heat-shock protein Hsp70 and 14-3-3 proteins (May and Soll, 2000). The guidance complex appears to stimulate the import: precursors complexed *in vitro* in the guidance complex are more efficiently imported into isolated chloroplasts than are noncomplexed precursors (May and Soll, 2000). Following release from the guidance complex, the precursor is transported across the outer membrane by the TOC complex. It is thought that at the late stage of the transport, the TOC complex comes at a contact site with the TIC complex and the translocating precursor passes across both membranes simultaneously at the contact site (Kouranov et al., 1998). The recognition and translocation of precursor proteins by the TOC and TIC complexes require hydrolysis of ATP and GTP. The transit peptide is removed by a stromal processing peptidase.

So far, four Toc proteins (Toc159, Toc75, Toc64 and Toc34, with a number referring to the molecular mass of the protein) and six Tic proteins (Tic110, Tic62, Tic55, Tic40, Tic22 and Tic20) have been identified from pea chloroplasts mainly by biochemical approaches. Several homologues of the Toc and Tic proteins are found in the genome of *Arabidopsis thaliana*, indicating that the chloroplast import systems are basically conserved among different plant species. The Toc and Tic proteins show no sequence similarity to the components of the TOM and TIM translocases of mitochondria. Therefore, it seems likely that the chloroplast translocases have evolved independently of the mitochondrial equivalents.

12.3.1 TOC Translocase

The Toc complex contains two GTP-binding receptors, Toc159 (originally identified as an 86-kDa proteolytic fragment called Toc86; Kessler et al., 1994; Hirsch et al., 1994; Perry and Keegstra, 1994; Bölter et al., 1998a) and Toc34 (Kessler et al., 1994), a putative receptor Toc64 (Sohrt and Soll, 2000) and a channel-forming protein Toc75 (Schnell et al., 1994; Perry and Keegstra, 1994). These Toc proteins are prominent constituents of the chloroplast outer envelope membrane. They interact with precursors during the import process and associate with each other even in the absence of the precursors.

12.3.1.1 Toc159 and Toc34

Toc159 consists of an amino-terminal acidic domain, a central GTP-binding domain and a carboxy-terminal membrane anchor domain (Chen et al., 2000). Toc34 consists of an amino-terminal GTP-binding domain and a carboxy-terminal membrane anchor domain (Kessler et al., 1994). The major part of these proteins, including the GTP-binding domains, projects into the cytosol (Seedorf et al., 1995; Chen et al., 2000; Schleiff et al., 2002). The GTP-binding domains of Toc159 and Toc34 are homologous and define a unique subgroup of the GTPase superfamily (Seedorf et al., 1995). Toc159 appears to be a major receptor for binding of precursors (Ma et al., 1996; Chen et al., 2000), whereas Toc34 plays a key role in the regulation by GTP of translocation of bound precursors across the outer envelope membrane (Chen et al., 2000; Sveshnikova et al., 2000; Schleiff et al., 2002). Toc159 and Toc34 can be phosphorylated by distinct protein kinases and the phosphorylation inhibits the precursor binding (Fulgosi and Soll, 2002), indicating that they are separately regulated by these kinases. Sun et al. (2002) have presented a crystal structure of the cytosolic part of Toc34 in complex with GDP and Mg^{2+}, which demonstrates that the GDP-bound Toc34 can form a homodimer *in vitro*.

12.3.1.2 *Toc75*

Toc75 is the most abundant protein in the chloroplast outer envelope membrane. The major part of this protein is deeply embedded within the outer envelope membrane, and topological studies suggest that it has a *b* barrel structure comprising 16 transmembrane β sheets (Hinnah et al., 1997; Sveshnikova et al., 2000). When reconstituted in proteoliposomes, Toc75 forms a voltage-sensitive, cation-selective channel (Hinnah et al., 1997). These data altogether indicate that Toc75 is a major constituent of the import channel of the TOC translocase.

12.3.1.3 *Toc64*

Toc64 is an integral membrane protein with a large portion of its carboxy terminus exposed to the cytosol. This protein shares a similarity with amidases, but it has a mutation in a conserved residue and appears to lack amidase activity (Sohrt and Soll, 2000). The carboxy-terminal cytosolic domain of Toc64 contains three TPR motifs, which are also found in the mitochondrial receptors (Tom70, Tom37, Tom22 and Tom20). Preliminary data indicate that the guidance complex interact with Toc64 (Sohrt and Soll, 2000), implying that Toc64 functions as a docking protein.

12.3.2 The TIC Translocase

The Tic proteins identified to date are a channel-forming protein Tic110 (Schnell et al., 1994; Wu et al., 1994), putative redox-sensing regulatory proteins Tic62 (Küchler et al., 2002) and Tic55 (Caliebe et al., 1997), membrane proteins Tic40 (Wu et al., 1994; Stahl et al., 1999) and Tic20 (Ma et al., 1996; Kouranov and Schnell, 1997; Kouranov et al., 1998) and a hydrophilic protein Tic22 (Ma et al., 1996; Kouranov and Schnell, 1997; Kouranov et al., 1998). *In vitro*, these Tic proteins can be cross-linked with precursors trapped at various stages during import in isolated chloroplasts. Except for Tic110, little is known about the roles these Tic proteins play in the import process, and there are some conflicting data on the participation of certain components in the TIC complex.

12.3.2.1 Tic110

Tic110 was the first component of the TIC complex to be identified (Schnell et al., 1994). When reconstituted in proteoliposomes, Tic110 forms a cation-selective, high-conductance channel and interacts specifically with precursors (Heins et al., 2002). Circular dichroism analysis reveals that Tic110 consists mainly of β strands (Heins et al., 2002). These data indicate that Tic110 constitutes a central part of the import channel of the TIC translocase. Tic110 is also implicated in the recruitment of stromal chaperones, such as a Hsp100 homologue, ClpC and a Cpn60 homologue (Kessler and Blobel, 1996; Nielsen et al., 1997). Because Tic110 can be found in a cross-link product with Toc75 and a precursor, it might be involved in the formation of the contact site of the TOC and TIC translocases.

12.3.2.2 *Tic55*

Tic55 can be detected in complexes containing an arrested precursor, Toc proteins (Toc159, Toc34 and Toc75) and Tic110. Tic55 is an integral membrane protein, only partially exposed to the intermembrane space. The predicted amino acid sequence of this protein contains a Rieske-type iron–sulfur cluster and a mononuclear iron-binding site (Caliebe et al., 1997). It has been demonstrated that translocation of a precursor across the inner envelope membrane was inhibited with diethyl pyrocarbonate, a reagent that interferes with histidine residues in the iron–sulfur cluster (Caliebe et al., 1997). This suggests a significance of Tic55 in the import process. Proteins that contain Rieske iron–sulfur clusters (Rieske iron–sulfur proteins) are usually involved in electron transfer chains, such as the cytochrome b_6f complex of the photosynthetic electron transport system. However, some Rieske iron–sulfur proteins function as redox sensors to regulate gene expression or programmed cell death (Hidalgo et al., 1997; Gray et al., 1997). By analogy with these sensors, Tic55 might also play a regulatory role during the import process by responding to changes in redox status within chloroplasts.

12.3.2.3 *Tic62*

Tic62 can be found in complexes containing other Tic proteins and be coimmunoprecipitated with antibodies raised against Tic110 or Tic55 (Küchler et al., 2002). The amino terminus of Tic62 contains a putative membrane anchor domain and a region of amino acid sequence similarities to eukaryotic NADH dehydrogeneases. The carboxy terminus of this protein is exposed to the stroma and contains a repetitive sequence module that interacts with a ferredoxin-NAD(P)+ oxidoreductase (FNR). It has been demonstrated that Tic62 binds NAD+, and the carboxy terminus of Tic62 interacts with FNR (Küchler et al., 2002). Furthermore, a competitor of NAD binding and a substrate that influences the ratio of NAD(P) to NAD(P)H in the stroma affect import of FNR isoforms into chloroplasts differently (Küchler et al., 2002). These results altogether indicate that Tic62, together with Tic55 and the FNR, functions as a redox-sensing regulator of the chloroplast protein import.

12.3.2.4 Tic22 and Tic20

Tic22 and Tic20 were identified by cross-linking experiments with arrested precursors (Ma et al., 1996; Kouranov and Schnell, 1997; Kouranov et al., 1998). These proteins associate with Tic110. Tic22 is a hydrophilic protein with no predicted transmembrane domains. It is peripherally associated with the outer surface of the inner envelope membrane, suggesting that it acts as a receptor for precursors as they emerge from the TOC complex (Kouranov et al., 1998). Tic20 is an integral membrane protein containing four predicted *a* helical transmembrane segments and short amino- and carboxy-terminal hydrophilic domains (Kouranov et al., 1998). Kouranov and Schnell (1997) provided evidence that precursors interact with Tic22 and Tic20 sequentially during the import process. Tic22 and Tic20 can associate with the major Toc proteins and Tic110 to form a TOC–TIC supercomplex that corresponds to the contact site, even in the absence of precursors, but do not associate with one another or with Tic110 in the absence of Toc proteins (Kouranov et al., 1998). This contrasts with Tic55, which was isolated as part of a stable complex containing Tic110 in the absence of Toc proteins (Caliebe et al., 1997). These observations suggest that the TIC complex is a dynamic structure in which the presence of one component can influence the association of others. It has been demonstrated that an *Arabidopsis* homologue of Tic20 plays a role in protein import (Chen et al., 2002; see later).

12.3.2.5 Tic40

Tic40 associates with precursors trapped during import and can be found in cross-linked complexes containing Tic110. Because Tic40 is resistant to proteases and is predicted to have a membrane-spanning region at its amino terminus, it seems likely that a large part of this protein is exposed on the stromal surface of the inner envelope membrane (Stahl et al., 1999). The role played by Tic40 in the import process is not known. Tic40 shares a limited sequence similarity with a protein called Hsp70-interacting protein (Hip) at its carboxy terminus. Hip is a mammalian cochaperone that regulates nucleotide exchange by Hsp70 proteins (Hohfeld et al., 1995). It is therefore tempting to speculate that Tic40 regulates the chaperones responsible for driving the protein import. However, the region of Hip that is similar to Tic40 is outside the Hsp70-interacting domain.

12.3.2.6 Arabidopsis *Homologues of the Toc and Tic Proteins*

A survey of the *Arabidopsis* genome sequence (Jackson-Constan and Keegstra, 2001) revealed several putative proteins whose deduced amino acid sequence shows a similarity to the pea Toc or Tic proteins. The Toc protein homologues include three Toc159 homologues (atToc159, atToc132 and atToc120), two Toc34 homologues (atToc33 and atToc34), three Toc75 homologues (atToc75-I, atToc75-III, atToc75-IV) and three Toc64 homologues (atToc64-III, atToc64-V, atToc64-I). Recently, the fourth Toc159 homologue (atToc90; Hiltbrunner et al., 2001a) and the fourth Toc75 homologue (atToc75-V; Eckart et al., 2002) have been described. The Tic homologues include a single homologue of each Tic110, Tic55, Tic62 and Tic40 (atTic110, atTic55, atTic62 and atTic40, respectively), two homologues of Tic22 (atTic22-I and atTic22-IV), and two homologues of Tic20 (atTic20-IV and atTic20-

III). Genetic strategies have been applied to characterize the functions of some of the Toc159, Toc34, Toc75, Tic40 and Tic20 homologues.

All the Toc159 homologues (atToc159, atToc132, atToc120 and atToc90) have the three-domain structure as pea Toc159 (psToc159). They share amino acid sequence similarities with psToc159 in the central GTP-binding and the carboxy-terminal membrane anchor domains (49%, 37%, 39% and 38% identity, respectively), whereas their amino-terminal acidic domains differ in length and share less or little sequence similarity with each other (Hiltbrunner et al., 2001a). AtToc159 is the most abundantly expressed and is therefore considered the orthologue of psToc159. atToc159 exists in both membrane-bound and cytosolic forms, and the cytosolic form binds to atToc33 (Hiltbrunner et al., 2001b). A knockout mutant of atToc159, *ppi2*, shows a seedling-lethal, albino phenotype, indicating that this protein is essential for chloroplast development (Bauer, J. et al., 2000). Expression of photosynthetic genes is downregulated in *ppi2*, whereas nonphotosynthetic genes appear to be expressed and imported into chloroplasts (Bauer, J. et al., 2000). It is therefore hypothesized that atToc159 is the main import receptor for photosynthetic proteins, which are expressed at high levels, and atToc132 and atToc120 are specific receptors for nonphotosynthetic proteins.

atToc33 and atToc34 share 61% and 64% amino acid sequence identity, respectively, with psToc34. atToc33 is the most abundantly expressed and is considered to be the orthologue of psToc34. A knockout mutant of atToc33, *ppi1*, shows a relatively weak phenotype and is able to produce seeds. Chloroplasts of *ppi1* import proteins with reduced efficiency (Jarvis et al., 1998), a demonstration that provided the first *in vivo* evidence of the role of a Toc protein in the chloroplast protein import machinery. It has also been shown that atToc34 has functional properties similar to those of atToc33 (Jarvis et al., 1998).

Among the first identified three Toc75 homologues, atToc75-III is expressed at a much higher level than the others; several expressed sequence tags (ESTs) can be found for atToc75-III, whereas no EST has been found for atToc75-I and atToc75-IV (Jackson-Constan and Keegstra, 2001). In addition, atToc75-III shows the highest level of identity with psToc75 (74%). It is therefore likely that atToc75-III is the major Toc75 isoform in *Arabidopsis* and the true orthologue of psToc75. atToc75-III and atToc75-I are similar in size and share >60% amino acid sequence identity; atToc75-IV, on the other hand, is smaller in size and shows a sequence similarity to the carboxy-terminal part of atToc75-III and atToc75-I. The fourth homologue, atToc75-V, expressed and localized in the outer envelope membrane (Eckart et al., 2002), exhibits a high sequence similarity to a cyanobacterial homologue, and is therefore proposed to be the most ancestral form of a Toc75-like protein (Eckart et al., 2002; see later). The functional and structural properties of the *Arabidopsis* Toc75 homologues remain to be studied.

atTic40 shares a high amino acid sequence similarity to psTic40 (53% identity). A knockout mutant of atTic40 was isolated during a large-scale genetic screening of tDNA insertional mutants (Budziszewski et al., 2001). The mutant shows a pale-green phenotype, a demonstration of defects in chloroplast development. Further characterization of this mutant will help establish the role of Tic40 in chloroplast protein import.

atTic20-I exhibits a high sequence similarity to psTic20 (>60% identity), whereas atTic20-IV shares a moderate similarity with psTic20 (33% identity). Therefore, it seems likely that atTic20-I is the functional counterpart of psTic20. In support of this, evidence has been presented that atTic20-I plays a role in protein import: *Arabidopsis* mutants in which levels of atTic20-I mRNA are reduced by antisense expression exhibit reduced accumulation of plastid proteins, a decrease in thylakoid membrane development in chloroplasts and severe growth defects (Chen et al., 2002).

The discovery of multiple homologues of the Toc and Tic proteins in the *Arabidopsis* genome opened up the possibility that different translocation complexes might exist in chloroplasts or other types of plastids. Different isoforms of Toc159 and Toc34, for instance, might associate with one another in different combinations to produce different translocases with different precursor or tissue specificity. However, certain components of the import machinery, such as Tic110, are invariant, indicating that all chloroplast precursors follow similar import pathways. The hypothesis that different translocation complexes exist would be in accordance with the conflicting data on the composition of the Tic complex (Caliebe et al., 1997; Kouranov et al., 1998), but has to be experimentally proven.

12.3.3 Origin of TOC and TIC Translocases

Cyanobacteria perform oxygenic photosynthesis by using photosystems similar to those found in plant chloroplasts. The cyanobacterial cell is enclosed by an outer membrane and a plasma membrane and contains thylakoid membranes, resembling the structure of chloroplasts. Based on these similarities between cyanobacteria and chloroplasts, it is assumed that cyanobacteria are the closest living relatives of the ancestral endosymbiont from which chloroplasts originate. This assumption brings up the question of whether there are homologues of Toc and Tic proteins in cyanobacteria. The genome sequences of several cyanobacterial strains have been determined to date, including those for a unicellular cyanobacterium, *Synechocystis* sp. PCC 6803 (hereafter *Synechocystis*) and a filamentous, heterocyst-forming cyanobacterium, *Anabaena* sp. PCC 7120 (hereafter *Anabaena*; Kaneko et al., 1996, 2001). A survey of amino acid sequences of their proteins, deduced from the nucleotide sequences of the *Synechocystis* and *Anabaena* genomes, revealed that both these cyanobacteria possess at least one putative homologue of each Toc75, Toc34, Tic55, Tic22 and Tic20. By contrast, no clear homologues of Toc159, Toc64, Tic110, Tic62 and Tic40 have been found in these cyanobacteria. These results suggest that some components of the TOC and TIC translocases might have been derived from preexisting cyanobacterial proteins.

Two *Synechocystis* homologues, synToc75 and synTic22, have been studied to some extent. synToc75 shows a low amino acid sequence similarity (22% identity) to Toc75. This protein is localized in the outer membrane, and, when reconstituted in proteoliposomes, forms an aqueous ion channel with properties similar to those of Toc75 (Bölter et al., 1998b). synToc75 is predicted to have a β barrel structure similar to Toc75 (Bölter et al., 1998b). Targeted inactivation of the synToc75 gene has failed, indicating that the gene is essential for the viability of *Synechocystis* (Reumann et al., 1999). synToc75 shares amino acid sequence similarities with outer membrane proteins of Gram-negative bacteria, which are implicated in hemolysin secretion (Bölter et al., 1998b; Reumann et al., 1999). Although the physiological role of synToc75 remains to be elucidated, it is hypothesized that the function of Toc75 as a protein-conducting channel has been maintained during the evolution of chloroplasts, whereas the preferred direction of permeation has been reversed. synTic22 shows a low amino acid sequence similarity to Tic22. Fulda et al. (2002) demonstrated that synTic22 is localized in the lumen of the thylakoid membranes. This contrasts that Tic22 is peripherally associated with the outer surface of the chloroplast inner envelope membrane. The *synTic22* gene is essential for the viability of *Synechocystis* sp. PCC 6803 (Fulda et al., 2002). The function and physiological role of synTic22 are not clear.

12.4 Intraorganellar Protein Translocation

Once inside the organelle, imported proteins have to be further sorted to their final destinations. Here we focus on the pathways for protein translocation across, or insertion into, the thylakoid membranes of chloroplasts. Four distinct pathways have been identified to date: the Sec, the SRP, the ΔpH-dependent or Tat and the YidC/Oxa1p/Alb3 pathways. In addition, insertion of some proteins into the thylakoid membrane seems to occur independently of the known pathways but spontaneously (Woolhead et al., 2001). This so-called spontaneous insertion pathway might be mediated by as yet unidentified proteinaceous factors. Translocation of proteins into or across the inner membrane of mitochondria also involves SRP components and Oxa1p. Bacteria, such as *Escherichia coli*, have systems homologous to the Sec, SRP, Tat and YidC/Oxa1p/Alb3 pathways for protein export to extracytoplasmic locations. This suggests that the intraorganellar pathways are derived from the preexisting prokaryotic systems.

12.4.1 The Sec Pathway

12.4.1.1 Bacterial Sec Pathway

In bacteria, the Sec translocase catalyzes export of virtually hundreds of different proteins across the plasma membrane and is responsible for most of the fundamental housekeeping secretion needs (Economou, 2000). The major components of the Sec translocase are the integral membrane proteins SecY and SecE and the peripheral membrane protein SecA ATPase. SecY and SecE form a firm dimeric complex (Joly et al., 1994), whereas SecA forms a homodimer and is only transiently associated to the SecYE core complex (Karamanou et al., 1999). A substantial portion of SecA exists in a soluble, cytoplasmic form. Further auxiliary components have been identified in association with the core complex, namely, SecG, SecD, SecF, YajC, and SecB.

SecG is a small membrane protein that stimulates translocation activity and has been postulated to enhance the membrane insertion–deinsertion cycle of SecA (Flower et al., 2000). SecG is apparently not present in all bacteria (Economou, 2000). SecD and SecF form an additional complex with the poorly investigated protein YajC. The SecDFYajC complex can substitute for the SecYEG complex, particularly when SecG is inactivated (Duong and Wickner, 1997b). In addition, YidC of *E. coli* can be copurified with the SecYEG complex and has been implicated in Sec-dependent protein integration into the plasma membrane (see Section 12.4.4; Scotti et al., 2000).

SecB acts as a molecular chaperone that prevents precursor proteins from folding and aggregation before the translocation step (Knoblauch et al., 1999). It recognizes stretches of aromatic and positively charged residues within the mature part of precursor proteins (Knoblauch et al., 1999). The complex of SecB and the precursor protein docks to the membrane via interaction of SecB with the carboxy-terminal portion of SecA (Fekkes et al., 1997). However, both the signal sequence and the mature part of the precursor protein can be directly recognized by SecA that is bound to the SecYE core complex (Wang et al., 2000; Fekkes et al., 1998). The initiation of translocation induces transfer of the mature part of the precursor protein from SecB to SecA, with a subsequent elimination of SecB from the complex (Fekkes et al., 1997, 1998).

SecA contains a sequence stretch called intramolecular regulator of ATP hydrolysis (IRA; Karamanou et al., 1999). The IRA functions as a molecular switch that prevents ATP hydrolysis in the cytoplasm and enables it when SecA is bound to the membrane and interacts with SecYEG. Because IRA lies within the region responsible for the binding to SecY, it has been proposed that SecY regulates the IRA switching by inducing a conformational change in SecA (Karamanou et al., 1999). ATP provides energy necessary for translocation, whereas an electrochemical proton motive force is required for additional efficiency (Schiebel et al., 1991).

Proteins are transported in an unfolded conformation through a channel consisting mainly of the SecYEG complex (Meyer et al., 1999; Manting et al., 2000). Substrate-laden SecA can insert in the membrane at the SecYEG complex (Economou and Wickner, 1994; Economou et al., 1995; Eichler et al., 1997; Eichler and Wickner, 1997; Matsumoto et al.,

1997; Ramamurthy and Oliver, 1997; Snyders et al., 1997; Chen et al., 1998). However, the binding energy of ATP promotes and stabilizes an even more integral membrane state (Economou et al., 1995; Economou and Wickner, 1994; Eichler and Wickner, 1997). Concomitant with the SecA insertion, ca. 25 amino acid residues of the translocating protein enter the membrane plane (Schiebel et al. 1991; van der Wolk et al. 1997). Threading of the protein through the channel is initiated in the vicinity of SecA and SecY (Joly and Wickner, 1993). Backward sliding is prevented by the fixation of SecA by SecD and SecF (Economou et al., 1995; Duong and Wickner, 1997a, b). Subsequent ATP hydrolysis is accompanied by release of the protein, deinsertion of SecA from the membrane and formation of stable association of the translocating protein with SecYEG (Schiebel et al., 1991; Joly and Wickner, 1993; Economou and Wickner, 1994). The completion of translocation requires a protein motive force (Schiebel et al., 1991), although its effect on translocation is most probably indirect (Nouwen et al., 1996). It has been proposed that multiple rounds of the membrane insertion–deinsertion cycle of SecA might propel the complete translocation of the entire polypeptide (Economou and Wickner, 1994).

Substrates of the Sec pathway that contain an amino-terminal signal sequence can be recognized by SRP (see Section 12.4.3). Because SRP can recognize highly hydrophobic stretches, it is particularly important to target a subset of polytopic membrane proteins (Ulbrandt et al., 1997; de Gier et al., 1998; Scotti et al., 1999). A membrane-bound receptor, FtsY, tethers SRP onto the membrane because of its high affinity for acidic phospholipids (de Leeuw et al., 2000). A further step includes removal of the signal sequence by a specific leader peptidase. The exact mechanism by which SRP-targeted substrates are further handed over to the Sec translocase remains elusive. Evidence from *in vitro* studies indicates that certain SRP substrates can insert into the membrane in the absence of SecA (Scotti et al., 1999). However, Kim et al. (2001) suggest that insertion of SRP substrates depends on interaction of Ffh, a component of SRP, with SecA. The latter mechanism appears more likely, because mutation of Ffh affects secretion of almost all proteins that are also affected by mutation of SecA (Hirose et al., 2000).

Homologues of SecA and SecY have been studied to some extent in a unicellular cyanobacterium, *Synechococcus* sp. PCC 7942 (hereafter *Synechococcus*). One gene for each SecA and SecY homologue has been found in *Synechococcus*. The SecA homologue is essential for the viability of *Synechococcus* and its transcription is controlled by the redox state of the cell (Nakai et al., 1994). The SecY homologue is localized in both the thylakoid membrane and the cytoplasmic membrane (Nakai et al., 1993). It is therefore likely that dual, but distinct, Sec pathways drive proteins into, and across, the thylakoid and cytoplasmic membranes (Barbrook et al. 1993; Steiner and Löffelhardt, 2002). The presence of a single SecA homologue was also shown in a thermophilic cyanobacterium, *Phormidium laminosum* (Barbrook et al., 1993). The genome of *Synechocystis* contains a homologue of each SecY, SecA and SecG, and three SecE homologues and two SecD homologues. This redundancy suggests that specificity of the Sec pathways might be determined by differential formation of the translocase between SecYG and various SecE or SecD proteins.

12.4.1.2 Chloroplast Sec Pathway

The majority of more than 80 proteins destined for the lumen of the thylakoid membrane in chloroplasts are synthesized in the cytosol as precursors with a bipartite signal sequence consisting of an amino-terminal stromal-targeting domain and a carboxy-terminal lumenal-targeting domain (Peltier et al., 2002). Once the stromal-targeting domain of the signal sequence is removed by the stromal processing peptidase, the intermediates are translocated across the thylakoid membrane via two different pathways, the Sec and the Tat. More than a dozen lumenal proteins, including the 33-kDa subunit of the oxygen-evolving complex

(OE33), plastocyanin and a prolyl-peptidyl isomerase (TLP40), are transported via the Sec pathway (Peltier et al., 2002). Although the substrates of the Sec translocase are mostly soluble proteins, integration of some proteins (cytochrome *f* and PsaF subunit of the photosystem I) into the thylakoid membrane is apparently Sec dependent. Following the translocation, the lumenal-targeting domain of the signal sequence is removed by the thylakoidal processing peptidase to yield the mature form of the protein. The peptidase shows high amino acid sequence similarity to signal peptidases of *Synechocystis*, suggesting that the processing mechanism in chloroplasts might have been derived from the endosymbiont (Chaal et al., 1998).

Akin to the bacterial counterpart, translocation of chloroplast Sec-dependent precursors is absolutely dependent on ATP and stromal factors and is stimulated by the pH gradient (ΔpH) of the thylakoid membrane (Bauerle and Keegstra, 1991; Mould et al., 1991; Karnauchov et al., 1994; Hulford et al., 1994 ; Yuan and Cline, 1994). Chloroplasts appear to possess only the crucial components of the Sec machinery. Genes encoding Sec components have first been found in algal plastid genomes (Scaramuzzi et al., 1992; Valentin, 1993) and their discovery has facilitated the identification of homologues in higher plants. A chloroplast SecA homologue (cpSecA) was identified from pea by using antibodies raised against an algal SecA and shown to substitute for stromal extracts and promote transport of OE33 and plastocyanin into the thylakoid membrane in an azide-sensitive manner (Yuan et al., 1994). In a later study, a chloroplast SecA homologue was identified by a genetic study of a maize mutant, *tha1* (Völker et al., 1997).

Chloroplast homologues of SecY (cpSecY) and SecE (cpSecE) were identified by using *Arabidopsis* EST and genomic sequences, respectively (Laidler et al., 1995; Schuenemann et al., 1999). Antibodies raised against cpSecY are able to selectively inhibit cpSecA-dependent protein translocation (Mori et al., 1999), whereas they affect neither ΔpH-dependent nor posttranslational SRP-dependent protein translocation (Mori et al., 1999; Schuenemann et al., 1999). cpSecY is a 50-kDa protein with 10 membrane-spanning domains (Laidler et al., 1995), whereas cpSecE is an integral membrane protein with 1 membrane-spanning domain (Schuenemann et al., 1999). cpSecY and cpSecE form the minimal translocase unit of 180 kDa in the thylakoid membrane (Schuenemann et al., 1999). The chloroplast Sec machinery lacks the chaperon subunit SecB; the *Arabidopsis* genome does not contain genes for SecB homologues. Consistently, cpSecA lacks a carboxy-terminal domain of 20 residues that binds to SecB (Fekkes and Driessen, 1999). The *Arabidopsis* genome does not contain genes with significant homology to SecG, SecD, SecF and YajC (Mori and Cline, 2001). Therefore, it seems that the chloroplast Sec machinery operates with the minimal number of Sec components. Alternatively, it might contain another yet unidentified components that have no bacterial counterparts. SecY appears to be duplicated in the *Arabidopsis* genome. It is not known whether the second protein is localized in the chloroplasts and associates with other members of cpSec machinery.

12.4.2 The Tat Pathway

A large number of precursors of thylakoid membrane proteins (Peltier et al., 2002) use another transport pathway that requires the ΔpH of the thylakoid membrane but neither stromal factors nor ATP (Cline et al., 1992; Henry et al., 1994). The existence of this pathway has been recognized at about the same time as the Sec (Cline et al., 1992), but recently it was found to be homologous to a protein transport pathway in bacteria called the Tat (Sargent et al., 1998; Santini et al., 1998; Weiner et al., 1998). The characteristic of the Tat pathway is the presence of a conserved sequence called twin-arginine motif, (S/T)-R-R-x-F-L-K, within the signal sequence (the twin-arginine signal sequence) of its substrates (Berks, 1996). Recent proteomic analysis of thylakoid lumen polypeptides indicates that the majority of the lumen proteins use the Tat pathway (Peltier et al., 2002). By contrast, substrates of the bacterial Tat pathway are much more scarce than the precursors transported by the Sec pathway (Berks, 1996). Targeting specificity of precursors to the Sec and Tat pathways is not well understood, but some chimeric precursors can be targeted to both pathways *in vitro* (Henry et al., 1997).

12.4.2.1 Bacterial Tat Pathway

The most outstanding characteristic of the Tat pathway is its capability of transporting folded proteins of variable dimensions across the membrane (Berks et al., 2000). In bacteria, the substrates of this pathway are mostly proteins that bind cytoplasmic cofactors and have to be delivered to the place of their action in a folded state. Such proteins are indispensable for bacterial energy metabolism because they participate in various respiratory electron transport chains. The mechanistic principles of the Tat machinery are much less understood than those of the Sec pathway. The most intriguing feature remains the principle by which mechanism the impermeability of the membrane to ions and solutes is maintained during transport of folded proteins. The bacterial Tat translocase can transport protein complexes of 142 kDa (FdnGH subcomplex of *E. coli* formate dehydrogenase-N; Berg et al., 1991), whereas the diameter of the translocation system ranges up to 7 nm (Berks et al., 2000). The Tat system is not present in bacteria that do not export proteins with the twin-arginine signal sequence, such as intracellular parasites, methanogens and bacteria with obligatory fermentative metabolism.

The Tat translocase in *E. coli* consists of four obligatory subunits, TatA, TatB, TatE and TatC (Bogsch et al., 1998; Sargent et al., 1998; Weiner et al., 1998). TatA, TatB and TatE are homologous membrane proteins (Chanal et al., 1998; Sargent et al., 1998; Weiner et al., 1998). TatA and TatE share ca. 60% amino acid sequence identity, whereas the TatB shares ca. 25% identity with TatA and TatE. The region that exhibits sequence similarity covers the amino-terminal domain consisting of a putative membrane-spanning helix followed by a cytoplasmically located amphipathic helix. Consistent with their structural similarity, TatA and TatE can functionally substitute one another (Sargent et al., 1998). TatC contains six transmembrane helices, with the amino and carboxy termini of the protein being exposed to the cytoplasm (Sargent et al., 1998).

In general, bacteria contain one gene for a TatC homologue and two or three genes for TatA/B/E homologues (Berks et al., 2000). The reasons behind this redundancy are poorly understood, but one possibility is that multiple homologues of TatA/B/E might interact with specific subsets of precursor proteins. It has been proposed that the Tat system could include an additional subunit, TatD, which is a hydrophilic cytoplasmic protein (Berks et al., 2000). However, TatD has been characterized as a magnesium-dependent DNase, which is not required for the proper function of the Tat machinery (Wexler et al., 2000). Both TatA/B/E homologues and TatC have been proposed as receptors for the twin-arginine signal sequence of precursor proteins (Settles et al., 1997; Chanal et al., 1998; Berks et al., 2000). It can be assumed that such function might be evolutionarily conserved and revealed as conserved amino acid residues that can participate in interaction with the twin-arginine motif. It has therefore been suggested that TatC might be a receptor because of its highly conserved nature (Berks et al., 2000). By contrast, TatA/B/E-like proteins do not fulfil the criterion; they contain only few highly conserved nonpolar amino acid residues.

In *Synechocystis*, homologues of TatC and TatD can be identified by BLAST search. Other subunits of the Tat system might be not present or more diverged in this organism. To date, no precursors of thylakoid (lumen) proteins with the twin-arginine signal sequence is known in *Synechocystis*, with exception of uncleavable signal sequences of the Rieske iron–sulfur proteins. However, proteomic analysis of the periplasm of *Synechocystis* has identified at least 10 candidates for the Tat pathway (Fulda et al., 2000).

12.4.2.2 Chloroplast Tat Pathway

The Tat pathway in the thylakoid membrane of chloroplasts includes Hcf106, Tha4 and cpTatC, which are orthologues of TatB, TatA/E and TatC of *E. coli*, respectively (Settles et al., 1997; Sargent et al., 1998; Bogsch et al., 1998; Mori et al., 1999, 2001; Walker et al., 1999). Hcf106 and Tha4 are homologous proteins, both consisting of an amino-terminal transmembrane domain, a putative amphipathic helix and an acidic carboxy-terminal domain exposed to the stroma. The amphipathic helix and the carboxy-terminal domain are much longer in Hcf106 than in Tha4, whereas the transmembrane domains share a high degree of sequence similarity. cpTatC is an integral membrane protein with topological features analogous to bacterial TatC, which are represented by six membrane-spanning helices and both amino and carboxyl termini being exposed to the stroma (Mori et al., 2001). Antibody inhibition studies have shown that Hcf106, Tha4 and cpTatC are directly involved in protein transport across the thylakoid membrane (Mori et al., 1999, 2001).

Several conserved structural features can be recognized in the components of the bacterial and chloroplast Tat translocases. Hcf106 and TatB proteins share a conserved glycine–proline dipeptide in a spacer region between the membrane-spanning and amphipathic helices (Mori and Cline, 2001). Both residues contribute to the flexibility of the interface segment. A proline–glutamic acid motif in the transmembrane domain of Hcf106 and Tha4 is also strictly conserved (Weiner et al., 1998), whereas TatB proteins have conserved glutamate residues in the corresponding position. Because maintenance of a charged residue in the transmembrane segment is coupled with considerable energy requirements, these glutamate residues are likely to play an important role in the translocation process (Mori and Cline, 2001). The glutamate residues of TatB/Hcf106 could, for example, interact with the transmembrane arginine residue of TatC/cpTatC to maintain or stabilize their complex formation. TatA, TatB and TatC are copurified as a complex of 600 kDa from *E. coli* (Bolhuis et al., 2001); in contrast, a 700-kDa complex purified from chloroplasts consists mostly of cpTatC and Hcf106 (Mori and Cline, 2001).

Experimentally determined substrates of the chloroplast Tat translocase range in size from 3.6 to 36 kDa. It has not yet been unequivocally demonstrated that the proteins maintain their folded conformation during the transport. If they do so, the putative channel of the Tat translocase would have to expand to about 5 nm. Binding of a precursor occurs in the absence of ΔpH, while the bound precursor is translocated to the lumen when the ΔpH is applied. As in the bacterial case, cpTatC has been suggested as the receptor for precursor proteins (Mori and Cline, 2001). The binding is strictly dependent on the twin-arginine motif of the lumenal-targeting domain of the signal sequence and is corroborated by the hydrophobic core.

Homologues of the Tat components are found in plant mitochondrial genomes, but are absent in animal and yeast mitochondria. Interestingly, the genome of *Arabidopsis* houses one chloroplast and two putative mitochondrial TatC homologues. Substrates of the plant mitochondrial Tat system have yet to be determined.

12.4.3 Signal Recognition Particle (SRP) Pathway

12.4.3.1 Cytoplasmic SRP Pathways

Prokaryotic polytopic proteins are targeted to the plasma membrane by using cotranslational targeting system composed of SRP and its cognate receptor component, FtsY (Rapoport et al., 1996; Keenan et al., 2001). Targeting of eukaryotic endomembrane proteins and secretory proteins to the endoplasmic reticulum is accomplished by using a similar principle (Rapoport et al., 1996). The core of cytoplasmic SRPs consists of a conserved RNA molecule and a 54-kDa GTPase, SRP54 [The corresponding protein in *E. coli* is called Ffh (fifty-four-kda protein homologue).) A key feature of the protein targeting by SRP is the ability of SRP54 to bind the hydrophobic hub of the signal sequence of precursors as they emerge from ribosomes. Binding of SRP to the signal sequence guides the entire ribosome-nascent chain-mRNA complex to the target membrane. Targeting fidelity can be at least partly accounted to the affinity of SRP54 for the membrane-bound receptor FtsY. In eukaryotes, binding of SRP to the ribosome induces translational arrest. At the membrane, SRP binds and hydrolyzes GTP and uses FtsY to coordinate release of the signal sequence and the subsequent handover of the nascent chain to a membrane translocase (primarily the Sec translocase). Homologues of *E. coli* Ffh are found in the *Synechocystis* and *Anabaena* genomes. However, little is known about the structure and function of the SRP pathway in cyanobacteria.

12.4.3.2 Chloroplast SRP Pathway

Chloroplast SRP (cpSRP) participates in a posttranslational targeting of a subset of nuclear-encoded thylakoid proteins (Eichacker and Henry, 2001). In addition, it appears to have retained a conserved role in a cotranslational targeting of organellar-encoded thylakoid proteins (Amin et al., 1999; Nilsson et al., 1999). The latter is accomplished by a similar mechanism as in bacteria, in which cpSRP54, the chloroplast homologue of SRP54, interacts only when the nascent polypeptide chain is attached to the ribosome (Eichacker and Henry, 2001).

To date, it seems likely that the posttranslational targeting by cpSRP is restricted to the light-harvesting chlorophyll *a/b*-binding proteins (LHCPs; Cline and Henry, 1996; Keegstra and Cline, 1999). LHCPs are synthesized in the cytosol as precursors. The amino-terminal signal sequence is removed on translocation across the envelope membranes through the TOC and TIC complexes. The apoprotein becomes engaged in a soluble transit complex and associates with the thylakoid membrane (Cline and Henry, 1996; Keegstra and Cline, 1999) and then it is integrated into the membrane in a GTP-dependent manner (Hoffman and Franklin, 1994). Li et al. (1995) demonstrated that the transit complex contains cpSRP54. Subsequent studies of cpSRP failed to identify an RNA component that is present in cytoplasmic SRPs, despite algal plastid genomes encoding genes for homologues of the RNA component (Packer and Howe, 1998), but instead revealed the existence of a dimer of a novel 43-kDa subunit, cpSRP43 (Schuenemann et al., 1998). No homologue of cpSRP43 has been found in prokaryotes, leaving its evolutionary origin uncertain. cpSRP54 can be found in two distinct pools in the stroma (Schuenemann et al., 1998), one associated with ribosomes and another with cpSRP43. LHCPs associate exclusively with the cpSRP43–cpSRP54 complex, which can also reconstitute membrane integration of LHCPs. The mechanistic reasons behind the existence of cpSRP43- and ribosome-bound pools of cpSRP54 are not understood. One possibility is that cpSRP43 functions as an uncoupling factor that releases ribosomes from cpSRP54. cpSRP43 would thus have to compete with ribosomes for the binding site on cpSRP54. cpSRP54 interacts with the so-called H domain within the mature portion of LHCPs (High and Dobberstein, 1991). The H domain has a unique recognition element that is used to promote posttranslational binding to cpSRP54. Additionally, an 18-amino acid hydrophilic domain (L18) situated between the second and third transmembrane domains of LHCPs is responsible for the interaction with cpSRP43 (Tu et al., 2000). Formation of the LHCP–cpSRP transit complex is apparently a synergistic process in which binding of L18 to cpSRP43 is required for subsequent interaction of LHCP with cpSRP54 (DeLille et al., 2000).

Kogata et al. (1999) isolated an *Arabidopsis* cDNA for cpFtsY, a homologue of FtsY. Akin to its bacterial counterpart, cpFtsY is a GTPase bound peripherally on the outer surface of the thylakoid membrane (Kogata et al., 1999; Tu et al., 1999). Based on antibody inhibition studies, cpFtsY is necessary for the cpSRP-dependent LHCP integration into the thylakoid membrane, suggesting its involvement in the pathway.

Yang and Mulligan (1996) isolated from maize mitochondria a SRP-like ribonucleoprotein complex that consists of a homologue of SRP54 and an RNA molecule with sequence similarity to the RNA component (4.5S RNA) of the bacterial SRP. The function of this SRP-like complex in mitochondria remains to be studied.

12.4.4 YidC/Oxa1p/Alb3 Translocase Family

12.4.4.1 Oxa1p and YidC

Oxa1p, first identified from yeast mitochondria (Kermorgant et al., 1997), is an inner membrane protein that plays a role in the integration of a subset of proteins from the matrix into the inner membrane (Hell et al., 2001). Oxa1p catalyzes its own integration in corroboration with the general import pathway (Stuart, 2002). Oxa1p homologues have been found in bacteria and plant chloroplasts and constitute a large family of protein translocases (Samuelson et al., 2000). The Oxa1p homologue of *E. coli*, YidC, can be associated with the SecYEG complex (Beck et al., 2001) and with the SecDFYajC complex (Nouwen and Driessen, 2002). It has been demonstrated that YidC, with SecY, mediates insertion of the so-called Type I transmembrane domains (Houben et al., 2002). YidC assists in sorting of some proteins that were previously believed to insert into the membrane without the aid of proteinaceous components. Integration of YidC itself into the plasma membrane of *E. coli* depends on the SRP, SecA and SecYEG (Koch et al., 2002). Therefore, in contrast to Oxa1p, YidC cannot catalyze its own integration as part of a Sec-independent protein transport pathway. YidC is a relatively abundant protein of the plasma membrane and it accumulates mostly at cell poles (Urbanus et al., 2002). Most eubacteria possess only one YidC homologue, but species of *Bacillus*, *Listeria* and *Streptomyces* contain an additional YidC-related protein (Luirink et al., 2001). The disruption of this additional protein in *B. subtilis* leads to a cell cycle arrest in the intermediate stage of spore formation. The protein has accordingly been designated SpoIIIJ (Stage III sporulation protein J; Errington et al., 1992). During vegetative growth, SpoIIIJ localizes to the cell membrane, whereas in sporulating cells it accumulates at polar zones and engulfment septa (Murakami et al., 2002).

12.4.4.2 Albino3 (Alb3) and ARTEMIS

Alb3 is a chloroplast homologue of Oxa1p located in the thylakoid membrane. The membrane integration of LHCPs depends on Alb3 (Moore et al., 2000). Inactivation of Alb3 in *Arabidopsis* leads to a severe damage of chloroplast ultrastructure and seedling lethality (Sundberg et al., 1997). Activity of Alb3 appears to be restricted to the integration of LHCPs only (Mant et al., 2001; Woolhead et al., 2001). Alb3 can functionally complement YidC-depleted strain of *E. coli* (Jiang et al., 2002).

The *Arabidopsis* genome contains several genes for YidC-like proteins (Luirink et al., 2001), but only three of them have so far been investigated. Apart from Alb3 and the plant mitochondrial Oxa1p (Sakamoto et al., 2000), chloroplasts possess an additional YidC-like protein in the inner envelope membrane, designated ARTEMIS (*Arabidopsis thaliana* envelope membrane integrase). The structure of ARTEMIS is quite unique among other members of the YidC/Oxa1p/Alb3 family. It consists of three distinct modules: an amino-terminal receptor-like region, a central glycine-rich stretch containing a nucleoside triphosphate-binding site and a carboxy-terminal YidC/Oxa1p/Alb3-like domain (Fulgosi et al., 2002). *Arabidopsis* plants in which ARTEMIS is inactivated by transposon insertion or with antisense suppression possess chloroplasts arrested in the late stages of division. This suggests that ARTEMIS might be involved in translocation or membrane tethering of a specific set of proteins necessary for chloroplast division or positioning of the organellar midpoint. *Synechocystis* has a gene that encodes a protein with sequence similarity to the YidC/Oxa1p/Alb3-like domain of ARTEMIS. Inactivation of this *Synechocystis* gene results in aberrant cell division (Fulgosi et al., 2002). This defect can be rescued by expression of the YidC/Oxa1p/Alb3-like domain of ARTEMIS, suggesting their evolutionary and functional relatedness.

Acknowledgment

This work was supported by grants from Deutsche Forschungsgemeinschaft, Sonderforschungsbereich TR-1.

References

Adam, A., Endres, M., Sirrenberg, C., Lottspeich, F., Neupert, W. and Brunner, M. (1999) Tim9, a new component of the TIM22.54 translocase in mitochondria. *EMBO J.* 18: 313–319.

Adams, M. D., Wagner, L. M., Graddis, T. J., Landick, R., Antonucci, T. K., Gibson, A. L. and Oxender, D. L. (1990) Nucleotide sequence and genetic characterization reveal six essential genes for the LIV-I and LS transport systems of *Escherichia coli*. *J. Biol. Chem.* 265: 11436–11443.

Ahting, U., Thieffry, M., Engelhardt, H., Hegerl, R., Neupert, W. and Nussberger, S. (2001) Tom40, the pore-forming component of the protein-conducting TOM channel in the outer membrane of mitochondria. *J. Cell Biol.* 153: 1151–1160.

Alconada, A., Kübrich, M., Moczko, M., Hönlinger, A. and Pfanner, N. (1995) The mitochondrial receptor complex: The small subunit Mom8b/Isp6 supports association of receptors with the general insertion pore and transfer of preproteins. *Mol. Cell Biol.* 15: 6196–6205.

Amin, P., Sy, D. A.C., Pilgrim, M. L., Parry, D. H., Nussaume, L. and Hoffman, N. E. (1999) *Arabidopsis* mutants lacking the 43- and 54-kilodalton subunits of the chloroplast signal recognition particle have distinct phenotypes. *Plant Physiol.* 121: 61–70.

Andersson, S. G. E., Zomorodipour, A., Andersson, J. O., Sicherits-Ponten, T., Alsmark, U. C. M., Podowski, R. M., Naslund, A. K., Eriksson, A. -S., Winkler, H. H. and Kurland, C. G. (1998) The genome sequence of *Rickettsia prowazekii* and the origin of mitochondria. *Nature* 396: 133–143.

Barbrook, A. C., Packer, J. C. and Howe, C. J. (1993) Components of the protein translocation machinery in the thermophilic cyanobacterium *Phormidium laminosum*. *Biochem. Biophys. Res. Commun.* 197: 874–877.

Bauer, J., Chen, K., Hiltbunner, A., Wehrli, E., Eugster, M., Schnell, D. and Kessler, F. (2000) The major protein import receptor of plastids is essential for chloroplast biogenesis. *Nature* 403: 203–207.

Bauer, M. F., Hofmann, S., Neupert, W. and Brunner, M. (2000) Protein translocation into mitochondria: The role of TIM complexes. *Trends Cell Biol.* 10: 25–31.

Bauerle, C. and Keegstra, K. (1991) Full-length plastocyanin precursor is translocated across isolated thylakoid membranes. *J. Biol. Chem.* 266: 5876–5883.

Beck, K., Eisner, G., Trescher, D., Dalbey, R. E., Brunner, J. and Müller, M. (2001) YidC, an assembly site for polytopic *Escherichia coli* membrane proteins located in immediate proximity to the SecYE translocon and lipids. *EMBO Rep.* 2: 709–714.

Berg, B. L., Li, J., Heider, J. and Stewart, V. (1991) Nitrate-inducible formate dehydrogenase in *Escherichia coli* K-12. I. Nucleotide sequence of the *fdnGHI* operon and evidence that opal (UGA) encodes selenocysteine. *J. Biol. Chem.* 266: 22380–22385.

Berks, B. C. (1996) A common export pathway for proteins binding complex redox cofactors? *Mol. Microbiol.* 22: 393–404.

Berks, B. C., Sargent, F. and Palmer, T. (2000) The Tat protein export pathway. *Mol. Microbiol.* 35: 260–274.

Blatch, G. L. and Lassle, M. (1999) The tetratricopeptide repeat: A structural motif mediating protein-protein interactions. *Bioessays* 21: 932–939.

Bölter, B., May, T. and Soll, J. (1998a) A protein import receptor in pea chloroplasts, Toc86, is only a proteolytic fragment of a larger polypeptid. *FEBS Lett.* 441: 59–62.

Bölter, B., Soll, J., Schulz, A., Hinnah, S. and Wagner, R. (1998b) Origin of a chloroplast protein importer. *Proc. Natl. Acad. Sci. USA* 95: 15831–15836.

Bogsch, E.G., Sargent, F., Stanley, N. R., Berks, B. C., Robinson, C. and Palmer, T. (1998) An essential component of a novel bacterial protein export system with homologues in plastids and mitochondria. *J. Biol. Chem.* 273: 18003–18006.

Bolhuis, A., Mathers, J. E., Thomas, J. D., Barrett, C. M. L. and Robinson, C. (2001) TatB and TatC form a functional and structural unit of the twin-arginine translocase from *Escherichia coli*. *J. Biol. Chem.* 276: 20213–20219.

Brix, J., Dietmeier, K. and Pfanner, N. (1997) Differential recognition of preproteins by the purified cytosolic domains of the mitochondrial import receptors Tom20, Tom22, and Tom70. *J. Biol. Chem.* 272: 20730–20735.

Budziszewski, G. J., Lewis, S. P., Glover, L. W., Reineke, J., Jones, G., Ziemnik, L. S., Lonowski, J., Nyfeler, B., Aux, G., Zhou, Q., McElver, J., Patton, D. A., Martienssen, R., Grossniklaus, U., Ma, H., Law, M. and Levin, J. Z. (2001) *Arabidopsis* genes essential for seedling viability: Isolation of insertional mutants and molecular cloning. *Genetics* 159: 1765–1778.

Caliebe, A., Grimm, R., Kaiser, G., Lubeck, J., Soll, J. and Heins, L. (1997) The chloroplastic protein import machinery contains a Rieske-type iron-sulfur cluster and a mononuclear iron-binding protein. *EMBO J.* 16: 7342–7350.

Chaal, B. K., Mould, R. M., Barbrook, A. C., Gray, J. C. and Howe, C. J. (1998) Characterization of a cDNA encoding the thylakoidal processing peptidase from *Arabidopsis thaliana*. *J. Biol. Chem.* 273: 689–692.

Chanal, A., Santini, C. L. and Wu, L. F. (1998) Potential receptor function of three homologous components, TatA, TatB and TatE, of the twin-arginine signal sequence-dependent metalloenzyme translocation pathway in *Escherichia coli*. *Mol. Microbiol.* 30: 674–676.

Chen, X. C., Brown, T. and Tai, P. C. (1998) Identification and characterization of protease-resistant SecA fragments: SecA has two membrane-integral forms. *J. Bacteriol.* 180: 527–537.

Chen, K., Chen, X. and Schnell, D. J. (2000) Initial binding of preproteins involving the Toc159 receptor can be bypassed during protein import into chloroplasts. *Plant Physiol.* 122: 813–822.

Chen, X., Smith, M. D., Fitzpatrick, L. and Schnell, D. J. (2002) In vivo analysis of the role of atTic20 in protein import into chloroplasts. *Plant Cell* 14: 641–654.

Cline, K., Ettinger, W. F. and Theg, S. M. (1992) Protein-specific energy requirements for protein transport across or into thylakoid membranes: Two lumenal proteins are transported in the absence of ATP. *J. Biol. Chem.* 267: 2688–2696.

Cline, K. and Henry, R. (1996) Import and routing of nucleus-encoded chloroplast proteins. *Annu. Rev. Cell Dev. Biol.* 12:1–26.

Curran, S. P., Leuenberger, D., Oppliger, W. and Koehler, C. M. (2002) The Tim9p-Tim10p complex binds to the transmembrane domains of the ADP/ATP carrier. *EMBO J.* 21: 942–953.

de Gier, J. W. L., Scotti, P. A., Saaf, A., Valent, Q. A., Kuhn, A., Luirink, J. and von Heijne, G. (1998) Differential use of the signal recognition particle translocase targeting pathway for inner membrane protein assembly in *Escherichia coli*. *Proc. Natl. Acad. Sci. USA* 95: 14646–14651.

de Leeuw, E., Kaat, T. K., Moser, C., Menestrina, G., Demel, R., de Kruijff, B., Oudega, B., Luirink, J. and Sinning, I. (2000) Anionic phospholipids are involved in membrane association of FtsY and stimulate its GTPase activity. *EMBO J.* 19: 531–541.

Dekker, P. J., Keil, P., Rassow, J., Maarse, A. C., Pfanner, N. and Meijer, M. (1993) Identification of MIM23, a putative component of the protein import machinery of the mitochondrial inner membrane. *FEBS Lett.* 330: 66–70.

Dekker, P. J., Martin, F., Maarse, A. C., Bömer, U., Müller, H., Guiard, B., Meijer, M., Rassow, J. and Pfanner, N. (1997) The Tim core complex defines the number of mitochondrial translocation contact sites and can hold arrested preproteins in the absence of matrix Hsp70-Tim44. *EMBO J.* 16: 5408–5419.

Dekker, P. J. T., Ryan, M. T., Brix, J., Müller, H., Hönlinger, A. and Pfanner, N. (1998) Preprotein translocase of the outer mitochondrial membrane: Molecular dissection and assembly of the general import core complex. *Mol. Cell Biol.* 18: 6515–6524.

DeLille, J., Peterson, E. C., Johnson, T., Moore. M., Kight, A. and Henry, R. (2000) A novel precursor recognition element facilitates posttranslational binding to the signal recognition particle in chloroplasts. *Proc. Natl. Acad. Sci. USA* 97: 1926–1931.

Diekert, K., Kispal, G., Guiard, B. and Lill, R. (1999) An internal targeting signal directing proteins into the mitochondrial intermembrane space. *Proc. Natl. Acad. Sci. USA* 96: 11752–11757.

Dietmeier, K., Hönlinger, A., Bömer, U., Dekker, P. J. T., Eckerskorn, C., Lottspeich, F., Kübrich, M. and Pfanner, N. (1997) Tom5 functionally links mitochondrial preprotein receptors to the general import pore. *Nature* 388: 195–200.

Duong, F. and Wickner, W. (1997a) Distinct catalytic roles of the SecYE, SecG and SecDFyajC subunits of preprotein translocase holoenzyme. *EMBO J.* 16: 2756–2768.

Duong, F. and Wickner, W. (1997b) The SecDFyajC domain of preprotein translocase controls preprotein movement by regulating SecA membrane cycling. *EMBO J.* 16: 4871–4879.

Eckart, K., Eichacker, L., Sohrt, K., Schleiff, E., Heins, L. and Soll, J. (2002) A Toc75-like protein import channel is abundant in chloroplasts. *EMBO Rep.* 3: 557–562.

Economou, A. (2000) Bacterial protein translocase: A unique molecular machine with an army of substrates. *FEBS Lett.* 476: 18–21.

Economou, A., Pogliano, J. A., Beckwith, J., Oliver, D. B. and Wickner W. (1995) SecA membrane cycling at SecYEG is driven by distinct ATP binding and hydrolysis events and is regulated by SecD and SecF. *Cell* 83: 1171–1181.

Economou, A. and Wickner, W. (1994) SecA promotes preprotein translocation by undergoing ATP-driven cycles of membrane insertion and deinsertion. *Cell* 78: 835–843.

Eichacker, L. A. and Henry, R. (2001) Function of a chloroplast SRP in thylakoid protein export. *Biochim. Biopys. Acta* 1541: 120–134.

Eichler, J., Brunner, J. and Wickner, W. (1997) The protease-protected 30 kDa domain of SecA is largely inaccessible to the membrane lipid phase. *EMBO J.* 16: 2188–2196.

Eichler, J. and Wickner, W. (1997) Both an N-terminal 65-kDa domain and a C-terminal 30-kDa domain of SecA cycle into the membrane at SecYEG during translocation. *Proc. Natl. Acad. Sci. USA* 94: 5574–5581.

Emtage, J. L. and Jensen, R. E. (1993) MAS6 encodes an essential inner membrane component of the yeast mitochondrial protein import pathway. *J. Cell. Biol.* 122: 1003–1012.

Errington, J., Appleby, L., Daniel, R. A., Goodfellow, H., Partridge, S. R. and Yudkin, M. D. (1992) Structure and function of the *spoIIIJ* gene of *Bacillus subtilis*: A vegetatively expressed gene that is essential for sigma G activity at an intermediate stage of sporulation. *J. Gen. Microbiol.* 138: 2609–2618.

Fekkes, P., van der Does, C. and Driessen, A. J. M. (1997) The molecular chaperone SecB is released from the carboxy-terminus of SecA during initiation of precursor protein translocation. *EMBO J.* 16: 6105–6113.

Fekkes, P., de Wit, J. G., van der Wolk, J. P. W., Kimsey, H. H., Kumamoto, C. A. and Driessen, A. J. (1998) Preprotein transfer to the *Escherichia coli* translocase requires the co-operative binding of SecB and the signal sequence to SecA. *Mol. Microbiol.* 29: 1179–1190.

Fekkes, P. and Driessen, A. J. (1999) Protein targeting to the bacterial cytoplasmic membrane. *Microbiol. Mol. Biol. Rev.* 63: 161–173.

Flower, A. M., Nines, L. L. and Pfenning, P. L. (2000) SecG is an auxiliary component of the protein export apparatus of *Escherichia coli*. *Mol. Gen. Genet.* 263: 131–136.

Fulda, S., Huang, F., Nilsson, F., Hagemann, M. and Norling, B. (2000) Proteomics of *Synechocystis* sp. strain PCC 6803: Identification of periplasmic proteins in cells grown at low and high salt concentrations. *Eur. J. Biochem.* 267: 5900–5907.

Fulda, S., Norling, B., Schoor, A. and Hagemann, M. (2002) The Slr0924 protein of *Synechocystis* sp. strain PCC 6803 resembles a subunit of the chloroplast protein import complex and is mainly localized in the thylakoid lumen. *Plant Mol. Biol.* 49: 107–118.

Fulgosi, H., Gerdes, L., Westphal, S., Glockmann, C. and Soll, J. (2002) Cell and organelle division requires ARTEMIS. *Proc. Natl. Acad. Sci. USA* 99: 11501–11506.

Fulgosi, H. and Soll, J. (2002) The chloroplast protein import receptors Toc34 and Toc159 are phosphorylated by distinct protein kinases. *J. Biol. Chem.* 277: 8934–8940.

Gabriel, K., Buchanan, S. K. and Lithgow, T. (2001) The alpha and the beta protein translocation across mitochondrial and plastid outer membranes. *Trends Biochem Sci.* 26: 36–40.

Gratzer, S., Lithgow, T., Bauer, R. E., Lamping, E., Paltauf, F., Kohlwein, S. D., Haucke, V., Junne, T., Schatz, G. and Horst, M. (1995) Mas37p, a novel receptor subunit for protein import into mitochondria. *J. Cell Biol.* 129: 25–34.

Gray, J., Close, P. S., Briggs, S. P. and Johal, G. S. (1997) A novel suppressor of cell death in plants encoded by the Lls1 gene of maize. *Cell* 89: 25–31.

Haucke, V., Ocana, C. S., Hönlinger, A., Tokatlidis, K., Pfanner, N. and Schatz, G. (1997) Analysis of the sorting signals directing NADH-cytochrome b_5 reductase to two locations within yeast mitochondria. *Mol. Cell Biol.* 17: 4024–4032.

Heins, L., Mehrle, A., Hemmler, R., Wagner, R., Küchler, M., Hörmann, F., Sveshnikov, D. and Soll, J. (2002) The preprotein conducting channel at the inner envelope membrane of plastids. *EMBO J.* 21: 2616–2625.

Hell, K., Neupert, W. and Stuart, R. A. (2001) Oxa1p acts as a general membrane insertion machinery for proteins encoded by mitochondrial DNA. *EMBO J.* 20: 1281–1288.

Henry, R., Kapazoglou, A., McCaffery, M. and Cline, K. (1994) Differences between lumen targeting domains of chloroplast transit peptides determine pathway specificity for thylakoid transport. *J. Biol. Chem.* 269: 10189–10192.

Henry, R., Carrigan, M., McCaffrey, M., Ma, X. and Cline, K. (1997) Targeting determinants and proposed evolutionary basis for the Sec and the Delta pH protein transport systems in chloroplast thylakoid membranes. *J. Cell. Biol.* 136: 823–832.

Hidalgo, E., Ding, H. and Demple, B. (1997) Redox signal transduction: Mutations shifting [2Fe-2S] centers of the SoxR sensor-regulator to the oxidized form. *Cell* 88: 121–129.

High, S. and Dobberstein, B. (1991) The signal sequence interacts with the methionine-rich domain of the 54-kD protein of signal recognition particle. *J. Cell Biol.* 113: 229–233.

Hill, K., Model, K., Ryan, M. T., Dietmeier, K., Martin, F., Wagner, R. and Pfanner, N. (1998) Tom40 forms the hydrophilic channel of the mitochondrial import pore for preproteins. *Nature* 395: 516–521.

Hiltbrunner, A., Bauer, J., Alvarez-Huerta, M. and Kessler, F. (2001a) Protein translocon at the *Arabidopsis* outer chloroplast membrane. *Biochem. Cell Biol.* 79: 1–7.

Hiltbrunner, A., Bauer, J., Vidi, P. A., Infanger, S., Weibel, P., Hohwy, M. and Kessler, F. (2001b) Targeting of an abundant cytosolic form of the protein import receptor atToc159 to the outer chloroplast membrane. *J. Cell Biol.* 154: 309–316.

Hines, V., Brandt, A., Griffiths, G., Horstmann, H., Brutsch, H. and Schatz, G. (1990) Protein import into yeast mitochondria is accelerated by the outer membrane protein MAS70. *EMBO J.* 9: 3191–3200.

Hinnah, S. C., Hill, K., Wagner, R., Schlicher, T. and Soll, J. (1997) Reconstitution of a chloroplast protein import channel. *EMBO J.* 16: 7351–7360.

Hirose, I., Sano, K., Shioda, I., Kumano, M., Nakamura, K. and Yamane, K. (2000) Proteome analysis of *Bacillus subtilis* extracellular proteins: A two-dimensional protein electrophoretic study. *Microbiology* 146: 65–75.

Hirsch, S., Muckel, E., Heemeyer, F., von Heijne, G. and Soll, J. (1994) A receptor component of the chloroplast protein translocation machinery. *Science* 266: 1989–1992.

Hoffman, N. E. and Franklin, A. E. (1994) Evidence for a stromal GTP requirement for the integration of a chlorophyll a/b-binding polypeptide into thylakoid membranes. *Plant Physiol.* 105: 295–304.

Hohfeld, J., Minami, Y. and Hartl, F. U. (1995) Hip, a novel cochaperone involved in the eukaryotic Hsc70/Hsp40 reaction cycle. *Cell* 83: 589–598.

Houben, E. N. G., Urbanus, M. L., van der Laan, M., ten Hagen-Jongman, C. M., Driessen, A. J. M., Brunner, J., Oudega, B. and Luirink, J. (2002) YidC and SecY mediate membrane insertion of a type I transmembrane domain. *J. Biol. Chem.* 277: 35880–35886.

Hönlinger, A., Bömer, U., Alconada, A., Eckerskorn, C., Lottspeich, F., Dietmeier, K. and Pfanner, N. (1996) Tom7 modulates the dynamics of the mitochondrial outer membrane translocase and plays a pathway-related role in protein import. *EMBO J.* 15: 2125–2137.

Hulford, A., Hazell, L. Mould, R. M. and Robinson, C. (1994) Two distinct mechanisms for the translocation of proteins across the thylakoid membrane, one requiring the presence of a stromal protein factor and nucleotide triphosphates. *J. Biol. Chem.* 269: 3251–3256.

Jackson-Constan, D. and Keegstra, K. (2001) *Arabidopsis* genes encoding components of the chloroplastic protein import apparatus. *Plant Physiol.* 125: 1567–1576.

Jarvis, P., Chen, L. J., Li, H., Peto, C. A., Fankhauser, C. and Chory, J. (1998). An *Arabidopsis* mutant defective in the plastid general protein import apparatus. *Science* 282: 100–103.

Jarvis, P. and Soll, J. (2001) Toc, Tic, and chloroplast protein import. *Biochim. Biophys. Acta* 1541: 64–79.

Jiang, F., Yi, L., Moore, M., Chen, M., Rohl, T., van Wijk, K. J., de Gier, J. W. L., Henry, R. and Dalbey, R. E. (2002) Chloroplast YidC homolog Albino3 can functionally complement the bacterial YidC depletion strain and promote membrane insertion of both bacterial and chloroplast thylakoid proteins. *J. Biol. Chem.* 277: 19281–19288.

Joly, J. C., Leonard, M. R. and Wickner, W. T. (1994) Subunit dynamics in *Escherichia coli* preprotein translocase. *Proc. Natl. Acad. Sci. USA* 91: 4703–4707.

Joly, J. C. and Wickner, W. (1993) The SecA and SecY subunits of translocase are the nearest neighbors of a translocating preprotein, shielding it from phospholipids. *EMBO J.* 12: 255–263.

Kanamori, T. Nishikawa, S., Shin, I., Schultz, P. G. and Endo, T. (1997) Probing the environment along the protein import pathways in yeast mitochondria by site-specific photocrosslinking. *Proc. Natl. Acad. Sci. USA* 94: 485–490.

Kaneko, T., Sato, S., Kotani, H., Tanaka, A., Asamizu, E., Nakamura, Y., Miyajima, N., Hirosawa, M., Sugiura, M., Sasamoto, S., Kimura, T., Hosouchi, T., Matsuno, A., Muraki, A., Nakazaki, N., Naruo, K., Okumura, S., Shimpo, S., Takeuchi, C., Wada, T., Watanabe, A., Yamada, M., Yasuda, M. and Tabata, S. (1996) Sequence analysis of the genome of the unicellular cyanobacterium *Synechocystis* sp. strain PCC6803. II. Sequence determination of the entire genome and assignment of potential protein-coding regions. *DNA Res.* 3: 109–136.

Kaneko, T., Nakamura, Y., Wolk, C. P., Kuritz, T., Sasamoto, S., Watanabe, A., Iriguchi, M., Ishikawa, A., Kawashima, K., Kimura, T., Kishida, Y., Kohara, M., Matsumoto, M., Matsuno, A., Muraki, A., Nakazaki, N., Shimpo, S., Sugimoto, M., Takazawa, M., Yamada, M., Yasuda, M. and Tabata, S. (2001) Complete genomic sequence of the filamentous nitrogen-fixing cyanobacterium *Anabaena* sp. strain PCC 7120. *DNA Res.* 8: 205–213.

Karamanou, S., Vrontou, E., Sianidis, G., Baud, C., Roos, T., Kuhn, A., Politou, A. S. and Economou, A. (1999) A molecular switch in SecA protein couples ATP hydrolysis to protein translocation. *Mol. Microbiol.* 34: 1133–1145.

Karlberg, O., Canback, B., Kurland, C. G. and Andersson, S. G. E. (2000) The dual origin of the yeast mitochondrial proteome. *Yeast* 17: 170–187.

Karnauchov, I., Cai, D., Schmidt, I., Herrmann, R. G. and Klösgen, R. B. (1994) The thylakoid translocation of subunit 3 of photosystem I, the *psaF* gene product, depends on a bipartite transit peptide and proceeds along an azide-sensitive pathway. *J. Biol. Chem.* 269: 32871–32878.

Kassenbrock, C. K., Cao, W. and Douglas, M. G. (1993) Genetic and biochemical characterization of ISP6, a small mitochondrial outer membrane protein associated with the protein translocation complex. *EMBO J.* 12: 3023–3034.

Keegstra, K. and Cline, K. (1999) Protein import and routing systems of chloroplasts. *Plant Cell* 11: 557–570.

Keenan, R. J., Freymann, D. M., Stroud, R. M. and Walter, P. (2001) The signal recognition particle. *Annu. Rev. Biochem.* 70: 755–775.

Kermorgant, M., Bonnefoy, N. and Dujardin, G. (1997) Oxa1p, which is required for cytochrome *c* oxidase and ATP synthase complex formation, is embedded in the mitochondrial inner membrane. *Curr. Genet.* 31: 302–307.

Kerscher, O., Holder, J., Srinivasan, M., Leung, R. S. and Jensen, R. E. (1997) The Tim54p-Tim22p complex mediates insertion of proteins into the mitochondrial inner membrane. *J. Cell Biol.* 139: 1663–1675.

Kerscher, O., Sepuri, N. B. and Jensen, R. E. (2000) Tim18p is a new component of the Tim54p-Tim22p translocon in the mitochondrial inner membrane. *Mol. Biol. Cell* 11: 103–116.

Kessler, F., Blobel, G., Patel, H. A. and Schnell, D. J. (1994) Identification of two GTP-binding proteins in the chloroplast protein import machinery. *Science* 266: 1035–1039.

Kessler, F. and Blobel, G. (1996) Interaction of the protein import and folding machineries of the chloroplast. *Proc. Natl. Acad. Sci. USA* 93: 7684–7689.

Kim, J., Rusch, S., Luirink, J. and Kendall, D. A. (2001) Is Ffh required for export of secretory proteins? *FEBS Lett.* 505: 245–248.

Knoblauch, N. T. M., Rudiger, S., Schonfeld, H. J., Driessen, A. J. M., Schneider-Mergener, J., and Bukau, B. (1999) Substrate specificity of the SecB chaperone. *J. Biol. Chem.* 274: 34219–34225.

Koch, H. G., Moser, M., Schimz, K. L. and Müller, M. (2002) The integration of YidC into the cytoplasmic membrane of *Escherichia coli* requires the signal recognition particle, SecA and SecYEG. *J. Biol. Chem.* 277: 5715–5718.

Koehler, C. M. (2000) Protein translocation pathways of the mitochondrion. *FEBS Lett.* 476: 27–30.

Koehler, C. M., Merchant, S., Oppliger, W., Schmid, K., Jarosch, E., Dolfini, L., Junne, T., Schatz, G. and Tokatlidis, K. (1998) Tim9p, an essential partner subunit of Tim10p for the import of mitochondrial carrier proteins. *EMBO J.* 17: 6477–6486.

Koehler, C. M., Leuenberger, D., Merchant, S., Renold, A., Junne, T. and Schatz, G. (1999) Human deafness dystonia syndrome is a mitochondrial disease. *Proc. Natl. Acad. Sci. USA* 96: 2141–2146.

Koehler, C. M., Murphy, M. P., Bally, N. A., Leuenberger, D., Oppliger, W., Dolfini, L., Junne, T., Schatz, G. and Or, E. (2000) Tim18p, a new subunit of the TIM22 complex that mediates insertion of imported proteins into the yeast mitochondrial inner membrane. *Mol. Cell Biol.* 20: 1187–1193.

Kogata, N., Nishio, K., Hirohashi, T., Kikuchi, S. and Nakai, M. (1999) Involvement of a chloroplast homologue of the signal recognition particle receptor protein, FtsY, in protein targeting to thylakoids. *FEBS Lett.* 447: 329–333.

Kouranov, A., Chen, X., Fuks, B. and Schnell, D. J. (1998) Tic20 and Tic22 are new components of the protein import apparatus at the chloroplast inner envelope membrane. *J. Cell Biol.* 143: 991–1002.

Kouranov, A. and Schnell, D. J. (1997) Analysis of the interactions of preproteins with the import machinery over the course of protein import into chloroplasts. *J. Cell Biol.* 139: 1677–1685.

Kovermann, P., Truscott, K. N., Guiard, B., Rehling, P., Sepuri, N. B., Müller, H., Jensen, R. E., Wagner, R. and Pfanner, N. (2002) Tim22, the essential core of the mitochondrial protein insertion complex, forms a voltage-activated and signal-gated channel. *Mol. Cell* 9: 363–373.
Krimmer, T., Geissler, A., Pfanner, N. and Rassow, J. (2001) Sorting of preproteins into mitochondria. *Chembiochem* 2: 505–512.
Kübrich, M., Dietmeier, K. and Pfanner, N. (1995) Genetic and biochemical dissection of the mitochondrial protein-import machinery. *Curr. Genet.* 27: 393–403.
Küchler, M., Decker, S., Soll, J. and Heins, L. (2002) Protein import into chloroplasts involves redox-regulated proteins. *EMBO J.* 21: 6136–6145.
Laidler, V., Chaddock, A. M., Knott, T. G., Walker, D. and Robinson, C. (1995) A SecY homolog in *Arabidopsis thaliana*: Sequence of a full-length cDNA clone and import of the precursor protein into chloroplasts. *J. Biol. Chem.* 270: 17664–17667.
Lang, B. F., Burger, G., O'Kelly, C. J., Cedergren, R., Golding, G. B., Lemieux, C., Sankoff, D., Turmel, M. and Gray, M. W. (1997) An ancestral mitochondrial DNA resembling a eubacterial genome in miniature. *Nature* 387: 493–497.
Leuenberger, D., Bally, N. A., Schatz, G. and Koehler, C. M. (1999) Different import pathways through the mitochondrial intermembrane space for inner membrane proteins. *EMBO J.* 18: 4816–4822.
Li, X., Henry, R., Yuan, J., Cline, K. and Hoffman, N. E. (1995) A chloroplast homologue of the signal recognition particle subunit SRP54 is involved in the posttranslational integration of a protein into thylakoid membranes. *Proc. Natl. Acad. Sci. USA* 92: 3789–3793.
Luirink, J., Samuelsson, T. and de Gier, J. W. L. (2001) YidC/Oxa1p/Alb3: Evolutionarily conserved mediators of membrane protein assembly. *FEBS Lett.* 501: 1–5.
Ma, Y., Kouranov, A., LaSala, S. E. and Schnell, D. J. (1996) Two components of the chloroplast protein import apparatus, IAP86 and IAP75, interact with the transit sequence during the recognition and translocation of precursor proteins at the outer envelope. *J. Cell Biol.* 134: 315–327.
Maarse, A. C., Blom, J., Grivell, L. A. and Meijer, M. (1992) MPI1, an essential gene encoding a mitochondrial membrane protein, is possibly involved in protein import into yeast mitochondria. *EMBO J.* 11: 3619–3628.
Maarse, A. C., Blom, J., Keil, P., Pfanner, N. and Meijer, M. (1994) Identification of the essential yeast protein MIM17, an integral mitochondrial inner membrane protein involved in protein import. *FEBS Lett.* 349: 215–221.
Mant, A., Woolhead, C. A., Moore, M., Henry, R. and Robinson, C. (2001) Insertion of PsaK into the thylakoid membrane in a "Horseshoe" conformation occurs in the absence of signal recognition particle, nucleoside triphosphates, or functional albino. *J. Biol. Chem.* 276: 36200–36206.
Manting, E. H., van Der Does, C., Remigy, H., Engel, A. and Driessen, A. J. M. (2000) SecYEG assembles into a tetramer to form the active protein translocation channel. *EMBO J.* 19: 852–861.
Martin, W., Stoebe, B., Goremykin, V., Hansmann, S., Hasegawa, M. and Kowallik, K. V. (1998) Gene transfer to the nucleus and the evolution of chloroplasts. *Nature* 393: 162–165.
Matsumoto, G., Yoshihisa, T. and Ito, K. (1997) SecY and SecA interact to allow SecA insertion and protein translocation across the *Escherichia coli* plasma membrane. *EMBO J.* 16: 6384–6393.
May, T. and Soll, J. (2000) 14-3-3 proteins form a guidance complex with chloroplast precursor proteins in plants. *Plant Cell* 12: 53–64.
Mayer, A., Nargang, F. E., Neupert, W. and Lill, R. (1995) MOM22 is a receptor for mitochondrial targeting sequences and cooperates with MOM19. *EMBO J.* 14: 4204–4211.
McFadden, G. I. (1999) Endosymbiosis and evolution of the plant cell. *Curr. Opin. Plant. Biol.* 2: 513–519.

Meyer, T. H., Menetret, J. F., Breitling, R., Miller, K. R., Akey, C. W. and Rapoport, T. A. (1999) The bacterial SecY/E translocation complex forms channel-like structures similar to those of the eukaryotic Sec61p complex. *J. Mol. Biol.* 285: 17789–17800.

Mitoma, J. and Ito, A. (1992) The carboxy-terminal 10 amino acid residues of cytochrome b_5 are necessary for its targeting to the endoplasmic reticulum. *EMBO J.* 11: 4197–4203.

Moczko, M., Dietmeier, K., Söllner, T., Segui, B., Steger, H. F., Neupert, W. and Pfanner, N. (1992) Identification of the mitochondrial receptor complex in *Saccharomyces cerevisiae. FEBS Lett.* 310: 265–268.

Model, K., Prinz, T., Ruiz, T., Radermacher, M., Krimmer, T., Kuhlbrandt, W., Pfanner, N. and Meisinger, C. (2002) Protein translocase of the outer mitochondrial membrane: Role of import receptors in the structural organization of the TOM complex. *J. Mol. Biol.* 316: 657–666.

Moore, M., Harrison, M. S., Peterson, E. C. and Henry, R. (2000) Chloroplast Oxa1p homolog albino3 is required for post-translational integration of the light harvesting chlorophyll-binding protein into thylakoid membranes. *J. Biol. Chem.* 275: 1529–1532.

Mori, H., Summer, E. J., Ma, X. and Cline, K. (1999) Component specificity for the thylakoidal Sec and Delta pH-dependent protein transport pathways. *J. Cell Biol.* 146: 45–56.

Mori, H., Summer, E.J. and Cline, K. (2001) Chloroplast TatC plays a direct role in thylakoid ΔpH-dependent protein transport. *FEBS Lett.* 501: 65–68.

Mori, H. and Cline, K. (2001) Post-translational protein translocation into thylakoids by the Sec and ΔpH-dependent pathways. *Biochim. Biophys. Acta* 1541: 80–90.

Mould, R. M., Shackleton, J. B. and Robinson, C. (1991) Transport of proteins into chloroplasts. Requirements for the efficient import of two lumenal oxygen-evolving complex proteins into isolated thylakoids. *J. Biol. Chem.* 266: 17286–17289.

Murakami, T., Haga, K., Takeuchi, M. and Sato, T. (2002) Analysis of the *Bacillus subtilis spoIIIJ* gene and its Paralogue gene, *yqjG*. *J. Bacteriol.* 184: 1998–2004.

Nakai, M., Sugita, D., Omata, T. and Endo, T. (1993) Sec-Y protein is localized in both the cytoplasmic and thylakoid membranes in the cyanobacterium *Synechococcus* PCC7942. *Biochem. Biophys. Res. Commun.* 193: 228–234.

Nakai, M., Nohara, T., Sugita, D. and Endo, T. (1994) Identification and characterization of the Sec-A protein homologue in the cyanobacterium *Synechococcus* PCC7942. *Biochem. Biophys. Res. Commun.* 200: 844–851.

Nielsen, E., Akita, M., Davila-Aponte, J. and Keegstra, K. (1997) Stable association of chloroplastic precursors with protein translocation complexes that contain proteins from both envelope membranes and a stromal Hsp100 molecular chaperone. *EMBO J.* 16: 935–946.

Nilsson, R., Brunner, J., Hoffman, N. E. and van Wijk, K. J. (1999) Interactions of ribosome nascent chain complexes of the chloroplast-encoded D1 thylakoid membrane protein with cpSRP54. *EMBO J.* 18: 733–742.

Nouwen, N., de Kruijff, B. and Tommassen, J. (1996) *prlA* suppressors in *Escherichia coli* relieve the proton electrochemical gradient dependency of translocation of wild-type precursors. *Proc. Natl. Acad. Sci. USA* 93: 5953–5957.

Nouwen, N. and Driessen, A. J. M. (2002) SecDFyajC forms a heterotetrameric complex with YidC. *Mol. Microbiol.* 44: 1397–1405.

Packer, J. C. L. and Howe, C. J. (1998) Algal plastid genomes encode homologues of the SRP-associated RNA. *Mol. Microbiol.* 27: 507–510.

Paschen, S. A., Rothbauer, U., Kaldi, K., Bauer, M. F., Neupert, W. and Brunner, M. (2000) The role of the TIM8-13 complex in the import of Tim23 into mitochondria. *EMBO J.* 19: 6392–6400.

Peltier, J. B., Emanuelsson, O., Kalume, D. E., Ytterberg, J., Friso, G., Rudella, A., Liberles, D. A., Soderberg, L., Roepstorff, P., von Heijne, G. and van Wijk, K. J. (2002) Central functions of the lumenal and peripheral thylakoid proteome of *Arabidopsis* determined by experimentation and genome-wide prediction. *Plant Cell* 14: 211–236.

Perry, S. E. and Keegstra, K. (1994) Envelope membrane proteins that interact with chloroplastic precursor proteins. *Plant Cell* 6: 93–105.

Pfanner, N., Douglas, M. G., Endo, T., Hoogenraad, N. J., Jensen, R. E., Meijer, M., Neupert, W., Schatz, G., Schmitz, U. K. and Shore, G. C. (1996) Uniform nomenclature for the protein transport machinery of the mitochondrial membranes. *Trends Biochem. Sci.* 21: 51–52.

Pohlmeyer, K., Soll, J., Steinkamp, T., Hinnah, S. and Wagner, R. (1997) Isolation and characterization of an amino acid-selective channel protein present in the chloroplastic outer envelope membrane. *Proc. Natl. Acad. Sci. USA* 94: 9504–9509.

Ramamurthy, V. and Oliver, D. (1997) Topology of the integral membrane form of *Escherichia coli* SecA protein reveals multiple periplasmically exposed regions and modulation by ATP binding. *J. Biol. Chem.* 272: 23239–23246.

Rapoport, T. A., Jungnickel, B. and Kutay, U. (1996) Protein transport across the eukaryotic endoplasmic reticulum and bacterial inner membranes. *Annu. Rev. Biochem.* 65: 271–303.

Rassow, J., Dekker, P. J. T., van Wilpe, S., Meijer, M. and Soll, J. (1999) The preprotein translocase of the mitochondrial inner membrane: Function and evolution. *J. Mol. Biol.* 286: 105–120.

Reichert, A. S. and Neupert, W. (2002) Contact sites between the outer and inner membrane of mitochondria: Role in protein transport. *Biochem. Biophys. Acta* 1592: 41–49.

Reumann, S., Davila-Aponte, J. and Keegstra, K. (1999) The evolutionary origin of the protein-translocating channel of chloroplastic envelope membranes: Identification of a cyanobacterial homolog. *Proc. Natl. Acad. Sci. USA* 96: 784–789.

Ryan, K. R., Menold, M. M., Garrett, S. and Jensen, R. E. (1994) SMS1, a high-copy suppressor of the yeast mas6 mutant, encodes an essential inner membrane protein required for mitochondrial protein import. *Mol. Biol. Cell* 5: 529–538.

Ryan, K. R., Leung, R. S. and Jensen, R. E. (1998) Characterization of the mitochondrial inner membrane translocase complex: the Tim23p hydrophobic domain interacts with Tim17p but not with other Tim23p molecules. *Mol. Cell Biol.* 18: 178–187.

Ryan, M. T., Müller, H. and Pfanner, N. (1999) Functional staging of ADP/ATP carrier translocation across the outer mitochondrial membrane. *J. Biol. Chem.* 274: 20619–20627.

Saeki, K., Suzuki, H., Tsuneoka, M., Maeda, M., Iwamoto, R., Hasuwa, H., Shida, S., Takahashi, T., Sakaguchi, M., Endo, T., Miura, Y., Mekada, E. and Mihara, K. (2000) Identification of mammalian TOM22 as a subunit of the preprotein translocase of the mitochondrial outer membrane. *J. Biol. Chem.* 275: 31996–32002.

Sakamoto, W., Spielewoy, N., Bonnard, G., Murata, M. and Wintz, H. (2000) Mitochondrial localization of AtOXA1, an *arabidopsis* homologue of yeast Oxa1p involved in the insertion and assembly of protein complexes in mitochondrial inner membrane. *Plant Cell Physiol.* 41: 1157–1163.

Samuelson, J. C., Chen, M., Jiang, F., Möller, I., Wiedmann, M., Kuhn, A., Phillips, G. J. and Dalbey, R. E. (2000) YidC mediates membrane protein insertion in bacteria. *Nature* 406: 637–641.

Santini, C. L., Ize, B., Chanal. A., Müller, M., Giordano, G. and Wu, L. F. (1998) A novel sec-independent periplasmic protein translocation pathway in *Escherichia coli*. *EMBO J.* 17: 101–112.

Sargent, F., Bogsch, E. G., Stanley, N. R., Wexler, M., Robinson, C., Berks B. C. and Palmer, T. (1998) Overlapping functions of components of a bacterial Sec-independent protein export pathway. *EMBO J.* 17: 3640–3650.

Sato, S., Nakamura, Y., Kaneko, T., Asamizu, E. and Tabata, S. (1999) Complete structure of the chloroplast genome of *Arabidopsis thaliana*. *DNA Res.* 29: 283–290.

Scaramuzzi, C. D., Hiller, R. G. and Stokes, H. W. (1992) Identification of a chloroplast-encoded *secA* gene homologue in a chromophytic alga: Possible role in chloroplast protein translocation. *Curr. Genet.* 22: 421–427.

Scherer, P. E., Manning-Krieg, U. C., Jenö, P., Schatz, G. and Horst, M. (1992) Identification of a 45-kDa protein at the protein import site of the yeast mitochondrial inner membrane. *Proc. Natl. Acad. Sci. USA* 89: 11930–11934.

Schiebel, E., Driessen, A. J. M., Hartl, F. U. and Wickner, W. (1991) $\Delta\mu H^+$ and ATP function at different steps of the catalytic cycle of preprotein translocase. *Cell* 64: 927–939.

Schleiff, E. and Klösgen, R. B. (2001) Without a little help from "my" friends: direct insertion of proteins into chloroplast membranes? *Biochim. Biophys. Acta* 1541: 22–33.

Schleiff, E., Soll, J., Sveshnikova, N., Tien, R., Wright, S., Dabney-Smith, C., Subramanian, C. and Bruce, B. D. (2002) Structural and guanosine triphosphate/diphosphate requirements for transit peptide recognition by the cytosolic domain of the chloroplast outer envelope receptor, Toc34. *Biochemistry* 41: 1934–1946.

Schnell, D. J., Kessler, F. and Blobel, G. (1994) Isolation of components of the chloroplast protein import machinery. *Science* 266: 1007–1012.

Schuenemann, D., Gupta, S., Persello-Cartieaux, F., Klimyuk, V. I., Jones, J. D. G., Nussaume, L. and Hoffman, N. E. (1998) A novel signal recognition particle targets light-harvesting proteins to the thylakoid membranes. *Proc. Natl. Acad. Sci. USA* 95: 10312–10316.

Schuenemann, D., Amin, P., Hartmann, E. and Hoffman, N. E. (1999) Chloroplast SecY is complexed to SecE and involved in the translocation of the 33-kDa but not the 23-kDa subunit of the oxygen-evolving complex. *J. Biol. Chem.* 274: 12177–12182.

Scotti, P. A., Valent, Q. A., Manting, E. H., Urbanus, M. L., Driessen, A. J. M., Oudega, B. and Luirink, J. (1999) SecA is not required for signal recognition particle-mediated targeting and initial membrane insertion of a nascent inner membrane protein. *J. Biol. Chem.* 274: 29883–29888.

Scotti, P. A., Urbanus, M. L., Brunner, J., de Gier, J. W. L., von Heijne, G., van der Does, C., Driessen, A. J. M., Oudega, B. and Luirink J. (2000) YidC, the *Escherichia coli* homologue of mitochondrial Oxa1p, is a component of the Sec translocase. *EMBO J.* 19: 542–549.

Seedorf, M., Waegemann, K. and Soll, J. (1995) A constituent of the chloroplast import complex represents a new type of GTP-binding protein. *Plant J.* 7: 401–411.

Settles, A. M., Yonetani, A., Baron, A., Bush, D.R., Cline, K. and Martienssen, R. (1997) Sec-independent protein translocation by the maize Hcf106 protein. *Science* 278: 1467–1470.

Sirrenberg, C., Bauer, M. F., Guiard, B., Neupert, W. and Brunner, M. (1996) Import of carrier proteins into the mitochondrial inner membrane mediated by Tim22. *Nature* 384: 582–585.

Sirrenberg, C., Endres, M., Folsch, H., Stuart, R. A., Neupert, W. and Brunner, M. (1998) Carrier protein import into mitochondria mediated by the intermembrane proteins Tim10/Mrs11 and Tim12/Mrs5. *Nature* 391: 912–915.

Snyders, S., Ramamurthy, V. and Oliver, D. (1997) Identification of a region of interaction between *Escherichia coli* SecA and SecY proteins. *J. Biol. Chem.* 272: 11302–11306.

Sohrt, K. and Soll, J. (2000) Toc64, a new component of the protein translocon of chloroplasts. *J. Cell Biol.* 148: 1213–1221.

Söllner, T., Griffiths, G., Pfaller, R., Pfanner, N. and Neupert, W. (1989) MOM19, an import receptor for mitochondrial precursor proteins. *Cell* 59: 1061–1070.

Söllner, T., Rassow, J., Wiedmann, M,, Schlossmann, J., Keil, P., Neupert, W. and Pfanner, N. (1992) Mapping of the protein import machinery in the mitochondrial outer membrane by crosslinking of translocation intermediates. *Nature* 355: 84–87.

Stahl, T., Glockmann, C., Soll, J. and Heins, L. (1999) Tic40, a new "old" subunit of the chloroplast protein import translocon. *J. Biol. Chem.* 274: 37467–37472.

Stan, T., Ahting, U., Dembowski, M., Kunkele, K. P., Nussberger, S., Neupert, W. and Rapaport, D. (2000) Recognition of preproteins by the isolated TOM complex of mitochondria. *EMBO J.* 19: 4895–4902.

Steiner, J. M., Löffelhardt, W. (2002) Protein import into cyanelles. *Trends Plant Sci.* 7: 72–77.

Strub, A., Lim, J. H., Pfanner, N. and Voos, W. (2000) The mitochondrial protein import motor. *Biol. Chem.* 381: 943–949.

Stuart, R. (2002) Insertion of proteins into the inner membrane of mitochondria: The role of the Oxa1p complex. *Biochim. Biophys. Acta* 1592: 79–87.

Sundberg, E., Slagter, J. G., Fridborg, I., Cleary, S. P., Robinson, C. and Coupland, G. (1997) ALBINO3, an *Arabidopsis* nuclear gene essential for chloroplast differentiation, encodes a chloroplast protein that shows homology to proteins present in bacterial membranes and yeast mitochondria. *Plant Cell* 9: 717–730.

Sveshnikova, N., Soll, J. and Schleiff, E. (2000) Toc34 is a preprotein receptor regulated by GTP and phosphorylation. *Proc. Natl. Acad. Sci. USA* 97: 4973–4978.

Sun, Y. -J., Forouhar, F., Li, H., Tu, S. -L., Yeh, Y. -H., Kao, S., Shr, H. -L., Chou, C. -C., Chen, C. and Hsiao, C. -D. (2002) Crystal structure of pea Toc34, a novel GTPase of the chloroplast protein translocon. *Nat. Struct. Biol.* 9: 95–100.

Tokatlidis, K., Junne, T., Moes, S., Schatz, G., Glick, B. S. and Kronidou, N. (1996) Translocation arrest of an intramitochondrial sorting signal next to Tim11 at the inner-membrane import site. *Nature* 384: 585–588.

Truscott, K. N., Kovermann, P., Geissler, A., Merlin, A., Meijer, M., Driessen, A. J., Rassow, J., Pfanner, N. and Wagner, R. (2001) A presequence- and voltage-sensitive channel of the mitochondrial preprotein translocase formed by Tim23. *Nat. Struct. Biol.* 8: 1074–1082.

Tu, C. J., Schuenemann, D. and Hoffman, N. E. (1999) Chloroplast FtsY, chloroplast signal recognition particle, and GTP are required to reconstitute the soluble phase of light-harvesting chlorophyll protein transport into thylakoid membranes. *J. Biol. Chem.* 274: 27219–27224.

Tu, C. J., Peterson, E. C., Henry, R. and Hoffman, N. E. (2000) The L18 domain of light-harvesting chlorophyll proteins binds to chloroplast signal recognition particle 43. *J. Biol. Chem.* 275: 13187–13190.

Ulbrandt, N. D., Newitt, J. A. and Bernstein, H. D. (1997) The *E. coli* signal recognition particle is required for the insertion of a subset of inner membrane proteins. *Cell* 88: 187–196.

Unseld, M., Marienfeld, J. R., Brandt, P. and Brennicke, A. (1997) The mitochondrial genome of *Arabidopsis thaliana* contains 57 genes in 366,924 nucleotides. *Nat. Genet.* 15: 57–61.

Urbanus, M. L., Fröderberg, L., Drew, D., Bjork, P., de Gier, J. W. L., Brunner, J., Oudega, B. and Luirink, J. (2002) Targeting, insertion, and localization of *Escherichia coli* YidC. *J. Biol. Chem.* 277: 12718–12723.

Valentin, K. (1993) SecA is plastid-encoded in a red alga: Implications for the evolution of plastid genomes and the thylakoid protein import apparatus. *Mol. Gen. Genet.* 236: 245–250.

van der Wolk, J. P. W., Fekkes, P., Boorsma, A., Huie, J. L., Silhavy, T. J. and Driessen, A. J. M. (1998) PrlA4 prevents the rejection of signal sequence defective preproteins by stabilizing the SecA-SecY interaction during the initiation of translocation. *EMBO J.* 17: 3631–3639.

Vestweber, D., Brunner, J., Baker, A. and Schatz, G. (1989) A 42K outer-membrane protein is a component of the yeast mitochondrial protein import site. *Nature* 341: 205–209.

Völker, J., Mendel-Hartvig, J. and Barkan, A. (1997) Transposon-disruption of a maize nuclear gene, *tha1*, encoding a chloroplast SecA homologue: *In vivo* role of cp-SecA in thylakoid protein targeting. *Genetics* 145: 467–478.

Waegemann, K. and Soll, J. (1996) Phosphorylation of the transit sequence of chloroplast precursor proteins. *J. Biol. Chem.* 271: 6545–6554.

Walker, M. B., Roy, L. M., Coleman, E., Voelker, R. and Barkan, A. (1999) The maize *tha4* gene functions in sec-independent protein transport in chloroplasts and is related to *hcf106*, *tatA*, and *tatB*. *J. Cell Biol.* 147: 267–276.

Wang, L., Miller, A. and Kendall, D. A. (2000) Signal peptide determinants of SecA binding and stimulation of ATPase activity. *J. Biol. Chem.* 275: 10154–10159.

Weiner, J. H., Bilous, P. T., Shaw, G. M., Lubitz, S. P., Frost, L., Thomas, G. H., Cole, J. A. and Turner, R. J. (1998) A novel and ubiquitous system for membrane targeting and secretion of cofactor-containing proteins. *Cell* 93: 93–101.

Werhahn, W., Niemeyer, A., Jansch, L., Kruft, V. V., Schmitz, U. K. and Braun, H. P. (2001) Purification and characterization of the preprotein translocase of the outer mitochondrial membrane from *Arabidopsis*: Identification of multiple forms of TOM20. *Plant Physiol.* 125: 943–954.

Wexler, M., Sargent, F., Jack, R. L., Stanley, N. R., Bogsch, E. G., Robinson, C., Berks, B. C. and Palmer, T. (2000) TatD is a cytoplasmic protein with DNase activity. No requirement for TatD family proteins in sec-independent protein export. *J. Biol. Chem.* 275: 16717–16722.

Woolhead, C. A., Thompson, S. J., Moore, M., Tissier, C., Mant, A., Rodger, A., Henry, R. and Robinson, C. (2001) Distinct Albino3-dependent and -independent pathways for thylakoid membrane protein insertion. *J. Biol. Chem.* 276: 40841–40846.

Wu, C., Seibert, F. S. and Ko, K. (1994) Identification of chloroplast envelope proteins in close physical proximity to a partially translocated chimeric precursor protein. *J. Biol. Chem.* 269: 32264–32271.

Yang, A. J. and Mulligan, R. M. (1996) Identification of a 4.5S-like ribonucleoprotein in maize mitochondria. *Nucleic Acids Res.* 24: 3601–3606.

Yuan, J. and Cline, K. (1994) Plastocyanin and the 33-kDa subunit of the oxygen-evolving complex are transported into thylakoids with similar requirements as predicted from pathway specificity. *J. Biol. Chem.* 269:18463–18467.

Yuan, J., Henry, R., McCaffery, M. and Cline, K. (1994) SecA homolog in protein transport within chloroplasts: Evidence for endosymbiont-derived sorting. *Science* 266: 796–798.

Chapter 13

Mitosomes, Hydrogenosomes and Mitochondria: Variations on a Theme?

Mark van der Giezen and Jorge Tovar

CONTENTS

Abstract

Mitochondria are considered defining features of eukaryotic cells, but many eukaryotic microorganisms do not possess these cellular organelles. Research on amitochondriate eukaryotes has revealed significant compartmentalization of metabolic functions in distinctive organelles that at first inspection appear to have little in common with mitochondria. However, more careful investigations have clearly linked these organelles to mitochondria and there is increasing evidence that they all likely represent evolutionary variants of a single microbial endosymbiosis that occurred before the divergence of all extant eukaryotes. The apparent reluctance of eukaryotic organisms to part with these seemingly valuable organelles points to the existence of a hitherto-unidentified organellar function essential for cell survival in both aerobic and anaerobic environments.

13.1 Introduction

Mutually beneficial endosymbiotic associations are commonly observed in nature between two primarily self-sufficient organisms (Brul and Stumm, 1994; Stumm and Zwart, 1986;

0-415-29904-7/04/$0.00+$1.50

Taylor, 1979, 1983). These interactions usually, but not exclusively, involve a phagocytic eukaryotic host and a prokaryotic or eukaryotic endosymbiont. Almost invariably, these interactions are either beneficial or selectively neutral, because interacting partners often fulfill complementary roles in resource recycling within an ecosystem. Most current theories of eukaryotic evolution propose that mitochondria and chloroplasts — organelles central to the evolution and diversification of eukaryotes — have arisen through endosymbiotic associations (Cavalier-Smith, 2002; Gray, 1999; Gray et al. 1999; Karlin et al., 1999; Kurland and Andersson, 2000; López-García and Moreira, 1999; Martin et al., 1998; Martin and Müller, 1998; Moreira et al., 2000; Moreira and López-García, 1998), although the identity of proposed interacting partners differs considerably between theories. Extensive molecular and phylogenetic analyses of genes and proteins from extant bacterial and eukaryotic cells support the view that mitochondria are remnants of an endosymbiosis established more than 1800 million years ago (Gray et al., 1999; Nisbet and Sleep, 2001). The current state of the nuclear, mitochondrial and plastid genomes is known to be the result of a complex process of reductive evolution, progressive loss of redundant genes from host and endosymbiont and of lateral gene transfer, mostly from the endosymbiont to the cell nucleus, further aided by mutation and recombination (Gray, 1999; Gray et al., 1999; Kurland and Andersson, 2000; Martin and Schnarrenberger, 1997).

Given the crucial role played by mitochondria in the evolution and diversification of eukaryotic cells, establishing unequivocally the timing of its endosymbiotic acquisition is central to our understanding of the natural history of eukaryotes. Eukaryotic organisms that lack typical mitochondria, such as the diplomonads (e.g., *Giardia*), parabasalids (e.g., *Trichomonas*), archamoebae (e.g., *Entamoeba*) and microsporidians (e.g., *Trachipleistophora*), were for many years considered representative examples of the primitive nucleated protoeukaryotic cell and were grouped together in the now abandoned subkingdom Archezoa (Cavalier-Smith, 1998, 2002). Most Archezoa members have since been shown to either harbor mitochondrion-related organelles (i.e., mitosomes, hydrogenosomes) or to contain in their nuclear genomes a handful of genes of mitochondrial ancestry, suggesting their postmitochondrial origins. This chapter reviews progress on the biochemical and structural characterization of mitosomes and hydrogenosomes of microaerophilic and anaerobic protists and fungi and discusses the contribution and implications of recent findings to our understanding of microbial eukaryotic evolution.

13.2 Derived Mitochondrial Organelles

Eukaryotic organisms that lack typical mitochondria can be conveniently assigned to one of three groups (modified from the classification proposed by Martin and Müller, 1998):

1. *Type I.* Organisms whose origins predate the endosymbiotic event that gave rise to mitochondria (primitively amitochondriate, no organelle)
2. *Type II.* Organisms that despite lacking most typical mitochondrial functions retain some form of compartmentalized energy metabolism in a mitochondrion-related organelle (hydrogenosome-containing organisms)
3. *Type III.* Organisms that have lost most mitochondrial functions secondarily and lack compartmentalized energy metabolism (a mitochondrion-derived organelle might or might not have been retained)

Current evidence suggests that most, if not all, amitochondriate eukaryotes belong to Types II and III, as mitochondrion-related organelles or genes of mitochondrial ancestry have been identified in all amitochondrial eukaryotes analyzed in sufficient detail thus far.

13.2.1 Hydrogenosomes

Type II amitochondriate organisms harbor hydrogenosomes. Hydrogenosomes are defined as organelles that, under anoxic conditions, oxidatively decarboxylate malate or pyruvate into acetate, carbon dioxide and hydrogen with the concomitant production of energy in the form of ATP (Müller, 1993). Hydrogen-producing organelles are found in a variety of eukaryotes, but never in mitochondrion-containing organisms, suggesting that hydrogenosomes and mitochondria are mutually exclusive. Hydrogenosomes were discovered in the early 1970s in the cattle parasite *Tritrichomonas foetus* (Cerkasovova et al., 1973; Lindmark and Müller, 1973) and in the human parasite *Trichomonas vaginalis* (Lindmark et al., 1975). Apart from parasitic hydrogenosome-containing organisms, these organelles are also found in some anaerobic ciliates (Berger and Lynn, 1992; Paul et al., 1990) and anaerobic fungi (Yarlett et al., 1986). As might be obvious from their distribution in phylogenetically distantly related organisms (Figure 13.1), hydrogenosomes are not vertically inherited but are an example of convergent evolution most likely driven by the similar environmental conditions encountered by these unrelated organisms.

An early theory, which was abandoned quickly, thought of hydrogenosomes as possible clostridial endosymbionts (Whatley et al., 1979) but it was soon realized that the mutual exclusion of mitochondria and hydrogenosomes might indicate a closer relationship between the two types of organelles (Cavalier-Smith, 1987). Although similarities between mitochondria and hydrogenosomes were observed repeatedly (Finlay and Fenchel, 1989; Humphreys et al., 1994; Lahti and Johnson, 1991; Marvin-Sikkema et al., 1994), a significant breakthrough in the interpretation of data came in 1996 when several groups reported that

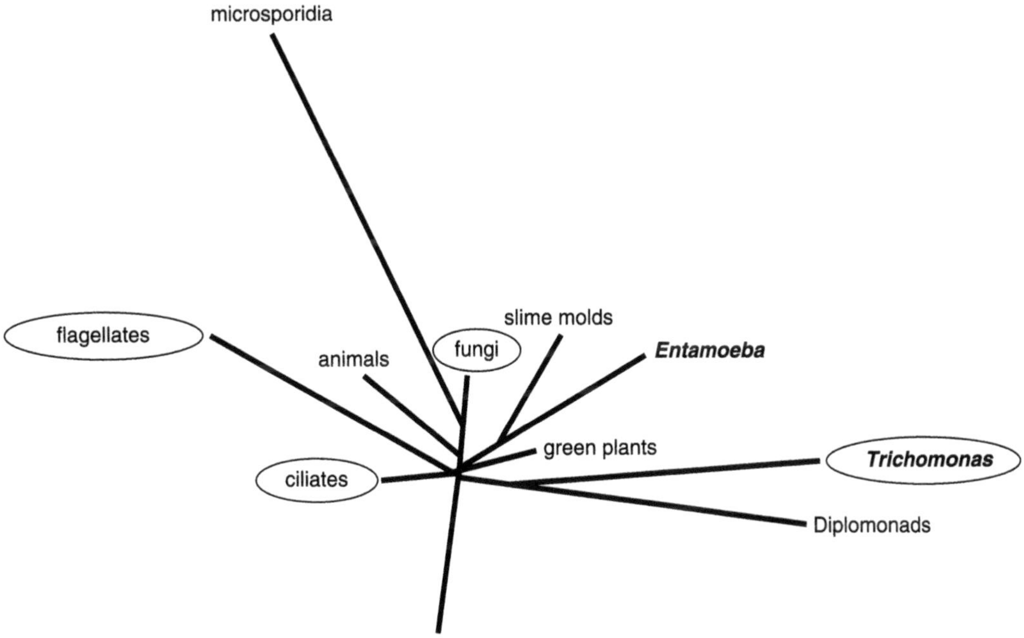

FIGURE 13.1 Schematic eukaryotic tree of life showing the distribution of hydrogenosomes. Groups that contain organisms with hydrogenosomes are indicated by an oval. The oval around the flagellates is based exclusively on the discovery of a putative hydrogenosome in *Carpediemonas membranifera* [Simpson, A.G. and Patterson, D.J. (1999) *Eur. J. Protistol.* 35: 353–370]. The tree has been corrected for major pitfalls affecting molecular phylogeny reconstructions. [Adapted from Gribaldo, S. and Philippe, H. (2002) *Theor. Popul. Biol.* 61: 391–408.]

trichomonad genes encoding hydrogenosomal heat-shock proteins were phylogenetically related to those of mitochondria (Bui et al., 1996; Germot et al., 1996; Horner et al., 1996; Roger et al., 1996). The traditional view that trichomonads were primitively amitochondriate made hydrogenosomes difficult to rationalize in the light of their mitochondrial features, but current evidence strongly supports the view that hydrogenosomes likely represent derived mitochondria (Embley et al., 1997, 2003).

13.2.2 Mitosomes

Over the past few years, the existence of highly derived mitochondrion-related organelles in amitochondrial protists lacking hydrogenase activity has been reported (Mai et al., 1999; Riordan et al., 2003; Tovar et al., 1999; Williams et al., 2002). It is thought that such organelles represent mitochondrial remnants. Although not much is known about the extent of their distribution among different amitochondriate taxa, current evidence suggests that mitochondrion-derived organelles may be widespread. Given the different environmental selective pressures experienced by different organisms, mitochondrion-derived organelles are likely to be at different stages of reductive evolution, and, as such, their proteome complements might vary in different organisms. This is readily illustrated by the presence of mitochondrial chaperonin 60 (Cpn60) in remnant mitochondrial organelles of *Entamoeba* and *Cryptosporidium*, but its apparent absence from those of microsporidia (Katinka et al., 2001; Mai et al., 1999; Riordan et al., 2003; Tovar et al., 1999; Williams et al., 2002). Two different names were originally proposed for the mitochondrial remnant organelles of *Entamoeba*: (1) mitosomes (Tovar et al., 1999) to indicate their relatedness to mitochondria and (2) cryptons (Mai et al., 1999) to indicate their cryptic nature and unknown functions. In the interest of readability and to avoid unnecessary confusion, all mitochondrion-related organelles that lack hydrogenase activity are collectively referred to as mitosomes in this chapter.

Mitosomes were first identified in the intestinal human pathogen *Entamoeba histolytica* as small cellular organelles housing the mitochondrial chaperonin Cpn60 (Mai et al., 1999; Tovar et al., 1999). Traditional ultrastructural and biochemical studies of *E. histolytica* failed to identify a specific cellular compartment of energy metabolism (Martínez-Palomo, 1993; McLaughlin and Aley, 1985). Such lack of recognizable mitochondria or hydrogenosomes together with the absence of tricarboxylic acid cycle enzymes and cytochromes and the detection and purification of enzymes of anaerobic metabolism from the cell cytoplasm (Reeves et al., 1977; Reeves, 1984; Takeuchi et al., 1975) appeared to confirm *E. histolytica* as a Type I amitochondriate organism (i.e., no organelle). However, the observation that this amoeboid eukaryote branches after the divergence of mitochondrion-bearing flagellates in various gene phylogenies suggested that *E. histolytica* might have undergone a process of reductive evolution from mitochondrion-bearing ancestors. Direct support for this hypothesis came from the work of Clark and Roger (1995), who identified in *E. histolytica* two nuclear genes of mitochondrial ancestry encoding the enzyme pyridine nucleotide transhydrogenase (PNT) and the molecular chaperonin Cpn60. Cloning and characterization of the Cpn60 gene and immunolocalization of its encoded product led to the identification of the mitosome, a small cytoplasmic structure characterized by a protein import mechanism functionally conserved with that of mitochondria and hydrogenosomes (Mai et al., 1999; Tovar et al., 1999).

Since then, mitochondrion-related organelles have also been identified in two other amitochondriate pathogens, the microsporidian *Trachipleistophora hominis* (Williams et al., 2002) and the apicomplexan *Cryptosporidium parvum* (Riordan et al., 2003), and the search for their presence in a number of other amitochondrial organisms — including members of the arguably early branching diplomonads *Hexamita, Spironucleus* and *Giardia* — is currently in progress in various laboratories worldwide. A recent report suggesting a close relationship between the excavate flagellate *Carpediemonas* and the diplomonad *Giardia* (Simpson et al., 2002) makes the presence of a mitochondrion-derived organelle in the latter all the more likely because *Carpediemonas* have been reported to contain an organelle resembling a hydrogenosome (Simpson and Patterson, 1999). Further support for this hypothesis comes from recent biochemical studies in *Giardia* trophozoites, in which the presence of a compartment with mitochondrial features in this organism has been suggested by using several cationic dyes (Lloyd et al., 2002; Lloyd and Harris, 2002). Together, these data suggest that all extant eukaryotes might have descended from ancestors that carried the mitochondrial endosymbiont.

13.3 Morphology

Most authors agree that mitochondria evolved as a result of an endosymbiotic event. There is enormous controversy about the actual players involved (Cavalier-Smith, 2002; Gray, 1999; Gray et al., 1999; Karlin et al., 1999; Kurland and Andersson, 2000; López-García and Moreira, 1999; Martin et al., 1998; Martin and Müller, 1998; Moreira et al., 2000) but it is thought that the endosymbiont that gave rise to the mitochondrion was a Gram-negative eubacterium (Gray et al., 1999). Comparative genomics indicates that many mitochondrial proteins are highly homologous to their α-proteobacterial counterparts, in particular to *Rickettsia* proteins (Andersson et al., 1998). One consequence of this is that mitochondria are surrounded by two bounding membranes, like the original symbiont (Cavalier-Smith, 1987). Although most textbooks present mitochondria in a classical cristated form, this oversimplified interpretation does not faithfully represent the true multifaceted mitochondrial morphology (Frey and Mannella, 2000). Mitochondria range from small elongated or elliptical structures ca. 1 to 2 μm in size to extended filamentous networks or clusters connected with intermitochondrial junctions. Moreover, their morphology seems to be dependent on cell type and physiological status (Hackenbrock, 1966; Prince, 1999; Scheffler, 2000). It is now apparent that mitochondrial morphology is also dependent on specific interactions between its outer membrane and various components of the cell cytoskeleton (Scheffler, 1999).

Having traditional mitochondria in mind, it was no surprise that hydrogenosomes were considered unrelated to mitochondria. Hydrogenosomes are ca.1-2 μm in diameter (similar to particulated mitochondria) and are without exception surrounded by two closely apposed membranes (Figure 13.2, Panels A and C; Benchimol et al., 1996, 1997; Benchimol and De Souza, 1983; Finlay and Fenchel, 1989; van der Giezen et al., 1997b). Inner membrane cristae are rarely seen, perhaps because the whole protein machinery that constitutes the electron transport chain is absent in hydrogenosome-containing eukaryotes. All these eukaryotes are facultative or strict anaerobes and therefore unable to use oxygen as terminal electron acceptor in a mitochondrial manner. This suggests that the large surface increase of the inner mitochondrial membrane to boost capacity of the electron transport chain, as seen in aerobic eukaryotes, might not be necessary for the inner hydrogenosomal membrane. In support of this suggestion is the observation that yeast mitochondria rapidly lose their cristae and change into a hydrogenosomal morphology when facing anoxic conditions (Lloyd, 1974).

Mitosomes appear to be less uniform in size. Preliminary estimates based on fluorescence microscopy imaging estimated the *E. histolytica* mitosomes at 1 to 2 μm in diameter (Tovar et al., 1999). In contrast, electron microscopy measurements of the *Trachipleistophora* organelles established their size at ca. 50 nm × 90 nm and those of *Cryptosporidium* appear to be ca. 150 to 300 nm in size, both significantly smaller than mitochondria and hydrogenosomes (Riordan et al., 2003; Williams et al., 2002). Like mitochondria and hydrogenosomes, *Cryptosporidium* and *Trachipleistophora* mitosomes appear to be surrounded by

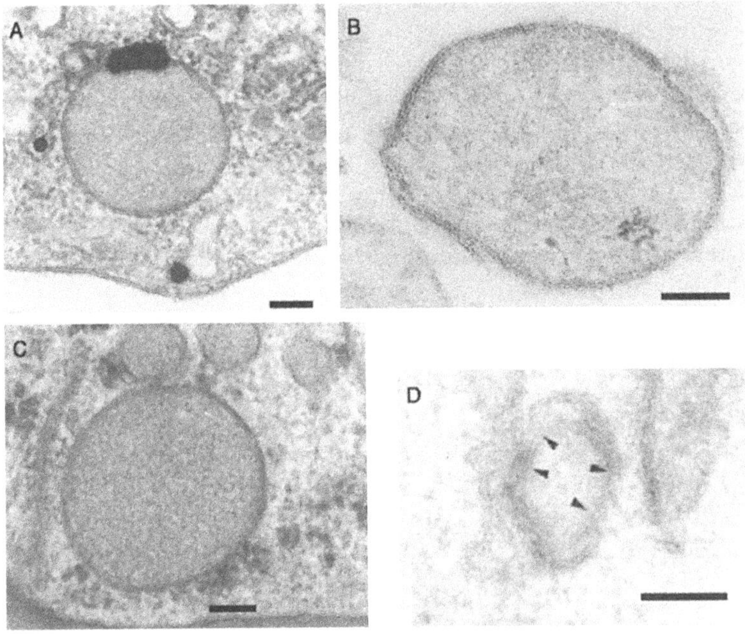

FIGURE 13.2 Electron micrographs of different mitochondrion-related organelles. (A) A hydrogenosome from the cattle parasite *Tritrichomonas foetus*. (B) A putative *Entamoeba histolytica* mitosome. (C) Hydrogenosomes from the anaerobic fungus *Neocallimastix patriciarum*. (D) Mitochondrial remnant organelle from the microsporidian *Trachipleistophora hominis*. Bars: 100 nm (A, B, C) or 50 nm (D). [(A) and (C) provided by Dr. Marlene Benchimol, Brazil; (B) reproduced with permission from the American Society for Microbiology; (D) reproduced with permission from *Nature*.]

double membranes (Figure 13.2D; Riordan et al., 2003; Williams et al., 2002). Although the *Entamoeba* mitosome has also been reported to be bounded by two closely apposed membranes, no mitosomal marker proteins were detected in these structures (Figure 13.2B; Ghosh et al., 2000). Instead, only partially purified fractions of propidium-iodide-stained organelles were used in electron microscopy studies, weakening the argument that these might represent bona fide mitosomes. The identity of these structures needs to be confirmed directly by *in situ* immuno-EM labeling by using antibodies specific to mitosomal proteins. Like most hydrogenosomes, neither *Trachipleistophora* nor *Entamoeba* mitosomes appear to contain inner membrane cristae.

13.4 Metabolic Capacity

The major functions of mitochondria are to convert pyruvate to acetyl-CoA and produce energy in the form of ATP (26 of the 30 molecules of ATP produced by the complete aerobic oxidation of 1 molecule of glucose generated in the mitochondrion). Pyruvate decarboxylation and ATP generation are also performed by all hydrogenosomes studied to date. The enzymes used and the amount of energy derived per mole of pyruvate varies among hydrogenosomes and is minimal compared to that generated by mitochondria. With a notable exception (Rotte et al., 2001), most aerobic mitochondria use the multienzyme complex pyruvate dehydrogenase for pyruvate decarboxylation and acetyl-CoA biosynthesis. In contrast, most hydrogenosomes use pyruvate:ferredoxin oxidoreductase (PFOR) for pyruvate decarboxylation. PFOR is an oxygen-sensitive enzyme of proposed eubacterial ancestry, but, although its precise phylogenetic origin has yet to be firmly established, current evidence does not support its direct descent from the mitochondrial endosymbiont (Embley et al., 2003; Horner et al., 1999; Rotte et al., 2001). The use of the enzyme pyruvate formate lyase (PFL) for pyruvate oxidation instead of PFOR in the hydrogenosomes of the anaerobic chytrid fungi *Piromyces* sp. and *Neocallimastix frontalis* has also been reported (Akhmanova et al., 1999). However, the PFL activity measured in this study is much lower than the previously reported hydrogenosomal PFOR activity in those organisms (Marvin-Sikkema et al., 1993; Yarlett et al., 1986).

A second major difference in hydrogenosomal metabolism compared with its mitochondrial counterpart is that reducing equivalents are not shunted to oxygen but are transferred to the canonical iron-dependent hydrogenase enzyme. Protons are therefore the terminal electron acceptors in hydrogenosomes and not oxygen as in mitochondria. Although there is some variation in hydrogenosomal metabolism, under anoxic conditions hydrogenosomes perform the oxidative decarboxylation of pyruvate, with the concomitant production of ATP via substrate-level phosphorylation, using an unusual acetate:succinate-CoA transferase (Müller, 2002). This enzyme has also been detected in functional anaerobic mitochondria, which can use inorganic nitrate or fumarate as the final electron acceptor (Tielens et al., 2002; van Hellemond et al., 1998). In addition, some ciliate hydrogenosomes produce ATP involving metabolism of butyrate or lactate (Paul et al., 1990; Williams and Harfoot, 1976; Yarlett et al., 1985).

An interesting development in the search for metabolic hydrogenosomal functions is the recent identification of two mitochondrial-type cysteine desulfurase genes (*isc*S) from *T. vaginalis* (Tachezy et al., 2001). IscS plays a central role in the biosynthesis of iron–sulfur (Fe–S) complexes (Figure 13.3), a critical function of the mitochondrion (Lill et al., 1999; Lill and Kispal, 2000). One of the trichomonad IscS proteins is predicted to contain a typical hydrogenosomal-targeting signal, suggesting its compartmentalized nature (Tachezy et al., 2001). Further, similar to hydrogenosomal chaperonins Cpn60 and Hsp70 (Bui et al., 1996; Germot et al., 1996; Horner et al., 1996; Roger et al., 1996), the mitochondrial ancestry of the trichomonad IscS proteins is strongly supported by phylogenetic analyses (Tachezy et al., 2001). Together, these findings suggest that *T. vaginalis* hydrogenosomes might be involved in the biosynthesis of Fe–S clusters in this organism.

Type III amitochondriate organisms lack compartmentalized energy metabolism, and so it is unclear what metabolic functions their mitochondrial remnant organelles might fulfill (Müller, 2000). Cpn60 and mitochondrial-type heat-shock protein 70 (mtHsp70) represent the only published examples of proteins whose mitosomal localization has been experimentally demonstrated (but see later), the former in *Entamoeba* and *Cryptosporidium* (Mai et al., 1999; Riordan et al., 2003; Tovar et al., 1999) the latter in *Trachipleistophora* (Williams et al., 2002). Genes for mtHsp70 have also been identified in *Entamoeba* and *Giardia*, but neither their cellular localization nor their function have been established (Arisue et al., 2002; Bakatselou et al., 2000; Morrison et al., 2001). Other putative mitosomal proteins have also been identified in *Entamoeba* (e.g., pyrimidine nucleotide transhydrogenase, PNT; Clark and Roger, 1995) and *Cryptosporidium* (e.g., adenylate kinase, valyl tRNA synthase, lactate and malate dehydrogenases; Zhu and Keithly, 2002), suggesting that their specific catalytic activities might represent mitosomal functions. But perhaps the most important window into the possible roles of mitosomes has been provided by the ongoing or recently completed genome sequencing projects of amitochondriate organisms, i.e., those of the microsporidian *E. cuniculi*, the diplomonad *G. lamblia*, the apicomplexan *C. parvum* and the archeoamebid *E. histolytica* (TIGR/Sanger Center, http://www.sanger.ac.uk/Projects/E_histolytica/, http://www.tigr.org/tdb/e2k1/eha1/; Katinka et al., 2001; McArthur et al., 2000; Strong and Nelson, 2000; TIGR/Sanger Center).

One of the most remarkable observations to come from the analysis of these genomes is the existence of a handful of genes phylogenetically related to mitochondrial and α-

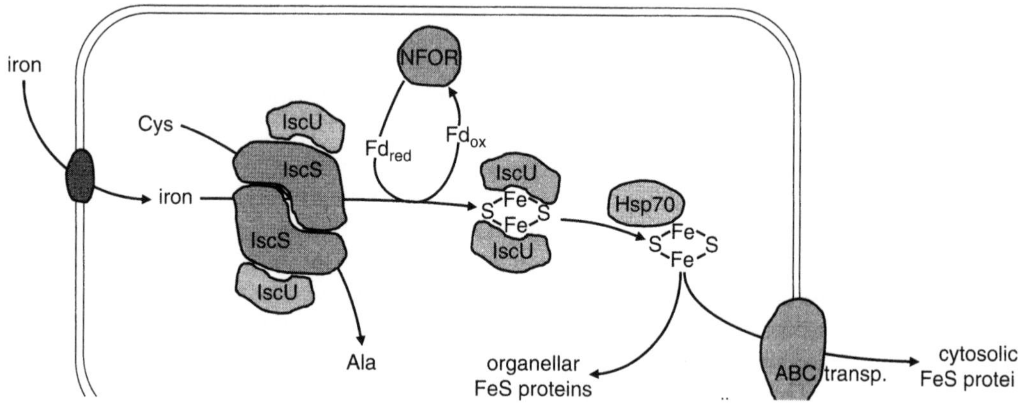

FIGURE 13.3 Simplified scheme showing the mechanism of maturation of Fe–S proteins in a eukaryotic cell. Functions of proteins (yeast homologues between brackets): ABC transp., ABC transporter (Atm1), export of Fe–S clusters; Fd, ferredoxin (Yah1), reduction of intermediate Fe–S clusters; Hsp70 (Ssq1), mitochondrial Hsp70 homologue involved in Fe–S assembly, function not clear; IscU (Isu1/2), scaffolding protein, binding of ferric iron and intermediate Fe–S clusters; IscS (Nfs1), cysteine desulfurase, transfer of sulfur into intermediate complex; NFOR (Arh1), NAD(P):ferredoxin oxidoreductase, recycling of ferredoxin. Distribution of proteins thus far identified in amitochondriate eukaryotes, *Cryptosporidium*: ABC transp., Fd, IscU, IscS, NFOR, Hsp70; *Encephalitozoon*: ABC transp., Fd, Hsp70, IscU, IscS, NFOR; *Entamoeba*: Fd, Hsp70, IscU; *Giardia*: Fd, Hsp70, IscU, IscS, NFOR; *Neocallimastix*: Fd, Hsp70, NFOR; *Trichomonas*: Fd, Hsp70, IscS, NFOR. N.B. The reported Hsp70 homologues detected in several species might not be the Ssq1 homologue involved in FeS assembly but the normal mitochondrial Ssc1 homologue.

proteobacterial sequences encoding Fe–S cluster assembly proteins (Figure 13.3). Genes for cysteine desulfurase (IscS), the intermediate scaffolding protein IscU, ferredoxin (Yah1), NAD(P)H:ferrodoxin reductase (Arh1), frataxin (Yfh1) and the ABC transporter Atm1 have been identified, wholly or in part, in the genomes of *Cryptosporidium, Trachipleistophora, Giardia* and *Entamoeba* (TIGR/Sanger Center, http://www.sanger.ac.uk/Projects/E_histolytica/, http://www.tigr.org/tdb/e2k1/eha1/; Katinka et al., 2001; McArthur et al., 2000; Strong and Nelson, 2000; Tachezy et al., 2001; Genbank entries AY040612, AY040613 and AY081205). Moreover, several of these sequences predict proteins with putative organelle-targeting peptides similar to those of hydrogenosomes and mitochondria, further suggesting their compartmentalized localization (Katinka et al., 2001; Strong and Nelson, 2000). The maturation of Fe–S proteins has only recently been recognized as an essential function of the mitochondrion (Lill et al., 1999) and, although preliminary, the available genome data strongly suggest that the assembly and maturation of Fe–S proteins are likely to be some of the main functions of mitochondrial remnant organelles. It is anticipated that the cellular localization of many of these putative mitosomal products will be unequivocally established within a short period of time as, to this aim, work is currently in progress in various laboratories worldwide.

This leaves us with the striking observation that the only unifying metabolic process of hydrogenosomes, mitosomes and mitochondria seems to be the biosynthesis of Fe–S clusters and subsequent maturation of Fe–S proteins. These essential metabolic functions are clearly more important than previously appreciated and might represent the driving force that led to the permanent establishment of the mitochondrial endosymbiont inside its host. Moreover, the requirement for Fe–S proteins in every conceivable environment inhabited by eukaryotic organisms — be it aerobic, anaerobic or microaerophilic — could have provided the selective pressure for the retention of the mitochondrial endosymbiont in all extant eukaryotes.

13.5 Biogenesis

13.5.1 Protein Import

One feature of the establishment of the endosymbiont is the lateral transfer of genes from the endosymbiont to the host's nucleus (Blanchard and Lynch, 2000; Gellissen and Michaelis, 1987). This possible means of a tighter control must have faced the cell with a problem, because the organellar proteins had to be transported back into the organelle to function. Thus, the transfer of organellar DNA must have been preceded or accompanied by the development of a protein import system. Some bacterial proteins have been put forward that might have evolved into this protein import system (Rassow et al., 1999). If, as it appears, mitochondria, hydrogenosomes and mitosomes are related, one should expect that this complex protein import system that comprises many interacting proteins would have been preserved through the process of reductive evolution; indeed, independent "invention" of such a complex system seems very unlikely (Cavalier-Smith, 1987). Most hydrogenosomal matrix proteins studied to date contain cleavable presequences that resemble, and, in some cases, have been proven to function as, typical mitochondrial targeting sequences (Table 13.1). The two hydrogenosomal inner membrane proteins studied thus far, Hmp31 in *T. vaginalis* (Dyall et al., 2001) and AAC in *Neocallimastix patriciarum* (van der Giezen et al., 2002; Voncken et al., 2002), use a second mitochondrial import pathway that relies on a cryptic signal which in yeast resides in the first one third of the protein (Sirrenberg et al., 1996, 1998). The conservation of both protein import pathways between mitochondria and hydrogenosomes strongly supports their common origin.

Information is still limited as to the mechanisms of protein import into mitosomes, but cross-functionality of mitochondrial and mitosomal presequences has been demonstrated in *E. histolytica* by deletion of amino acid residues 2 to 15 from the mitosomal chaperonin Cpn60. Removal of the amino terminal leader sequence led to the accumulation of Cpn60 in the cell cytoplasm, a mutant phenotype that could be reversed by adding the *Trypanosoma cruzi* mtHsp70 mitochondrial targeting signal to the truncated protein (Tovar et al., 1999). *Entamoeba* PNT and mtHsp70 genes also seem to contain targeting presequences (Bakatselou et al., 2000; Clark and Roger, 1995). Although sequence analysis of the *E. cuniculi* genome predicts the existence of several mitosomal proteins with putative mitochondrial presequences (Katinka et al., 2001), the only protein so far demonstrated to reside in microsporidian mitosomes (i.e., mtHsp70) lacks a recognizable organelle-targeting sequence (Williams et al., 2002). This result strongly suggests that unidentified internal sequences must function during mitosomal mtHsp70 import in *Trachipleistophora*. Whether the conventional amino terminal presequence import pathway is still operational in microsporidia remains to be demonstrated experimentally. *Cryptosporidium* Cpn60 contains a putative mitochondrial targeting signal, and its functionality was recently demonstrated by targeting a reporter protein (GFP) into yeast mitochondria. Together with the demonstrated mitosomal localization of Cpn60 by immunoelectron microscopy, this result suggests that a functional protein import mechanism that relies on mitochondrial-like targeting presequences might still be operational in *Cryptosporidium* (Riordan et al., 2003).

Although the existence of mitochondrion-related organelles has not yet been reported in the diplomonads, at least two of their putative mitochondrial proteins contain recognizable organelle-targeting signals, i.e., *Giardia* ferredoxin and *Spironucleus* Cpn60 (Table 13.1; Horner and Embley, 2001; Nixon et al., 2002). Whether these putative targeting

TABLE 13.1 Presequences of Hydrogenosomal and Mitosomal Proteins

Organism	*Localization*	*Protein*	*Presequence*	*References*
Eh	Mi	Cpn60	MLSSSSHYNKLLSLN[a]	Tovar et al., 1999
Eh	Mi	Hsp70	MFVSQPARS	Bakatselou et al., 2000
Eh	Mi	PNT	MSTSSSIEEEVFNYM	Clark and Roger, 1995
Gi	?	Ferredoxin	MSLLSSIRRFITFRVVQQ	Nixon et al., 2002
Sb	?	Cpn60	MHFSLYKITYHIYSIILILRK	Horner and Embley, 2001
Nf	Hy	Beta-SCS	MLANVTRSTSKAAPALASIAQTAQKRF	Brondijk et al., 1996
Nf	Hy	Hydrogenase	MLSSVLNKAVVNPKLTRSLATAAAEK	Davidson et al., 2002
Nf	Hy	Malic enzyme	MLAPIQTIARPVSSILPATGALAAKRT[a]	van der Giezen et al., 1997a
Np	Hy	Cpn60	MLSARSLICKSMIKSGFRRAVAPSVAMA ASSMTLTARRNY[a]	van der Giezen et al., 2003
Np	Hy	Hsp70	MFLSTLAKKSTTFGVSNVVKNALSSKVMRTTPRMFQRF[#]	van der Giezen et al., 2003
No	Hy	Hydrogenase	MISRLIAKKAPLFLRTFATSE	Akhmanova et al., 1998)
Pl	Hy	Ferredoxin	MVSGVSRN	Brul et al., 1994
Tv	Hy	Adenylate kinase	MLSTLAKRF[a]	Länge et al., 1994
Tv	Hy	Alpha-SCS 1	MLAGDFSRN[a]	Lahti et al., 1994
Tv	Hy	Alpha-SCS 2	MLSSSFERN[a]	Lahti et al., 1994
Tv	Hy	Beta-SCS 1	MLSSSFARN[a]	Lahti et al., 1992
Tv	Hy	Beta-SCS 2	MLSASSNFARN[a]	Lahti et al., 1992
Tv	Hy	Ferredoxin	MLSQVCRF[a]	Johnson et al., 1990

Tv	Hy	Hydrogenase	MLASSSRA	Bui and Johnson, 1996
Tv	Hy	HMP31	MAQPAEQILIAT[a]	Dyall et al., 2001
Tv	Hy	Cpn10	MLATFARN	Bui et al., 1996
Tv	Hy	Cpn60	MSLIEAAKHFTRAF[a]	Bui et al., 1996
Tv	Hy	Hsp70	MLKMFNSIFARE	Bui et al., 1996
Tv	Hy	Malic enzyme A	MLTSSVSVPVRN[a]	Hrdý and Müller, 1995b
Tv	Hy	Malic enzyme B	MLTSSVNFPARE[a]	Hrdý and Müller, 1995b
Tv	Hy	Malic enzyme C	MLTSVSYPVRN[a]	Hrdý and Müller, 1995b
Tv	Hy	Malic enzyme D	MLTSVSLPVRN[a]	Hrdý and Müller, 1995b
Tv	Hy	PFO A	MLRSF[a]	Hrdý and Müller, 1995a
Tv	Hy	PFO B	MLRNF[a]	Hrdý and Müller, 1995a

Note: Eh, *Entamoeba histolytica*; Gi, *Giardia intestinalis*; Nf, *Neocallimastix frontalis*; No, *Nyctotherus ovalis*; Np, *Neocallimastix patriciarum*; Pl, *Psalteriomonas lanterna*; Sb, *Spironucleus barkhanus*; Tv, *Trichomonas vaginalis*; Hy, hydrogenosome; Mi, mitosome; Cpn, chaperonin; HMP31, 31-kDa hydrogenosomal membrane protein; Hsp, heat-shock protein; PFO, pyruvate:ferredoxin oxidoreductase; PNT, pyridine nucleotide transhydrogenase; SCS, succinyl-CoA synthetase; ?, presence of mitochondrion-related organelles not known.

[a] Biochemically or functionally confirmed cleaved presequences.

signals indeed deliver ferredoxin and Cpn60 to a hitherto-unknown cellular compartment remains to be demonstrated. However, that *Giardia* Cpn60 does not contain a recognizable organelle-targeting sequence (Roger et al., 1998) would suggest that if a remnant mitochondrial compartment exists, import of matrix proteins in *Giardia* might not be solely dependent on amino terminal presequences. Interestingly, an intron separates the ferredoxin putative targeting signal from the predicted mature protein in the gene sequence, a phenomenon that has been proposed to explain the evolution of targeting signals (Watanabe and Ohama, 2001). Because not all putative mitochondrial proteins present in amitochondriate microbial organisms contain typical amino-terminal targeting domains, their cellular localization and the functionality of each putative organelle-targeting sequence will need to be demonstrated experimentally in a case-by-case basis.

Analogous to the situation with hydrogenosomes, very little is known about the functionality of membrane translocases (TOMs and TIMs) that participate in mitosomal protein import pathways. Analysis of the *E. cuniculi* genome revealed the existence of putative TOM70 and TIM22 homologues, whereas a putative TIM17 homologue has been identified in the *C. parvum* genome (Katinka et al., 2001; Strong and Nelson, 2000). It is likely that more mitosomal translocases will be identified in the near future, as the proteomes of these organisms become known, aiding our understanding of organelle protein import pathways retained by amitochondrial eukaryotes.

13.5.2 Organellar Dynamics

Unlike mitochondria, almost no work has been reported on organellar dynamics for hydrogenosomes or mitosomes. Our knowledge on the biogenesis of trichomonad hydrogenosomes is limited to the fact that *T. vaginalis* hydrogenosomal proteins are synthesized on free polyribosomes and posttranslationally translocated into the organelle by mechanisms akin to those that operate in mitochondria (Bradley et al., 1997; Dyall and Johnson, 2000; Johnson et al., 1993). Mitosomal proteins can also be translocated into the organelle via mitochondrion-related pathways, but details on the precise mechanisms involved are still very limited (Katinka et al., 2001; Mai et al., 1999; Tovar et al., 1999). There are no published studies on hydrogenosomal or mitosomal dynamics, like segregation, fusion, migration or their interaction with cytoskeletal proteins. Much work has been done on mitochondrial dynamics and division in yeast, *Drosophila*, *Ceanorhabditis elegans* and mammalian cells (Shaw and Nunnari, 2002), and this should fuel future related work on amitochondrial eukaryotes. We have recently identified two dynamin-like proteins of *Entamoeba* thought to be involved in mitosomal fission and fusion (van der Giezen and Tovar, unpublished data). Work is in progress in our laboratory to test their involvement in mitosome dynamics.

13.6 Evolutionary Considerations: A Common Selective Force?

Although mitochondria are usually seen as efficient energy-generating organelles central to the diversification and evolution of the eukaryotic lineage, mitosomes and hydrogenosomes tend to be regarded as mere oddities of anaerobiosis, or, at best, as derived microbial adaptations to anaerobic environments. Understanding the natural history of the mitochondrion and related organelles — mitosomes and hydrogenosomes — will require answers to two important questions: (1) What was the main driving force that led to the establishment of a permanent endosymbiosis in a primitive, highly anoxic environment? (2) What is the selective pressure that has operated throughout millions of years of evolution to prevent the loss of the original endosymbiont?

The first question remains one of the most hotly debated topics of early eukaryotic evolution (Andersson et al., 1998; Andersson and Kurland, 1999; Castresana and Saraste, 1995; Cavalier-Smith, 2002; Dyall and Johnson, 2000; Gray, 1999; Gray et al., 1999; Karlin et al., 1999; Kurland and Andersson, 2000; López-García and Moreira, 1999; Margulis, 1996; Martin and Müller, 1998; Moreira and López-García, 1998; Müller and Martin, 1999; Rotte et al., 2000; Sogin, 1991; Vellai et al., 1998; Vellai and Vida, 1999). Energy generation coupled to aerobic respiration has traditionally been regarded as the most important functional advantage conferred by the mitochondrial endosymbiont on its host (Andersson and Kurland, 1999; Cavalier-Smith, 2002; Kurland and Andersson, 2000; Margulis, 1996; Sogin, 1991; Vellai et al., 1998; Vellai and Vida, 1999). But this view has been robustly contended by proposers of the anaerobic origin of mitochondria who suggest primitive anaerobic microbial communities (mats) as the most likely ecological niche for the origin of the first eukaryote, with metabolic syntrophy as the driving force for a successful and permanent endosymbiosis (López-García and Moreira, 1999; Martin and Müller, 1998; Moreira and López-García, 1998; Müller and Martin, 1999). The arguments for and against these two contrasting views are beyond the scope of this review but are by their very nature complex, multifaceted and at times incomplete and speculative. Undoubtedly, comparative genomics, paleobiology and molecular ecology will have much to contribute over the next few years to the consolidation of current arguments and to the development of alternative hypotheses.

The second question concerning the selective pressures required for organelle retention might be more tractable. Characterizing the proteomes of extant endosymbiosis-derived organelles should provide clues as to their metabolic functions and to their *raison d' être*. Energy generation and recycling of resources are likely to have driven the establishment of the mitochondrial endosymbiosis. Regardless of whether energy generation in the original heterotrophic endosymbiont was coupled to the reduction of NO_2, H^+, SO_3 or O_2 (or a combination of them), most electron transport chains and modes of heterotrophic organic matter oxidation have one thing in common: several of their key effectors (e.g., ferredoxins, succinate dehydrogenase, PFOR, cytochrome *b*-associated Rieske protein, hydrogenase) rely on Fe–S centers for their catalytic functions (Castresana and Saraste, 1995). It follows that the original endosymbiont must have possessed the capacity to synthesize Fe–S clusters and to assemble them into functional redox proteins. The striking observation that all mitosomes and hydrogenosomes studied in sufficient detail thus far contain identifiable components of the mitochondrial pathway for the biosynthesis and maturation of Fe–S clusters (Katinka et al., 2001; McArthur et al., 2000; Strong and Nelson, 2000; Tachezy et al., 2001; *Entamoeba* genome project, TIGR/Sanger Centre) strongly suggests that this metabolic function could have played an important role in securing the permanent establishment of the protomitochondrion in the newly formed protoeukaryotic cell. Despite the apparently limited biochemical repertoire so far identified or predicted in mitosomes and hydrogenosomes, Fe–S cluster assembly appears to be the only biochemical feature of the heterotrophic protomitochondrion that proved essential for eukaryotic cell survival in both anaerobic and oxygenic environments. Thus, the requirement for the assembly and maturation of Fe–S proteins might have provided, for millions of years, the selective pressure required to retain the mitochondrial endosymbiont in all its current manifestations, i.e., aerobic and anaerobic mitochondria, hydrogenosomes and mitosomes.

13.7 Conclusions

It is apparent from the information presented that mitochondria, hydrogenosomes and mitosomes have much more in common than seems evident at first sight. Although *in vivo* they might display different shapes and sizes, these organelles share a common morphological denominator: two closely apposed membranes surround them. Transport of proteins across these doubly membraned organelles is conducted through functionally conserved import pathway mechanisms. However, there is evidence that some of the requirements of matrix protein import (e.g., positively charged amino-terminal targeting sequences) might have been partially lost or modified in the recently discovered and highly derived mitosomes. In terms of their biochemistry, significant diversity is apparent between and within these three types of organelles. In mitochondrial and hydrogenosomal organisms, pyruvate oxidation and electron transport linked to ATP biosynthesis are performed inside their respective organelles, but in mitosomal organisms these functions have been relocated, at least in part, to the cytoplasm. Despite no longer being compartmentalized, key redox enzymes of mitosomal organisms (e.g., PFOR, ferredoxin, hydrogenase) are structurally and functionally homologous to their organellar counterparts in hydrogenosomes or anaerobic mitochondria, or both, with strong phylogenetic support for a common ancestry. Current evidence suggests that Fe–S cluster metabolism might be the only physiological function common to all endosymbiosis-derived organelles.

Finally, although other mitochondrial functions (e.g., apoptosis, calcium storage) might also be carried out by mitochondrion-derived organelles of amitochondriate protists, data from completed or ongoing genome sequencing projects do not predict an extensive biochemical repertoire for these organelles. Much remains to be tested experimentally to ascertain the interrelationships, differences and similarities between mitochondria, mitosomes and hydrogenosomes, but well-defined threads have emerged. A significant body of evidence points to a common ancestry for mitochondria, mitosomes and hydrogenosomes; all these organelles might be evolutionary variations of a theme. Different environmental conditions and selective pressures experienced by diverse organisms might have tailored their mitochondrial organelles to suit different life strategies. The observed diversity in terms of their morphological appearance, metabolic capacity and remnant protein import mechanisms should come as no surprise. The enormous flexibility of eukaryotic organisms to adapt to ever-changing environmental conditions in order to survive is well represented in these enigmatic organelles.*

Acknowledgments

We thank Marlene Benchimol, Martin Embley and John Samuelson for permission to reproduce electron microscopy micrographs; Janet Keithly for communicating experimental data before publication; and Graham Clark for his critical reading of the manuscript. The use of information from the various amitochondrial genome projects (as referenced in the text) is also acknowledged. Work on this topic in our laboratory is supported by grants from the Wellcome Trust (059845) and from the BBSRC (111/C13820) to J.T.

References

Akhmanova, A., Voncken, F., van Alen, A.T., van Hoek, H.A., Boxma, B., Vogels, G., Veenhuis, M. and Hackstein, J.H. (1998) A hydrogenosome with a genome. *Nature* 396: 527–528.

* *Note*: The study of early eukaryotic evolution is a fast-moving area of research. After acceptance of this review article for publication, the existence of mitosomes in the diplomonad *Giardia intestinalis* was reported (Tovar et al., 2003). As suggested in this book chapter, the work by Tovar et al. demonstrated that *Giardia* is not primitively amitochondrial and furnished direct evidence that Fe-S cluster metabolism, an essential mitochondrial function, is carried out in mitochondrial remnant organelles.

Akhmanova, A., Voncken, F.G., Hosea, K.M., Harhangi, H., Keltjens, J.T., op den Camp, H.J., Vogels, G.D. and Hackstein, J.H. (1999) A hydrogenosome with pyruvate formate-lyase: Anaerobic chytrid fungi use an alternative route for pyruvate catabolism. *Mol. Microbiol.* 32: 1103–1114.

Andersson, S.G. and Kurland, C.G. (1999) Origins of mitochondria and hydrogenosomes. *Curr. Opin. Microbiol.* 2: 535–541.

Andersson, S.G., Zomorodipour, A., Andersson, J.O., Sicheritz-Ponten, T., Alsmark, U.C., Podowski, R.M., Naslund, A.K., Eriksson, A.S., Winkler, H.H. and Kurland, C.G. (1998) The genome sequence of *Rickettsia prowazekii* and the origin of mitochondria. *Nature* 396: 133–140.

Arisue, N., Sánchez, L.B., Weiss, L.M., Müller, M. and Hashimoto, T. (2002) Mitochondrial-type hsp70 genes of the amitochondriate protists *Giardia intestinalis, Entamoeba histolytica* and two microsporidians. *Parasitol. Int.* 51: 9–16.

Bakatselou, C., Kidgell, C. and Clark, C.G. (2000) A mitochondrial-type hsp70 gene of *Entamoeba histolytica. Mol. Biochem. Parasitol.* 110: 177–182.

Benchimol, M. and De Souza, W. (1983) Fine structure and cytochemistry of the hydrogenosome of *Tritrichomonas foetus. J. Protozool.* 30: 422–425.

Benchimol, M., Durand, R. and Almeida, J.C.A. (1997) A double membrane surrounds the hydrogenosomes of the anaerobic fungus *Neocallimastix frontalis. FEMS Microbiol. Lett.* 154: 277–282.

Benchimol, M., Johson, P.J. and De Souza, W. (1996) Morphogenesis of the hydrogenosomes: An ultrastructural study. *Biol. Cell* 87: 197–205.

Berger, J. and Lynn, D.H. (1992) Hydrogenosome-methanogen assemblages in the echinoid endocommensal plagiopylid ciliates *Lechriopyla mystax* Lynch, 1930, and *Plagiopyla minuta* Powers, 1933. *J. Protozool.* 39: 4–8.

Blanchard, J.L. and Lynch, M. (2000) Organellar genes: Why do they end up in the nucleus? *Trends Genet.* 16: 315–320.

Bradley, P.J., Lahti, C.J., Plümper, E. and Johnson, P.J. (1997) Targeting and translocation of proteins into the hydrogenosome of the protist *Trichomonas*: Similarities with mitochondrial protein import. *EMBO J.* 16: 3484–3493.

Brondijk, T.H.C., Durand, R., van der Giezen, M., Gottschal, J.C., Prins, R.A. and Fèvre, M. (1996) *scsB*, a cDNA encoding the hydrogenosomal protein b-succinyl-CoA synthetase from the anaerobic fungus *Neocallimastix frontalis. Mol. Gen. Genet.* 253: 315–323.

Brul, S. and Stumm, C.K. (1994) Symbionts and organelles in anaerobic protozoa and fungi. *Trends Ecol. Evol.* 9: 319–324.

Brul, S., Veltman, R.H., Lombardo, M.C. and Vogels, G.D. (1994) Molecular cloning of hydrogenosomal ferredoxin cDNA from the anaerobic amoeboflagellate *Psalteriomonas lanterna. Biochem. Biophys. Acta* 1183: 544–546.

Bui, E.T.N., Bradley, P.J. and Johnson, P.J. (1996) A common evolutionary origin for mitochondria and hydrogenosomes. *Proc. Natl. Acad. Sci. USA* 93: 9651–9656.

Bui, E.T.N. and Johnson, P.J. (1996) Identification and characterization of [Fe]-hydrogenases in the hydrogenosome of *Trichomonas vaginalis. Mol. Biochem. Parasitol.* 76: 305–310.

Castresana, J. and Saraste, M. (1995) Evolution of energetic metabolism: The respiration-early hypothesis. *Trends Biochem. Sci.* 20: 443–448.

Cavalier-Smith, T. (1987) The simultaneous symbiotic origin of mitochondria, chloroplasts, and microbodies. *Ann. N. Y. Acad. Sci.* 503: 55–71.

Cavalier-Smith, T. (1998) A revised six-kingdom system of life. *Biol. Rev. Camb. Philos. Soc.* 73: 203–266.

Cavalier-Smith, T. (2002) The phagotrophic origin of eukaryotes and phylogenetic classification of Protozoa. *Int. J. Syst. Evol. Microbiol.* 52: 297–354.

Cerkasovova, A., Lukasova, G., Cerkasov, J. and Kulda, J. (1973) Biochemical characterization of large granule fraction of *Tritrichomonas foetus* (strain KV1). *J. Protozool.* 20: 525.

Clark, C.G. and Roger, A.J. (1995) Direct evidence for secondary loss of mitochondria in *Entamoeba histolytica. Proc. Natl. Acad. Sci. USA* 92: 6518–6521.

Davidson, E.A., van der Giezen, M., Horner, D.S., Embley, T.M. and Howe, C.J. (2002) An [Fe] hydrogenase from the anaerobic hydrogenosome-containing fungus *Neocallimastix frontalis* L2. *Gene* 296: 45–52.

Dyall, S.D. and Johnson, P.J. (2000) Origins of hydrogenosomes and mitochondria: Evolution and organelle biogenesis. *Curr. Opin. Microbiol.* 3: 404–411.

Dyall, S.D., Koehler, C.M., Delgadillo-Correa, M.G., Bradley, P.J., Plümper, E., Leuenberger, D., Turck, C.W. and Johnson, P.J. (2001) Presence of a member of the mitochondrial carrier family in hydrogenosomes: Conservation of membrane-targeting pathways between hydrogenosomes and mitochondria. *Mol. Cell Biol.* 20: 2488–2497.

Embley, T.M., Horner, D.S. and Hirt, R.P. (1997) Anaerobic eukaryote evolution: Hydrogenosomes as biochemically modified mitochondria? *Trends Ecol. Evol.* 12: 437–441.

Embley, T.M., van der Giezen, M., Horner, D.S., Dyal, P.L. and Foster, P. (2003) Mitochondria and hydrogenosomes are two forms of the same fundamental organelle. *Phil. Trans. R. Soc. Lond.* 358: 191–204.

Finlay, B.J. and Fenchel, T. (1989) Hydrogenosomes in some anaerobic protozoa resemble mitochondria. *FEMS Microbiol. Lett.* 65: 311–314.

Frey, T.G. and Mannella, C.A. (2000) The internal structure of mitochondria. *Trends Biochem. Sci.* 25: 319–324.

Gellissen, G. and Michaelis, G. (1987) Gene transfer: Mitochondria to nucleus. *Ann. N. Y. Acad. Sci.* 503: 391–401.

Germot, A., Philippe, H. and Le Guyader, H. (1996) Presence of a mitochondrial-type 70-kDa heat shock protein in *Trichomonas vaginalis* suggests a very early mitochondrial endosymbiosis in eukaryotes. *Proc. Natl. Acad. Sci. USA* 93: 14614–14617.

Ghosh, S., Field, J., Rogers, R., Hickman, M. and Samuelson, J. (2000) The *Entamoeba histolytica* mitochondrion-derived organelle (crypton) contains double-stranded DNA and appears to be bound by a double membrane. *Infect. Immun.* 68: 4319–4322.

Gray, M.W. (1999) Evolution of organellar genomes. *Curr. Opin. Genet. Dev.* 9: 678–687.

Gray, M.W., Burger, G. and Lang, B.F. (1999) Mitochondrial evolution. *Science* 283: 1476–1481.

Gribaldo, S. and Philippe, H. (2002) Ancient phylogenetic relationships. *Theor. Popul. Biol.* 61: 391–408.

Hackenbrock, C.R. (1966) Ultrastructural bases for metabolically linked mechanical activity in mitochondria. I. Reversible ultrastructural changes with change in metabolic steady state in isolated liver mitochondria. *J. Cell Biol.* 30: 269–297.

Horner, D.S. and Embley, T.M. (2001) Chaperonin 60 phylogeny provides further evidence for secondary loss of mitochondria among putative early-branching eukaryotes. *Mol. Biol. Evol.* 18: 1970–1975.

Horner, D.S., Hirt, R.P. and Embley, T.M. (1999) A single eubacterial origin of eukaryotic pyruvate: ferredoxin oxidoreductase genes: Implications for the evolution of anaerobic eukaryotes. *Mol. Biol. Evol.* 16: 1280–1291.

Horner, D.S., Hirt, R.P., Kilvington, S., Lloyd, D. and Embley, T.M. (1996) Molecular data suggest an early acquisition of the mitochondrion endosymbiont. *Proc. R. Soc. Lond. B. Biol. Sci.* 263: 1053–1059.

Hrdý, I. and Müller, M. (1995a) Primary structure and eubacterial relationships of the pyruvate: ferredoxin oxidoreductase of the amitochondriate eukaryote *Trichomonas vaginalis. J. Mol. Evol.* 41: 388–396.

Hrdý, I. and Müller, M. (1995b) Primary structure of the hydrogenosomal malic enzyme of *Trichomonas vaginalis* and its relationship to homologous enzymes. *J. Euk. Microbiol.* 42: 593–603.

Humphreys, M.J., Ralphs, J., Durrant, L. and Lloyd, D. (1994) Hydrogenosomes in trichomonads are calcium stores and have a transmembrane electrochemical potential. *Biochem. Soc. Trans.* 22: 324S.

Johnson, P.J., D'Oliveira, C.E., Gorrell, T.E. and Müller, M. (1990) Molecular analysis of the hydrogenosomal ferredoxin of the anaerobic protist *Trichomonas vaginalis. Proc. Natl. Acad. Sci. USA* 87: 6097–6101.

Johnson, P.J., Lahti, C.J. and Bradley, P.J. (1993) Biogenesis of the hydrogenosome in the anaerobic protist *Trichomonas vaginalis. J. Parasitol.* 79: 664–670.

Karlin, S., Brocchieri, L., Mrazek, J., Campbell, A.M. and Spormann, A.M. (1999) A chimeric prokaryotic ancestry of mitochondria and primitive eukaryotes. *Proc. Natl. Acad. Sci. USA* 96: 9190–9195.

Katinka, M.D., Duprat, S., Cornillot, E., Metenier, G., Thomarat, F., Prensier, G., Barbe, V., Peyretaillade, E., Brottier, P., Wincker, P., Delbac, F., El Alaoui, H., Peyret, P., Saurin, W., Gouy, M., Weissenbach, J. and Vivares, C.P. (2001) Genome sequence and gene compaction of the eukaryote parasite *Encephalitozoon cuniculi. Nature* 414: 450–453.

Kurland, C.G. and Andersson, S.G. (2000) Origin and evolution of the mitochondrial proteome. *Microbiol. Mol. Biol. Rev.* 64: 786–820.

Lahti, C.J., Bradley, P.J. and Johnson, P.J. (1994) Molecular characterization of the a-subunit of *Trichomonas vaginalis* hydrogenosomal succinyl CoA synthetase. *Mol. Biochem. Parasitol.* 66: 309–318.

Lahti, C.J., D'Oliveira, C.E. and Johnson, P.J. (1992) b-Succinyl-coenzyme A synthetase from *Trichomonas vaginalis* is a soluble hydrogenosomal protein with an amino-terminal sequence that resembles mitochondrial presequences. *J. Bacteriol.* 174: 6822–6830.

Lahti, C.J. and Johnson, P.J. (1991) *Trichomonas vaginalis* hydrogenosomal proteins are synthesized on free polyribosomes and may undergo processing upon maturation. *Mol. Biochem. Parasitol.* 46: 307–310.

Länge, S., Rozario, C. and Müller, M. (1994) Primary structure of the hydrogenosomal adenylate kinase of *Trichomonas vaginalis* and its phylogenetic relationships. *Mol. Biochem. Parasitol.* 66: 297–308.

Lill, R., Diekert, K., Kaut, A., Lange, H., Pelzer, W., Prohl, C. and Kispal, G. (1999) The essential role of mitochondria in the biogenesis of cellular iron-sulfur proteins. *Biol. Chem.* 380: 1157–1166.

Lill, R. and Kispal, G. (2000) Maturation of cellular Fe-S proteins: An essential function of mitochondria. *Trends Biochem. Sci.* 25: 352–356.

Lindmark, D.G. and Müller, M. (1973) Hydrogenosome, a cytoplasmic organelle of the anaerobic flagellate *Tritrichomonas foetus*, and its role in pyruvate metabolism. *J. Biol. Chem.* 248: 7724–7728.

Lindmark, D.G., Müller, M. and Shio, H. (1975) Hydrogenosomes in *Trichomonas vaginalis. J. Parasitol.* 63: 552–554.

Lloyd, D. (1974) *The Mitochondria of Microorganisms,* Academic Press, London.

Lloyd, D. and Harris, J.C. (2002) *Giardia*: Highly evolved parasite or early branching eukaryote? *Trends Microbiol.* 10: 122–127.

Lloyd, D., Harris, J.C., Maroulis, S., Wadley, R., Ralphs, J.R., Hann, A.C., Turner, M.P. and Edwards, M.R. (2002) The "primitive" microaerophile *Giardia intestinalis* (syn. *lamblia, duodenalis*) has specialized membranes with electron transport and membrane-potential-generating functions. *Microbiology* 148: 1349–1354.

López-García, P. and Moreira, D. (1999) Metabolic symbiosis at the origin of eukaryotes. *Trends Biochem. Sci.* 24: 88–93.

Mai, Z., Ghosh, S., Frisardi, M., Rosenthal, B., Rogers, R. and Samuelson, J. (1999) Hsp60 is targeted to a cryptic mitochondrion-derived organelle ("crypton") in the microaerophilic protozoan parasite *Entamoeba histolytica. Mol. Cell Biol.* 19: 2198–2205.

Margulis, L. (1996) Archaeal-eubacterial mergers in the origin of Eukarya: Phylogenetic classification of life. *Proc. Natl. Acad. Sci. USA* 93: 1071–1076.

Martin, W. and Müller, M. (1998) The hydrogen hypothesis for the first eukaryote. *Nature* 392: 37–41.

Martin, W. and Schnarrenberger, C. (1997) The evolution of the Calvin cycle from prokaryotic to eukaryotic chromosomes: A case study of functional redundancy in ancient pathways through endosymbiosis. *Curr. Genet.* 32: 1–18.

Martin, W., Stoebe, B., Goremykin, V., Hapsmann, S., Hasegawa, M. and Kowallik, K.V. (1998) Gene transfer to the nucleus and the evolution of chloroplasts. *Nature* 393: 162–165.

Martínez-Palomo, A. (1993) Parasitic amoebas of the intestinal tract. In *Parasitic Protozoa* (Kreier, J. P. and Baker, J. R., Eds.) Academic Press, New York, pp. 65–141.

Marvin-Sikkema, F.D., Driessen, A.J.M., Gottschal, J.C. and Prins, R.A. (1994) Metabolic energy generation in hydrogenosomes of the anaerobic fungus *Neocallimastix*: Evidence for a functional relationship with mitochondria. *Mycol. Res.* 98: 205–212.

Marvin-Sikkema, F.D., Pedro Gomes, T.M., Grivet, J.P., Gottschal, J.C. and Prins, R.A. (1993) Characterization of hydrogenosomes and their role in glucose metabolism of *Neocallimastix* sp. L2. *Arch. Microbiol.* 160: 388–396.

McArthur, A.G., Morrison, H.G., Nixon, J.E., Passamaneck, N.Q., Kim, U., Hinkle, G., Crocker, M.K., Holder, M.E., Farr, R., Reich, C.I., Olsen, G.E., Aley, S.B., Adam, R.D., Gillin, F.D. and Sogin, M.L. (2000) The *Giardia* genome project database. *FEMS Microbiol. Lett.* 189: 271–273.

McLaughlin, J. and Aley, S. (1985) The biochemistry and functional morphology of the *Entamoeba*. *J. Protozool.* 32: 221–240.

Moreira, D., Le Guyader, H. and Philippe, H. (2000) The origin of red algae and the evolution of chloroplasts. *Nature* 405: 69–72.

Moreira, D. and López-García, P. (1998) Symbiosis between methanogenic archaea and delta-proteobacteria as the origin of eukaryotes: The syntrophic hypothesis. *J. Mol. Evol.* 47: 517–530.

Morrison, H.G., Roger, A.J., Nystul, T.G., Gillin, F.D. and Sogin, M.L. (2001) *Giardia lamblia* expresses a proteobacterial-like DnaK homolog. *Mol. Biol. Evol.* 18: 530–541.

Müller, M. (1993) The hydrogenosome. *J. Gen. Microbiol.* 139: 2879–2889.

Müller, M. (2000) A mitochondrion in *Entamoeba histolytica*? *Parasitol. Today* 16: 368–369.

Müller, M. (2002) Energy metabolism. Part I: Anaerobic protozoa. In *Molecular Medical Parasitology* (Marr, J., Nilsen, T. and Komuniecki, R., Eds.), Academic Press, New York, pp. 125–139.

Müller, M. and Martin, W. (1999) The genome of *Rickettsia prowazekii* and some thoughts on the origin of mitochondria and hydrogenosomes. *Bioessays* 21: 377–381.

Nisbet, e.g., and Sleep, N.H. (2001) The habitat and nature of early life. *Nature* 409: 1083–1091.

Nixon, J.E.J., Wang, A., Morrison, H.G., McArthur, A.G., Sogin, M.L., Loftus, B.J. and Samuelson, J. (2002) A spliceosomal intron in *Giardia lamblia*. *Proc. Natl. Acad. Sci. USA* 99: 3701–3705.

Paul, R.G., Williams, A.G. and Butler, R.D. (1990) Hydrogenosomes in the rumen entodiniomorphid ciliate *Polyplastron multivesiculatum*. *J. Gen. Microbiol.* 136: 1981–1989.

Prince, F.P. (1999) Mitochondrial cristae diversity in human Leydig cells: A revised look at cristae morphology in these steroid-producing cells. *Anat. Rec.* 254: 534–541.

Rassow, J., Dekker, P.J., van Wilpe, S., Meijer, M. and Soll, J. (1999) The preprotein translocase of the mitochondrial inner membrane: Function and evolution. *J. Mol. Biol.* 286: 105–120.

Reeves, R.E. (1984) Metabolism of *Entamoeba histolytica* Schaudinn, 1903. *Adv. Parasitol.* 23: 105–142.

Reeves, R.E., Warren, L.G., Susskind, B. and Lo, H.S. (1977) An energy-conserving pyruvate-to-acetate pathway in *Entamoeba histolytica*. Pyruvate synthase and a new acetate thiokinase. *J. Biol. Chem.* 252: 726–731.

Riordan, C.E., Ault, J.G., Langreth, S.G. and Keithly, J.S. (2003) *Cryptosporidium parvum* Cpn60 targets a relict organelle. *Curr. Genet.* 44: 138–147.

Roger, A.J., Clark, C.G. and Doolittle, W.F. (1996) A possible mitochondrial gene in the early-branching amitochondriate protist *Trichomonas vaginalis*. *Proc. Natl. Acad. Sci. USA* 93: 14618–14622.

Roger, A.J., Svärd, S.G., Tovar, J., Clark, C.G., Smith, M.W., Gillin, F.D. and Sogin, M.L. (1998) A mitochondrial-like chaperonin 60 gene in *Giardia lamblia*: Evidence that diplomonads once harbored an endosymbiont related to the progenitor of mitochondria. *Proc. Natl. Acad. Sci. USA* 95: 229–234.

Rotte, C., Henze, K., Müller and Martin, W. (2000) Origins of hydrogenosomes and mitochondria. *Curr. Opin. Microbiol.* 3: 481–486.

Rotte, C., Stejskal, F., Zhu, G., Keithly, J.S. and Martin, W. (2001) Pyruvate:NADP+ oxidoreductase from the mitochondrion of *Euglena gracilis* and from the apicomplexan *Cryptosporidium parvum*: A biochemical relic linking pyruvate metabolism in mitochondriate and amitochondriate protists. *Mol. Biol. Evol.* 18: 710–720.

Scheffler, I.E. (1999) *Mitochondria,* Wiley Liss, New York.

Scheffler, I.E. (2000) A century of mitochondrial research: Achievements and perspectives. *Mitochondrion* 1: 3–31.

Shaw, J.M. and Nunnari, J. (2002) Mitochondrial dynamics and division in budding yeast. *Trends Cell Biol.* 12: 178–184.

Simpson, A.G. and Patterson, D.J. (1999) The ultrastructure of *Carpediemonas membranifera* (Eukaryota) with reference to the "Excavate Hypothesis." *Eur. J. Protistol.* 35: 353–370.

Simpson, A.G., Roger, A.J., Silberman, J.D., Leipe, D.D., Edgcomb, V.P., Jermiin, L.S., Patterson, D.J. and Sogin, M.L. (2002) Evolutionary history of "early-diverging" eukaryotes: The excavate taxon *Carpediemonas* is a close relative of *Giardia*. *Mol. Biol. Evol.* 19: 1782–1791.

Sirrenberg, C., Bauer, M.F., Guiard, B., Neupert, W. and Brunner, M. (1996) Import of carrier proteins into the mitochondrial inner membrane mediated by Tim22. *Nature* 384: 582–585.

Sirrenberg, C., Endres, M., Folsch, H., Stuart, R.A., Neupert, W. and Brunner, M. (1998) Carrier protein import into mitochondria mediated by the intermembrane proteins Tim10/Mrs11 and Tim12/Mrs5. *Nature* 391: 912–915.

Sogin, M.L. (1991) Early evolution and the origin of eukaryotes. *Curr. Opin. Genet. Dev.* 1: 457–463.

Strong, W.B. and Nelson, R.G. (2000) Preliminary profile of the *Cryptosporidium parvum* genome: An expressed sequence tag and genome survey sequence analysis. *Mol. Biochem. Parasitol.* 107: 1–32.

Stumm, C.K. and Zwart, K.B. (1986) Symbiosis of protozoa with hydrogen-utilizing methanogens. *Microbiol. Sci.* 3: 100–105.

Tachezy, J., Sánchez, L.B. and Müller, M. (2001) Mitochondrial type iron-sulfur cluster assembly in the amitochondriate eukaryotes *Trichomonas vaginalis* and *Giardia intestinalis*, as indicated by the phylogeny of IscS. *Mol. Biol. Evol.* 18: 1919–1928.

Takeuchi, T., Weinbach, E.C. and Diamond, L.S. (1975) Pyruvate oxidase (CoA acetylating) in *Entamoeba histolytica*. *Biochem. Biophys. Res. Commun.* 65: 591–596.

Taylor, F.J. (1979) Symbionticism revisited: A discussion of the evolutionary impact of intracellular symbioses. *Proc. R. Soc. Lond. B. Biol. Sci.* 204: 267–286.

Taylor, F.J.R. (1983) Some eco-evolutionary aspects of intracellular symbiosis. *Int. Rev. Cytol.* S14: 1–28.

Tielens, A.G., Rotte, C., van Hellemond, J.J. and Martin, W. (2002) Mitochondria as we don't know them. *Trends Biochem. Sci.* 27: 564–572.

Tovar, J., Fischer, A. and Clark, C.G. (1999) The mitosome, a novel organelle related to mitochondria in the amitochondrial parasite *Entamoeba histolytica*. *Mol. Microbiol.* 32: 1013–1021.

Tovar, J., Leon-Avila, G., Sánchez, L.B., Sutak, R., Tachezy, J., van der Giezen, M., Hernández, M., Müller, M. and Lucocq, J.M. (2003) Mitochondrial remnant organelles of *Giardia* function in iron-sulfur protein maturation. *Nature* 426: 172–176.

van der Giezen, M., Birdsey, G.M., Horner, D.S., Lucocq, J., Dyal, P.L., Benchimol, M., Danpure, C.J. and Embley, T.M. (2003) Fungal hydrogenosomes contain mitochondrial heat-shock proteins. *Mol. Biol. Evol.* 20: 1051–1061.

van der Giezen, M., Rechinger, K.B., Svendsen, I., Durand, R., Hirt, R.P., Fèvre, M., Embley, T.M. and Prins, R.A. (1997a) A mitochondrial-like targeting signal on the hydrogenosomal malic enzyme from the anaerobic fungus *Neocallimastix frontalis*: Evidence for the hypothesis that hydrogenosomes are modified mitochondria. *Mol. Microbiol.* 23: 11–21.

van der Giezen, M., Sjollema, K.A., Artz, R.R.E., Alkema, W. and Prins, R.A. (1997b) Hydrogenosomes in the anaerobic fungus *Neocallimastix frontalis* have a double membrane but lack an associated organelle genome. *FEBS Lett.* 408: 147–150.

van der Giezen, M., Slotboom, D.J., Horner, D.S., Dyal, P.L., Harding, M., Xue, G.-P., Embley, T.M. and Kunji, E.R.S. (2002) Conserved properties of hydrogenosomal and mitochondrial ADP/ATP carriers: A common origin for both organelles. *EMBO J.* 21: 572–579.

van Hellemond, J.J., Opperdoes, F.R. and Tielens, A.G.M. (1998) Trypanosomatidae produce acetate via a mitochondrial acetate:succinate CoA-transferase. *Proc. Natl. Acad. Sci. USA* 95: 3036–3041.

Vellai, T., Takacs, K. and Vida, G. (1998) A new aspect to the origin and evolution of eukaryotes. *J. Mol. Evol.* 46: 499–507.

Vellai, T. and Vida, G. (1999) The origin of eukaryotes: The difference between prokaryotic and eukaryotic cells. *Proc. R. Soc. Lond. B. Biol. Sci.* 266: 1571–1577.

Voncken, F., Boxma, B., Tjaden, J., Akhmanova, A., Huynen, M., Verbeek, F., Tielens, A.G., Haferkamp, I., Neuhaus, H.E., Vogels, G., Veenhuis, M. and Hackstein, J.H. (2002) Multiple origins of hydrogenosomes: Functional and phylogenetic evidence from the ADP/ATP carrier of the anaerobic chytrid *Neocallimastix* sp. *Mol. Microbiol.* 44: 1441–1454.

Watanabe, K.I. and Ohama, T. (2001) Regular spliceosomal introns are invasive in *Chlamydomonas reinhardtii*: 15 introns in the recently relocated mitochondrial cox2 and cox3 genes. *J. Mol. Evol.* 53: 333–339.

Whatley, J.M., John, P. and Whatley, F.R. (1979) From extracellular to intracellular: The establishment of mitochondria and chloroplasts. *Proc. R. Soc. Lond. B.* 204: 165–187.

Williams, A.G. and Harfoot, C.G. (1976) Factors affecting the uptake and metabolism of soluble carbohydrates by the rumen ciliate *Dasytricha ruminantium* isolated from ovine rumen contents by filtration. *J. Gen. Microbiol.* 96: 125–136.

Williams, B.A.P., Hirt, R.P., Lucocq, J.M. and Embley, T.M. (2002) A mitochondrial remnant in the microsporidian *Trachipleistophora hominis*. *Nature* 418: 865–869.

Yarlett, N., Lloyd, D. and Williams, A.G. (1985) Butyrate formation from glucose by the rumen protozoon *Dasytricha ruminantium*. *Biochem. J.* 228: 187–192.

Yarlett, N., Orpin, C.G., Munn, E.A. and Greenwood, C. (1986) Hydrogenosomes in the rumen fungus *Neocallimastix patriciarum*. *Biochem. J.* 236: 729–739.

Zhu, G. and Keithly, J.S. (2002) Alpha-proteobacterial relationship of apicomplexan lactate and malate dehydrogenases. *J. Euk. Microbiol.* 49: 255–261.

Chapter 14

Eukaryotic Cell Evolution from a Comparative Genomic Perspective: The Endomembrane System

Joel B. Dacks and Mark C. Field

CONTENTS

Abstract

Comparative genomics provides a powerful tool for both evolutionary and cellular biology. As an example of how comparative genomics can be used in these fields, we examine how

0-415-29904-7/04/$0.00+$1.50

cell biological studies in model systems, together with the rapidly accumulating genomic data from diverse taxa, can be applied to reconstruct complex aspects of the biosynthetic-secretory and endocytic pathways in eukaryotes. The near-universal presence within eukaryotes of the core features of an intracellular transport system serves to highlight the vital role that this elaborate system must play in cell function. The evolution of this system is non-obvious, as prokaryotes have been generally considered to lack primitive or precursor structures that could have given rise to an endomembrane system. We consider, in detail, the proteins involved in vesicular transport, emphasizing a number of insights from selected divergent systems and comparing these with crown eukaryotes. We highlight possible prokaryotic precursors, survey the eukaryotic diversity of vesicular transport machinery and discuss how genomics initiatives have helped push forward cell biological studies of the endomembrane system in diverse organisms. Importantly, the mechanistic details of the transport systems are essentially conserved, indicating an ancient origin for these processes. All the while, increasing complexity in the sense of pathway multiplicity is observed in the vesicular transport system when comparing unicellular eukaryotes to more complex multicellular organisms.

14.1 Introduction

One of the most profound divisions in the biological world is between eukaryotic and non-eukaryotic cells. Prokaryotic organisms exhibit a huge diversity in biochemical and metabolic processes, but are underpinned with a comparatively simple cellular structure. Eukaryotes, in contrast, have massively expanded structural diversity and complexity in cell biological systems. Features such as the membrane-bound nucleus, the cytoskeleton, mitochondria, plastids and a system of functionally connected membrane-bound compartments (collectively referred to as the endomembrane system) are among the major characteristics that set eukaryotes apart from prokaryotes. This division is bridgeable though. Sophisticated homology searches and structural examination have identified prokaryotic homologues for proteins once thought to be strictly eukaryotic (Addinall and Holland, 2002; Kasinsky et al., 2001; van den Ent et al., 2001). Eukaryotes are also not as uniform in their cellular organization as once imagined. Organelles such as mitochondria, peroxisomes and stacked Golgi complexes have been lost or transformed many times in the course of eukaryotic history (Roger, 1999; see also Chapter 2), whereas plastid evolution is an even more sordid tale of theft, kidnapping and metamorphosis (Delwiche, 1999; see also Chapter 3 and Chapter 4). A broad comparative approach across a wide range of taxa is therefore key to making any generalizations about eukaryotic cell biology or evolution.

Two major advances have made the study of eukaryotic cell evolution more tractable. The first is the increasing wealth of molecular information about eukaryote-specific features. For example, it is no longer the cytoskeleton or even microtubules that are the defining characteristic, but the microtubule proteins (tubulins), and their genes, that can be compared (Addinall and Holland, 2002). This is important, because it facilitates a more objective analysis rather than a dependence on morphology or pharmacology, which cannot be quantified accurately in terms of genetic distance or functional divergence. Additionally, significant advances in molecular cell biology have identified many of the gene products involved in meiosis, chromatin organization and the endomembrane system, among other functions. The second advance is genomics, particularly comparative genomics. Prokaryotic genomes are being released at a tremendous rate, and eukaryotic genome initiatives are becoming more common whether as draft or full genomes, expressed sequence tag (EST) or genome sequence survey (GSS) projects. These sequences, organized and annotated into databases, will offer up gene sequences useful for addressing all aspects of eukaryotic evolution. Many of the advances made in molecular cell biology are because of progress in the genome projects of models systems such as mice, *Caenorhabditis*, *Drosophila* and yeasts.

It is likely that eukaryotes emerged from a single common prokaryote-like ancestor. A comparative genomics examination of the transition then from that prokaryote-like state and the emergence of the cellular systems that define eukaryotes can be approached in a number of ways. Prokaryotic homologues of proteins thought to be uniquely eukaryotic can provide hints as to a system's origin. A survey of the components of a system present in a wide diversity of eukaryotes allows an estimate of the complexity already established in the last common eukaryotic ancestor, as well as opens the door to more detailed questions of evolution and function in that system. Such an examination, however, requires methods for searching the genomic databases, candidate proteins to search for and a cell biological system to investigate.

14.2 The Endomembrane System

14.2.1 Evolution of an Endomembrane System: An Important Transition

One of the features that most distinguishes eukaryotes from prokaryotes is the assemblage of internal membrane-bound compartments for protein trafficking that constitutes the biosynthetic-secretory and endocytic pathways. This organellar system sorts, modifies, transports and even captures material (Alberts et al., 1994). Evolving this endomembrane system would have been a crucial step in the transition from the prokaryotic to the eukaryotic condition. The typical eukaryotic cell is 10- to 30-fold larger in linear dimensions than that of a prokaryote, with a consequent volume increase of 10^3- to 10^4-fold, and the result that simple diffusion of macromolecules through the cytoplasm is too slow for biological processes. An efficient transport system allows for increased cellular size, making available novel ecological niches. Endocytosis allows for more efficient heterotrophy and sets the stage for the acquisition of mitochondria and plastids. Targeted protein transport confers the ability to modify and control the composition of the cell surface, which is likely to have been an essential aspect of constructing a complex multicellular state. The evolution of the endomembrane system has been proposed to be the key step in the evolution of eukaryotes (Stanier, 1970).

In the past 5 to 10 years, there has been a huge increase in molecular data accumulated on the endomembrane system and membrane trafficking from cell biological studies. The picture that emerges is of a highly complicated and dynamic network of assembling and disassembling protein complexes required for the transport of material from one compartment to another. Although most of the molecular data have been obtained from animal and fungal model systems (Jahn and Sudhof, 1999), more limited studies in selected organisms will also be outlined.

14.2.2 Organelles of the Endomembrane System

The endoplasmic reticulum (ER), which is contiguous with the nuclear envelope (Figure 14.1), can be considered to be the beginning of the endomembrane system, on account of this being the point of insertion of polypeptides into the secretory pathway. Rough ER (rER) has a studded appearance, because of bound ribosomes, and is the site of synthesis for proteins destined to travel via vesicular transport. At the very heart of this transport process lies the mechanism by which polypeptides are translocated across the ER membrane (cotranslational translocation). The ribosome is targeted to the ER membrane via the signal sequence on the nascent chain of the polypeptide being translated. The information in the signal sequence is read by the signal recognition particle (SRP), a cytoplasmic riboprotein complex consisting of six polypeptides together with a core RNA (Rapoport et al., 1996). Transport vesicles bud from ribosome-free regions of the rER, called transitional elements (Klumperman, 2000), and quickly fuse with other vesicles derived from the same source. They might also fuse with a network of tubules termed the vesicular-tubular compartment (VTC) or ER

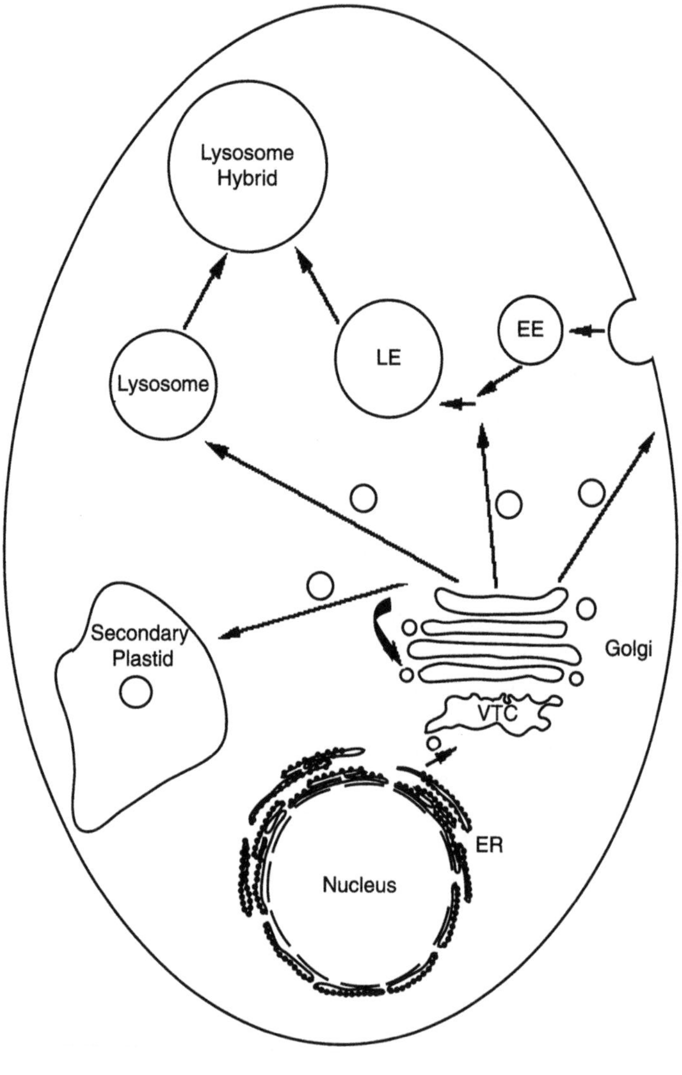

FIGURE 14.1 Organelles and direction of vesicular transport in a hypothetical eukaryotic cell. Straight arrows in this cartoon show the anterograde movement of vesicles between membrane compartments. The curved arrow illustrates retrograde transport in this case alone, as retrograde transport is the only non-controversial function of COPI vesicles. Small circles represent transport vesicles and large ones represent digestive organelles. EE and LE denote early and late endosomes, respectively. ER = endoplasmic reticulum, VTC = vesicular-tubular compartment. Much of the information shown here was derived from studies on mammalian and yeast cells, but the movement of Golgi-derived vesicles to a secondary plastid (as in *Euglena*) is also depicted.

Golgi intermediate compartment (ERGIC). Transport intermediates migrate from the VTC to a complex network of structures at the *cis*-face of the Golgi complex, referred to as the *cis*-Golgi network.

Most familiar as parallel stacks of flattened membrane-bound compartments (cisternae), the Golgi apparatus is the next distinct organelle in the endomembrane system. The Golgi apparatus receives material from the ER and is responsible for modification of proteins and sorting for later transport to various organelles. The compartments receiving material from the ER are called *cis*-Golgi, the main portions of the stack are the medial-Golgi and stacks that subsequently receive material are termed the *trans*-Golgi cisternae. From these last cisternae, vesicles bud for further transport from an elaborate network of membranes termed the *trans*-Golgi network (TGN). The morphology of the Golgi complex is quite varied among eukaryotes, with distinct flattened stacks in animals, plants and many protozoa; punctate vesicles in most fungi (but not chytrids); and smaller but numerous stacks in algae (Becker and Melkonian, 1996). This structural variation does not necessarily correlate with phylogeny. Although, in general, higher plants tend to have large numbers of small stacks, the organization, number and location is different in comparatively closely related yeasts [*Saccharomyces cerevisiae* and *Picia pastoris* (Glick, 2000)]. Metazoans have typically a single contiguous Golgi ribbon (Shorter and Warren, 2002); however, *Drosophila* can exhibit distinct Golgi morphologies in different life stages (Stanley et al., 1997). Given this diversity, a definition that depends on function and not classical stacked morphology is appropriate.

The progression of material becomes less linear on exiting the Golgi stack (Figure 14.1). In mammals and yeast, vesicles emerge from the TGN and might travel in at least four possible directions: retrograde, i.e., backward to previous compartments within the Golgi or to the ER; anterograde to the plasma membrane; intersect with the endocytic pathway; or be targeted to the lysosome. Each pathway is accompanied by distinct protein factors responsible for sorting, targeting and transporting the vesicle (Pryer et al., 1992). The secondary plastids of some organisms, such as *Euglena*, also receive Golgi-derived vesicles (Sulli and Schwartzbach, 1995; Sulli et al., 1999).

The plasma membrane represents the end point of the secretion–biosynthetic pathway and the beginning of the endocytic system. Vesicles leaving the TGN for the plasma membrane travel to the surface, where they fuse, either releasing their soluble contents or presenting their membrane-bound cargo. At the plasma membrane, endocytic vesicles are created to entrap food, internalize ligand-bound cell surface receptors or take up fluid phase material (Figure 14.1). Vesicles derived from the TGN destined for intracellular compartments fuse either with endocytic vesicles derived from the plasma membrane, called early or sorting endosomes, or with a preexisting late endosome (Figure 14.1). These pathways seem to differ in the components required for vesicle budding from the TGN, but share much of the same machinery for fusion at the lysosome (Bryant and Stevens, 1998; Luzio et al., 2000). The late endosome then fuses with lysosomes to create a hydrolytic organelle involved in degradation. This late endosome–lysosomal compartment is considered by some authors as a separate organelle, the term *lysosome* being reserved for the organelle that contains concentrated hydrolytic enzymes (Figure 14.1; Luzio et al., 2000).

14.2.3 Steps in the Transport Reaction

Regardless of the donating and receiving organelles, the mechanistic process of vesicular transport has many shared features and can therefore be described in a generalized model with three basic steps: vesicle formation and budding from the donor organelle, vesicle movement and finally fusion of the vesicle with the target organelle. The machinery used for vesicular transport between the different organelles is a mixture of components common to a reaction, regardless of location; members of protein families with paralogues specific for transport between two given organelles; and organellar specific complexes.

14.2.3.1 Vesicle Budding

The process begins by recruitment of a small GTPase to the cytosolic side of the membrane at the site of vesicle formation. Initially, the GTPase is GDP-bound, but a guanine exchange factor (GEF) protein catalyzes exchange of GDP for GTP. The GTPase regulates vesicle formation by recruiting the cytosolic coat proteins required for vesicle budding. Cargo proteins to be transported by the vesicle can be packaged by bulk flow, direct interaction with coat proteins via amino acid motifs in the cargo or via adaptor proteins. After cargo selection, the protein coat polymerizes, deforming the membrane and the ensuing vesicle buds. Figure 14.2 illustrates this generalized model (Springer et al., 1999).

Although the well-characterized types of vesicles built within the cell all conform to the generalized model of vesicle formation, their protein components differ significantly. In anterograde ER to Golgi transport, vesicles are coated with a complex COPII (Kaiser and Ferro-Novick, 1998; Springer et al., 1999). In the creation of COPII vesicles, the GTPase is called Sar1, which binds to the cytosolic face of the ER, with Sec12 acting as its GEF. The Sec23/24 protein complex interacts with the target membrane, in part through Sar1 (the complex is a regulator of the GTPase), and probably also through the cargo and putative cargo receptors. Cargo is concentrated into these exit regions and incorporated into transport vesicles, presumably via retention though proteins of the Emp24 family (Muniz et al., 2000) and also through bulk flow (Klumperman, 2000). The Sec23/24 complex recruits the Sec13/31 coat complex, which, by self-assembly, acts to coat the cytoplasmic surface of the nascent vesicle.

COPI vesicles recycle material from the Golgi apparatus back to the ER. In formation of the COP I complex, a distant paralogue to Sar1, called Arf, binds to the cytosolic portion of the membrane in GDP bound form. The nucleotide diphosphate is then exchanged for a GTP moiety by an Arf GEF in a manner similar to Sar1 activation. Membrane-bound cargo might interact with a preassembled vesicle coat-forming coatomer complex. Coatomer, Arf and ArfGAP (GAP for GTPase-activating protein) complex to form the polymeric coat and vesicle budding occurs (Springer et al., 1999). The formation of COPI vesicles is clearly involved in retrograde transport from the Golgi back to the ER, but it is probable that it is also important in anterograde transport within the Golgi stack (Orci et al., 1997; Schekman and Mellman, 1997).

Many of the remaining vesicles formed in the cell are coated with clathrin or clathrin-related proteins and include vesicles destined for both the endosome and the plasma membrane. In the formation of clathrin-coated vesicles, Arf also acts as the GTPase, with an Arf GEF again providing the GTP exchange (Kirchhausen, 2000; Springer et al., 1999). Heterotetrameric adaptin complexes (AP) bind cargo and provide specificity for particular organellar destinations. AP1 and AP3 at the TGN are involved in targeting material to the late endosome and lysosome, respectively. AP2 is involved in cargo selection for plasma-membrane-derived vesicles entering the endocytic pathway (Kirchhausen, 2000; Robinson and Bonifacino, 2001). Adaptors bind via *cis*-acting amino acid motifs in the cargo or via additional adaptor proteins, such as the mannose-6-phosphate receptors in mammals for the transport of material to the late endosome (Kirchhausen, 2000). Clathrin itself acts as the protein coat, polymerizing and forming the vesicle. In the case of AP3, clathrin is not involved but VPS41 acts as the protein coat polymer and appears to have a domain homologous to clathrin (Kirchhausen, 2000; Robinson and Bonifacino, 2001). Additional coat systems have also been reported, e.g., the GGA family (Golgi-associated, gamma adaptin ear-containing Arf-binding protein). They are also important in mannose-6-phosphate receptor trafficking (Doray et al., 2002) but, at present, it is unclear how these are recruited or function in detail.

14.2.3.2 Vesicle Translocation

After vesicle formation and budding, the vesicle is transported to its eventual target. At some point after vesicle formation, the GTP on the GTPase is hydrolyzed back to GDP via the action of an ArfGAP homologue. The role of this hydrolysis is unclear, although it has

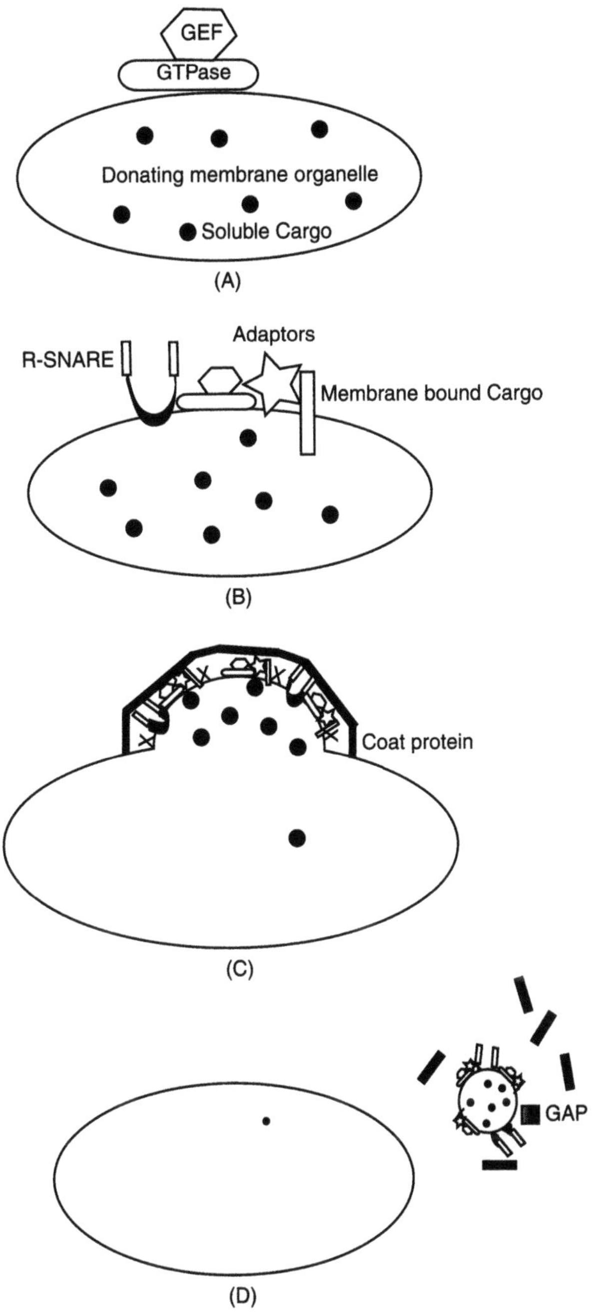

FIGURE 14.2 Generalized cartoon of vesicle formation, budding and movement. (A) GTPase attaches to the membrane and a GEF swaps GDP for GTP. (B) Adaptor proteins and cargo attach to the nucleating site of vesicle formation. (C) Coat proteins arrive and form a scaffolding complex for vesicle formation. Soluble cargo might be incorporated into the vesicle via adaptors or by bulk flow. (D) Vesicle has budded away from the donating membrane, a GAP hydrolyzed GTP and the vesicle uncoats. *Note:* All shapes once named in a panel retain their assignment in subsequent panels.

been suggested that ArfGAPs are involved in signaling and interactions with the cytoskeleton (Donaldson and Lippincott-Schwartz, 2000). It has also been demonstrated that when GTP hydrolysis is blocked, intracellular transport vesicles are unable to uncoat (Tanigawa et al., 1993). Regardless of whether this is a causal relationship, uncoating of the transport vesicles occurs after leaving the donor membrane and before vesicle fusion. Critically, vesicles are transported by interaction with microtubules and appear to be driven through interaction with kinesins (Girod et al., 1999).

14.2.3.3 Vesicle Fusion

The final stage of vesicular transport is the fusion of the vesicle with its target (Figure 14.3). An R-SNARE (soluble NSF attachment protein receptor) is incorporated as membrane cargo during vesicle formation. SNARE proteins are characterized by the presence of extensive coiled-coil regions and can form both *cis-* and *trans-*complexes. SNAREs are classed as Q or R, based on the position of an arginine (R) or glutamine (Q) at a highly conserved position within the coiled-coil domain (Antonin et al., 2002; Fasshauer et al., 1998). On the target membrane, at least one member of the Q-SNARE protein family, syntaxin (Edwardson, 1998), is present. This protein is complexed by a Sec1/Munc18 homologue, which appears to inhibit syntaxin from promiscuous interaction before the appropriate fusion event (Schulze et al., 1994). As the incoming vesicle reaches the target membrane (Figure 14.3B), syntaxin, a SNAP-25 homologue (which also possesses a Q within the critical region) and the R-SNARE form a coiled-coil, four-helix bundle (Hay, 2001). The SNARE complex has been implicated in docking, tethering (Ungermann et al., 1998, 2000) and physical fusion (Nickel et al., 1999) of the membranes. SNARE–SNARE interaction might also provide some specificity for vesicular transport (McNew et al., 2000). After fusion (Figure 14.3D), the *trans*-SNARE bundle is disassembled and recycled via the action of an ATPase, either NSF or p97. Whereas NSF acts in vesicular transport (Edwardson, 1998), its paralogue p97 plays a similar role in the postmitotic reassembly of organelles (Rabouille et al., 1998).

14.2.3.4 Regulation and Specificity

GTPases of the Rab protein family are essentially involved in the vesicular transport process in a variety of steps. They interact with a large number of proteins, both physically and genetically. These include SNARES, docking factors and the cytoskeleton. Most significantly, Rabs act in regulation of the overall process (Armstrong, 2000). Similarly to Sar1 and Arf, Rab proteins are recruited to the membrane as the GDP form, but are rapidly converted to the GTP state, which is the active form. The best-characterized function for Rab proteins is control of vesicle fusion, which is dependent on the GTP form being present on the vesicle membrane. Significantly, target membranes appear to contain a GAP activity, which results in rapid inactivation of the G protein and control of fusion efficiency (Rybin et al., 1996). Rab function is extensive, as the protein interacts with a large number of effector molecules. In the case of mammalian Rab5, at least 20 of these effectors have been identified, including EEA1 (a tethering factor) and several kinases, which most likely influence lipid structure and dynamics (Christoforidis et al., 1999). A highly significant aspect of the Rab gene family is specificity: members of the family are targeted precisely to endomembrane subcompartments. This feature makes these proteins attractive as molecular flags for pathways within the membrane transport system and has been exploited in several systems discussed next.

14.2.4 Vesicular Transport in Non-Opisthokont Lineages

Whereas much of the generalized information given has been developed in yeast and mammals, several divergent systems have been studied at the molecular level in some detail,

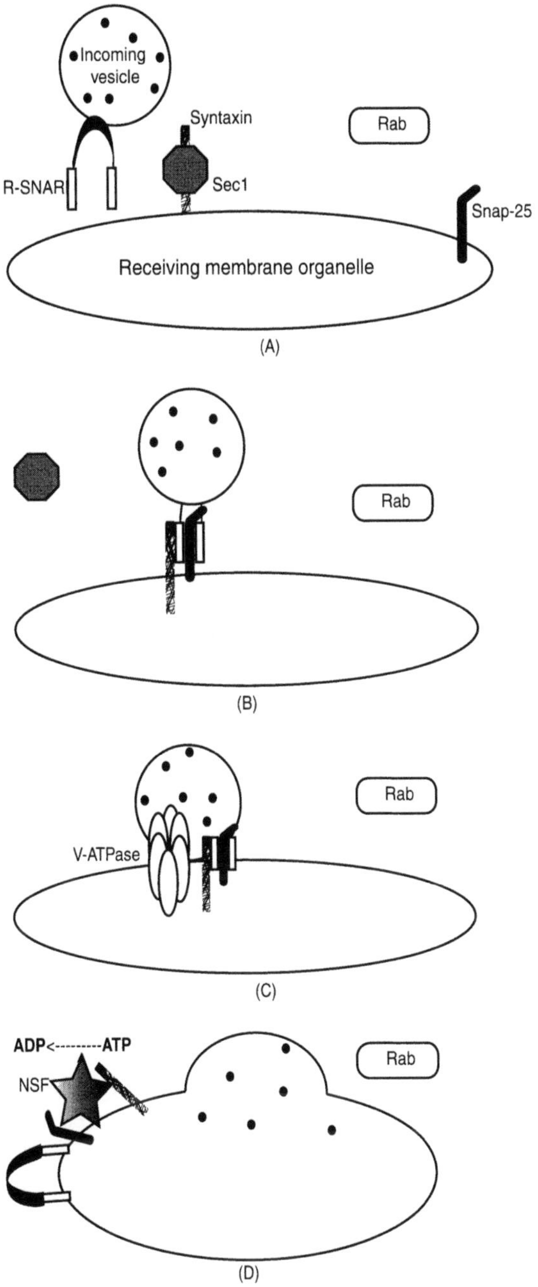

FIGURE 14.3 Generalized cartoon of vesicle fusion. (A) Incoming vesicle, containing cargo and R-SNARE homologue, approaches receiving organelle possessing Snap-25 and syntaxin homologue complexed with a Sec1 homologue. (B) Sec1 releases syntaxin, which forms a coiled coil with the R-SNARE and Snap25 homologues, prompting vesicle docking and tethering. (C) Vesicle fusion begins with the SNARE complex and other proteins (V-ATPase subunits) being implicated in creating the fusion pore. (D) NSF hydrolyzes ATP to dissociate the SNARE complex and recycle components for future rounds of vesicle fusion. Rab is implicated at various steps in the process. *Note:* All shapes once named in a panel retain their assignment in subsequent panels.

primarily because of the importance of these organisms as major disease agents. The best examples are among the apicomplexans *Plasmodium falciparum* and *Toxoplasma gondii* and the kinetoplastid *Trypanosoma brucei*. The last two are the best understood, in part for technical reasons, but also because of the advanced stage of their respective genome projects. The functional analyses in these taxa provide some information that is not forthcoming from sequence analysis alone, and might be informative of the manner in which the endomembrane system has evolved. It must be recognized that these systems are highly developed and are as far from the eukaryote common ancestor as metazoans. Some aspects of these systems have likely arisen from adaptation rather than being a reflection of a true basal or pleisiomorphic state. However, the presence of such systems provides a triangulation point for reconstructing evolutionary processes and importantly can demonstrate whether differences between higher eukaryotes reflect fundamental changes or simply specialization.

In *T. gondii*, members of the ARF family are present, and at least one member (ARF1) shares a clear role in transport through the Golgi complex and secretion with higher eukaryotes (Liendo et al., 2001). A number of Rabs have also been identified and their functions described. *T. gondii* homologues of Rab4, 5, 7 and 11 are present, and data suggest that these proteins also have conserved functions (Stedman et al., 2003). Interestingly, two Rab5 isoforms have been identified, but only one has been studied in detail (Robibaro et al., 2002). In addition, a clear Rab6 homologue is also present; this protein serves to define a retrograde transport pathway that delivers material to the *Toxoplasma* Golgi complex (Stedman et al., 2003). Therefore, once more the basic pathway of the endomembrane system is easily detectable and characterized, suggesting that these pathways emerged rapidly after eukaryotes evolved. Furthermore, indirect biochemical analysis demonstrates that the NSF/SNAP/Rab system is present in this organism (Chaturvedi et al., 1999).

A particularly interesting aspect of the Apicomplexa is the presence of specialized organelles, the rhoptries, dense granules and micronemes, all of which appear to play a role in invasion and establishment of the intracellular environment that the parasite requires for replication. Significantly, targeting of proteins to the rhoptries is dependent on recognition of sequences homologous to lysosomal targeting signals, and it is likely that rhoptry biogenesis is derived from targeting of rhoptry proteins via the early endosome (Robibaro et al., 2001).

T. brucei can be considered representative of the kinetoplastida, and all evidence suggests that other organisms of this order conform to the overall morphology and level of intracellular complexity seen in this paradigm organism. Clathrin has been identified in *Trypanosoma* (Morgan et al., 2001), as has the entire COPI coat (Maier et al., 2001). Other protein families have also been identified, including the subunits for three adaptin complexes; interestingly, the AP2 complex involved in recognition of cargo molecules at the cell surface was not identified (Morgan et al., 2002).

The best-studied family of endomembrane proteins in trypanosomes are the Rab proteins. As in mammals, several Rab proteins are likely involved in ER to Golgi transport: Rab1, 2A and 2B. These three proteins are well conserved, as evidenced by BLAST, and functional data are in good agreement with this assignment (Field et al., 1999; MCF and V. Dhir, unpublished). This is in contrast to *S. cerevisiae*, in which only a single Rab protein appears involved in ER to Golgi transport (the Rab1 homologue, Ypt1p). This suggests that for ER to Golgi transport, yeast is less complex than the trypanosome, and might indicate a loss of function during fungal evolution. For intra-Golgi transport, a similar level of complexity is apparent between yeast (Ypt31p and 32p) and trypanosomes (Rab18 and Rab31). In contrast, mammals have several Rabs that are likely responsible for regulating intra-Golgi transport, consistent with increased pathway complexity among the metazoans.

14.3 Mining the Databases

A comparative genomic study of the evolution of the endomembrane system requires three major pieces: genomic databases (details in Section 14.7.3), specific proteins to search for and methods to do those searches.

14.3.1 Candidate Proteins To Be Used as Representative Queries

It is clear that key proteins or protein families are involved in the generalized steps of the vesicular transport. The common machinery components involved in vesicle formation are Arf/Sar1 GTPases, GAPs and GEFs. As there are at least three major types of vesicles that share some of these common components, it is possible to search for these types of vesicles by looking for a representative component of their respective coat polymers. Clathrin (heavy chain) is the obvious representative for clathrin-coated vesicles, whereas α-COP and Sec31 are used as representatives of COPI and COPII vesicles, respectively. The fusion machinery also provides several attractive search query candidates. Sec1 and SNAREs are excellent examples of protein families with multiple paralogues involved in the same role at various steps of transport. NSF and p97 play keys roles in membrane fusion events and, as such, are good query candidates. Finally, Rab essentiality is undeniable and will also be included. Representatives of each of these protein families or protein complexes were assembled (Table 14.1) and used as query sequences for the subsequent comparative genomic surveys.

TABLE 14.1 Genes Used as Queries for Comparative Genomic Survey

Query	*Gene Family*	*Paralogue*	*Taxon*	*Accession No.*
Arf	ADP ribosylating factor	ARF1	Homo	P32889
Sar1	Secretion-associated, Ras-related	Sar1p	Saccharomyces	NP_015106
ArfGEF	Arf-GTP exchange factor	GEA1	Saccharomyces	P47102
AP	Adaptin	AP2 alpha subunit	Homo	NP_055018
COPII	COP II vesicle coat	Sec31p	Homo	NP_055748
COPI	Coatomer alpha	Alpha-COP	Saccharomyces	P53622
Clathrin	Clathrin	Chc1p	Saccharomyces	NP_011309
ArfGAP	Arf-GTP activating factor	Gcs1	Saccharomyces	NP_010055
R-SNARE	Synaptobrevin	Ykt6p	Saccharomyces	NP_012725
Syntaxin	Syntaxin	Sso1p	Saccharomyces	NP_015092
Sec1	Sec1	Syntaxin-binding protein 2	Homo	XP_008937
Rab	Rab	Ypt52p	Saccharomyces	P36018
NSF	N-ethylmaleimide-sensitive factor	NSF	Homo	XP_032173
p97	Transitional ER ATPase	TERA	Homo	P55072

Note: Homo: *Homo sapiens*; Saccharomyces: *Saccharomyces cerevisiae*.

14.3.2 Search Methods

The BLAST search algorithm (Altschul et al., 1997) can be used to find homologues of either DNA or protein sequences (queries) by searching genomic databases containing either sequence type. This algorithm aligns the query sequence with others in the database and assigns it a score based on how similar two sequences are. The reliability of a match by BLAST is measured in expectation (E) values and is usually expressed as a negative exponent.

This corresponds to the probability of observing an alignment that scores the same as the alignment between the query and a retrieved database entry, based on chance alone. This value is also corrected for the size of the database. The lower the *E* value, the more significant the match. At some point, the *E* value drops so low that the server may merely state the value as 0. PSI-BLAST is an iterative BLAST program that uses a scoring matrix based on a consensus of retrieved homologues to increase the sensitivity of the subsequent search. This method can also counteract lineage-specific peculiarities for a given search query, such as amino acid compositional bias, rapid evolutionary rate or divergence of a key motif (Altschul et al., 1997).

There are a variety of reasons why a particular protein might not appear in a genomic initiative database, other than its true absence from a genome. EST projects are a snapshot of genes expressed at a given time. If a gene is not expressed at that life cycle stage, or is expressed in low abundance, then it might not be represented. GSS surveys are random samplings of a genome, and so, by chance, a gene of interest simply might not have been encountered when the search was performed. Finally, when looking among diverse eukaryotes, the gene of interest might have diverged so much in that taxon that it is unrecognizable by a BLAST search; this can only be reliably confirmed by functional analysis. If no homologue can be identified in response to a particular query, then stating simply that a homologue was not identified is the most prudent response in the majority of cases.

The conservative nature of the "not identified" label released us from having to use a method that rigorously excludes claims about the lack of a homologue in a genome. Instead, we were able to use a search strategy that was biased against the other major pitfall, false positive identification of homologues. For each protein component, the relevant query sequence was used in a BLAST search against a given database. All sequences retrieved as possible homologues, given a generous cut-off value for significance, were then reciprocally used as queries for a BLAST search. Only those that retrieved the query sequence, and other defined orthologues of it, were noted as true homologues. This struck a balance between allowing for divergent sequence in distantly related taxa (i.e., weak but real BLAST hits) and caution in assigning homology. In cases where the retrieved sequence was a named homologue of the query (implying that functional characterization or at least BLAST identification had already been done), reciprocal BLAST analysis was not performed. For details of the search methodology, refer to Section 14.7.2.

Bioinformatic surveys of diverse genomes were performed by using homologues of the above components as queries in order to examine the origin and evolution of the vesicular transport machinery.

14.4 Endomembrane System Component Homologues in Diverse Genomes

The origins of a eukaryotic cellular system can be approached from the bottom up, looking to diverse prokaryotes for homologues of components of complex eukaryotic cellular machinery, or from the top down by reconstructing from extant taxa a consensus of commonly held and therefore likely ancestral machinery. Both approaches will be taken, using the components listed in Table 14.1, together with BLAST analysis (Altschul et al., 1997).

14.4.1 Bottom Up: Prokaryotic Homologues of Endomembrane System Components

Several of the vesicular transport components have clear homologues in prokaryotic genomes, but these are highly significant and have narrow taxonomic distributions. When the proposed prokaryotic homologue is reciprocally used as a query for a BLAST search (reciprocal BLAST analysis), the sequences retrieved are not prokaryotic homologues but rather eukaryotic proteins. This pattern seems indicative of a lateral gene transfer (LGT) event, transferring the eukaryotic gene to the prokaryote rather than the gene being the progenitor of a novel eukaryotic gene family. For example, when using ArfGEF as a query in a BLAST search, the RalF protein from *Legionella* is returned at 5e-19. This protein has been shown to have Arf-modulating activity *in vivo* (Nagai et al., 2002). Another protein from *Rickettsia*, identified as a further ArfGEF homologue, is also returned at 2e-15. On reciprocal BLAST analysis, both return each other with E values in the range of e-78 and eukaryotic ArfGEFs (e-30), but do not seem to have a wide distribution among prokaryotes. This likely represents a lateral transfer to either *Legionella* or *Rickettsia* and subsequent transfer to the other (Nagai et al., 2002). Similarly, the RecO protein from *Deinococcus* has an identifiable GAP domain at its C-terminal end (E = 0.051), but other RecO homologues do not. This might be a case of LGT and domain fusion specifically in this taxon.

The cotranslational system is highly conserved and clearly has its origin with the prokaryotes; in these organisms, the system is used for direct export across the plasma membrane into the periplasmic space (reviewed in Rapoport et al., 1996). Significantly, the minimal functional core of the RNA is conserved in both prokaryotes and eukaryotes. As well, the protein responsible for recognizing the signal sequence (SRP54) in eukaryotes has a bacterial homologue. The remainder of the eukaryotic SRP protein components are absent in prokaryotes. Furthermore, the SRP receptor, responsible for recognizing the ribosome–SRP complex, is also highly conserved between the kingdoms. In addition, in prokaryotes, the nascent protein is translocated through a proteinaceous channel composed of SecY, which exhibits clear homology to a protein performing an analogous function in eukaryotes, Sec61. Interestingly, Sec61 functions in collaboration with several additional polypeptides, including Sec62/63, which are responsible for mediating interactions with machinery in the ER lumen. These are absent from the prokaryotic system (although additional non-conserved polypeptides are also present in the bacterial pathway), highlighting that the basic machinery appears conserved but a number of important (and frequently essential) functions are eukaryote specific. The high degree of conservation seen in the polypeptide translocation system provides a rare and focused insight into evolution of the endomembrane system.

A number of important components of the vesicular transport machinery belong to larger gene families, each having intriguing putative prokaryotic homologues (Table 14.2).

TABLE 14.2 Comparison of Eukaryotic vs. Prokaryotic Endomembrane Component Homologues

Component[a]	*Eukaryotic E value*[b]	*Prokaryotic Homologue*[c]	*Prokaryotic E value*[d]
Rab/Sar/Arf	E-05 to E-98	Putative GTPases	Ψ^e I2 = E-06 to E-11
α-COP	E-26 to E-101	WD-40 proteins	E-20
Sec31	E-100	WD-40 proteins	E-40 to E-79
p97	E-130 to E-0.0	cdc48 homologues	E-180
NSF	E-50 to E-0.0	cdc48 homologues	E-50

[a]This column lists the protein family, or families, used as queries with specific queries matching their family designation in Table 14.1.

[b]This column lists the range of expectation value scores seen for retrieved eukaryotic homologues.

[c]This column lists the general assignment of prokaryotic sequences assigned as putative homologues.

[d]This column lists the range of expectation value scores seen for putative prokaryotic homologues.

[e]In the case of the Arf/Sar1/Rab searches, two iterations of PSI-BLAST were done before a significant prokaryotic homologue was retrieved.

A BLAST search using Arf, Sar1 or Rab queries produces similarity to each other and a number of other GTPases, mostly eukaryotic. No single, clear, prokaryotic homologue can be said to have given rise to endomembrane system GTPases. Nonetheless, position-specific-interated (PSI)-BLAST searches with Arf, Sar1 or Rab retrieved several GTPases with moderate prokaryote taxonomic distribution. These hit eukaryotic GTPases (*E* = 5e-13) and elongation factors (*E* = 4e-04) in reciprocal BLAST analysis. Most likely, an ancestral GTPase gave rise to the eukaryotic GTPases, but which is unclear. Eukaryotic small GTPases being more closely related to each other (*E* = e-05 to e-98) than to any given prokaryotic homologue suggests that the common ancestor had a very simple GTPase composition, which was expanded following the initial diversification of eukaryotes.

Proteins of both COPI and COPII vesicles (α-COP and Sec31, respectively) possess WD-40 domains. This domain, present in a wide variety of functionally unrelated proteins, is implicated as a scaffolding domain that facilitates protein–protein interactions (Smith et al., 1999). A number of bacterial and archael proteins also have very clear WD-40 domains. BLAST analysis of Sec31 retrieved eukaryotic homologues scoring in the range of E = e–100. In the same BLAST search, multiple cyanobacterial sequences were retrieved that, when used as queries in reciprocal BLAST analysis, retrieved diverse prokaryotic sequences from Bacteria and Archaea (*E* = e-40 to e-77). Use of α-COP as a query in BLAST analysis retrieved eukaryotic sequences from various taxa (*E* = e-26 to e-101), whereas prokaryotic sequences were obtained with expectation values in the range of e-20. Many of the putative prokaryotic homologues are simply assigned as WD-40 proteins without further functional prediction. This suggests that only the WD-40 domain is conserved and not necessarily homologous functionality. Clearly, Sec31 and α-COP have arisen from one or more ancestral proteins containing such a domain, but this analysis suggests that a true functional homologue is not likely present within the sampled prokaryotic taxa.

The AAA-type ATPase family is a well-defined group of proteins associated with a wide variety of cellular functions (Ye et al., 2001). One member of this family, p97, has been shown to be involved in homotypic membrane fusion events, such as postmitotic reassembly of ER (Latterich et al., 1995) and Golgi (Rabouille et al., 1998). It has also been implicated in a number of additional functional processes, including ubiquitin-dependant protein degradation (Ghislain et al., 1996) and the cell cycle (Moir et al., 1982). A second AAA-type ATPase paralogue, NSF, on the other hand, is known to be involved only in SNARE complex disassembly and recycling (Edwardson, 1998). Clear homologues of AAA-type ATPases can be found in both Bacteria and Archaea (Pamnani et al., 1997). A BLAST search with p97 as the query sequence yields eukaryotic homologues with expectation values ranging from e-130 to 0.0 and prokaryotic homologues with scores of ca. *E* = e-150. BLAST analysis of NSF retrieves eukaryotic NSF homologues (*E* = e-50 to 0.0) and prokaryotic sequences in the *E* = e-50 range as well. A number of indications suggest that p97 might be the ancestral and pleisiomorphic form of the protein (Zhang et al., 2000). BLAST values for prokaryotic homologues are higher when using the p97 version than with the NSF query. Also, the broad spectrum of cellular processes with which p97 is involved suggests that NSF might have been a specialized offshoot. However, because BLAST values might be affected by evolutionary rate, this should be examined by phylogenetic analysis.

Although not a prokaryotic connection, it is still possible to find some evolutionary affinities that go beyond simple paralogue expansion. For example, the adaptin complex has common origins with a subcomplex of COPI (Duden et al., 1991; Schledzewski et al., 1999). Both complexes interact with, and help to form, vesicle coats at the Golgi. The large subunits of each complex are clearly paralogues as are the medium and small subunits. It is proposed that what began as a heterodimer of a large and small subunit duplicated and differentiated to form an ancestral heterotetramer. Further duplications later produced the F-COP complex and then subsequently the AP3, 2 and 1 adaptin complexes (Schledzewski et al., 1999). Recent data also suggest that the regulatory V1H subunit of the V-ATPase is homologous to the N-terminus of the β-COP and μ-adaptin subunits (Geyer et al., 2002). These studies, however, are restricted in their taxonomic sampling, primarily including sequences from animals, fungi and land plants. A broader sampling would be useful to verify and expand this proposed evolutionary scheme. No clear prokaryotic homologues were identifiable for SNAREs, adaptins, Sec1 and clathrin by BLAST or PSI-BLAST analyses.

14.4.2 Top Down: Reconstructing the Vesicular Transport Machinery of the Last Common Eukaryotic Ancestor

Reconstructing the evolution of a cell biological system is done, in the ideal case, by deducing which components are present in the common ancestor of the group of organisms under consideration. Features found in all descendants of a common ancestor must, barring lateral gene transfer, have been present in that ancestor. Features found in most taxa, and in the deepest branch of a resolved phylogeny, are also most likely to have been present in the common ancestor. To make such deductions, knowledge of the cell biology of the taxa under examination, plus their phylogeny, is required. Systems central to eukaryotic cellular evolution will require a deeply branched phylogeny.

Initially, small subunit ribosomal DNA sequences were used alone to resolve eukaryotic relationships (Sogin, 1991), but for multiple reasons these data are insufficient to construct robust phylogenies and must be supplemented by protein and morphological data (Dacks and Doolittle, 2001; Embley and Hirt, 1998; Philippe and Adoutte, 1998; Philippe et al., 2000; Chapter 2). Issues relating to construction of phylogenies are discussed elsewhere in the book (Chapter 6). Despite analyses showing that the major eukaryotic lineages might have diverged from one another rapidly (Philippe and Adoutte, 1998; Philippe et al., 2000), evidence from improved species sampling, increased availability of molecular data and optimized computational analysis to suggest phylogenetic structure in eukaryotic relationships (Figure 14.4 and references therein). Genomics can further provide us with some of the data needed to better resolve that structure (Dacks and Doolittle, 2001).

Despite these advances, a resolved, broadly sampled and rooted eukaryotic phylogeny does not seem imminent. Nonetheless, reconstructing the evolution of eukaryotic cell biology is still possible in its absence. Rather than looking at a designated deepest taxon, diversity can be used to approximate the last common eukaryotic ancestor. Using comparative genomics to search, among diverse taxa, for proteins known to be functionally important in a particular system allows estimation of the minimal protein machinery present in the last common ancestor of the taxa sampled. The wider the diversity of sampling, the better their last common ancestor approximates the last common ancestor of all eukaryotes. The scheme in Figure 14.4 shows the state of knowledge ca. 2002 regarding eukaryotic relationships. Coded in this picture are major lineages with publicly accessible genomics initiatives. Although there are areas of the eukaryotic tree vastly undersampled, the current diversity of genome initiatives allows us to examine a last common ancestor that is a crude but reasonable approximation of the last common eukaryotic ancestor.

Eukaryotic genomic initiative databases publicly accessible as of September 2002 were searched for homologues of the protein family queries listed in Table 14.1. Most genomes examined have, at least, one member of the protein families identified as important components of the vesicle transport machinery (Table 14.3 and Table 14.4). Several queries were not identified in the *Paramecium* (ciliate) genome, but this project was in its earliest stages at the time of this survey. Given that the apicomplexans (sisters to the ciliates in the alveolate supergroup) possess all the components, it is likely that the scarcity of *Paramecium* components is due to sampling. The shared presence of the basic vesicular transport machinery in most of the surveyed genomes implies that it was already established in an early eukaryotic

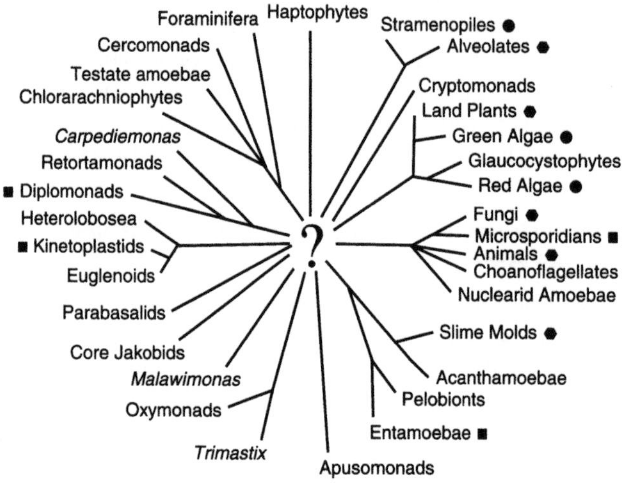

FIGURE 14.4 Schematic of proposed eukaryotic relationships, ca. 2002. This unrooted star phylogeny incorporates references fully cited in Dacks and Doolittle (2001) 419–425, as well as new morphological SSUrDNA and protein data [Archibald et al. (2003) 62–66; Arisue et al. (2002) 1–10; Bapteste et al. (2002) 1414–1419; Silberman et al. (2002) 777–786; Simpson et al. (2002a) 239–248; Simpson et al. 1782–1791]. Taxa with publicly available EST projects are noted with a circle. Those with GSS or genome projects are noted with a square. Taxa with both are marked with a hexagon.

TABLE 14.3 Comparative Genomic Survey of Vesicle Formation and Movement Proteins in Diverse Eukaryotic Genomes

Higher Taxon	*Organism*	*Arf*	*Sar1*	*ArfGEF*	*AP*	*COPII*	*COPI*	*Clathrin*	*ArfGAP*
Fungi	*Saccharomyces*	A	A	A	A	A	A	A	A
Land plants	*Arabidopsis*	A	C	A	A	C	A	C	A
Animal	*Homo*	A	A	A	A	A	A	A	A
Diplomonad	*Giardia*	A	D	D	C	E	B	D	D
Kinetoplastid	*Trypanosoma*	C	E	E	C	E	B	C	E
Apicomplexa	*Plasmodium*	A	E	E	E	C	B	E	C
Slime molds	*Dictyostelium*	C	E	E	D	NI	D	A	C
Entamoebae	*Entamoeba*	C	E	E	E	E	B	E	E
Red Algae	*Porphyra*	D	D	D	D	E	B	D	NI
Stramenopiles	*Phytophthora*	D	E	D	D	D	B	D	D
Green algae	*Chlamydomonas*	A	E	NI	D	NI	B	D	A
Ciliates	*Paramecium*	NI	NI	A	NI	NI	C	C	NI

Note: A: homologues published in separate analyses; B: homologues previously identified in Dacks and Doolittle (2001) 419–425; C: genes not yet published but found in Genbank; D: a gene listed on the respective genome initiative web site; E: homologue found by reciprocal BLAST analysis in this study; NI: a clear homologue not reliably identified by any of the given criteria. Table last updated in September 2002.

TABLE 14.4 Comparative Genomic Survey of Vesicle Fusion Proteins in Diverse Eukaryotic Genomes

Higher Taxon	*Organism*	*R-SNARE*	*Syntaxin*	*Sec1*	*Rab*	*NSF*	*p97*
Fungi	*Saccharomyces*	A	A	A	A	A	A
Land plants	*Arabidopsis*	A	A	A	A	C	C
Animal	*Homo*	A	A	A	A	A	A
Diplomonad	*Giardia*	B	A	B	A	D	D
Kinetoplastid	*Trypanosoma*	B	A	B	A	C	C
Apicomplexa	*Plasmodium*	B	C	B	B	C	C
Slime molds	*Dictyostelium*	B	A	B	A	A	C
Entamoebae	*Entamoeba*	B	B	B	A	E	C
Red algae	*Porphyra*	B	A	NI	B	NI	D
Stramenopiles	*Phytophthora*	B	A	B	B	E	D
Green algae	*Chlamydomonas*	B	A	B	A	E	D
Ciliates	*Paramecium*	NI-a	NI	NI	C	C	NI

Note: A: homologues published in separate analyses; B: homologues previously identified in Dacks and Doolittle (2001) 419–425; C: genes not yet published but found in Genbank; D: a gene listed on the respective genome initiative web site; E: homologue found by reciprocal BLAST analysis in this study; NI: a clear homologue not reliably identified by any of the given criteria. In the case of NI-a, an *Euplotes* (ciliate) homologue has been identified. Table last updated in September 2002.

ancestor and that the basic mechanism of vesicular transport has also been conserved. Similarly, the common presence of clathrin, Sec31, and α-COP homologues in the various genomes suggests that the last common ancestor also had the ability to form the three classes of vesicles seen at present.

Many of the questions surrounding the evolution and complexification of eukaryotic systems are ones of duplications. Some queries involve multiple duplications of closely related proteins, which might be difficult to assess by BLAST alone (see Section 14.5). Others, however, involve deep duplications, yielding paralogues with divergent function such as Sar1/Arf. The Arf protein family is composed of several paralogous subfamilies, each playing a similar role in the formation of clathrin and COPI vesicles as Sar1 does for COPII. The majority of taxa examined have at least one homologue of both Arf and Sar1 present in their genomes (Table 14.3 and Table 14.4), and thus the duplication that gave rise to Arf and Sar1 is likely to have occurred before the divergence of the taxa examined.

The situation with NSF vs. p97 is slightly more complicated. Both proteins are members of a larger AAA-type ATPase family. Most taxa examined seem to have at least one copy of both genes. However, both proteins also retrieve eukaryotic cdc48 homologues as well as a number of uncharacterized cdc48-like ORFs with significant BLAST scores. This makes it quite difficult to distinguish the presence of p97 vs. that of NSF. As well, although the biological function of NSF is well established, p97 seems to have multiple roles in the cell, membrane fusion being only one of them (Ye et al., 2001). As such, the biological significance of the duplication is difficult to assess. Although the story is likely to be infinitely more complex, it is possible to deduce, at a minimum, that the duplication which gave rise to p97 and NSF occurred before the last common ancestor of the taxa tested.

Comparative genomic surveys have also been used to demonstrate a trend of expansion in some of the families involved in vesicular transport in unicellular organisms as compared with multicellular taxa. This is seen quite strikingly in the Rab proteins when comparing *Saccharomyces* (11 Rabs), *Plasmodium* (11) and *Trypanosoma* (16) to *Caenorhabditis* (29), *Homo* (60) and *Arabidopsis* (57); (Bock et al., 2001; Rutherford and Moore, 2002).

14.5 Beyond BLAST: Examples from Functional Studies

BLAST can survey for the presence of relevant protein families in diverse eukaryotic genomes. However, many more detailed questions are beyond the scope of sequence analysis alone and require additional data.

Questions of detailed evolutionary history might require the identification of a gene sequence at the level of its paralogue subfamily within a larger gene family, ideally within a phylogenetic framework. The reliability of such an assignment by BLAST might be compromised, because the algorithm does not take into account evolutionary rate and so a sequence from an organism with a rapid rate might be misidentified. Additionally, many of the databases provide only partial (end reads of cDNAs) or poor-quality gene sequence (single-pass reads of genomic fragments). These should provide enough conserved sequence to yield a broad gene family assignment (such as in the case of Sar1 vs. Arf), but a subfamily identification might be beyond the boundaries of reliability. For identification of close paralogue affiliation, molecular biology, phylogeny and functional assignment might be required. Other detailed questions of evolution within a gene family might involve establishing the relationship of paralogues, and the timing of their expansion relative to various lineage divergences or the relationship of various paralogues relative to an outgroup.

14.5.1 Genomics and Phylogeny

Several phylogenetic studies have used genomic databases for the initial identification of partial sequences, which were then confirmed and expanded through standard molecular biological means. Phylogenetic analysis of the various subunits of the adaptin and COP complexes, as obtained in part by sequencing cDNAs, revealed not only homologies between the seemingly unrelated endomembrane components but also some internal paralogue relationships (Chow et al., 2001; Schledzewski et al., 1999). Similar studies have also been performed on the syntaxin gene family. Syntaxin genes were identified from a variety of protist EST and GSS surveys and further characterized by molecular biological means. Phylogenetic analysis determined that the duplication giving rise to the syntaxin gene families must have occurred early on in eukaryotic evolution (Dacks and Doolittle, 2002), as well as identifying several lineage specific paralogue expansions within the gene family. The Rab protein family has been studied in perhaps the most diverse array of taxa (Bush et al., 1993; Janoo et al., 1999; Langford et al., 2002; Morgan et al., 2002; Rutherford and Moore, 2002; Saito-Nakano et al., 2001; Stedman et al., 2003). Various Rab homologues have been identified from genomics initiatives in *Entamoeba* (Saito-Nakano et al., 2001) and more recently from *Giardia* (Langford et al., 2002). The phylogenetic analysis in this study shows a family with multiple deep duplications giving rise to several clades early on in the history of eukaryotes. On the other hand, phylogenetic analysis of the Rab5 families from mammals, yeast and trypanosomes indicates that the evolution of multiple Rab5 genes postdates the common eukaryotic ancestor as the Rab5 genes for each organism segregate into separate clades (Field et al., 1998). This implies that the common ancestor had a single Rab5 and hence most likely a simplified endocytic system. Interestingly, in yeast there is a high degree of redundancy between the three Rab5 isoforms. In the simpler trypanosomal system, wherein there are only two Rab5 family members, these functions appear fully distinct. The status of the two *Toxoplasma* gene products awaits functional analysis (Robibaro et al., 2002).

Phylogenetic analysis has also shed light on the story of paralogue expansion in both SNAREs and Rab proteins. There is a clear story of convergent, lineage-specific expansion in the plasma-membrane-localized syntaxin families of both metazoa and plants (Dacks and Doolittle, 2002). The *Arabidopsis* genome, in particular, shows a heavily expanded SNARE complement (Sanderfoot et al., 2000). A similar story is seen with the Rab proteins. A phylogenetic study of Rabs in yeast, mammals and *Arabidopsis* showed not only an expanded Rab content in *Arabidopsis*, as compared with yeast, but also that the mammalian and plant Rabs have expanded separately (Rutherford and Moore, 2002).

14.5.2 Genomics and Cell Biology

As for phylogenetic studies, genomics has allowed for identification of novel vesicular transport component homologues that can then be studied functionally. This has yielded both comforting underlying generalities to the model of endomembrane system organization as well as some surprising differences.

A number of significant finds regarding functional vs. *in silico* assignment of paralogues have emerged so far. A BLAST result alone might not be sufficient to assign functional homology. For example, one trypanosome Rab homologue clearly belongs to the Rab 18 family, based on BLAST. Paradoxically, the trypanosomal protein localizes to the Golgi complex whereas in metazoans Rab18 is associated with an endosomal compartment (Jeffries et al., 2002). Further, for a protein finally assigned as a Rab31 homologue, sequence comparisons were unable to discriminate its assignment between several Rab subfamilies. Functional analysis indicated trypanosomal Rab31 as a Golgi protein, in common with human Rab31 (Field et al., 2000). These observations indicate that detailed reconstruction of pathways within the endomembrane system, based purely on the presence of similar protein factors, is likely to be inaccurate; they probably also reflect the strong emphasis on functional data from higher eukaryotes. Most likely, as studies progress in divergent systems, the information will be of major utility for improving the accuracy of *in silico* assignments. For a second family of proteins, the SNAREs, this will be even more critical as these proteins contain extensive coiled-coil regions and retain limited sequence homology. Assignment of the full SNARE complement took several years, together with functional analysis, in the accessible *S. cerevisiae* system (Lewis and Pelham, 2002). It is clear that BLAST alone will be unable to even identify all SNAREs in divergent systems, let alone assign them a specific function. Phylogenetic analysis and functional cell biology will be even more important to fully understand the role of SNAREs in eukaryotes.

Nonetheless, the generalities of function for both these protein families have been confirmed in diverse taxa. Studies of Rabs have been done in a wide array of organisms, helped in part by sequences derived from genomic initiatives. Syntaxins too, albeit to a lesser extent, have benefited from the genomic windfall (Bogdanovic et al., 2000, 2002; Zhu et al., 2002). This will help establish a generalized model of how the endomembrane system functions have evolved.

14.6 Conclusions

The analysis of prokaryotic homologues raises several points. It appears likely that direct prokaryotic homologues of the proteins involved in vesicle transport are absent. Proteins containing the domains from which the eukaryotic components are built are present, but the occurrence of multiple GTPase or WD-40 domains in prokaryotes is perhaps unsurprising. For the prokaryotic taxa with genomes currently in hand, it is unlikely that any function homologous to vesicle transport is present. However, a recent characterization of the archeon *Ignicoccus* revealed an intracellular vesicle (Rachel et al., 2002). On the other hand, there are well-characterized prokaryotic homologues for various pieces of the protein translocation machinery, which serve similar if not identical roles in the cell (Rapoport et al., 1996). A recent functional study has even shown that when this system is blocked in *E. coli*, stacks of internal membranes with attached ribosomes accumulate in the cell (Herskovits et al., 2002), eerily reminiscient of ER. Although these and the archael vesicle might only be superficially similar and not truly homologous to the eukaryotic systems, they provide examples of structures similar to those in the endomembrane system arising in a prokaryotic context. Having more than one example of this makes any suggested models of the process more plausible.

From the current survey of eukaryotes, it is clear that the majority of the vesicular transport protein machinery that is well characterized in model systems is present in diverse taxa. This indicates that the entire system is relatively conserved and that the models for the mechanisms of vesicular transport are broadly applicable to eukaryotes beyond yeast and humans. The mere presence of a homologue does not necessarily imply the same function, but the presence of multiple interacting components makes the conservation of mechanism the most parsimonious working hypothesis. This mechanism needs to be tested *in vivo*, however, in diverse eukaryotes. The difference in function will tell exactly how the overall model must be modified to be universally applicable, as well as provide insight into specific evolutionary modifications. The last common eukaryotic ancestor appears to have had a complex endomembrane system. If generalities can be drawn from the evolution of the syntaxin and Rab families, then the elaboration of the vesicular transport components is likely to have begun early in eukaryotic evolution and ballooned on the various incidents of multicellularity. Nonetheless, other protein families will also have to be examined and this detailed picture of the evolution of the protein machinery will ultimately flesh out our understanding of the evolution of the endomembrane system in eukaryotes and provide deeper insight into organisms that are a threat or a benefit to us.

14.7 Materials and Methods

14.7.1 Search Queries

Either animal or fungal representatives of each protein family identified in the introduction were retrieved from Genbank and used as queries for the BLAST analyses. Table 14.1 gives the full listing of queries with their accession numbers. The protein representatives from these taxa were used as queries because the functional characterization of the protein families occurred in these model systems.

14.7.2 Search Methods

Keyword searching was performed at all databases that supported this option, in order to retrieve identified homologues. BLAST analysis was performed at the NCBI BLAST server (http://www.ncbi.nlm.nih.gov/BLAST/). Both the BLASTp algorithm and PSI-BLAST algorithm when necessary were used to search the protein databases. The tBLASTn algorithm was used when searching nucleotide databases. A cut-off value of 0.05 was used when selecting potential homologues, and each retrieved sequence was reciprocally used as a query back to the nr database. Only sequences that retrieved the initial query sequence were deemed legitimate homologues.

Two sets of searches were performed. In September 2001, searches were performed for a subset of the vesicular transport proteins, the results of which were published in November 2001 (Dacks and Doolittle, 2001). A second search was performed in September 2002 for an expanded set of vesicular transport proteins and to search for prokaryotic homologues of the queries listed in Table 14.1. Therefore, the homologues identified as B in Table 14.3 and Table 14.4 were identified in the September 2001 search, and all results are current as of September 2002.

14.7.3 Databases

The nonredundant (nr) database at Genbank was the only database searched when attempting to find prokaryotic homologues. The search for eukaryotic orthologues was also primarily performed in the nr database. However, the others ESTs database, HTGS and the GSS databases were also searched. Searches were also performed at a number of genome project Web sites, including the *Dictyostelium* cDNA project (http://www.csm.biol.tsukuba.ac.jp/cDNAproject.html), the *Giardia* genome project (http://jbpc.mbl.edu/Giardia-HTML/index2.html), the *Phytophthora* Genome Consortium (https://xgi.ncgr.org/pgc/) as well as the *Chlamydomonas* and *Porphyra* genome projects (http://www.kazusa.or.jp/en/plant/database.html).

Acknowledgments

JBD thanks W. F. Doolittle for helpful discussion, critical reading of the manuscript and supervision. Various discussions with T. Cavalier-Smith and members of the Doolittle and Roger labs have helped to shape many of the ideas in this chapter. P. P. Poon, M. Sogin, L. A. M. Davis and A. McArthur are also thanked for critical reading of this manuscript. MCF thanks members of the Field Laboratory for comments on the manuscript. We acknowledge the various genome projects that have made data publicly available; without these data such discussion would be impossible. Work in this chapter was supported by a grant to W. F. Doolittle (MT4467), a Canadian Institutes of Health Research doctoral research award to JBD and the Wellcome Trust (program grant to MCF).

References

Addinall, S. G. and Holland, B. (2002) The tubulin ancestor, FtsZ, draughtsman, designer and driving force for bacterial cytokinesis. *J. Mol. Biol.* 318: 219–236.

Alberts, B., Bray, D., Lewis, J., Raff, M., Roberts, K. and Watson, J. D. (1994) *Molecular Biology of the Cell,* Garland Publishers, New York.

Altschul, S. F., Madden, T. L., Schaffer, A. A., Zhang, J., Zhang, Z., Miller, W. and Lipman, D. J. (1997) Gapped BLAST and PSI-BLAST: A new generation of protein database search programs. *Nucleic Acids Res.* 25: 3389–3402.

Antonin, W., Fasshauer, D., Becker, S., Jahn, R. and Schneider, T. R. (2002) Crystal structure of the endosomal SNARE complex reveals common structural principles of all SNAREs. *Nat. Struct. Biol.* 9: 107–111.

Archibald, J. M., Longet, D., Pawlowski, J. and Keeling, P. J. (2003) A novel polyubiquitin structure in cercozoa and foraminifera: Evidence for a new eukaryotic supergroup. *Mol. Biol. Evol.* 20: 62–66.

Arisue, N., Hashimoto, T., Lee, J. A., Moore, D. V., Gordon, P., Sensen, C. W., Gaasterland, T., Hasegawa, M. and Muller, M. (2002) The phylogenetic position of the pelobiont *Mastigamoeba balamuthi* based on sequences of rDNA and translation elongation factors EF-1alpha and EF-2. *J. Euk. Microbiol.* 49: 1–10.

Armstrong, J. (2000) Membrane traffic between genomes. *Genome Biol.* 1: REVIEWS104.

Bapteste, E., Brinkmann, H., Lee, J. A., Moore, D. V., Sensen, C. W., Gordon, P., Durufle, L., Gaasterland, T., Lopez, P., Muller, M. and Philippe, H. (2002) The analysis of 100 genes supports the grouping of three highly divergent amoebae: *Dictyostelium*, *Entamoeba*, and *Mastigamoeba*. *Proc. Natl. Acad. Sci. USA* 99: 1414–1419.

Becker, B. and Melkonian, M. (1996) The secretory pathway of protists: Spatial and functional organization and evolution. *Microbiol. Rev.* 60: 697–721.

Bock, J. B., Matern, H. T., Peden, A. A. and Scheller, R. H. (2001) A genomic perspective on membrane compartment organization. *Nature* 409: 839–841.

Bogdanovic, A., Bruckert, F., Morio, T. and Satre, M. (2000) A syntaxin 7 homologue is present in *Dictyostelium discoideum* endosomes and controls their homotypic fusion. *J. Biol. Chem.* 275: 36691–36697.

Bogdanovic, A., Bennett, N., Kieffer, S., Louwagie, M., Morio, T., Garin, J., Satre, M. and Bruckert, F. (2002) Syntaxin 7, syntaxin 8, Vti1 and VAMP7 (vesicle-associated membrane protein 7) form an active SNARE complex for early macropinocytic compartment fusion in *Dictyostelium discoideum*. *Biochem. J.* 368: 29–39.

Bryant, N. J. and Stevens, T. H. (1998) Vacuole biogenesis in *Saccharomyces cerevisiae*: Protein transport pathways to the yeast vacuole. *Microbiol. Mol. Biol. Rev.* 62: 230–247.

Bush, J., Franek, K., Daniel, J., Spiegelman, G. B., Weeks, G. and Cardelli, J. (1993) Cloning and characterization of five novel *Dictyostelium discoideum* rab-related genes. *Gene* 136: 55–60.

Chaturvedi, S., Qi, H., Coleman, D., Rodriguez, A., Hanson, P. I., Striepen, B., Roos, D. S. and Joiner, K. A. (1999) Constitutive calcium-independent release of *Toxoplasma gondii* dense granules occurs through the NSF/SNAP/SNARE/Rab machinery. *J. Biol. Chem.* 274: 2424–2431.

Chow, V. T., Sakharkar, M. K., Lim, D. P. and Yeo, W. M. (2001) Phylogenetic relationships of the seven coat protein subunits of the coatomer complex, and comparative sequence analysis of murine xenin and proxenin. *Biochem. Genet.* 39: 201–211.

Christoforidis, S., Miaczynska, M., Ashman, K., Wilm, M., Zhao, L., Yip, S. C., Waterfield, M. D., Backer, J. M. and Zerial, M. (1999) Phosphatidylinositol-3-OH kinases are Rab5 effectors. *Nat. Cell Biol.* 1: 249–252.

Dacks, J. B. and Doolittle, W. F. (2001) Reconstructing/deconstructing the earliest eukaryotes: How comparative genomics can help. *Cell* 107: 419–425.

Dacks, J. B. and Doolittle, W. F. (2002) Novel syntaxin gene sequences from *Giardia*, *Trypanosoma* and algae: Implications for the ancient evolution of the eukaryotic endomembrane system. *J. Cell Sci.* 115: 1635–1642.

Delwiche, C. F. (1999) Tracing the thread of plastid diversity through the tapestry of life. *Am. Nat.* 154: S164–S177.

Donaldson, J. G. and Lippincott-Schwartz, J. (2000) Sorting and signaling at the Golgi complex. *Cell* 101: 693–696.

Doray, B., Ghosh, P., Griffith, J., Geuze, H. J. and Kornfeld, S. (2002) Cooperation of GGAs and AP-1 in packaging MPRs at the trans-Golgi network. *Science* 297: 1700–1703.

Duden, R., Griffiths, G., Frank, R., Argos, P. and Kreis, T. E. (1991) Beta-COP, a 110 kDa protein associated with non-clathrin-coated vesicles and the Golgi complex, shows homology to beta-adaptin. *Cell* 64: 649–665.

Edwardson, J. M. (1998) Membrane fusion: All done with SNAREpins? *Curr. Biol.* 8: R390–393.

Embley, T. M. and Hirt, R. P. (1998) Early branching eukaryotes? *Curr. Opin. Genet. Dev.* 8: 624–629.

Fasshauer, D., Sutton, R. B., Brunger, A. T. and Jahn, R. (1998) Conserved structural features of the synaptic fusion complex: SNARE proteins reclassified as Q- and R-SNAREs. *Proc. Natl. Acad. Sci. USA* 95: 15781–15786.

Field, H., Farjah, M., Pal, A., Gull, K. and Field, M. C. (1998) Complexity of trypanosomatid endocytosis pathways revealed by Rab4 and Rab5 isoforms in *Trypanosoma brucei*. *J. Biol. Chem.* 273: 32102–32110.

Field, H., Sherwin, T., Smith, A. C., Gull, K. and Field, M. C. (2000) Cell-cycle and developmental regulation of TbRAB31 localisation, a GTP-locked Rab protein from *Trypanosoma brucei*. *Mol. Biochem. Parasitol.* 106: 21–35.

Field, H., Ali, B. R., Sherwin, T., Gull, K., Croft, S. L. and Field, M. C. (1999) TbRab2p, a marker for the endoplasmic reticulum of *Trypanosoma brucei*, localises to the ERGIC in mammalian cells. *J. Cell Sci.* 112: 147–156.

Geyer, M., Yu, H., Mandic, R., Linnemann, T., Zheng, Y. H., Fackler, O. T. and Peterlin, B. M. (2002) Subunit H of the V-ATPase binds to the medium chain of adaptor protein complex 2 and connects Nef to the endocytic machinery. *J. Biol. Chem.* 277: 28521–28529.

Ghislain, M., Dohmen, R. J., Levy, F. and Varshavsky, A. (1996) Cdc48p interacts with Ufd3p, a WD repeat protein required for ubiquitin-mediated proteolysis in *Saccharomyces cerevisiae*. *EMBO J.* 15: 4884–4899.

Girod, A., Storrie, B., Simpson, J. C., Johannes, L., Goud, B., Roberts, L. M., Lord, J. M., Nilsson, T. and Pepperkok, R. (1999) Evidence for a COP-I-independent transport route from the Golgi complex to the endoplasmic reticulum. *Nat. Cell Biol.* 1: 423–430.

Glick, B. S. (2000) Organization of the Golgi apparatus. *Curr. Opin. Cell Biol.* 12: 450–456.

Hay, J. C. (2001) SNARE complex structure and function. *Exp. Cell Res.* 271: 10–21.

Herskovits, A. A., Shimoni, E., Minsky, A. and Bibi, E. (2002) Accumulation of endoplasmic membranes and novel membrane-bound ribosome-signal recognition particle receptor complexes in *Escherichia coli*. *J. Cell Biol.* 159: 403–410.

Jahn, R. and Sudhof, T. C. (1999) Membrane fusion and exocytosis. *Annu. Rev. Biochem.* 68: 863–911.

Janoo, R., Musoke, A., Wells, C. and Bishop, R. (1999) A Rab1 homologue with a novel isoprenylation signal provides insight into the secretory pathway of *Theileria parva*. *Mol. Biochem. Parasitol.* 102: 131–143.

Jeffries, T. R., Morgan, G. W. and Field, M. C. (2002) TbRAB18, a developmentally regulated Golgi GTPase from *Trypanosoma brucei*. *Mol. Biochem. Parasitol.* 121: 63–74.

Kaiser, C. and Ferro-Novick, S. (1998) Transport from the endoplasmic reticulum to the Golgi. *Curr. Opin. Cell. Biol.* 10: 477–482.

Kasinsky, H. E., Lewis, J. D., Dacks, J. B. and Ausio, J. (2001) Origin of H1 linker histones. *FASEB J.* 15: 34–42.

Kirchhausen, T. (2000) Clathrin. *Annu. Rev. Biochem.* 69: 699–727.

Klumperman, J. (2000) Transport between ER and Golgi. *Curr. Opin. Cell Biol.* 12: 445–449.

Langford, T. D., Silberman, J. D., Weiland, M. E., Svard, S. G., McCaffery, J. M., Sogin, M. L. and Gillin, F. D. (2002) *Giardia lamblia*: Identification and characterization of Rab and GDI proteins in a genome survey of the ER to Golgi endomembrane system. *Exp. Parasitol.* 101: 13–24.

Latterich, M., Frohlich, K. U. and Schekman, R. (1995) Membrane fusion and the cell cycle: Cdc48p participates in the fusion of ER membranes. *Cell* 82: 885–893.

Lewis, M. J. and Pelham, H. R. (2002) A new yeast endosomal SNARE related to mammalian syntaxin 8. *Traffic* 3: 922–929.

Liendo, A., Stedman, T. T., Ngo, H. M., Chaturvedi, S., Hoppe, H. C. and Joiner, K. A. (2001) *Toxoplasma gondii* ADP-ribosylation factor 1 mediates enhanced release of constitutively secreted dense granule proteins. *J. Biol. Chem.* 276: 18272–18281.

Luzio, J. P., Rous, B. A., Bright, N. A., Pryor, P. R., Mullock, B. M. and Piper, R. C. (2000) Lysosome-endosome fusion and lysosome biogenesis. *J. Cell Sci.* 113: 1515–1524.

Maier, A. G., Webb, H., Ding, M., Bremser, M., Carrington, M. and Clayton, C. (2001) The coatomer of *Trypanosoma brucei*. *Mol. Biochem. Parasitol.* 115: 55–61.

McNew, J. A., Parlati, F., Fukuda, R., Johnston, R. J., Paz, K., Paumet, F., Sollner, T. H. and Rothman, J. E. (2000) Compartmental specificity of cellular membrane fusion encoded in SNARE proteins. *Nature* 407: 153–159.

Moir, D., Stewart, S. E., Osmond, B. C. and Botstein, D. (1982) Cold-sensitive cell-division-cycle mutants of yeast: Isolation, properties, and pseudoreversion studies. *Genetics* 100: 547–563.

Morgan, G. W., Allen, C. L., Jeffries, T. R., Hollinshead, M. and Field, M. C. (2001) Developmental and morphological regulation of clathrin-mediated endocytosis in *Trypanosoma brucei*. *J. Cell Sci.* 114: 2605–2615.

Morgan, G. W., Hall, B. S., Denny, P. W., Field, M. C. and Carrington, M. (2002) The endocytic apparatus of the kinetoplastida. Part II: Machinery and components of the system. *Trends Parasitol.* 18: 540–546.

Muniz, M., Nuoffer, C., Hauri, H. P. and Riezman, H. (2000) The Emp24 complex recruits a specific cargo molecule into endoplasmic reticulum-derived vesicles. *J. Cell Biol.* 148: 925–930.

Nagai, H., Kagan, J. C., Zhu, X., Kahn, R. A. and Roy, C. R. (2002) A bacterial guanine nucleotide exchange factor activates ARF on *Legionella* phagosomes. *Science* 295: 679–682.

Nickel, W., Weber, T., McNew, J. A., Parlati, F., Sollner, T. H. and Rothman, J. E. (1999) Content mixing and membrane integrity during membrane fusion driven by pairing of isolated v-SNAREs and t-SNAREs. *Proc. Natl. Acad. Sci. USA* 96: 12571–12576.

Orci, L., Stamnes, M., Ravazzola, M., Amherdt, M., Perrelet, A., Sollner, T. H. and Rothman, J. E. (1997) Bidirectional transport by distinct populations of COPI-coated vesicles. *Cell* 90: 335–349.

Pamnani, V., Tamura, T., Lupas, A., Peters, J., Cejka, Z., Ashraf, W. and Baumeister, W. (1997) Cloning, sequencing and expression of VAT, a CDC48/p97 ATPase homologue from the archaeon *Thermoplasma acidophilum. FEBS Lett.* 404: 263–268.

Philippe, H. and Adoutte, A. (1998) The molecular phylogeny of Eukaryota: Solid facts and uncertainties. In *Evolutionary Relationships among Protozoa* (Coombs, G., Vickerman, K., Sleigh, M. and Warren, A., Eds.), Kluwer, Dordrecht, pp. 25–56.

Philippe, H., Lopez, P., Brinkmann, H., Budin, K., Germot, A., Laurent, J., Moreira, D., Muller, M. and Le Guyader, H. (2000) Early-branching or fast-evolving eukaryotes? An answer based on slowly evolving positions. *Proc. R. Soc. Lond. B Biol. Sci.* 267: 1213–1221.

Pryer, N. K., Wuestehube, L. J. and Schekman, R. (1992) Vesicle-mediated protein sorting. *Annu. Rev. Biochem.* 61: 471–516.

Rabouille, C., Kondo, H., Newman, R., Hui, N., Freemont, P. and Warren, G. (1998) Syntaxin 5 is a common component of the NSF- and p97-mediated reassembly pathways of Golgi cisternae from mitotic Golgi fragments *in vitro*. *Cell* 92: 603–610.

Rachel, R., Wyschkony, I., Riehl, S. and Huber, H. (2002) The ultrastructure of *Ignicoccus*: Evidence for a novel outer membrane and for intracellular vesicle budding in an archeon. *Archaea* 1: 9–18.

Rapoport, T. A., Jungnickel, B. and Kutay, U. (1996) Protein transport across the eukaryotic endoplasmic reticulum and bacterial inner membranes. *Annu. Rev. Biochem.* 65: 271–303.

Robibaro, B., Hoppe, H. C., Yang, M., Coppens, I., Ngo, H. M., Stedman, T. T., Paprotka, K. and Joiner, K. A. (2001) Endocytosis in different lifestyles of protozoan parasitism: Role in nutrient uptake with special reference to *Toxoplasma gondii*. *Int. J. Parasitol.* 31: 1343–1353.

Robibaro, B., Stedman, T. T., Coppens, I., Ngo, H. M., Pypaert, M., Bivona, T., Nam, H. W. and Joiner, K. A. (2002) *Toxoplasma gondii* Rab5 enhances cholesterol acquisition from host cells. *Cell Microbiol.* 4: 139–152.

Robinson, M. S. and Bonifacino, J. S. (2001) Adaptor-related proteins. *Curr. Opin. Cell Biol.* 13: 444–453.

Roger, A. J. (1999) Reconstructing early events in eukaryotic evolution. *Am. Nat.* 154: S146–S163.

Rutherford, S. and Moore, I. (2002) The *Arabidopsis* Rab GTPase family: Another enigma variation. *Curr. Opin. Plant Biol.* 5: 518–528.

Rybin, V., Ullrich, O., Rubino, M., Alexandrov, K., Simon, I., Seabra, M. C., Goody, R. and Zerial, M. (1996) GTPase activity of Rab5 acts as a timer for endocytic membrane fusion. *Nature* 383: 266–269.

Saito-Nakano, Y., Nakazawa, M., Shigeta, Y., Takeuchi, T. and Nozaki, T. (2001) Identification and characterization of genes encoding novel Rab proteins from *Entamoeba histolytica*. *Mol. Biochem. Parasitol.* 116: 219–222.

Sanderfoot, A. A., Assaad, F. F. and Raikhel, N. V. (2000) The *Arabidopsis* genome: An abundance of soluble N-ethylmaleimide-sensitive factor adaptor protein receptors. *Plant Physiol.* 124: 1558–1569.

Schekman, R. and Mellman, I. (1997) Does COPI go both ways? *Cell* 90: 197–200.

Schledzewski, K., Brinkmann, H. and Mendel, R. R. (1999) Phylogenetic analysis of components of the eukaryotic vesicle transport system reveals a common origin of adaptor protein complexes 1, 2, and 3 and the F subcomplex of the coatomer COPI. *J. Mol. Evol.* 48: 770–778.

Schulze, K. L., Littleton, J. T., Salzberg, A., Halachmi, N., Stern, M., Lev, Z. and Bellen, H. J. (1994) rop, a *Drosophila* homolog of yeast Sec1 and vertebrate n-Sec1/Munc-18 proteins, is a negative regulator of neurotransmitter release in vivo. *Neuron* 13: 1099–1108.

Shorter, J. and Warren, G. (2002) Golgi architecture and inheritance. *Annu. Rev. Cell Dev. Biol.* 18: 379–420.

Silberman, J. D., Simpson, A. G., Kulda, J., Cepicka, I., Hampl, V., Johnson, P. J. and Roger, A. J. (2002) Retortamonad flagellates are closely related to diplomonads: Implications for the history of mitochondrial function in eukaryote evolution. *Mol. Biol. Evol.* 19: 777–786.

Simpson, A. G., Radek, R., Dacks, J. B. and O'Kelly, C. J. (2002a) How oxymonads lost their groove: An ultrastructural comparison of *Monocercomonoides* and excavate taxa. *J. Euk. Microbiol.* 49: 239–248.

Simpson, A. G., Roger, A. J., Silberman, J. D., Leipe, D. D., Edgcomb, V. P., Jermiin, L. S., Patterson, D. J. and Sogin, M. L. (2002b) Evolutionary history of "early-diverging" eukaryotes: The excavate taxon *Carpediemonas* is a close relative of *Giardia*. *Mol. Biol. Evol.* 19: 1782–1791.

Smith, T. F., Gaitatzes, C., Saxena, K. and Neer, E. J. (1999) The WD repeat: A common architecture for diverse functions. *Trends Biochem. Sci.* 24: 181–185.

Sogin, M. L. (1991) Early evolution and the origin of eukaryotes. *Curr. Opin. Genet. Dev.* 1: 457–463.

Springer, S., Spang, A. and Schekman, R. (1999) A primer on vesicle budding. *Cell* 97: 145–148.

Stanier, R. (1970) Some aspects of the biology of cells and their possible evolutionary significance. In *Organization and Control in Prokaryotic and Eukaryotic Cells* (Charles, H. and Knight, B., Eds.), Cambridge University Press, Cambridge, pp. 1–38.

Stanley, H., Botas, J. and Malhotra, V. (1997) The mechanism of Golgi segregation during mitosis is cell type-specific. *Proc. Natl. Acad. Sci. USA* 94: 14467–14470.

Stedman, T. T., Sussmann, A. R. and Joiner, K. A. (2003) *Toxoplasma gondii* Rab6 mediates a retrograde pathway for sorting of constitutively secreted proteins to the Golgi complex. *J. Biol. Chem.* 278: 5433–5443.

Sulli, C. and Schwartzbach, S. D. (1995) The polyprotein precursor to the *Euglena* light-harvesting chlorophyll a/b-binding protein is transported to the Golgi apparatus prior to chloroplast import and polyprotein processing. *J. Biol. Chem.* 270: 13084–13090.

Sulli, C., Fang, Z., Muchhal, U. and Schwartzbach, S. D. (1999) Topology of *Euglena* chloroplast protein precursors within endoplasmic reticulum to Golgi to chloroplast transport vesicles. *J. Biol. Chem.* 274: 457–463.

Tanigawa, G., Orci, L., Amherdt, M., Ravazzola, M., Helms, J. B. and Rothman, J. E. (1993) Hydrolysis of bound GTP by ARF protein triggers uncoating of Golgi-derived COP-coated vesicles. *J. Cell Biol.* 123: 1365–1371.

Ungermann, C., Price, A. and Wickner, W. (2000) A new role for a SNARE protein as a regulator of the Ypt7/Rab-dependent stage of docking. *Proc. Natl. Acad. Sci. USA* 97: 8889–8891.

Ungermann, C., Sato, K. and Wickner, W. (1998) Defining the functions of trans-SNARE pairs *Nature* 396: 543–548.

van den Ent, F., Amos, L. A. and Lowe, J. (2001) Prokaryotic origin of the actin cytoskeleton. *Nature* 413: 39–44.

Ye, Y., Meyer, H. H. and Rapoport, T. A. (2001) The AAA ATPase Cdc48/p97 and its partners transport proteins from the ER into the cytosol. *Nature* 414: 652–656.

Zhang, X., Shaw, A., Bates, P. A., Newman, R. H., Gowen, B., Orlova, E., Gorman, M. A., Kondo, H., Dokurno, P., Lally, J., Leonard, G., Meyer, H., van Heel, M. and Freemont, P. S. (2000) Structure of the AAA ATPase p97. *Mol. Cell* 6: 1473–1484.

Zhu, J., Gong, Z., Zhang, C., Song, C. P., Damsz, B., Inan, G., Koiwa, H., Zhu, J. K., Hasegawa, P. M. and Bressan, R. A. (2002) OSM1/SYP61: A syntaxin protein in *Arabidopsis* controls abscisic acid-mediated and non-abscisic acid-mediated responses to abiotic stress. *Plant Cell* 14: 3009–3028.

Chapter 15

The Membranome and Membrane Heredity in Development and Evolution

*Thomas Cavalier-Smith**

CONTENTS

Abstract

Cells are organisms created by cooperating membranes, chromosomes, skeletons and gene-encoded catalysts. Genetic information is transmitted intramolecularly by DNA replication and repair and supramolecularly by membrane heredity. All genes and membranes stem ultimately from those of precellular organisms, probably inside-out cells or obcells, the putative ancestors of the last common ancestor of life, probably a green bacterium with a murein exoskeleton responsible for osmotic stability and helping faithful DNA segregation and cell division. Membrane heredity is mediated by molecular complementarity of proteins that define membrane identity through their key roles in targeting membrane proteins. Membrane polarity and topology and membrane protein and lipid composition are the key features transmitted by membrane heredity. Loss and gain of discrete genetic membranes are very rare evolutionary events that change the membranome — the set of topologically and compositionally distinct membranes of a cell. Gain involves origin of novel targeting

* Tel.: +44-1865-281-065; fax: +44-1865-281-310; e-mail: tom.cavalier-smith@zoology.oxford.ac.uk

0-415-29904-7/04/$0.00+$1.50

machinery and can occur autogenously by membrane subdivision or symbiogenetically by enslaving a foreign cell and recruiting some or all its genetic membranes and genomes. Intricately interconnected features of multiple membrane morphogenesis impose exceptional epigenetic constraints on cell evolution. Cell architecture, anatomy and genetic viability depend on proper targeting of integral membrane proteins that bind to the bacterial exoskeleton (wall) or eukaryotic cytoskeleton.

15.1 Introduction: Supramolecular Preformed Structure Is Important in Cell Heredity

The fundamental unit of life is the cell cycle. For the simplest organisms there is nothing more. More complex ones add on cell differentiation. In unicells, this typically takes the form of resting spores or cysts or an alternation between motile and nonmotile forms; in multicells, extracellular glue creates temporary multicellular bodies, which either lack internal differentiation (e.g., simple filaments or balls of cells) or have varying degrees of cell differentiation, as in cellular slime molds, trees or dinosaurs. These general principles apply equally to the two major forms of life, bacteria and eukaryotes, but there are fundamental differences between them in the structural organization of their cells, which profoundly affect their evolutionary potential (Cavalier-Smith, 1991b, c; *bacteria* refers to all prokaryotes, both archaebacteria and eubacteria; a recent tendentious and harmful fashion to use *bacteria* as a synonym for eubacteria alone is highly confusing). These differences in supramolecular structural organization are as profoundly important as the internal structure of genes and molecules in determining the nature of living organisms.

It is unfortunate that chemists have hijacked the term *structural biology* to refer primarily to the study of three-dimensional structure of protein molecules and their interactions with their immediate neighbors. Structural biology must also include the study of how molecules interact to make supramolecular assemblies and higher-order structures — membranes, organelles, cells and multicellular bodies — that are the traditional domain of the morphologist. This chapter examines some basic principles of how membranes and cell skeletons interact with each other and with genes, catalysts, small molecules, ions and water to make cells that can grow and divide. It emphasizes that genes do not make organisms. DNA itself is chemically inert and of limited morphogenetic potential. As Sonneborn (1963) first clearly emphasized, preformed supramolecular structure plays a fundamental role in cell heredity. Cells are organisms that grow and divide; unlike complex self-assembly systems such as ribosomes and microtubules, they cannot be assembled simply from homogeneous solutions of their molecular constituents. As Sonneborn (1963) put it, there is an important difference between a chicken and chicken soup.

Viruses are not organisms but genetic parasites. As viruses differ fundamentally from cells not only in lacking ribosomes but also in lacking genetic membranes, they are very inadequate models for understanding the molecular basis of life; unfortunately, early over-reliance on them as conceptual models, though immediately highly beneficial, helped foster a longer-term serious neglect by classical molecular biologists of the fundamental roles of membranes and skeletons in biology, including genetics itself (Cavalier-Smith, 2001).

15.2 DNA and Membrane Heredity Compared

The core essentials of DNA heredity and gene expression and of biological catalysis by enzymes and ribozymes (e.g. ribosomes) are fundamentally the same in bacteria and eukaryotes. Genes are derived from preexisting genes by DNA replication and by the more innovatory processes of gene duplication, divergence, fusion, chimaerization and fission. All descend from the genes present in the last common ancestor of life, which was almost certainly a complex Gram-negative photosynthetic eubacterium — quite possibly a green nonsulfur bacterium (Cavalier-Smith, 2002b) — with at least 1500 different genes and a highly developed ultrastructure and well-ordered cell cycle. Such negibacterial cells, unlike our own, are bounded by an envelope of two dissimilar lipid membranes: (1) the inner cytoplasmic membrane that is the osmotic barrier to inorganic ions, and (2) the outer membrane (OM) through which they pass freely because of large central hydrophilic pores in cylindrical macromolecular protein complexes (porins) embedded in the lipid bilayer (Figure 15.1c). All membranes, whether of structurally simpler bacteria (e.g., mycoplasmas or archaebacteria) or structurally more complex eukaryotes, have probably descended by successive growth and division from one or other of these two negibacterial membranes (Blobel, 1980; Cavalier-Smith, 1980, 1982, 1983, 1987a, b, 1991b, c, 1995, 2000, 2002d, 2003). Thus membranes, like genes and chromosomes, never form *de novo*, but always arise from preexisting structures of a related kind. Like chromosomes, they carry genetic information in their preexisting structure. Although the information transmitted by membrane heredity differs from that of genes in certain key respects, both membrane heredity and DNA heredity are essential for the development and evolution of organisms.

In genes, the information is intramolecular and essentially linear — the sequence of the nucleotides, which is replicated precisely, except for rare errors, many of which can be repaired. Membranes, in marked contrast, are fundamentally supramolecular structures held together by weaker noncovalent forces; unlike a nucleic acid or protein molecule, they do not have a unique atomic structure. Their capacity for stably transmitting and storing genetic information over billenia is therefore more limited than it is for genes, which can be multiplied almost without limit (Maynard Smith and Szathmáry, 1995). They consist of a fluid bilayer (or, in many archaebacteria, a monolayer) of amphipathic lipid molecules in which a great variety of amphipathic proteins are also embedded; some membrane proteins are free to diffuse laterally, others are anchored to the cell skeleton. Unlike chromosomes, membranes do not replicate. They grow by insertion of new lipid and protein molecules into preexisting membranes. Thus, they do not have a constant size or composition. After growth, they multiply by division into two roughly equal or very unequal daughter membranes. Even in the most equal divisions, which are the basis for cell division in all organisms and organelle division in eukaryotes, the two daughter membranes will almost never have exactly the same number of constituent molecules of each type. In the grossly unequal divisions that occur in eukaryotes when vesicles bud off and separate topologically from endomembranes or during endocytosis from the plasma membrane, the composition of the daughter membranes is dramatically different, because the budding process is almost invariably highly selective. Such selectivity is a fundamental basis for the greater intracellular differentiation of eukaryotes compared with that of prokaryotes. The origin of coated vesicle budding (of three main kinds: clathrin, copI and copII), which is unique to eukaryotes, is one of the most profound differences between bacteria and eukaryotes and was a prerequisite for the origin of the nucleus (Figure 15.1g to i; Cavalier-Smith, 2002d, 1987b).

It is also interesting to contrast membrane heredity with epigenetic inheritance (Jablonka and Lamb, 1995). Epigenetic inheritance refers to the stable inheritance of differentiated cell states by somatic cell lineages in multicellular eukaryotes. It probably comes about by somatically heritable modifications to chromatin, by altering its composition or the actual structure of its DNA or protein in heritable ways. Epigenetic and membrane inheritance are both examples of inheritance not mediated simply by nucleotide sequences, but they differ in specific mechanisms and in the fact that epigenetic inheritance is essentially somatic whereas all known examples of membrane heredity involve the germ line (see also Chapter 16). It is an interesting question whether any epigenetic inheritance is mediated by somatic membrane heredity rather than simply by chromatin modification as generally assumed. For example, one could imagine that the differentiation of animal epithelial cells might involve

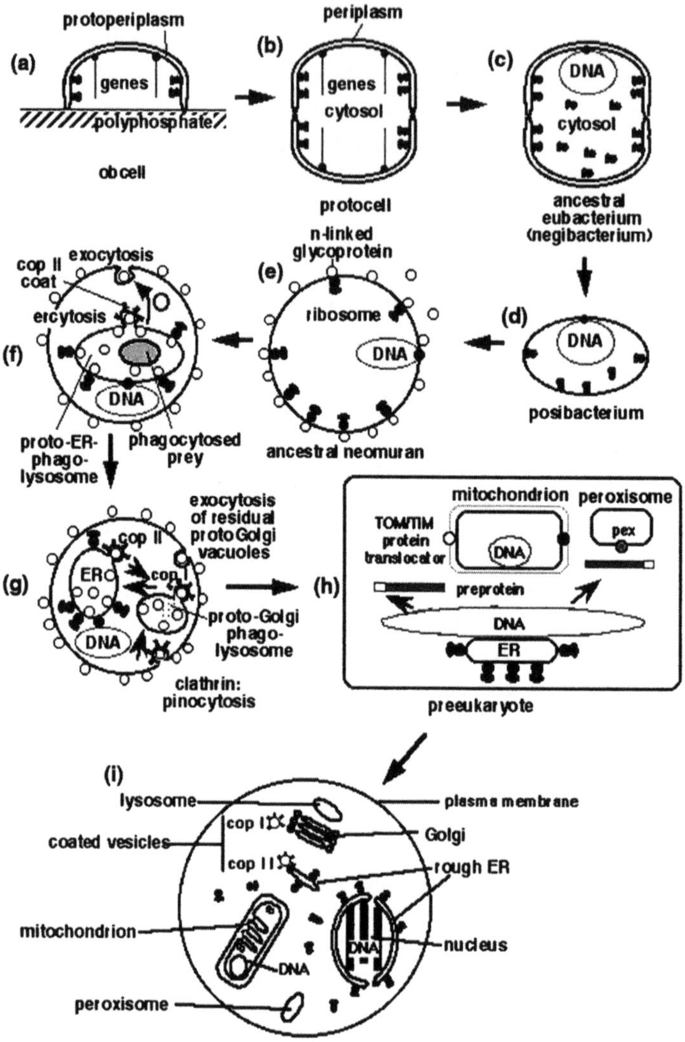

FIGURE 15.1

a purely somatic epigenetic version of membrane inheritance ensuring the stable transmission down cell lineages of membrane receptors responsible for the differential targeting of endomembrane vesicles to the apical and basolateral domains of the plasma membrane.

Epigenetic and membrane inheritance are not the only forms of heredity not directly mediated by nucleic acid replication — for other types see Szathmáry (2000) and (1999) — but together with cultural inheritance they are probably those of most general evolutionary and developmental importance. In contrast to nucleic acid, epigenetic and cultural inheritance, all relatively unlimited in information-carrying capacity, membrane heredity is limited in variational scope, only ca. 18 different kinds of genetic membrane having arisen in the history of life (Table 15.1). DNA and membranes have both genetic and structural roles in the cell. Just as the genetic role of membranes is limited compared with that of DNA, the structural roles of DNA are limited compared with the multifarious ones of membranes; yet DNA's structural roles are as functionally crucial as its genetic ones. In bacteria DNA supercoiling, bulk, and membrane attachments probably play key physiological and integrative roles in the cell cycle (Errington et al., 2003; Wu and Errington, 2002), whereas in

FIGURE 15.1 Major changes to the membranome from the origin of cells to the origin of eukaryotes. (a) Obcells or inside-out-cells with RNA genes attached to the topologically external face of the single genetic membrane by covalently attached hydrophobic peptides might have been the first living organism. Its energy might have come from inorganic polyphosphate and its protoribosomes inserted primitive proteins directly into the membrane. (b) Fusion of two obcells generated the first protocell with a double envelope; this internalized the genes and created a closed intraorganismal cytosol, enabling water-soluble (nonmembrane attached) metabolic enzymes to evolve therein for the first time; the cytoplasmic (CM) and outer membrane (OM) differentiated into two compositionally distinct genetic membranes as special protein-targeting machinery arose for the OM. (c) Origin of a circular DNA genome and of the murein peptidoglycan wall within the periplasm created the first bacterium, a negibacterial eubacterium. (d) Loss of the OM (possibly caused by murein hypertrophy) generated the Gram-positive bacteria, leading to greater emphasis on cotranslational synthesis of numerous secretory digestive enzymes. (e) Adaptation to thermophily during the neomuran revolution that created the common ancestor of archaebacteria and eukaryotes led to the loss of posttranslational secretion mechanisms and the evolution of cotranslational synthesis by oligosaccharide transfer of N-linked cell surface glycoproteins (open circles). (f) Eukaryotes alone evolved phagotrophy and an endoskeleton of actin and microtubules (plus associated molecular motors myosin, dynein and kinesin) and centrosomes for mitosis. Phagotrophy internalized surface membranes bearing ribosome/SRP receptors and DNA-attachment proteins; these internalized membranes were stabilized as a protoER/nuclear envelope by evolution of ercytosis (budding from them of copII-coated vesicles that selectively exclude DNA and ribosome receptors) coupled with the origin of exocytosis of these vesicles alone. (g) Origin of copI- and clathrin-coated vesicles for membrane retrieval from the cell surface, the phagosome and other internal membranes led to the origin of distinct Golgi and lysosomal/endosomal/digestive compartments of the endomembrane system. (h) This preeukaryote cell permanently enslaved an α-proteobacterium by inserting an ADP/ATP translocator (black blob) in its inner membrane and Tom/Tim translocons (open circle) into its outer and inner membranes (for import of preproteins with N-terminal transit sequence), converting them into the mitochondrial envelope, thereby acquiring two extra genetic membranes and an extra genome. Later the genome was lost in secondarily anaerobic lineages, but the two membranes were retained as the envelope of hydrogenosomes or mitosomes. The novel genetic membrane of the peroxisomes might have originated similarly by enslaving a single-membraned posibacterium and losing its genome (Cavalier-Smith, 1990) or (more simply?) autogenously by modifying an endomembrane (Cavalier-Smith, 2002d), enabling it to divide; either mechanism must have involved the novel insertion of pex protein receptor/translocators into the membrane, which thereafter were perpetually inherited, maintaining its unique identity. The relative timing of the origin of mitochondria, peroxisomes and nuclear pore complexes, which enabled the ER cisternae to fuse to form the nuclear envelope without a lethal total separation of cytosol and nucleoplasm, is unclear, but all three were perfected (like the tripartite subdivision of the endomembrane system (g) prior to the eukaryote cenancestor. (i) The major membrane types present in the ancestral eukaryote (an aerobic nonphotosynthetic protozoan), animals and fungi; probably all six types other than lysosomes and coated vesicles are distinct genetic membranes.

eukaryotes DNA probably has a key structural role in nuclear assembly and size determination (Cavalier-Smith and Beaton, 1999; Cavalier-Smith, 2003, 2004).

15.3 The Membranome: Genetic Membranes and the Determinants of Membrane Identity

Organisms differ greatly in the diversity of the different types of membranes that they possess, which I collectively designate their *membranome*. Some of the most basic concepts of membrane heredity (Cavalier-Smith, 1995, 2000) are most simply explained by contrasting the rough endoplasmic reticulum (ER) membranes and the plasma membrane of eukary-

TABLE 15.1 The 18 Kinds of Genetic Membranes

Bacteria (= prokaryotes)
1. Cytoplasmic membrane of all bacteria
2. Outer membrane of all Negibacteria
3. Thylakoid membrane of all cyanobacteria except *Gloeobacter*

Eukaryotes
4. Plasma membrane
5. Rough endoplasmic reticulum/nuclear envelope membrane
6. Peroxisome membrane (?)
7. Golgi membranes (?)
8. Inner membrane of mitochondria, hydrogenosomes and mitosomes
9. Outer membrane of mitochondria, hydrogenosomes and mitosomes
10. Inner membrane of all chloroplast envelopes
11. Outer membrane of plant, chromist and sporozoan chloroplast/plastid envelopes (= middle membrane of euglenoid and dinoflagellate chloroplast envelopes)
12. Epiplastid membrane of chlorarachnean and euglenoid protozoa (cabozoa)
13. Epiplastid membrane of dinoflagellate and sporozoan protozoa
14. Periplastid membrane of chlorarachneans
15. Periplastid membrane of chromalveolates (i.e., chromists + alveolate protozoa)
16. Nucleomorph membrane of chlorarachneans
17. Nucleomorph membrane of cryptomonads and periplastid reticulum of chromobiotes and putatively Sporozoa
18. Thylakoid membranes of chloroplasts

Note: Other genetic membranes possibly exist; the membrane surrounding the β-glucan storage vesicles of photosynthetic cabozoa (euglenoids and chlorarachneans) might be one. The status of Golgi and peroxisome membranes is unclear, as they might in some circumstances be capable of regeneration from vesicles derived from the endoplasmic reticulum, even though they normally are formed by growth and division of their own membrane type.

ote cells (Figure 15.1i). Although these two membranes have many lipids in common, there are major quantitative and some qualitative differences in lipid composition between them. There are even more dramatic differences in protein composition, many types of protein occurring in one but not the other. The unique composition of these two membranes has been maintained for the more than 800 million years since it first arose (Cavalier-Smith, 2002d), because each membrane arises by growth and division of the same kind of membrane and because exchange of lipids and proteins between the two types of membrane is selective and therefore fails ever to homogenize them.

Although the cell has some phospholipid-exchange proteins that can move certain lipids to and fro between topologically distinct membranes, the primary mode of growth of the plasma membrane is by the fusion with it of vesicles (in the process of exocytosis) that have budded off from the Golgi apparatus (or of much larger cisternae or vacuoles also of Golgi origin). In turn, Golgi membranes grow by fusion with preexisting Golgi membranes of vesicles budded off from the ER. Virtually all proteins specific to Golgi or plasma membranes are not inserted directly into those membranes. Instead, they are inserted cotranslationally into rough ER membranes by ribosomes that attach to those membranes with the help of a ribonucleoprotein particle known as the signal recognition particle (SRP; Pool et al., 2002). The SRP can be thought of as an optional third ribosomal subunit that recognizes the hydrophobic signal at the N-terminus of the nascent membrane protein as it emerges from the cavity in the large ribosomal subunit and specifically binds to it. The SRP also recognizes a docking protein embedded in the RER membrane and binds to it, causing the ribosome to become attached to the membrane by means of ribosome receptors (Herskovits et al.,

2001). The membrane protein is then extruded from the ribosome into the ER lumen through a gated pore, but remains partially trapped within the membrane by its hydrophobic signal sequence (in cases when this is not removed) or by one or more distinct hydrophobic stop-transfer sequences, or both. This mode of insertion ensures that every protein is inserted into the membrane with a characteristic polarity and with specific regions of its surfaces embedded within the membrane or projecting from one or both surfaces. This conserved polarity is vital for subsequent function and a key aspect of the conservation of supramolecular organization within each discrete class of membrane. Water-soluble proteins destined for the cytosol, nucleus or mitochondria lack the N-terminal signal sequence, and therefore the ribosomes making them do not dock onto ER membrane but typically associate instead with other structures, notably the cytoskeleton, and liberate their proteins into the cytosol from where some of them are subsequently targeted into other organelles by distinct topogenic (targeting) sequences. Water-soluble secretory proteins, like membrane proteins, have signal sequences, but remain in the vesicles budded serially from ER and Golgi until they fuse with the plasma membrane. This exocytosis liberates them to the exterior because they lack stop-transfer sequences that trap them in the membrane or (unlike resident soluble ER proteins) retrieval sequences or others that trap them in other cell compartments on their way to the cell surface.

Ribosomes cannot bind to the plasma membrane or Golgi membranes and insert proteins directly into them because there have no docking proteins or ribosome receptors. Docking proteins themselves are inserted cotranslationally into the rough ER by the cotranslational SRP machinery (Herskovits et al., 2001). This ensures that they can be inserted only into membranes in which they themselves are already present. Thus, the docking protein is a key molecular label that provides the molecular basis for ER identity together with the complementary specificity of the SRP. If one were able to remove all docking protein molecules from the ER and place them instead in the outer membrane (OM) of the mitochondrion or in the plasma membrane, then ribosomes would cease to bind to the ER and attach to the mitochondrial OM or the plasma membrane instead and insert the ER, Golgi and plasma membrane proteins there, thus dramatically changing the composition of the various cell membranes in the complete absence of any DNA mutations. Either experiment (in effect an artificial membrane mutation) would soon kill the cell, because it would totally disrupt function. The reader can easily deduce the reasons for this in the two instances given. The key point is that plasma membranes remain genetically distinct from ER because their contrasting identities are perpetually propagated through growth and division by invariably continued presence of key proteins or protein assemblies whose presence in the membrane is partially autocatalytic.

The identity of the ER depends on the location in it of docking protein, whereas that of the plasma membrane depends on the correct targeting to it of vesicles bearing plasma membrane proteins (Archer et al., 2002). In general, specific vesicle targeting depends on the presence of protein labels on the cytosolic face of the vesicle (v-SNARES) and complementary receptors (t-SNARES) on the cytosolic face of their proper target membrane (Parlati et al., 2002; Rein et al., 2002; Bi et al., 2002; Chen et al., 2002). Thus, if in a thought experiment one were (without mutating any DNA) to move all the t-SNARES specific for the plasma membrane from it to the OM of the mitochondrion, the latter would incorporate plasma membrane proteins and the plasma membrane could not. In practice, the situation would be more complex if a particular membrane has two distinct labels ensuring its identity. The genetic identity of the OM of the mitochondrion depends on the continued presence in it of the outer membrane translocon (Tom), which binds the usually removable N-terminal peptide that characterizes most nuclear-coded mitochondrial proteins, thus ensuring that they are imported into the correct organelle (Rapaport, 2002). In plants, chromists and photosynthetic protozoa, the translocon of the OM of the plastid envelope (Toc) plays a similar role. Tom and Toc are both autocatalytically self-inserting; therefore, their self-complementarity is the essential molecular basis of mitochondrial and plastid OM heredity.

Topologically and compositionally distinct membranes such as the RER, plasma membrane and OM of mitochondria and plastids that always grow only from preexisting membranes of the same kind are called genetic membranes and bear a specific identity assembly like the docking protein, t-SNAREs, Tom and Toc. Other membranes, such as the coated vesicle membranes or food vacuole membranes, that have no permanent identity but are generated by budding from compositionally different membranes are nongenetic membranes. Genetic membranes are in effect the germ line of the membranome; if one kind of genetic membrane is lost but all the genes encoding its proteins and lipid-synthesizing machinery remain, it would be highly improbable that it could be replaced, because the requisite molecules would be most unlikely to be capable of self-assembly or self-insertion into another type of membrane so as to regenerate it. For practical purposes, it would be gone forever. Nongenetic membranes are part of the soma; even if one class is totally eliminated, the cell should be able to regenerate it by budding from heterologous membranes so long as all relevant genes remain intact.

15.4 Stasis and Quantum Changes in the Membranome during Megaevolution

Changes in the number of genetic membranes in the membranome are among the most important megaevolutionary changes in the history of life, as they have been responsible for much of its structural diversity. If Golgi membranes constitute at least one kind of genetic membrane, which is probable but debatable (Shorter and Warren, 2002; Pelletier et al., 2002), then all eukaryote cells have at least three different kinds of genetic membranes: ER, plasma membrane and Golgi. Possibly, however, all eukaryote cells also have two further genetic membranes homologous with the inner and outer membranes of the mitochondria. These two mitochondrial membranes are probably the direct descendants of the two membranes of the envelope of the α-proteobacterium that was enslaved by the prekaryote cell to form mitochondria (Cavalier-Smith, 1983), probably very soon after the origin of phagotrophy and the endomembrane system (Cavalier-Smith, 2002d). It is now generally accepted that there are no living eukaryotes that primitively lack mitochondria (Silberman et al., 2002). All anaerobic eukaryotes are now thought to have evolved by the secondary loss of mitochondria or by their conversion into hydrogenosomes or mitosomes. Hydrogenosomes are respiratory organelles found in a variety of protozoa and fungi that lack cytochromes and any capacity for oxidative phosphorylation, but can generate ATP by substrate-level phosphorylation, disposing of waste electrons and protons by combining them to make molecular hydrogen; for these functions, they have two characteristic enzymes, hydrogenase (Horner et al., 2002) and pyruvate-ferredoxin oxidoreductase (Horner et al., 1999). Even though (in all thoroughly studied examples) they have entirely lost the mitochondrial genome, they have retained both membranes of the mitochondrial envelope and its protein-targeting machinery and the mitochondrial chaperones needed for protein import (van der Giezen et al., 2002, 2003). Many amitochondrial groups lack hydrogenosomes, but three of these have been shown to retain even simpler double-membrane organelles, almost certainly also of mitochondrial origin, i.e., the mitosomes. Mitosomes were first discovered in *Entamoeba* (Tovar et al., 1999), but there is now evidence for similar organelles in microsporidian fungi (Williams et al., 2002; Katinka et al., 2001) and diplomonads (Tovar et al., 2003). As mitosomes are thought to be needed for assembling iron–sulfur proteins, a mitochondrial function that appears to be less dispensable than oxidative phosphorylation, it is likely that all amitochondrial eukaryotes that lack mitochondria or hydrogenosomes still retain mitosomes. Regardless of whether this turns out to be true, mitosomes and hydrogenosomes are both remarkable examples of the persistence of a complex structure embodying two distinct membranes for hundreds of millions of years after the genome of the α-proteobacterium they once enclosed has been entirely lost or partially transferred to the nucleus. Other eukaryotic genetic membranes probably exist, notably peroxisome membranes (Purdue and Lazarow, 2001; Brosius and Gartner, 2002). Defining them is a key task for cell biologists.

If the tree of life is really rooted among Gram-negative eubacteria, as the fossil record strongly indicates when critically interpreted (Cavalier-Smith, 2002b), then the two membranes of the mitochondrial and negibacterial envelope originated before the cenancestor of all life. I have argued that these two genetic membranes originated when the ancestral cell was formed by the fusion of two separate obcells or inside-out cells (Figure 15.1a); according to this theory, they differentiated from the single membrane of the hypothetical obcell that had already been divided into two domains by a band of integral membrane proteins that restricted the diffusion of other membrane proteins (Cavalier-Smith, 2001). In bacteria, the distinctiveness of the two membranes is maintained by the OM proteins having additional targeting signals that ensure that they are transferred from the cytoplasmic membrane or periplasm into the outer membrane. I have also argued that the OM was lost only once in the history of life during the origin of Posibacteria, by the hypertrophy of the periplasmic peptidoglycan murein causing the OM to break away from the adhesion sites with the cytoplasmic membrane, through which the lipids necessary for OM growth must move (Cavalier-Smith, 1987a, b, 2002b). If the OM had never been lost, it is unlikely that eukaryotes would ever have evolved, because a single bounding membrane is much more conducive to the origin of phagotrophy than is the double negibacterial envelope. Both the secondary origin of the single posibacterial membrane (at least 2.8 Gy ago) and the much later origin (ca. 850 My ago) of cotranslational synthesis of surface N-linked glycoproteins, which occurred during the neomuran revolution that generated the common ancestor of eukaryotes and archaebacteria (Cavalier-Smith, 2002b), were probably prerequisites for the origin of phagotrophy and the endomembrane system, during which new genetic membranes evolved as a result of the successive origins of novel types of coated vesicle budding (Cavalier-Smith, 2002d).

During the entire history of bacteria, only cyanobacteria certainly evolved novel genetic membranes: the ancestral cyanobacteria almost certainly had no thylakoids; their phycobilisomes associated with the cytoplasmic membrane as in *Gloeobacter.* As cells grew larger, invaginations similar to the chromatophores of purple bacteria and mitochondrial cristae probably evolved to increase the amount of photosynthetic machinery. At some stage they must have become topologically separated from the cytoplasmic membrane. However, such separated cisternae would only be a novel genetic membrane once they evolved a specific protein-insertion mechanism for the photosystems and other thylakoid complexes that was distinct from the various generalized mechanisms used in the cytoplasmic membrane. Currently, the mechanisms of differential targeting to the thylakoids and cytoplasmic membrane are poorly understood. In chloroplasts, thylakoid membrane proteins are targeted either by the plastid SRP (which unlike other SRPs has lost the RNA component) or insert spontaneously; two other pathways (Sec and TAT) are used for thylakoid lumenal proteins. When a cyanobacterium was enslaved to form the first plastid, the negibacterial OM was also retained as its OM (Cavalier-Smith, 1982, 2000); thylakoids are similarly retained — even in nonphotosynthetic plastids there are typically internal membranes likely to be their homologues.

All three membranes were also retained when plastids were transferred during secondary symbiogenesis when eukaryote algae were enslaved by protozoan hosts and converted to organelles (Figure 15.2) by evolving novel coated vesicles budded from the endomembrane system and targeted to the food vacuole membrane, which thereby became converted into a novel genetic membrane, the epiplastid membrane. [For detailed discussion and how the algal plasma membrane was probably converted into the periplastid membrane (PPM) by acquiring and modifying Tocs see Cavalier-Smith (2003).] This happened independently to create chromalveolates when a red alga was enslaved, probably by the origin of a novel ER-(copII-) coated vesicle (Figure 15.2; see Chapter 4), and in euglenoids and chlorarachneans when a green alga was enslaved by the origin of a novel Golgi-coated vesicle [probably copI; possibly in their cabozoan common ancestor (Cavalier-Smith, 2003)].

Purple bacterial chromatophores and other apparently internal membranes of Proteobacteria, e.g., methylotrophs, are invaginations of the cytoplasmic membrane, of which they are specialized domains. The recently discovered anammoxosomes of certain Planctomycetes that engage in ammonia oxidation by using nitrite as the terminal electron acceptor (Sinninghe Damste et al., 2002), which are assumed to be unusually impermeable bilayers of unique ladderene monoether lipids, are possible candidates for distinct genetic membranes, but nothing is known of their protein composition and still less of their biogenesis. In the absence of thorough serial section electron microscopy, it is unclear whether they (and other complex apparently internal membranes of Planctomycetes) are topologically connected to the cytoplasmic membrane like chromatophores or distinct like thylakoids; their status and biogenesis need clarification.

15.5 Lipid Targeting and Membrane Polarity

Not only proteins but also lipids are differentially targeted to separate genetic membranes. In eukaryotes, the ER is the sole site of synthesis of the simplest phospholipids, presumably because the acyl transferase that esterifies glycerol phosphate by adding acyl groups from fatty acids is an integral membrane protein targeted to it by an SRP-recognized signal and trapped in the ER by exclusion from the copII-coated vesicles that bud from it. The later stages of sterol biosynthesis are similarly localized. Thus, cholesterol and most phospholipids reach the Golgi indirectly via such vesicles and pass on to the plasma membrane even more indirectly. Sphingolipids are made in the Golgi because their biosynthetic proteins are targeted there, and are therefore abundant in the plasma membrane but rarer in the ER. (Some get there because of membrane retrieval from Golgi to ER by retrograde vesicle budding.) In eukaryote–eukaryote chimaeras that converted the algal plasma membrane into the PPM (chromistan and chlorarachnean algae and Sporozoa), it has been postulated that a capacity for phospholipid synthesis still remains in the nucleomorph envelope (ER homologue) or the periplastid reticulum (PPR) of chromobiotes and Sporozoa (probably a relic of the nucleomorph/ER membrane), and that periplastid vesicles bud from it and fuse with the PPM in a relic of exocytosis (Cavalier-Smith, 2003). If this is correct, then the PPR has been retained with the same topology and biogenetic function as the ancestral algal nuclear envelope half a billion years after the genome it once enclosed was lost or transferred to the host nucleus. Like the hydrogenosome/mitosome membrane, this attests to the remarkable longevity and stasis in the biogenetic mechanisms underlying membrane heredity. In negibacteria lacking thylakoids, the cytoplasmic membrane is the sole site of phospholipid synthesis; phospholipids are transferred secondarily to the OM by extra mechanisms. Virtually all genetic membranes have qualitatively or quantitatively different lipid composition in their two leaflets, which in part reflects that, as for proteins, new lipids are inserted unilaterally; for example, in green plant chloroplast OMs glycolipids come from within the plastid, whereas phosphatidylcholine comes from the cytosol by lipid-exchange proteins and remains concentrated in the cytosolic face. Primary growth membranes such as the ER and

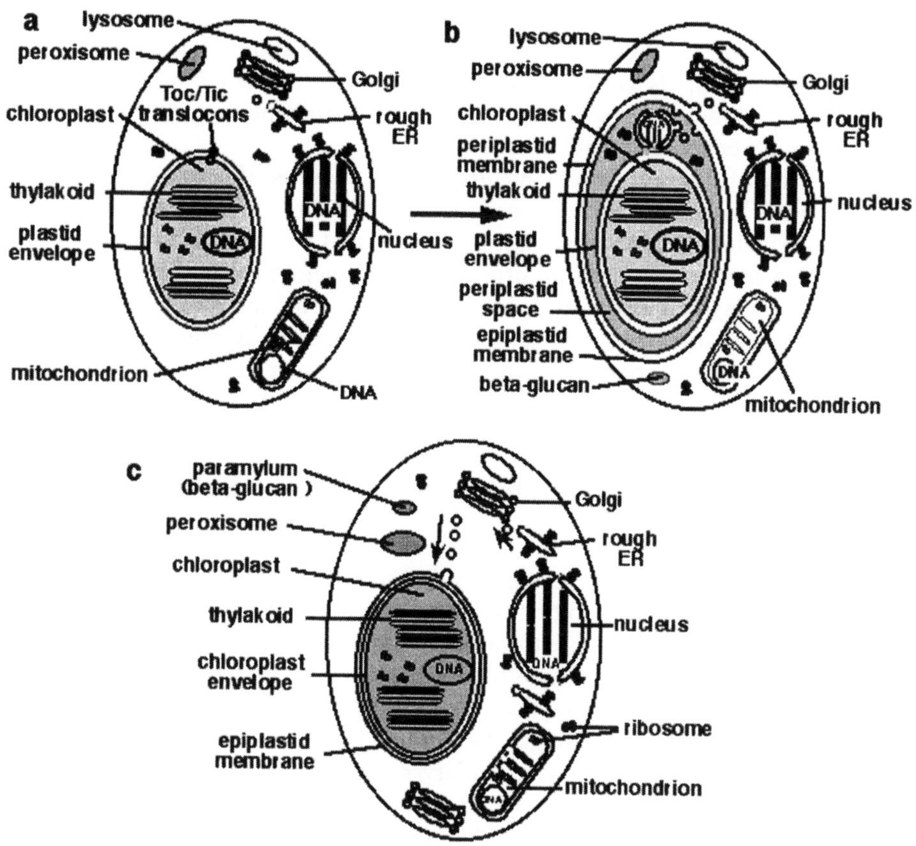

FIGURE 15.2 Additional genetic membranes present in plants (a) and photosynthetic protozoa (b and c). (a) A biciliate eukaryote became the ancestral plant by enslaving a cyanobacterium by converting its double envelope into the plastid double envelope by inserting energy translocators and the Toc/Tic translocons for protein import. (b) After plants diverged into three lineages, a red alga was enslaved to form chromalveolates (see Chapter 4 for details of their novel genetic membranes); independently, a green algal enslavement generated the equally complex chlorarachnean algae (shown here; chlorarachneans probably have the most complex membranome of any organism — at least 11 distinct genetic membranes); euglenoid plastids (c) which have a triple chloroplast envelope of three distinct genetic membranes evolved either from the same chimaeric cabozoan ancestor by losing the periplastid membrane and nucleomorph or, less likely, independently; chlorarachneans and euglenoids store the β-glucan paramylum within a cytosolic membrane that might be another genetic membrane. In both cases, enslavement was initiated by the budding of novel vesicles targeted to the perialgal vacuole, converting it into a novel genetic membrane, the epiplastid membrane. (The presumably coated transport vesicles came from ER in chromalveolates, from Golgi in euglenoids, and, hypothetically, in chlorarachneans.)

bacterial cytoplasmic membrane have flippases (Kol et al., 2002; Doerrler and Raetz, 2002; Nicolson and Mayinger, 2000; Hrafnsdottir and Menon, 2000; Daleke and Lyles, 2000) that can flip selected lipids between the two leaflets of membrane bilayers (essential for membrane growth), but their selectivity plus that of lipid binding to polarized proteins prevents the total abolition of lipid polarity.

Thus, polarity of different genetic membranes is conserved throughout evolution by the great stasis in their biogenetic mechanisms, notably in the directionality, mode and machinery of membrane protein and lipid targeting. What is passed on through membrane heredity is this polarity, the specific composition of each kind of membrane and its topology, i.e., the nature of what lies inside it and outside it, which is positional information at the subcellular level. This preformed supramolecular structure is not encoded by genes, but must be passed on directly by membrane growth and division, though many genes are also necessary to control these processes and to encode the targeting machinery and the proteins that it directs to their correct location in the cell.

15.6 Biogenesis of Multiple Membrane Systems as an Epigenetic Constraint on Evolution

The nature of membrane biogenesis imposes epigenetic constraints on what can evolve. In principle, a chloroplast can function physiologically with only a single bounding membrane. But no organism has ever been able to dispense with the plastid OM, presumably because to do so would eliminate the Toc receptors for protein import. An all-powerful designer or creator could have chosen such simplicity by putting Tocs and inner membrane permeases and biosynthetic enzymes in a single membrane from the start. Likewise, in eukaryote–eukaryote chimaeras created by secondary symbiogenesis (Cavalier-Smith, 2003), the extra (third and fourth) membranes are an even greater physiological encumbrance and energetic burden imposed by the inability of mutations to simplify their biogenesis sufficiently to dispense with them. Occasionally, functionally superfluous genetic membranes acquired by symbiogenesis have been lost, e.g., the PPM in euglenoids and dinoflagellates [which unlike all other organisms have plastids with triple envelopes, itself a character stemming from historical epigenetic constraints not functional optimality (Cavalier-Smith, 2003)], but its rarity shows that it is not easy when they form part of the biogenetic mechanisms of other functionally necessary membranes. Complete organelles, e.g., mitochondria and peroxisomes, that were present in the algae enslaved by the chromalveolate and cabozoan ancestors can more readily be lost, showing that it is not loss per se that is difficult but the loss of one of several membranes that share a tightly integrated mutually dependent biogenesis. Thus, the persistence of the great variety of membrane topologies in eukaryotic algae (Cavalier-Smith, 1995, 2000, 2002a, c, 2003) is one of the most striking examples of the importance of epigenetic constraints in evolution, showing that evolution cannot be understood simply in terms of general concepts such as gene mutation and selection; the physical processes that build cells and how they are both limiting and enabling for evolution must also be understood.

15.7 Membranes, Cell Skeletons and Genomes Interact to Build and Maintain Organisms

Although membranes and genes are the primary joint carriers of long-term genetic information, they cannot build organisms without the help of a cell skeleton and catalysts. Because the essential role of catalysts is well known to all (biological catalysts select reactants as well as accelerate reactions), I emphasize the cell skeleton here. That the cell skeleton is largely nonhomologous between bacteria and eukaryotes has helped conceal its fundamental and universal importance in the fidelity of genetic transmission as well as in the mechanical stability and function of organisms. In all cells, the cell skeleton and the molecular motors associated with it or the chromosomes provide the mechanical basis for faithful DNA segregation and cell division and in the vital coupling between them. In bacteria, this genetic and mechanical function is achieved by an exoskeleton — the cell wall, ancestrally by murein peptidoglycan (in most eubacteria), alternatively by protein (most Planctobacteria), glycoprotein (ancestrally for archaebacteria), pseudomurein (a few derived methanogenic archaebacteria) or carbohydrates (mycoplasmas and a few euryarchaeotes). By contrast, in eukaryotes it is provided by an endoskeleton: the microtubules, with molecular motors dynein and kinesin that mediate mitosis, and actin microfilaments and myosin motors for cytokinesis. Attachment of chromosomes to the skeleton by centromeres or their bacterial equivalent (Yamaichi and Niki, 2000; Moller-Jensen et al., 2002), and probably also of replisomes that mediate replication (Lemon and Grossman, 2000; Jensen et al., 2001), plays a fundamental role in the cell cycle and genetic persistence of all organisms.

The architecture, anatomy and motility of cells depend fundamentally on the mechanical attachment of the skeleton, whether exo or endo, to integral membrane proteins embedded in their membranes. It is the genes for such proteins, rather than those for catalysts or transcriptional switches, which biochemists traditionally emphasize, that play the most decisive role in making organisms in all their remarkable structural diversity. But they cannot do this without the perpetuation by a combination of membrane heredity and DNA heredity of the distinctive genetic membranes and their somatic derivatives into which these key links between membrane and skeleton become embedded. Much of the control of diverse morphology lies in the choice of which membranes they insert into and how these are organized into differentiated domains in association with cytoskeletal elements.

Membrane-cytoskeleton attachments are equally important for ensuring faithful DNA segregation and division and the viability of all daughter cells, and thus genetic stasis and the ability to conserve valuable evolutionary advances over hundreds of millions of years. In eukaryotes, the DNA of nuclear genomes probably has a structural, nongenic and sequence-independent role, as a nuclear skeleton enabling nuclear volume to be easily adjusted (by deletions and insertions of purely skeletal DNA) to changing cell volumes during evolution, thus ensuring balanced growth through maintaining an optimal ratio of nuclear and cytoplasmic volumes (Cavalier-Smith, 1985, 1991a, 1999, 2003, 2004). An often very small fraction *also* has the same genic roles in specifying RNAs and encoding proteins as does most of the genome of bacteria, mitochondria and chloroplasts, which lack a skeletal DNA function as they have no ER/nuclear envelope requiring such nucleation for its assembly. The origin of the nuclear envelope required the prior origin of the novel ER genetic membrane plus that of the nuclear pore complexes assembled on it through attachment to novel integral membrane nucleoporin proteins (Cavalier-Smith, 1988, 2002d); indirectly, therefore, this innovation in the membranome led to the novel nucleoskeletal function for DNA, thus explaining why lilies of the field (that do toil and spin) have so much larger genomes than we do — because they need very large cells in their bulbs to allow rapid escape from dormancy through cell expansion without growth and division (Cavalier-Smith, 1985).

15.8 Envoi

Although the cell skeleton plays an essential genetic role in ensuring accurate DNA segregation, it does not itself normally carry genetic information, as it can generally be assembled *de novo* from its molecular components. One interesting exception is cortical inheritance in ciliate protozoa, where the preformed pattern of kineties is probably determined by the nucleation of centrioles in precise positions in the cell cortex by apparently amorphous cytoskeletal nucleating sites arranged in a preexisting pattern through mutual binding of the cortical cytoskeleton, plasma membranes and cortical alveolar membranes (Frankel, 1989). Although it is sometimes postulated that centrioles themselves might be replicators (Maynard Smith and Szathmáry, 1995), there is no necessity to postulate this to explain cortical inheritance; centrioles themselves are almost certainly not replicators, though the possibility that tiny amorphous nucleators of them might be has not been excluded (Cavalier-Smith, 1974). It is also possible that the initial polarity and asymmetry of animal eggs are influenced by the asymmetry of surrounding follicle cells, so that supramolecular inheritance might play a role in the hereditary transmission of our own bilaterality (Gardner, 2001). But neither of these interesting exceptions has the same generality for understanding the basic nature of life as does membrane heredity. I suggest that Sonneborn's term *cytotaxis* (Sonneborn, 1963) be reserved for such higher-level cases of a genetic role for supramolecular structure.

For membrane heredity itself, the basic mechanism ensuring genetic continuity is molecular self-complementarity between proteins (components of the identity-determining integral membrane proteins such as ER docking proteins, the mitochondrial Tom or the chloroplast Toc) or molecular complementarity between t-SNAREs and v-SNAREs. For DNA heredity it is self-complementarity of nucleotides. For genomes, their location within a particular cell compartment (e.g., the four distinct genomes of a cryptomonad cell — Chapter 4) is determined by the conserved location of the preexisting genome that is being replicated; thus, even for effective DNA heredity, preformed structure of much of the rest of the cell, not only DNA itself, is important. For membranes, the location and polarity of membrane proteins are determined by the location of docking proteins, Tocs, Toms, t-SNAREs and so on in preexisting membranes. Perpetual maintenance of this preexisting structure through interactions among the membranome, genome and cell skeleton is as important for DNA as for membrane heredity, not only in the most complex cells like cryptomonads and chlorarachneans but even in the most drastically simplified ones such as *Mycoplasma* or *Thermoplasma*, which have only a single genetic membrane.

The popular notion that the genome contains "all the information needed to make a worm" (Sulston and Ferry, 2002) is simply false. Membrane heredity, by providing chemically specific two-dimensional surfaces with mutually conserved topological relationships in the three spatial dimensions, plays a key role in the mechanisms that convert the linear information of DNA into the three-dimensional shapes of single cells and multicellular organisms. Animal development creates a complex three-dimensional multicellular organism not by starting from the linear information in DNA (Edelman, 1988) but always starting from an already highly complex three-dimensional unicellular organism, the fertilized egg, which membrane and DNA heredity together have perpetuated.

References

Archer, D. A., Graham, M. E. and Burgoyne, R. D. (2002) Complexin regulates the closure of the fusion pore during regulated vesicle exocytosis. *J. Biol. Chem.* 277: 18249–18252.

Bi, X., Corpina, R. A. and Goldberg, J. (2002) Structure of the Sec23/24-Sar1 pre-budding complex of the COPII vesicle coat. *Nature* 419: 271–277.

Blobel, G. (1980) Intracellular protein topogenesis. *Proc. Natl. Acad. Sci. USA* 77: 1496–1500.

Brosius, U. and Gartner, J. (2002) Cellular and molecular aspects of Zellweger syndrome and other peroxisome biogenesis disorders. *Cell Mol. Life Sci.* 59: 1058–1069.

Cavalier-Smith, T. (1974) Basal body and flagellar development during the vegetative cell cycle and sexual cycle of *Chlamydomonas reinhardii*. *J. Cell Sci.* 16: 529–556.

Cavalier-Smith, T. (1980) Cell compartmentation and the origin of eukaryote membranous organelles. In *Endocytobiology: Endosymbiosis and Cell Biology, A Synthesis of Recent Research* (Schwemmler, W. and Schenk, H. E. A., Eds.), deGruyter, Berlin, pp. 893–916.

Cavalier-Smith, T. (1982) The origins of plastids. *Biol. J. Linn. Soc.* 17: 289–306.

Cavalier-Smith, T. (1983) Endosymbiotic origin of the mitochondrial envelope. In *Endocytobiology II* (Schwemmler, W. and Schenk, H. E. A, Eds.) deGruyter, Berlin, pp. 265–279.

Cavalier-Smith, T. (1985) Cell volume and the evolution of genome size. In *The Evolution of Genome Size* (Cavalier-Smith, T., Ed.), Wiley, Chichester, pp. 105–184.

Cavalier-Smith, T. (1987a) The origin of cells: A symbiosis between genes, catalysts, and membranes. *Cold Spring Harb. Symp. Quant. Biol.* 52: 805–824.

Cavalier-Smith, T. (1987b) The origin of eukaryotic and archaebacterial cells. *Ann. N. Y. Acad. Sci.* 503: 17–54.

Cavalier-Smith, T. (1988) Origin of the cell nucleus. *Bioessays* 9: 72–78.

Cavalier-Smith, T. (1990) Symbiotic origin of peroxisomes. In *Endocytobiology IV* (Nardon, P., Gianinazzi-Pearson, V., Grenier, A. M., Margulis, L. and Smith, D. C., Eds.), Institut National de la Recherche Agronomique, Paris, pp. 515–521.

Cavalier-Smith, T. (1991a) Coevolution of vertebrate genome, cell and nuclear sizes. In *Symposium on the Evolution of Terrestrial Vertebrates: Selected Symposia and Monographs U.Z.I., 4* (Ghiara, G., Eds.), Muchi, Modena, pp. 51–88.

Cavalier-Smith, T. (1991b) The evolution of cells. In *Evolution of Life* (Osawa, S. and Honjo, T., Eds.), Springer-Verlag, Tokyo, pp. 271–304.

Cavalier-Smith, T. (1991c) The evolution of prokaryotic and eukaryotic cells. In *Fundamentals of Medical Cell Biology* (Bittar, G. E., Ed.), JAI Press, Greenwich, CT, pp. 217–272.

Cavalier-Smith, T. (1995) Membrane heredity, symbiogenesis, and the multiple origins of algae. In *Biodiversity and Evolution* (Arai, R., Kato, M. and Doi, Y., Eds.), The National Science Museum Foundation, Tokyo, pp. 75–114.

Cavalier-Smith, T. (1999) Principles of protein and lipid targeting in secondary symbiogenesis: Euglenoid, dinoflagellate, and sporozoan plastid origins and the eukaryotic family tree. *J. Euk. Microbiol.* 46: 347–366.

Cavalier-Smith, T. (2000) Membrane heredity and early chloroplast evolution. *Trends Plant Sci.* 5: 174–182.

Cavalier-Smith, T. (2001) Obcells as proto-organisms: Membrane heredity, lithophosphorylation, and the origins of the genetic code, the first cells, and photosynthesis. *J. Mol. Evol.* 53: 555–595.

Cavalier-Smith, T. (2002a) Chloroplast evolution: Secondary symbiogenesis and multiple losses. *Curr. Biol.* 12: R62–64.

Cavalier-Smith, T. (2002b) The neomuran origin of archaebacteria, the negibacterial root of the universal tree and bacterial megaclassification. *Int. J. Syst. Evol. Microbiol.* 52: 7–76.

Cavalier-Smith, T. (2002c) Nucleomorphs: Enslaved algal nuclei. *Curr. Opin. Microbiol.* 5: 612–619.

Cavalier-Smith, T. (2002d) The phagotrophic origin of eukaryotes and phylogenetic classification of Protozoa. *Int. J. Syst. Evol. Microbiol.* 52: 297–354.

Cavalier-Smith, T. (2003) Genomic reduction and evolution of novel genetic membranes and protein-targeting machinery in eukaryote-eukaryote chimaeras (meta-algae). *Phil. Trans. Roy. Soc. B.* 358: 109–134.

Cavalier-Smith, T. (2004) Economy, speed and size matter: Evolutionary forces driving nuclear genome miniaturisation and expansion. *Ann. Bot.*, in press.

Cavalier-Smith, T. and Beaton, M. J. (1999) The skeletal function of non-genic nuclear DNA: New evidence from ancient cell chimaeras. *Genetica* 106: 3–13.

Chen, X., Tomchick, D. R., Kovrigin, E., Arac, D., Machius, M., Sudhof, T. C. and Rizo, J. (2002) Three-dimensional structure of the complexin/SNARE complex. *Neuron* 33: 397–409.

Daleke, D. L. and Lyles, J. V. (2000) Identification and purification of aminophospholipid flippases. *Biochim. Biophys. Acta* 1486: 108–127.

Doerrler, W. T. and Raetz, C. R. (2002) ATPase activity of the MsbA lipid flippase of *Escherichia coli*. *J. Biol. Chem.* 277: 36697–36705.

Edelman, G. (1988) *Topobiology: An Introduction to Molecular Embryology*, HarperCollins, New York.

Errington, J., Daniel, R. A. and Scheffers, D. J. (2003) Cytokinesis in bacteria. *Microbiol. Mol. Biol. Rev.* 67: 52–65.

Frankel, J. (1989) *Pattern Formation: Ciliate Studies and Models*, Oxford University Press, Oxford.

Gardner, R. L. (2001) The initial phase of embryonic patterning in mammals. *Int. Rev. Cytol.* 203: 233–290.

Herskovits, A. A., Seluanov, A., Rajsbaum, R., ten Hagen-Jongman, C. M., Henrichs, T., Bochkareva, E. S., Phillips, G. J., Probst, F. J., Nakae, T., Ehrmann, M., Luirink, J. and Bibi, E. (2001) Evidence for coupling of membrane targeting and function of the signal recognition particle (SRP) receptor FtsY. *EMBO Rep.* 2: 1040–1046.

Horner, D. S., Heil, B., Happe, T. and Embley, T. M. (2002) Iron hydrogenases: Ancient enzymes in modern eukaryotes. *Trends Biochem. Sci.* 27: 148–153.

Horner, D. S., Hirt, R. P. and Embley, T. M. (1999) A single eubacterial origin of eukaryotic pyruvate:ferredoxin oxidoreductase genes: Implications for the evolution of anaerobic eukaryotes. *Mol. Biol. Evol.* 16: 1280–1291.

Hrafnsdottir, S. and Menon, A. K. (2000) Reconstitution and partial characterization of phospholipid flippase activity from detergent extracts of the *Bacillus subtilis* cell membrane. *J. Bacteriol.* 182: 4198–4206.

Jablonka, E. and Lamb, M. (1995) *Epigenetic Inheritance and Evolution*, Oxford University Press, Oxford.

Jensen, R. B., Wang, S. C. and Shapiro, L. (2001) A moving DNA replication factory in *Caulobacter crescentus*. *EMBO J.* 20: 4952–4963.

Katinka, M. D., Duprat, S., Cornillot, E., Metenier, G., Thomarat, F., Prensier, G., Barbe, V., Peyretaillade, E., Brottier, P., Wincker, P., Delbac, F., El Alaoui, H., Peyret, P., Saurin, W., Gouy, M., Weissenbach, J. and Vivares, C. P. (2001) Genome sequence and gene compaction of the eukaryote parasite *Encephalitozoon cuniculi*. *Nature* 414: 450–453.

Kol, M. A., de Kruijff, B. and de Kroon, A. I. (2002) Phospholipid flip-flop in biogenic membranes: What is needed to connect opposite sides. *Semin. Cell Dev. Biol.* 13: 163–170.

Lemon, K. P. and Grossman, A. D. (2000) Movement of replicating DNA through a stationary replisome. *Mol. Cell* 6: 1321–1330.

Maynard Smith, J. and Szathmáry, E. (1995) *The Major Transitions in Evolution*, University Press, Oxford.

Moller-Jensen, J., Jensen, R. B., Lowe, J. and Gerdes, K. (2002) Prokaryotic DNA segregation by an actin-like filament. *EMBO J.* 21: 3119–3127.

Nicolson, T. and Mayinger, P. (2000) Reconstitution of yeast microsomal lipid flip-flop using endogenous aminophospholipids. *FEBS Lett.* 476: 277–281.

Parlati, F., Varlamov, O., Paz, K., McNew, J. A., Hurtado, D., Sollner, T. H. and Rothman, J. E. (2002) Distinct SNARE complexes mediating membrane fusion in Golgi transport based on combinatorial specificity. *Proc. Natl. Acad. Sci. USA* 99: 5424–5429.

Pelletier, L., Stern, C. A., Pypaert, M., Sheff, D., Ngo, H. M., Roper, N., He, C. Y., Hu, K., Toomre, D., Coppens, I., Roos, D. S., Joiner, K. A. and Warren, G. (2002) Golgi biogenesis in *Toxoplasma gondii*. *Nature* 418: 548–552.

Pool, M. R., Stumm, J., Fulga, T. A., Sinning, I. and Dobberstein, B. (2002) Distinct modes of signal recognition particle interaction with the ribosome. *Science* 297: 1345–1348.

Purdue, P. E. and Lazarow, P. B. (2001) Peroxisome biogenesis. *Annu. Rev. Cell Dev. Biol.* 17: 701–752.

Rapaport, D. (2002) Biogenesis of the mitochondrial TOM complex. *Trends Biochem. Sci.* 27: 191–197.

Rein, U., Andag, U., Duden, R., Schmitt, H. D. and Spang, A. (2002) ARF-GAP-mediated interaction between the ER-Golgi v-SNAREs and the COPI coat. *J. Cell Biol.* 157: 395–404.

Shorter, J. and Warren, G. (2002) Golgi architecture and inheritance. *Annu. Rev. Cell Dev. Biol.* 18: 379–420.

Silberman, J. D., Simpson, A. G., Kulda, J., Cepicka, I., Hampl, V., Johnson, P. J. and Roger, A. J. (2002) Retortamonad flagellates are closely related to diplomonads: Implications for the history of mitochondrial function in eukaryote evolution. *Mol. Biol. Evol.* 19: 777–786.

Sinninghe Damste, J. S., Strous, M., Rijpstra, W. I., Hopmans, E. C., Geenevasen, J. A., Van Duin, A. C., Van Niftrik, L. A. and Jetten, M. S. (2002) Linearly concatenated cyclobutane lipids form a dense bacterial membrane. *Nature* 419: 708–712.

Sonneborn, T. M. (1963) Does preformed structure play an essential role in cell heredity? In *The Nature of Biological Diversity* (Allen, J. M., Ed.), McGraw-Hill, New York, pp. 165–221.

Sulston, J. and Ferry, G. (2002) *The Common Thread*, Bantam, London.

Szathmáry, E. (1999) Chemes, genes, memes: A revised classification of replicators. *Lect. Math. Life Sci.* 26: 1–10.

Szathmáry, E. (2000) The evolution of replicators. *Phil. Trans. Roy. Soc. Lond. B* 335: 1669–1676.

Tovar, J., Fischer, A. and Clark, C. G. (1999) The mitosome, a novel organelle related to mitochondria in the amitochondrial parasite *Entamoeba histolytica. Mol. Microbiol.* 32: 1013–1021.

van der Giezen, M., Dyal, P. L., Birdsey, G. M., Horner, D. S., Lucocq, J., Benchimol, M., Danpure, C. J. and Embley, T. M. (2003) Hydrogenosomal chaperones of the anaerobic fungus *Neocallimastix patriciarum*; characterization, localization and mitochondrial import. *Mol. Biol. Evol.* 20: 1051–1061.

Tovar, J., Leon-Avila, G., Sánchez, L.B., Sutak, R., Tachezy, J. van der Giezen, M., Hernandez, M. Müller, M. and Lucocq, J.M. (2003) Mitochondrial remnant organelles of *Giardia* function in iron-sulphur protein metabolism. *Nature* 426: 172–176.

van der Giezen, M., Slotboom, D. J., Horner, D. S., Dyal, P. L., Harding, M., Xue, G. P., Embley, T. M. and Kunji, E. R. (2002) Conserved properties of hydrogenosomal and mitochondrial ADP/ATP carriers: A common origin for both organelles. *EMBO J.* 21: 572–579.

Williams, B. A., Hirt, R. P., Lucocq, J. M. and Embley, T. M. (2002) A mitochondrial remnant in the microsporidian *Trachipleistophora hominis. Nature* 418: 865–869.

Wu, L. J. and Errington, J. (2002) A large dispersed chromosomal region required for chromosome segregation in sporulating cells of *Bacillus subtilis. EMBO J.* 21: 4001–4011.

Yamaichi, Y. and Niki, H. (2000) Active segregation by the *Bacillus subtilis* partitioning system in *Escherichia coli. Proc. Natl. Acad. Sci. USA* 97: 14656–14661.

Chapter 16

Epigenetic Inheritance and Evolutionary Adaptation

Csaba Pál and Laurence D. Hurst

CONTENTS

Abstract

Although epigenetic inheritance has been recognized to be crucial to maintain different cellular states during development, it is still unclear whether and how often epigenetic marks can be important in adaptation. As epigenetic inheritance encapsulates a wide range of phenomena, we first briefly describe the mechanisms behind the heritable potential of (1) metabolic steady-state systems, (2) cellular structural elements and (3) chromatin marks including DNA methylation. Next, we discuss the experimental evidences for the transmission of chromatin marks

0-415-29904-7/04/$0.00+$1.50

through meiosis. Although these results provide a clear mechanistic basis for heritable epigenetic variation, an important possible objection is that they might not be stably transmitted through meiosis (and hence not between generations). Moreover, in many cases, epigenetic marks are not inherited in a Mendelian fashion; they are either transmitted to too many progeny, in which case they can also be deleterious, or to too few, in which case even if advantageous they will often be lost. This suggests that under sexual reproduction — possibly associated with cell fusion — epigenetic inheritance is unlikely to play a fair Mendelian game, which is a prerequisite for adaptive evolution.

16.1 Introduction

There is a growing recognition that there is more to heredity than DNA. That this might be so has come to prominence through the study of how it is that the genetically identical cells of an embryo come to have, and stably maintain, different fates (Holliday, 1987). This non-DNA-based heredity, called epigenetic inheritance, although initially a vague concept has gained a secure mechanistic basis and is now crucial to understand development. In principle, the same sorts of epigenetic modifications if transferable between organism generations could be a source of heritable variation (Jablonka and Lamb, 1989, 1995; Maynard Smith, 1990). It is tempting therefore to speculate that there is more to adaptation than the fixation of point mutations, deletions or insertions (Jablonka and Lamb, 1995). But is this speculation reasonable? We argue that although a mechanistic basis removes epigenetic inheritance from the realms of vague speculation, it is still unclear whether and how often epigenetic marks can be important in adaptation.

It should be recognized that epigenetic inheritance encapsulates a broad range of phenomena. Jablonka and Lamb (1995, 1998) recognized three main types of systems that can contribute to cellular inheritance: (1) steady-state systems, (2) structural inheritance and (3) chromatin marking systems. We first briefly describe the mechanisms and evolutionary potential of the first two systems. Next we describe the mechanisms behind the inheritance of chromatin marks and how from this basis we can understand the nature of heritable epimutations (Kermicle, 1978). An important possible objection to the idea that epimutations might be important is that they might not be stably transmitted through meiosis (and hence not between generations). We therefore review what is known about this process. Finally, we point to what we believe is the more serious objection, this being not that epimutations are not transferable between sexual generations but rather that the rate at which it occurs must exist in a very particular window (neither too high nor too low) and that only some of the known cases fulfill this requirement.

16.2 Steady-State Systems

It has long been noted that some regulatory and metabolic patterns might have hereditary potentials (Novick and Weiner, 1957). A simple example is autoregulatory genes, which regulate their own transcription by positive feedback. Once turned on, the gene activates its own transcription. If cell division is more or less equal and the concentration of the gene product is high enough in the cytoplasm, the daughter cells might inherit the active state of the gene. In multicellular organisms, many positive gene regulatory feedbacks have been found (Serfling, 1989), supporting the notion that these loops have an important role in maintaining active gene states during somatic development.

Owing to recent advances in genome projects, we now have a detailed knowledge of regulatory networks in *E. coli* (Shen-Orr et al., 2002) and baker's yeast (Guelzim et al., 2002). The networks of these organisms also contain several positive regulatory feedbacks at either the transcriptional or the post-transcriptional level. The mechanism also works at the posttranslational level: it has been shown that some enzymes are needed for their own assembly. For example, Hsp60p (Cheng et al., 1990) and Yah1p proteins (Lange et al., 2000) are found active in the mitochondrial matrix in yeast and are needed to produce additional active enzymes. The loss of proper localization of these enzymes is therefore a potentially irreversible change.

There is also evidence that positive regulatory feedback loops contribute to heritable clonal variation in the expression of the lac operon (Novick and Weiner, 1957). When *E. coli* is cultured under low concentrations of the galactose, two cell types can be distinguished, those with fully active or fully repressed lac operons, and these cell states can be stably inherited for hundreds of generations. The difference is due to the initial stochastic fluctuation in the intracellular concentration of the transport inducer of galactose (permease). This gene is part of the lac operon, and therefore its own induction also depends on the intracellular concentration of galactose. Hence, once the permease is above a critical concentration because of stochastic events, it will increase the concentration of galactose in the cell and hence activate its own transcription. Remarkably, after removal of the nutrient, the cell still maintains the activated state of the lac operon for some generations. It is tempting to speculate that this mechanism is an adaptive response to short spatial or temporal fluctuations in nutrient concentration. It is possible that cells cannot find the nutrient in a given microenvironment even when the extracellular concentration of galactose is still relatively high. In this case, it would be unfavorable to shut down the operon immediately; instead, it would be better to wait for some generations and ensure that the nutrient is no longer present. Hence, as the cells cannot judge the outer concentration of the nutrient with 100% accuracy, some lag in response to changing environmental conditions might sometimes be favorable.

The inheritance of phenotypic state in this example critically depends on the assumption that the permease is part of the operon and is therefore induced by lactose. Therefore, the comparison of the lac operon structure with the ecological conditions in different bacterial species might shed some light on the possible advantages of this feedback loop.

16.3 Structural Inheritance

The common feature of the following examples is that preexisting cell structures can be used as templates for the assembly of new structures (Jablonka and Lamb, 1995). We consider three examples: cortical inheritance, genetic membranes and prions.

16.3.1 Inheritance of Cell Surface Structure

Some of the earliest examples of nongenetic heredity come from ciliates. It has been known since the early 1960s that the large-scale structure of cortical surface of ciliates shows nongenetic inheritance (Jablonka and Lamb, 1995). The cell surface of these protozoa is covered by thousands of cilia arranged in longitudinal rows and the same orientation. In a series of classical experiments on ciliates, Sonneborn showed that variants of structural organization of the cell surface are stably inherited (Tamm et al., 1975; see also Frankel, 1989). The interpretation for cortical inheritance is that the old kinetid (cilia plus associated structures) serves as a molecular scaffold on which the new one is built. When the orientation of the kinetid is inverted, the new kinetids show the same orientation. Although the example is suggestive, it is unclear whether any of the structural variants could confer advantage over the wild-type organization. Nevertheless, it is likely that structural inheritance associated with cytoskeleton structure is more frequent than what was previously thought. The positional inheritance of the flagellum in trypanosomes is another example of such a phenomenon. It has recently been shown that the old flagellum directs the morphogenesis and position of the new flagellum in relation to the cell body, and consequently of the internal cytoskeleton (Moreira-Leite et al., 2001). A similar phenomenon has been observed in maintaining the bipolar pattern of budding in yeast (Jablonka and Lamb, 1995).

16.3.2 Genetic Membranes

There are also claims that membranes have some hereditary capabilities (Cavalier-Smith, 2000, 2001; Szathmary, 2000). It has been recognized that mitochondrial membranes are autocatalytic for the incorporation of some proteins. The majority of mitochondrial proteins are encoded by the nucleus, and hence they must somehow be imported to the mitochondria. This process is facilitated by import proteins, which are themselves encoded by the nucleus. However, these import proteins must also be imported, leading to a positive feedback loop for the incorporation of these proteins. Although the idea of membrane inheritance is purely theoretical at this stage, some new experimental results are consistent with this notion. At least two mitochondrial matrix proteins — Hsp60p and Yah1p — are known to be essential for their own assembly. Moreover, some yeast strains seem to be the result of a heritable structural alteration in mitochondria and are not simply genetic mutations in the mitochondria (Lockshon, 2002). These strains are known to be defective in certain steps of leucine biosynthesis that involve transport across mitochondrial membranes.

16.3.3 Prions

Prions provide another example of structural inheritance (Prusiner, 1998). Prions are generally known as infectious agents widely implicated in a variety of mammalian neurodegenerative diseases, e.g., bovine spongiform encephalopathy (BSE), scrapie of sheep and Creutzfeldt–Jakob disease (CJD). The infectious nature of these proteins comes from the ability of the prion protein to catalyze its own propagation. These proteins can have at least two stable conformations: normal and prion. The prion form might rarely arise as a spontaneous misfolding event, possibly facilitated by translational errors. The prion form induces the normal protein to adopt the altered prion conformation, probably leading to the accumulation of amyloid protein aggregates.

Prions are also found in baker's yeast and in the fungus *Podospora* (Wickner et al., 1999). In contrast to mammalian prions, some evidences suggest that the prion form is not fatal to the organism, although the exact biological function is generally unknown. In *Podospora*, the prion form influences cell fusion incompatibility. In the case of yeast, at least three different proteins with prion forms are known (Bradley et al., 2002): (1) [Psi$^+$], the prion form of a translational termination factor protein (Sup35); (2) [URE3], the prion form of a regulator of nitrogen metabolism (Ure2); and (3) [Pin$^+$], the prion form of Rnq1. Remarkably, in all cases, the prion forms can be stably propagated through asexual cell division for hundreds of generations: the frequency of loss is less than 0.8%. In contrast, they are more frequently eliminated during meiosis (2%). Notably, a heat-shock protein (HSP104) is very highly expressed in sporulating cultures (Sanchez et al., 1992). As overexpression of this protein is known to eliminate [PSI$^+$], it is suggested that overexpression might be an evolved response against the rapid spread of prions under sexual reproduction.

The different prions influence the *de novo* appearance of each other in both positive and negative manner. [PSI$^+$] and [Pin$^+$] catalyze the formation of each other, and hence their relationship can be considered mutualistic. In contrast, although [PSI$^+$] facilitates the *de novo* appearance of [URE3], [URE3] inhibits [PSI$^+$].

Even more remarkably, numerous distinct strains of the [PSI$^+$] prion form exist (Bradley et al., 2002). These strains are likely to be alternative prion conformations of the same protein (Sup35) and differ in their mitotic stabilities (frequency of [PSI+] loss) and translational termination efficiencies. In contrast to the stable maintenance of different prion types, coexistence is not possible between two variants of the same prion. This makes sense, as the two variants of the same prion compete for the same pool of newly synthesized proteins to reproduce, and the faster growing prion strain eventually outcompetes the slower, less-stable variant.

Why are prions found in yeast? Are they simply epigenetic parasites, or do they confer any advantage to the host? There are some suggestions that the [PSI+] prions facilitate adaptation to stressful environmental conditions by producing new phenotypic variants (True and Lindquist, 2000). True and Lindquist (2000) showed that this prion has a strong and diverse effect on colony growth and morphology, and sometimes confers advantage over isogenic strains that lack the prion form. The prion reduces the fidelity of translation termination process in a heritable manner (Serio et al., 2001). It causes the read-through of stop codons, leading to an abnormally extended peptide. By reducing the fidelity of protein synthesis, the prion generates enhanced variation at the proteomic level (Pal, 2001). Some of the variants produced by the prion might permit survival under fluctuating environmental conditions.

Previous work has also demonstrated that the N-terminal region of the protein (prion determinant) is essential for converting normal proteins to the prion form (Serio et al., 2001). Prion determinant regions have similar very unusual amino acid composition and imperfect oligopeptide repeats, suggesting that these properties might underlie prion-based inheritance. Similar domains are widely found in eukaryotes (Michelitsch and Weissman, 2000), suggesting that prions might occur more frequently than generally thought. Deletion of this region does not have any harmful effect on colony growth under normal conditions. Remarkably, the region is well conserved across species (Santoso et al., 2000) and is under stabilizing selection (Jensen et al., 2001). However, this is not proof that this region is maintained to provide the prion conformation: there is some evidence that the prion determinant interacts with cytoskeletal proteins (Bailleul et al., 1999).

16.4 Inheritance of Chromatin Marks: Some Mechanistical Considerations

The most prominent examples of epigenetic inheritance are based on the transmission of specific patterns of chromatin structure, or chromatin marks (Jablonka and Lamb, 1995). Chromatin marks consist of complexes of DNA-binding proteins, RNAs and chemical modifications of the DNA (e.g., DNA methylation). The presence of certain marks does not change the coding property of the gene. Rather, it influences the rate and long-term stability of gene expression. Comparable to the semiconservative replication of DNA, these marks are also frequently inherited after cell division. We first illustrate this mechanism by reference to DNA methylation, which is the best-described type of chromatin mark.

16.4.1 DNA Methylation

In many eukaryotes (Regev et al., 1998), some of the cytosines are methylated. DNA methylation contributes to the control of gene expression, parental imprinting (Nicholls and Knepper, 2001), X-chromosome inactivation in mammals and protection of the genome against selfish DNA (Yoder et al., 1997). Methylation usually inhibits transcription initiation, although it is also known that DNA methylation affects transcript elongation in fungi (Martienssen and Colot, 2001). In numerous cases, the DNA methylation pattern of given genomic regions is faithfully transmitted to daughter cells after cell division. How is this achieved?

According to the most prominent model, methylation patterns can be inherited if cytosines of palindromic sequences (such as CpG or CNG triplets, where N denotes any of the four base pairs) are involved. Because of complementary base pairing, the daughter strand also has the same sequence. Following DNA replication, the parental strand is methylated whereas the daughter strand is not. An enzyme complex including methyltransferase recognizes the hemimethylated state and appropriately methylates the daughter strand. By this mechanism, a pattern of methylated and nonmethylated cytosines is copied, leading to inheritance of silenced or expressed states.

Although many experiments support this model, it must be emphasized that it cannot provide the whole picture. Most importantly, cytosine methylation is not confined to CpG or other symmetrical sequences in plants and fungi (Martienssen and Colot, 2001). It is also known that after treating with demethylating agents, the genomic level of DNA methylation is drastically reduced. However, following removal of the drug, methylation level slowly recovers (Bird, 2002). This result can hardly be explained without assuming some *de novo* methylation process. Indeed, in mammals, an extensive demethylation process occurs in the primordial germ-cell stage and during early development (Yoder et al., 1997). We also have good evidence for the enzymes with some *de novo* activity and others responsible for removing methyl groups (Bird, 2002). It seems that methylation patterns are maintained at a genomic domain level, even if some of the constituting cytosines do not reside at symmetrical sites (Bird, 2002).

There is also hope for a better understanding how DNA methylation affects gene expression level. In an elegant study, Amedeo et al. (2002) identified a gene in *Arabidopsis* whose product is required to maintain transcriptional gene silencing. Mutation of this gene leads to reactivation of several genes, even though these genes remain heavily methylated. Possibly, this gene is involved downstream of methylation in epigenetic regulation.

16.4.2 Chromatin Remodeling

Inheritance of silenced epigenetic state can occur without DNA methylation. DNA methylation is completely absent in numerous model organisms, including *C. elegans* and fission yeast (Regev et al., 1998), and it is also rare in *Drosophia* (Lyko, 2001). Nevertheless, these organisms provide examples of mitotic and meiotic transmission of epigenetic silencing by DNA-protein and protein–protein interactions. The proposed mechanism is remarkably similar to the classic model of DNA methylation, and it is largely inspired by results on the interaction of polycomb-group response elements in *Drosophila* (Lyko and Paro, 1999). Assume that certain chromatin proteins have an ability to bind to certain DNA regions and also to each other. If the association of these chromatin proteins is facilitated by cooperative interactions, then the more proteins found on a certain genomic region, the higher the possibility that a new protein can be attached. After replication, the semibound sites of the new DNA molecules could be preferential sites for the assembly of new complexes on the daughter strand.

16.4.3 Histone Acetylation

Histone acetylation is another heritable modification of chromatin structure and like DNA methylation is also involved in such processes as genomic imprinting (Turner, 2000). Acetylation reduces the affinity of the H4 histone protein to DNA, leading to relaxed chromatin structure and higher transcription rate. In contrast, deacetylation of H4 is associated with highly condensed DNA, with low or no transcription. Several mechanisms have been proposed to explain how the acetylation pattern can be transmitted after DNA replication. One possibility is that the enzymes responsible for acetylation process form part of a complex that remains associated with its target DNA throughout the cell cycle. Alternatively, the enzyme responsible for histone acetylation affects genomic regions with some DNA methylation. Indeed, it is known that the histone deacetylase enzyme and methyltransferase interact with each other (Fuks et al., 2000).

16.4.4 RNA-Mediated Gene Silencing

It has recently become obvious that in many organisms, including fungi, higher plants and animals, small RNAs derived from cleavage of double-stranded RNA are involved in post-transcriptional gene silencing (Kooter et al., 1999; Matzke et al., 2001). Although the details are somewhat obscure, these diverse silencing mechanisms have some common features, indicating an ancient origin. It is claimed that RNA silencing evolved to counter the spread of viruses and transposable elements, many of which produce double-stranded RNAs during their replication. It has also been suggested that host defense mechanisms provided a raw material for the evolution of new regulatory mechanisms for host genes required during development. Many genes are known to contain TE insertions, which might have imposed changes in the regulatory control of the genes.

There is also evidence that small RNAs can guide *de novo* methylation of homologous DNA sequences. Another interesting feature of RNA-mediated gene silencing is that they produce mobile signals (small RNAs) that can potentially induce silencing in cells distant from the origin. These results open an intriguing possibility that RNA-mediated gene silencing can provide a feedback from somatic to germ cells (E. Jablonka, personal communication). Small RNAs derived from somatic cells might move to germ cells and induce *de novo* methylation.

16.4.5 Fidelity of Transmission and Epimutations

Whatever the exact mechanism by which DNA methylation pattern and chromatin marks are transmitted, it is clear that the copying process has limited fidelity. The infidelity of replication of methylation patterns has the capacity to generate heritable phenotypic diversity among genetically identical cells. In almost all somatic cells, illegitimate transcripts occur as a result of spontaneous reactivation (Chelly et al., 1989; McAdams and Arkin, 1999), and this process is especially pronounced during ageing (Brown and Rastan, 1988; Catania and Fairweather, 1991). There is also increasing evidence that methylation changes are involved in cancer initiation (Jones and Laird, 1999; Ohlsson et al., 1999), although it is less clear whether these changes are the result of somatic mutations or they precede it. Kermicle (1978) has termed randomly produced modifications of epigenetic silencing *epimutations* and the possible variants at a given locus *epialleles*.

16.5 Inheritance of Chromatin Marks through Meiosis

It has long been known that the cortical surface structure in asexual ciliates and metabolic states in *E. coli* show nongenetic inheritance (see previously). In these cases, genetically identical populations show clonal heritable variation in these traits. However, this variation may be maintained only because these organisms are unicellular and lack specialized gametes. It has long been argued that in multicellular organisms developing from a single cell, resetting epigenetic information during gametogenesis is necessary to restore totipotency. Does this also imply that *all* epigenetic marks are erased during gametogenesis? Were this so, then clearly, epimutation cannot be important to the process of adaptation. We briefly review some of the best examples for the inheritance of chromatin marks through meiosis to establish that this objection is not terminal for the speculation that epimutations are involved in adaptation.

It is important to emphasize that we consider only epigenetic traits that are inherited for numerous organism generations; therefore, we do not discuss many other epigenetic phenomena, such as genetic imprinting. Imprints are established and erased every generation in a parental-specific manner, and hence these marks cannot be inherited in the long term and are instead under genetic (DNA-based) control and are better regarded simply as the mechanism by which the imprinting genes exercise their effects.

Consider first a unicellular organism in which chromatin marks are inherited through mitosis and meiosis. In the fission yeast *Schizosaccharomyces pombe*, the two mating cell types (plus and minus) switch efficiently by interchanging alleles at the mating type locus (mat1). This process occurs by directed gene conversion of the information located at the silent mat2 and mat3 loci. These loci are located 11 kb away from the active loci. When a reporter gene is inserted within the regions between the two silent loci, expression from this gene is also greatly reduced (Grewal and Klar, 1996; Grewal, 2000). The authors also identified *cis*-acting regions and *trans*-acting factors responsible for the inheritance of silent states. A partial deletion of these regions results in variegated expression of the inserted gene. More precisely, some of the cell lines with the genetic modification express the transgene whereas others do not. Even more remarkably, the silent or active state of the reporter gene is stably inherited through mitosis and meiosis. The rate of switching between the two states is relatively low (once in every 30 to 100 generations).

Heritable epigenetic silencing is observed not only at the mating type region in fission yeast but also at centromeric regions (Nonaka et al., 2002). Remarkably, in both cases, a chromodomain protein (swi6) is involved, suggesting a connection between silencing at these loci.

Epigenetic inheritance through meiosis is not restricted to unicellular organisms. For example, Kakutani and colleagues investigated an *Arabidopsis* mutant (ddm1) defective in maintaining methylation patterns (Kakutani et al., 1999). This leads to a reduced methylation level along the genome and consequently to numerous developmental changes. These changes are stably inherited even when segregated from the mutant genetic background. Similar examples of heritable epigenetic variation in plants were described for laboratory strains, as in the case of transposons of maize (Martienssen and Baron, 1994) and some repeated transgenes of tobacco (Park et al., 1996). In these examples, mutations causing genome-wide demethylation unleashed numerous heritable developmental abnormalities, even if the mutant gene is no longer present. Stable inheritance of methylation changes for numerous generations enables the identification of many controlled genes by conventional linkage analysis and cloning (Habu et al., 2001). This approach appeared to be especially fruitful: heritable epigenetic silencing associated with locus-specific DNA methylation changes have been documented for numerous genes involved in plant developments, including superman (Jacobsen and Meyerowitz, 1997), agamous (Jacobsen et al., 2000) and the flowering locus WA (Soppe et al., 2000).

There is also evidence for the occurrence of epigenetic variation in nature (Cubas et al., 1999). More than 250 years ago, Linneaus described natural variation of flower symmetry in *Linaria vulgaris*. Mostly, the flower of this plant is bilaterial, but radial flowers also occur. It has recently been revealed that epigenetic changes are responsible for this natural variation. The authors have investigated a gene (Lcyc) that controls floral dorsoventral symmetry in this and many other plant species. They failed to find specific genetic changes responsible for the variation. Rather, they found that the gene is extensively methylated and silent in radial variants. The epimutation is transmitted to future plant generations, although much less efficiently than genetic mutations: it was reported that demethylation during somatic development did sometimes occur, reverting to flowers with radial symmetry.

It has long been argued that because of early separation of germline and soma, meiotically heritable epigenetic variation cannot occur in most animal species. On a similar vein, others argued that epigenetic modifications that suppress gene activity in mammals are cleared in the mammalian germline, restoring totipotency of the genome. However, there is direct evidence for the transmission of chromatin marks through meiosis in *Drosophila* (Cavalli and Paro, 1998) and mammals (Morgan et al., 1999; Sutherland et al., 2000).

For example, in an elegant study, Cavalli and Paro (1998) demonstrated that a DNA regulatory motif (Fab-7) confers heritable states of expression and repression during somatic cell divisions. This motif is also known to be located in the homeotic gene cluster, and it maintains the expression state of developmental genes by being the target of protein complexes that organize heritable chromatin structures. Strikingly, the derepressed and depressed states of reporter genes are transmitted to the progeny through the female germline.

In another study, Morgan et al. (1999) described epigenetic inheritance at the agouti gene of mice. Insertion of a retrotransposon upstream of the gene results in ectopic expression of its gene product, with characteristic phenotypic changes (e.g., yellow skin color and obesity). The phenotype also shows variable expression in isogenic mice and it is maternally heritable. Hence, isogenic strains (all containing the inserted retrotransposon) differ only in the expression of the agouti locus, and these changes are also transmitted to the progeny. The possibility of a simple maternal effect was excluded. These results prompted the authors to suggest that mobile genetic elements can produce heritable epigenetic modifications (Whitelaw and Martin, 2001). In a similar vein, Sutherland et al. (2000) have revealed that silent state of the transgene in mammals is inherited for multiple generations irrespective of the sex of the parent, implying maintenance of the epigenetic state through meiosis. Furthermore, silencing is transcriptional and correlates with methylation of the transgene as well as an inaccessible chromatin structure; these changes are reversed when expression is reactivated. For other examples of epigenetic inheritance in mammals (including human), see Rakyan et al. (2001).

Although these studies suggest that epigenetic inheritance can occur in organisms with early separation of germline and soma, there are many more examples of the phenomenon in plants. This might simply reflect the advance of certain genetic techniques in plants (such as forward genetics), but differences in the role of DNA methylation might also have some role (Habu et al., 2001). For example, in contrast to plants and fungi, mutants defective in DNA methylation are generally inviable in mammals (Li et al., 1992). Hence, heritable epigenetic variants that could co-segregate with mutants may rarely arise in mammals, as these mutants have more serious fitness consequences.

It is important to emphasize that transgenerational epigenetic inheritance is not restricted to organisms with extensive DNA methylation. In *Drosophila*, DNA methylation is barely detectable (Lyko, 2001), and it is likely to be completely absent in fission yeast. Nevertheless, some of the most detailed studies on epigenetic inheritance are from these organisms. Hence, although DNA methylation can have an important contribution, it is neither necessary nor sufficient to maintain expression states.

Nevertheless, it is still possible that there is a common mechanism behind epigenetic inheritance; for example, the chromatin proteins responsible for heritable gene silencing in fission yeast, *Drosophila* and plants are related to each other (Klar, 1998; Habu et al., 2001).

16.5.1 Is Epigenetic Inheritance a Specific Property of Certain Genomic Regions?

Although the previous examples indicate that epigenetic marks can be transmitted between generations, an important caveat is that all regions of the genome need not be equally likely to permit such inheritance, i.e., it is conceivable that epigenetic inheritance through meiosis is restricted to certain genomic regions, which are somehow safeguarded from the erasure of chromatin marks. For example, in contrast to housekeeping genes, methylation patterns of certain retrotransposons (e.g., Alu) remain relatively unchanged in the female genome during mammalian development (Yoder et al., 1997). Some other facts point in the same direction. First, epigenetic inheritance of certain traits is often associated with transposable elements (Fedoroff et al., 1995; Whitelaw and Martin, 2001). Second, there is also evidence that heritable gene silencing of the Mu transposon and paramutation in maize are mechanistically linked (Lisch et al., 2002). Hence, housekeeping genes that happen to be close to regions with many transposable elements might have a relatively high chance to show epigenetic inheritance.

This regionality concept may help to reconcile the controversial findings on the stability and transmissibility of chromatin marks during mammalian gametogenesis (Yoder et al. 1997, but see Whitelaw and Martin, 2001).

16.6 Evolutionary Potential

Is heritable epigenetic variation merely an aberrant manifestation of developmental processes, as often claimed, or can it also have important roles during evolution? Jablonka and Lamb (1995) argue that epigenetic inheritance provides heritable phenotypic variation, which can be exploited during adaptation to new environmental conditions. Although theoretical models that assume asexuality support their verbal argument (Lachmann and Jablonka, 1996; Pal and Miklos, 1999), the case for adaptation in sexuals must be different.

Consider a simple model (Keller, 1995). Assume that an epimutation has arisen that is neutral. Initially, it will be present in heterozygotes only. What proportion of the progeny of the individual with the epimutation will also have the epimutation? Were the epimutation a point mutation of another DNA-based mutation then, owing to Mendelian inheritance, the answer would be 1/2. Therefore, there would be no deterministic force acting to reduce the allele's frequency, because it must be assumed that the bearer of the mutation has the average number of progeny (i.e., 2). But with epimutations Mendelian inheritance need not be supposed. Many are removed when passaged through meiosis. If k is the proportion of progeny of a heterozygote bearing the epimutation, then $k < 1/2$ might well be true; alternatively, in some instances, $k > 1/2$ might be true. In either case, such epimutations are unlikely to provide raw material for adaptation. In the former case ($k < 1/2$), an advantageous mutation can easily be deterministically lost owing to the failure to be transmitted. If the beneficial effect of the allele increases fitness by s in the heterozygotes, then $s > (1 - 2k)/2k$ must hold for the selection effects to permit spread. Much as with Jeffries objection to Darwinism under blending inheritance, the advantageous mutation must be very advantageous to counteract the transmission system. In the latter case ($k > 1/2$), a deleterious epimutation (with fitness effects $1 - s$) can spread so long as $2k(1 - s) > 1$; that is, the spread need not be associated with advantageous alleles. This does not mean that all such traits will be deleterious. However, if most epialleles are deleterious, then if $k > 1/2$, epimutations are expected to act against the process of adaptation. Only when $k = 1/2$ does s — and s alone — matter. In a finite population, however, there is probably a domain around $k = 1/2$ for which an epimutation can be regarded as effectively Mendelian (much as with s, $s < 1/2$ Ne ensures a mutation to be effectively neutral, where Ne denotes the effective population size).

However, such considerations add an extra wrinkle to the analysis. It must be supposed that for a small Ne, the zone of effective Mendelian transmission is the largest. But with a low Ne, the zone for s being effectively neutral is also larger and hence the zone for adaptation also shrinks. Clearly, one needs to ask what k is for epimutations.

16.6.1 Why k<1/2 Might Often Be the Case

For an epimutation to be transmitted with $k = 1/2$, it must not be lost in cells that are not dividing, or when cells are dividing or through meiosis. What is the fidelity of the molecular mechanisms responsible to maintain epigenetic marks? Methylated GCs can spontaneously arise and be lost, partly as a result of the imperfect copying process and because there is evidence of *de novo* activity of certain enzymes (Bird, 2002). Although the replication accuracy of methylation patterns *in vivo* is unknown, *in vitro* studies suggest that it is 95 to 99% per site per cell generation (Holliday, 1987). Obviously, the fidelity of this process is lower than that of DNA replication. However, it is unclear what fraction of the methylated CGs is necessary to maintain the silenced state of a given gene. There are some claims that even if numerous methylated GCs are lost around the genic regions, the silenced state of the gene can remain.

For these reasons, it is important to ask about the fidelity by which the silenced state of a given gene remains through cell divisions. Works on mammalian and plant cell cultures have revealed a remarkably stable inheritance of epigenetic variants. For example, in plant cell cultures, the rate of such changes is ca. 10^{-4} to 10^{-6}, although lower fidelity has been observed in other cases (Jablonka and Lamb, 1995).

Most importantly, what is the transmission fidelity of epigenetic marks through organismal generations? Unfortunately, there is almost no systematic approach to investigate this question. Mostly, the fates of the epigenetic variants were followed only for one or two generations. Therefore, we have only very crude estimates of the proportion of progeny to which it can be transmitted. However, there exists the strong generality that marks might be transmitted fairly well through female meiosis but not through male meiosis (Cavalli and Paro, 1998; Morgan et al., 1999). This suggests that often $k \approx 1/4$ must be the case. (Half the time the epimutation will be in a male.) In some cases, the rate is lower still. For example, inheritance of the expression state at the reporter gene in *Drosophila* was possible only through maternal germline, and only 30% of the descendants down the maternal line inherited the new trait (Cavalli and Paro, 1998) suggests k is around 0.15.

However, in other cases, a highly reliable transmission was observed. For example, changes in the protein complexes bound to the Y chromosome in *Drosophila* have affected gene silencing in nearby genomic regions and these changes have been faithfully transmitted for 11 generations (Dorn et al., 1993). Silencing of the mating type locus in fission yeast provides another example of a stable inheritance of chromatin marks (Grewal and Klar, 1996).

Furthermore, as pointed out by Maynard Smith and Szathmáry (1995), for epigenetic inheritance to be important in multicellular organisms, it is not enough that chromatin marks are stably inherited through the germline. It is also important that they have reproducible phenotypic consequences in somatic cells. Hence, if a silent epigenetic state is transmitted through the germline, the silent state is expected to be maintained in somatic cells in a similar manner. In the examples of the previous section, this seems to be generally the case. However, there are also examples of partial reactivation of the silenced epigenetic state in somatic cell lineages, resulting in a variegated phenotype (Rakyan et al., 2001). The descendants generally show a full spectrum of phenotypes (showing more or less somatic expression of the gene), with increased frequency of those that are similar to the mother.

16.6.2 Why $k > 1/2$ Might Be Found

Epigenetic marks can receive transmission at rates higher than those found in Mendelian inheritance. Paramutation is an allelic interaction in higher plants that results in segregationally biased, meiotically heritable changes in expression (Hollick et al., 1997). The best-studied system showing paramutation involves the R locus in maize, which affects pigment intensity in the plant (Chandler et al., 2000). Alleles sensitive to paramutation are called paramutable, and alleles that initiate paramutation are paramutagenic. Under heterozygotic conditions when a paramutable allele meets a paramutagenic one, the expression of the paramutable allele is decreased through exposure to the paramutagenic allele. After meiotic segregation, the former paramutable allele retains its lowered expression state and itself becomes paramutagenic. Remarkably, in some cases, the two — paramutable and paramutagenic — alleles are genetically identical to each other and differ only in chromatin structure (Hollick et al., 1997). In contrast to Mendel's first law, the lowly and highly expressed alleles do not segregate unchanged from heterozygotes. The low expression state is heritably transmitted through the formerly highly expressed allele. Hence, irrespective of the selective advantage or disadvantage of the paramutagenic allele, it might deterministically spread in the population because of its overepresentation among the products of meiosis.

A similar problem is likely to arise with structural inheritance and other examples that involve alternative states of the cytoplasm. This problem is elucidated with prions. Under sexual reproduction, prion strains in yeast receive transmission at much higher rates than those found in Mendelian inheritance (Tuite and Lindquist, 1996). Under heterozygotic conditions, when strains differing in the presence of the prion meet, nearly all daughter cells will inherit the prion form. A similar problem is likely to arise with other examples of structural inheritance and steady-state systems.

From an evolutionary viewpoint, paramutation and prions are reminiscent of classical examples of meiotic drive. In the latter case, one of the chromosomes has a more than 50% chance of ending up in functional gametes (Hurst et al., 1996). Biased gene conversion has comparable dynamics, but this process affects only one gene instead of a whole chromosome (Hurst and Werren, 2001). Gene conversion is a nonreciprocal recombination event resulting in the alteration of one allele with the other. It routinely occurs during meiosis and is sometimes biased in one direction, leading to one of the two alleles being overrepresented after segregation. The difference between paramutation and biased gene conversion is that instead of a genetic variant, an expression state is transmitted to the other allele, leading to its increased frequency in gametes of heterozygotes.

From these considerations, it can be concluded that there exists a mechanistic basis for epimutations and that sometimes epimutations are transmissable through meiosis and contribute to standing variation, but these facts alone are not adequate to defend the conjecture that epimutations are an important source of heritable variation that might lead to adaptation. Only a limited subset will have transmission rates within the relatively small window ca. $k = 1/2$ to allow selection — and selection alone — to determine their fate. Although there are some cases for near-Mendelian segregation of the epialleles, these are more likely to be the exception rather than the rule. The opposite is true for most DNA-based mutations.

16.6.3 Range of Heritable Epigenetic Variation

However, suppose that $k = 1/2$. Would then epi- and DNA-based mutations be equally likely to be raw material for adaptation? A further difference between the two is the extent to which they permit variation.

Maynard Smith and Szathmáry (1995) have noticed that heredity systems can have limited and unlimited repertoire of hereditary variants. Genetic systems provide an example of unlimited inheritance, as the possible number of genetic variants is practically unlimited and only a very tiny fraction is actually realized in the population. In contrast, in a system with limited range of possible variants, the same few variants emerge repeatedly. In this case, the frequency of backward changes is much higher than that of a system with unlimited heredity.

At first sight, epigenetic inheritance seems to be a system with limited hereditary potential. There are some claims that epigenetic silencing is frequently an all-or-none phenomenon; that is, if a single locus is considered, it can have only two heritable states: expressed and silenced. Although it might be generally so, there are some examples with multiple heritable epialleles. For example, in *Arabidopsis*, seven heritable epialleles have been observed at the

SUP locus. These epialleles are associated with some differences in excess cytosine methylation within the SUP gene and a decreased level of gene expression (Flavell and O'Dell, 1990). Another study has revealed at least three meiotically heritable epigenetic states of a transposable element in maize, and all these states correspond with slight DNA methylation differences in nearby genomic regions (Fedoroff et al., 1995).

However, it is obvious that when a single locus is considered, the number of epialleles is far less than the practically infinite number of mutational variants a gene can have. A much better figure can be obtained if a combinatorial approach across numerous loci is used (see also (Jablonka and Lamb, 1998). Consider an organism with 10, 000 different loci. If only two possible epialleles per locus are considered (active vs. inactive), the number of possible epigenetic variants is $2^{10,000}$, an enormously large number.

The basic assumption behind this calculation is that silenced and expressed states can be achieved independently for all loci considered. One can argue that only a tiny fraction of these possible variants can be achieved because of various regulatory constraints. It is unlikely that heritable epigenetic variants can arise on a gene-by-gene basis. DNA methylation patterns are likely to be maintained through cell divisions at the genomic domain level instead of at the genic level (Bird, 2002). Therefore, the size of independently methylated domains would be of special importance. Assume that these domains generally include 100 genes. In this case, the possible number of epigenetic variants reduces to $2^{1000/100} = 2^{100}$, still a very large number. Hence, it is safe to conclude that even if the vast space of variation is constrained, epigenetic inheritance could theoretically provide variation on which selection could act.

16.7 Discussion

Owing to the advances in molecular techniques, there are many good examples of the inheritance of epigenetic traits through both mitosis and meiosis. Although the role of epigenetic inheritance during development is generally accepted, it is much less clear whether heritable epigenetic variation can play a significant role during evolution. We have argued that epimutations are unlikely to be the source of heritable variation that is formed by selection into adaptations: they are either transmitted to too many progeny, in which case they can also be deleterious, or to too few, in which case even if advantageous they will often be lost. This suggests that under sexual reproduction — possibly associated with cell fusion — epigenetic inheritance is unlikely to play a fair Mendelian game, which is a prerequisite for adaptive evolution. Only when transmissibility from a heterozygote (k) is half will selective advantage of the epimutation alone specify the fate of the variant. In all other cases ($k < 1/2$ or $k > 1/2$), the transmission capability of the system interferes with its selective advantage.

But what if k depends on the environment? It has been known that most chromatin marks are labile and transmission accuracy depends on the environment considered. Therefore, recurrent induction might increase the otherwise low transmissibility (E. Jablonka, personal communication). However, we think that it is not enough if recurrent environmental changes influence the average value of transmissibility over generations. It is hard to imagine how environmental changes could provide compensation for the otherwise low transmissibility, leading to an average $k = 0.5$ transmissibility. Consider, for example, in a population a rare epigenetic variant that is generally transmitted to 49% of the offspring. (Note that this is a very slight bias.) After only 35 generations its frequency will be halved, and not even a complete induction ($k = 1$) that lasts for one generation can restore its original frequency.

In sum, epigenetic inheritance is expected to have more profound effect under asexual, unicellular conditions. Most theoretical models on epigenetic inheritance consider asexual populations (Pal and Miklos, 1999; Lachmann and Jablonka, 1996). However, we do not consider our argument decisive in this debate. It is possible that our knowledge is biased. After all, most examples of nongenetic inheritance were discovered because of the suspiciously non-Mendelian segregation of the trait considered. Further, we have not considered the possibility that group selection might play a role. We have argued that $k > 1/2$ will allow easier invasion of both advantageous and deleterious epialleles. As most epialleles can be assumed to be deleterious, this process acts to degrade populations. But if we allow for between-group competition (with minimal mixing or sex between the groups), then selection might favor those groups with epimutations (with $k > 1/2$) over those without, as some groups might, by chance, have more advantageous alleles. A detailed model would be needed to see whether this speculation might be supportable even in theory.

Some facts suggest that fair segregation of chromatin marks during meiosis might occur more often than previously thought. Recently, Riddle and Richards (2002) found significant natural variation in cytosine methylation in *Arabidopsis*, particularly in the nucleolus organizer regions, which constitute ca. 6% of the genome. Their results indicate that besides the differences in *trans*-acting modifier genes and rRNA copy number among the populations, epigenetic inheritance of methylation patterns also contributes to the variation. The authors also noted that gene methylation in F1 hybrids created by reciprocal crosses was intermediate between the two parents (high and low methylation). This could potentially be due to a homogenization of methylation at intermediate value on both parental chromosomes in the hybrids. In this case, the variation would disappear and provide an example of blending inheritance. Alternatively, parental methylation patterns might be preserved and inherited in the hybrids. The latter explanation turned out to be true. Some further examples show that hypomethylation of genomic sequences is inherited in a Mendelian manner.

Although it is hard to judge how often Mendelian segregation of an epigenetic trait can occur, it does not follow that epigenetic inheritance cannot have important evolutionary consequences. One need, for example, consider only genomic imprinting. Marks are put on or taken off genes in one germline, effects that are reversed in the germline of the opposite sex. But this is not the independent evolution of a heritable system that runs in parallel to DNA-based polymorphisms. Rather, DNA-based mutations direct the placement and removal of the marks, much as coding sequences determine which amino acids are to be employed.

Further, we should consider some of the consequences of having epigenetic systems. Most notably, transcriptionally active regions are more prone to mutations (Datta and Jinksrobertson, 1995), have higher recombination rates (Gerton et al., 2000) and gene silencing mechanisms influence the invasion possibility of transposable elements and foreign sequences (Yoder et al., 1997). Restriction modification systems in bacteria (a system that involves DNA methylation) might itself be considered as selfish genetic elements (Kobayashi, 2001).

It is also known that highly expressed genes are under more stringent selection pressure (Pal et al., 2001), possibly to reduce metabolic costs of amino acid biosynthesis (Akashi and Gojobori, 2002). Gene silencing mechanisms could also influence the formation of viable hybrids between related species (Pikaard 2000, 2001). For example, in newly formed allopolyploids of plant species, substantial DNA methylation changes are observed in genes and transposable elements.

We can then be sure that epigenetics is of importance, but it is unclear whether when uncoupled from DNA-based inheritance (i.e., when considered as a system parallel to DNA-based inheritance) it will be of importance in the process of adaptation. Possibly, only under unusual circumstances can epimutations have any lasting role.

References

Akashi, H. and Gojobori, T. (2002) Metabolic efficiency and amino acid composition in the proteomes of *Escherichia coli* and *Bacillus subtilis*. *Proc. Natl. Acad. Sci. USA* 99: 3695–700.

Amedeo, P., Habu, Y., Afsar, K., Scheid, O. M. and Paszkowski, J. (2000) Disruption of the plant gene MOM releases transcriptional silencing of methylated genes. *Nature* 405: 203–206.

Bailleul, P. A., Newnam, G. P., Steenbergen, J. N. and Chernoff, Y. O. (1999) Genetic study of interactions between the cytoskeletal assembly protein sla1 and prion-forming domain of the release factor Sup35 (eRF3) in Saccharomyces cerevisiae. *Genetics* 153: 81–94.

Bird, A. (2002) DNA methylation patterns and epigenetic memory. *Genes Dev.* 16: 6–21.

Bradley, M. E., Edskes, H. K., Hong, J. Y., Wickner, R. B. and Liebman, S. W. (2002) Interactions among prions and prion "strains" in yeast. *Proc. Natl. Acad. Sci. USA* 99: 16392–16399.

Brown, S. and Rastan, S. (1988) Age-related reactivation of an X-linked gene close to the inactivation centre in the mouse. *Genet Res.* 52: 151–154.

Catania, J. and Fairweather, D. S. (1991) DNA methylation and cellular ageing. *Mutat. Res.* 256: 283–293.

Cavalier-Smith, T. (2000) Membrane heredity and early chloroplast evolution. *Trends Plant Sci.* 5: 174–182.

Cavalier-Smith, T. (2001) Obcells as proto-organisms: Membrane heredity, lithophosphorylation, and the origins of the genetic code, the first cells, and photosynthesis. *J. Mol. Evol.* 53: 555–595.

Cavalli, G. and Paro, R. (1998) The *Drosophila* Fab-7 chromosomal element conveys epigenetic inheritance during mitosis and meiosis. *Cell* 93: 505–518.

Chandler, V. L., Eggleston, W. B. and Dorweiler, J. E. (2000) Paramutation in maize. *Plant Mol. Biol.* 43: 121–145.

Chelly, J., Concordet, J. P., Kaplan, J. C. and Kahn, A. (1989) Illegitimate transcription: Transcription of any gene in any cell type. *Proc. Natl. Acad. Sci. USA* 86: 2617–2621.

Cheng, M. Y., Hartl, F. U. and Horwich, A. L. (1990) The mitochondrial chaperonin hsp60 is required for its own assembly. *Nature* 348: 455–458.

Cubas, P., Vincent, C. and Coen, E. (1999) An epigenetic mutation responsible for natural variation in floral symmetry. *Nature* 401: 157–161.

Datta, A. and Jinksrobertson, S. (1995) Association of increased spontaneous mutation-rates with high-levels of transcription in yeast. *Science* 268: 1616–1619.

Dorn, R., Krauss, V., Reuter, G. and Saumweber, H. (1993) The enhancer of position-effect variegation of *Drosophila*, E(var)3-93D, codes for a chromatin protein containing a conserved domain common to several transcriptional regulators. *Proc. Natl. Acad. Sci. USA* 90: 11376–11380.

Fedoroff, N., Schlappi, M. and Raina, R. (1995) Epigenetic regulation of the maize Spm transposon. *Bioessays* 17: 291–297.

Flavell, R. B. and O'Dell, M. (1990) Variation and inheritance of cytosine methylation patterns in wheat at the high molecular weight glutenin and ribosomal RNA gene loci. *Dev. Suppl.*: 15–20.

Frankel, J. (1989) *Pattern Formation: Ciliate Studies and Models*, Oxford University Press, New York.

Fuks, F., Burgers, W. A., Brehm, A., Hughes-Davies, L. and Kouzarides, T. (2000) DNA methyltransferase Dnmt1 associates with histone deacetylase activity. *Nat. Genet.* 24: 88–91.

Gerton, J. L., DeRisi, J., Shroff, R., Lichten, M., Brown, P. O. and Petes, T. D. (2000) Inaugural article: Global mapping of meiotic recombination hotspots and coldspots in the yeast Saccharomyces cerevisiae. *Proc. Natl. Acad. Sci. USA* 97: 11383–11390.

Grewal, S. I. (2000) Transcriptional silencing in fission yeast. *J. Cell Physiol.* 184: 311–318.

Grewal, S. I. and Klar, A. J. (1996) Chromosomal inheritance of epigenetic states in fission yeast during mitosis and meiosis. *Cell* 86: 95–101.
Guelzim, N., Bottani, S., Bourgine, P. and Kepes, F. (2002) Topological and causal structure of the yeast transcriptional regulatory network. *Nat. Genet.* 31: 60–63.
Habu, Y., Kakutani, T. and Paszkowski, J. (2001) Epigenetic developmental mechanisms in plants: Molecules and targets of plant epigenetic regulation. *Curr. Opin. Genet. Dev.* 11: 215–220.
Hollick, J. B., Dorweiler, J. E. and Chandler, V. L. (1997) Paramutation and related allelic interactions. *Trends Genet.* 13: 302–308.
Holliday, R. (1987) The inheritance of epigenetic defects. *Science* 238: 163–170.
Hurst, G. D. and Werren, J. H. (2001) The role of selfish genetic elements in eukaryotic evolution. *Nat. Rev. Genet.* 2: 597–606.
Hurst, L. D., Atlan, A. and Bengtsson, B. O. (1996) Genetic conflicts. *Q. Rev. Biol.* 71: 317–364.
Jablonka, E. and Lamb, M. J. (1995) *Epigenetic Inheritance and Evolution*, Oxford University Press, Oxford.
Jablonka, E. and Lamb, M. J. (1989) The inheritance of acquired epigenetic variations. *J. Theor. Biol.* 139: 69–83.
Jablonka, E. and Lamb, M. J. (1998) Epigenetic inheritance in evolution (A target article). *J. Evol. Biol.* 11: 159–183.
Jacobsen, S. E. and Meyerowitz, E. M. (1997) Hypermethylated SUPERMAN epigenetic alleles in *Arabidopsis*. *Science* 277: 1100–1103.
Jacobsen, S. E., Sakai, H., Finnegan, E. J., Cao, X. and Meyerowitz, E. M. (2000) Ectopic hypermethylation of flower-specific genes in *Arabidopsis*. *Curr. Biol.* 10: 179–186.
Jensen, M. A., True, H. L., Chernoff, Y. O. and Lindquist, S. (2001) Molecular population genetics and evolution of a prion-like protein in Saccharomyces cerevisiae. *Genetics* 159: 527–535.
Jones, P. A. and Laird, P. W. (1999) Cancer epigenetics comes of age. *Nat. Genet.* 21: 163–167.
Kakutani, T., Munakata, K., Richards, E. J. and Hirochika, H. (1999) Meiotically and mitotically stable inheritance of DNA hypomethylation induced by ddm1 mutation of *Arabidopsis thaliana*. *Genetics* 151: 831–838.
Keller, A. D. (1995) Fixation of epigenetic states in a population. *J. Theor. Biol.* 176: 211–219.
Kermicle, J. L. (1978) Imprinting of gene action in maize endosperm. In *Maize Breeding and Genetics* (Walden D. B., Ed.), Wiley, New York, pp. 357–371.
Klar, A. J. (1998) Propagating epigenetic states through meiosis: Where Mendel's gene is more than a DNA moiety. *Trends Genet.* 14: 299–301.
Kobayashi, I. (2001) Behavior of restriction-modification systems as selfish mobile elements and their impact on genome evolution. *Nucleic Acids Res.* 29: 3742–3756.
Kooter, J. M., Matzke, M. A. and Meyer, P. (1999) Listening to the silent genes: Transgene silencing, gene regulation and pathogen control. *Trends Plant Sci.* 4: 340–347.
Lachmann, M. and Jablonka, E. (1996) The inheritance of phenotypes: An adaptation to fluctuating environments. *J. Theor. Biol.* 181: 1–9.
Lange, H., Kaut, A., Kispal, G. and Lill, R. (2000) A mitochondrial ferredoxin is essential for biogenesis of cellular iron-sulfur proteins. *Proc. Natl. Acad. Sci. USA* 97: 1050-1055.
Li, E., T. Bestor, H. and Jaenisch, R. (1992) Targeted mutation of the DNA methyltransferase gene results in embryonic lethality. *Cell* 69: 915–926.
Lisch, D., Carey, C. C., Dorweiler, J. E., Chandler, V. L. (2002). A mutation that prevents paramutation in maize also reverses Mutator transposon methylation and silencing. *Proc. Natl. Acad. Sci. USA* 99: 6130–6135.
Lockshon, D. (2002) A heritable structural alteration of the yeast mitochondrion. *Genetics* 161: 1425–1435.
Lyko, F. (2001) DNA methylation learns to fly. *Trends Genet.* 17: 169–172.
Lyko, F. and Paro, R. (1999) Chromosomal elements conferring epigenetic inheritance. *Bioessays* 21: 824–832.

Martienssen, R. and Baron, A. (1994) Coordinate suppression of mutations caused by Robertson's mutator transposons in maize. *Genetics* 136: 1157–1170.

Martienssen, R. A. and Colot, V. (2001) DNA methylation and epigenetic inheritance in plants and filamentous fungi. *Science* 293: 1070–1074.

Matzke, M. A., Matzke, A. J., Pruss, G. J. and Vance, V. B. (2001).RNA-based silencing strategies in plants. *Curr. Opin. Genet. Dev.* 11: 221–227.

Maynard Smith, J. (1990) Models of a dual inheritance system. *J. Theor. Biol.* 143: 41–53.

Maynard Smith, J. and Szathmary, E. (1995) *The Major Transitions in Evolution*, W.H. Freeman, Spektrum, Oxford.

McAdams, H. H. and Arkin, A. (1999) It's a noisy business! Genetic regulation at the nanomolar scale. *Trends Genet.* 15: 65–69.

Michelitsch, M. D. and Weissman, J. S. (2000) A census of glutamine/asparagine-rich regions: Implications for their conserved function and the prediction of novel prions. *Proc. Natl. Acad. Sci. USA* 97: 11910–11915.

Moreira-Leite, F. F., Sherwin, T. Kohl, L. and Gull, K. (2001) A trypanosome structure involved in transmitting cytoplasmic information during cell division. *Science* 294: 610–612.

Morgan, H. D., Sutherland, H. G., Martin, D. I. and Whitelaw, E. (1999) Epigenetic inheritance at the agouti locus in the mouse. *Nat. Genet.* 23: 314–318.

Nicholls, R. D. and Knepper, J. L. (2001) Genome organization, function, and imprinting in Prader-Willi and Angelman syndromes. *Annu. Rev. Genomics Hum. Genet.* 2: 153–175.

Nonaka, N., Kitajima, T., Yokobayashi, S., Xiao, G., Yamamoto, M., Grewal, S. I. and Watanabe, Y. (2002) Recruitment of cohesin to heterochromatic regions by Swi6/HP1 in fission yeast. *Nat. Cell Biol.* 4: 89–93.

Novick, A. and Weiner, M. (1957) Enzyme induction as an all-or-none phenomenon. *Proc. Natl. Acad. Sci. USA* 43: 553–566.

Ohlsson, R., Cui, H., He, L., Pfeifer, S., Malmikumpu, H., Jiang, S., Feinberg, A. P. and Hedborg, F. (1999) Mosaic allelic insulin-like growth factor 2 expression patterns reveal a link between Wilms' tumorigenesis and epigenetic heterogeneity. *Cancer Res.* 59: 3889–3892.

Pal, C. (2001) Yeast prions and evolvability. *Trends Genet.* 17: 167–169.

Pal, C. and Miklos, I. (1999) Epigenetic inheritance, genetic assimilation and speciation. *J. Theor. Biol.* 200: 19–37.

Pal, C., Papp, B. and Hurst, L. D. (2001) Highly expressed genes in yeast evolve slowly. *Genetics* 158: 927–931.

Park, Y. D., Papp, I., Moscone, E. A., Iglesias, V. A., Vaucheret, H., Matzke, A. J. and Matzke, M. A. (1996) Gene silencing mediated by promoter homology occurs at the level of transcription and results in meiotically heritable alterations in methylation and gene activity. *Plant J.* 9: 183–194.

Pikaard, C. S. (2000) The epigenetics of nucleolar dominance. *Trends Genet.* 16: 495–500.

Pikaard, C. S. (2001) Genomic change and gene silencing in polyploids. *Trends Genet.* 17: 675–677.

Prusiner, S. B. (1998) Prions. *Proc. Natl. Acad. Sci. USA* 95: 13363–13383.

Rakyan, V. K., Preis, J., Morgan, H. D. and Whitelaw, E. (2001) The marks, mechanisms and memory of epigenetic states in mammals. *Biochem. J.* 356: 1–10.

Regev, A., Lamb, M. J. and Jablonka, E. (1998) The role of DNA methylation in invertebrates: Developmental regulation or genome defense? *Mol. Biol. Evol.* 15: 880–891.

Riddle, N. C. and Richards, E. J. (2002) The control of natural variation in cytosine methylation in *Arabidopsis*. *Genetics* 162: 355–363.

Sanchez, Y., Taulien, J., Borkovich, K. A. and Lindquist, S. (1992) Hsp104 is required for tolerance to many forms of stress. *EMBO J.* 11: 2357–2364.

Santoso, A., Chien, P., Osherovich, L. Z. and Weissman, J. S. (2000) Molecular basis of a yeast prion species barrier. *Cell* 100: 277–288.

Serfling, E. (1989) Autoregulation — a common property of eukaryotic transcription factors? *Trends Genet.* 5: 131–133.

Serio, T. R., Cashikar, A. G., Kowal, A. S., Sawicki, G. J. and Lindquist, S. L. (2001) Self-perpetuating changes in Sup35 protein conformation as a mechanism of heredity in yeast. *Biochem. Soc. Symp.* 68: 35–43.
Shen-Orr, S. S., Milo, R., Mangan, S. and Alon, U. (2002) Network motifs in the transcriptional regulation network of *Escherichia coli. Nat. Genet.* 31: 64–68.
Soppe, W. J., Jacobsen, S. E., Alonso-Blanco, C., Jackson, J. P., Kakutani, T., Koornneef, M. and Peeters, A. J. (2000) The late flowering phenotype of fwa mutants is caused by gain-of-function epigenetic alleles of a homeodomain gene. *Mol. Cell* 6: 791–802.
Sutherland, H. G., Kearns, M., Morgan, H. D., Headley, A. P., Morris, C., Martin, D. I. and Whitelaw, E. (2000) Reactivation of heritably silenced gene expression in mice. *Mamm. Genome* 11: 347–355.
Szathmary, E. (2000) The evolution of replicators. *Philos. Trans. R. Soc. Lond. B Biol. Sci.* 355: 1669–1676.
Tamm, S. L., Sonneborn, T. M. and Dippell, R. V. (1975) The role of cortical orientation in the control of the direction of ciliary beat in *Paramecium. J. Cell Biol.* 64: 98–112.
True, H. L. and Lindquist, S. L. (2000) A yeast prion provides a mechanism for genetic variation and phenotypic diversity. *Nature* 407: 477–483.
Tuite, M. F. and Lindquist, S. L. (1996) Maintenance and inheritance of yeast prions. *Trends Genet.* 12: 467–471.
Turner, B. M. (2000) Histone acetylation and an epigenetic code. *Bioessays* 22: 836–845.
Whitelaw, E. and Martin, D. I. (2001) Retrotransposons as epigenetic mediators of phenotypic variation in mammals. *Nat. Genet.* 27: 361–365.
Wickner, R. B., Taylor, K. L., Edskes, H. K., Maddelein, M. L., Moriyama, H. and Roberts, B. T. (1999) Prions in *Saccharomyces* and *Podospora* spp.: Protein-based inheritance. *Microbiol. Mol. Biol. Rev.* 63: 844–861.
Yoder, J. A., Walsh, C. P. and Bestor, T. H. (1997) Cytosine methylation and the ecology of intragenomic parasites. *Trends Genet.* 13: 335–340.

Index

B

C

D

E

F

G

H

I

J

K

L

M

R

T

U

V

W

X

Y

Z

Systematics Association Publications

1. Bibliography of Key Works for the Identification of the British Fauna and Flora, 3rd edition (1967)†
Edited by G.J. Kerrich, R.D. Meikie and N. Tebble
2. Function and Taxonomic Importance (1959)†
Edited by A.J. Cain
3. The Species Concept in Palaeontology (1956)†
Edited by P.C. Sylvester-Bradley
4. Taxonomy and Geography (1962)†
Edited by D. Nichols
5. Speciation in the Sea (1963)†
Edited by J.P. Harding and N. Tebble
6. Phenetic and Phylogenetic Classification (1964)†
Edited by V.H. Heywood and J. McNeill
7. Aspects of Tethyan biogeography (1967)†
Edited by C.G. Adams and D.V. Ager
8. The Soil Ecosystem (1969)†
Edited by H. Sheals
9. Organisms and Continents through Time (1973)†
Edited by N.F. Hughes
10. Cladistics: A Practical Course in Systematics (1992)*
P.L. Forey, C.J. Humphries, I.J. Kitching, R.W. Scotland, D.J. Siebert and D.M. Williams
11. Cladistics: The Theory and Practice of Parsimony Analysis (2nd edition)(1998)*
I. J. Kitching, P.L. Forey, C.J. Humphries and D.M. Williams
*Published by Oxford University Press for the Systematics Association
†Published by the Association (out of print)

Systematics Association Special Volumes

1. The New Systematics (1940)
Edited by J.S. Huxley (reprinted 1971)
2. *Chemotaxonomy and serotaxonomy (1968)**
Edited by J.C. Hawkes
3. Data Processing in Biology and Geology (1971)*
Edited by J.L. Cutbill
4. Scanning Electron Microscopy (1971)*
Edited by V.H. Heywood
5. Taxonomy and Ecology (1973)*
Edited by V.H. Heywood
6. The Changing Flora and Fauna of Britain (1974)*
Edited by D.L. Hawksworth
7. Biological Identification with Computers (1975)*
Edited by R.J. Pankhurst
8. Lichenology: Progress and Problems (1976)*
Edited by D.H. Brown, D.L. Hawksworth and R.H. Bailey
9. Key Works to the Fauna and Flora of the British Isles and Northwestern Europe, 4th edition (1978)*
Edited by G.J. Kerrich, D.L. Hawksworth and R.W. Sims
10. Modern Approaches to the Taxonomy of Red and Brown Algae (1978)
Edited by D.E.G. Irvine and J.H. Price

11. Biology and Systematics of Colonial Organisms (1979)*
Edited by C. Larwood and B.R. Rosen
12. The Origin of Major Invertebrate Groups (1979)*
Edited by M.R. House
13. Advances in Bryozoology (1979)*
Edited by G.P. Larwood and M.B. Abbott
14. Bryophyte Systematics (1979)*
Edited by G.C.S. Clarke and J.G. Duckett
15. The Terrestrial Environment and the Origin of Land Vertebrates (1980)
Edited by A.L. Pachen
16. Chemosystematics: Principles and Practice (1980)*
Edited by F.A. Bisby, J.G. Vaughan and C.A. Wright
17. The Shore Environment: Methods and Ecosystems (2 volumes)(1980)*
Edited by J.H. Price, D.E.C. Irvine and W.F. Farnham
18. The Ammonoidea (1981)*
Edited by M.R. House and J.R. Senior
19. Biosystematics of Social Insects (1981)*
Edited by P.E. House and J.-L. Clement
20. Genome Evolution (1982)*
Edited by G.A. Dover and R.B. Flavell
21. Problems of Phylogenetic Reconstruction (1982)
Edited by K.A. Joysey and A.E. Friday
22. Concepts in Nematode Systematics (1983)*
Edited by A.R. Stone, H.M. Platt and L.F. Khalil
23. Evolution, Time And Space: The Emergence of the Biosphere (1983)*
Edited by R.W. Sims, J.H. Price and P.E.S. Whalley
24. Protein Polymorphism: Adaptive and Taxonomic Significance (1983)*
Edited by G.S. Oxford and D. Rollinson
25. Current Concepts in Plant Taxonomy (1983)*
Edited by V.H. Heywood and D.M. Moore
26. Databases in Systematics (1984)*
Edited by R. Allkin and F.A. Bisby
27. Systematics of the Green Algae (1984)*
Edited by D.E.G. Irvine and D.M. John
28. The Origins and Relationships of Lower Invertebrates (1985)‡
Edited by S. Conway Morris, J.D. George, R. Gibson and H.M. Platt
29. Infraspecific Classification of Wild and Cultivated Plants (1986)‡
Edited by B.T. Styles
30. Biomineralization in Lower Plants and Animals (1986)‡
Edited by B.S.C. Leadbeater and R. Riding
31. Systematic and Taxonomic Approaches in Palaeobotany (1986)‡
Edited by R.A. Spicer and B.A. Thomas
32. Coevolution and Systematics (1986)‡
Edited by A.R. Stone and D.L. Hawksworth
33. Key Works to the Fauna and Flora of the British Isles and Northwestern Europe, 5th edition (1988)‡
Edited by R.W. Sims, P. Freeman and D.L. Hawksworth
34. Extinction and Survival in the Fossil Record (1988)‡
Edited by G.P. Larwood
35. The Phylogeny and Classification of the Tetrapods (2 volumes)(1988)‡
Edited by M.J. Benton

36. Prospects in Systematics (1988)‡
Edited by J.L. Hawksworth
37. Biosystematics of Haematophagous Insects (1988)‡
Edited by M.W. Service
38. The Chromophyte Algae: Problems and Perspective (1989)‡
Edited by J.C. Green, B.S.C. Leadbeater and W.L. Diver
39. Electrophoretic Studies on Agricultural Pests (1989)‡
Edited by H.D. Loxdale and J. den Hollander
40. Evolution, Systematics, and Fossil History of the Hamamelidae (2 volumes)(1989)‡
Edited by P.R. Crane and S. Blackmore
41. Scanning Electron Microscopy in Taxonomy and Functional Morphology (1990)‡
Edited by D. Claugher
42. Major Evolutionary Radiations (1990)‡
Edited by P.D. Taylor and G.P. Larwood
43. Tropical Lichens: Their Systematics, Conservation and Ecology (1991)‡
Edited by G.J. Galloway
44. Pollen and Spores: Patterns and Diversification (1991)‡
Edited by S. Blackmore and S.H. Barnes
45. The Biology of Free-Living Heterotrophic Flagellates (1991)‡
Edited by D.J. Patterson and J. Larsen
46. Plant–Animal Interactions in the Marine Benthos (1992)‡
Edited by D.M. John, S.J. Hawkins and J.H. Price
47. The Ammonoidea: Environment, Ecology and Evolutionary Change (1993)‡
Edited by M.R. House
48. Designs for a Global Plant Species Information System (1993)‡
Edited by F.A. Bisby, G.F. Russell and R.J. Pankhurst
49. Plant Galls: Organisms, Interactions, Populations (1994)‡
Edited by M.A.J. Williams
50. Systematics and Conservation Evaluation (1994)‡
Edited by P.L. Forey, C.J. Humphries and R.I. Vane-Wright
51. The Haptophyte Algae (1994)‡
Edited by J.C. Green and B.S.C. Leadbeater
52. Models in Phylogeny Reconstruction (1994)‡
Edited by R. Scotland, D.I. Siebert and D.M. Williams
53. The Ecology of Agricultural Pests: Biochemical Approaches (1996)**
Edited by W.O.C. Symondson and J.E. Liddell
54. Species: the Units of Diversity (1997)**
Edited by M.F. Claridge, H.A. Dawah and M.R. Wilson
55. Arthropod Relationships (1998)**
Edited by R.A. Fortey and R.H. Thomas
56. Evolutionary Relationships among Protozoa (1998)**
Edited by G.H. Coombs, K. Vickerman, M.A. Sleigh and A. Warren
57. Molecular Systematics and Plant Evolution (1999)
Edited by P.M. Hollingsworth, R.M. Bateman and R.J. Gornall
58. Homology and Systematics (2000)
Edited by R. Scotland and R.T. Pennington
59. The flagellates: Unity, Diversity and Evolution (2000)
Edited by B.S.C. Leadbeater and J.C. Green
60. Interrelationships of the Platyhelminthes (2001)
Edited by D.T.J. Littlewood and R.A. Bray
61. Major Events in Early Vertebrate Evolution (2001)

Edited by P.E. Ahlberg
62. The Changing Wildlife of Great Britain and Ireland (2001)
Edited by D.L. Hawksworth
63. Brachiopods Past and Present (2001)
Edited by H. Brunton, L.R.M. Cocks and S.L. Long
64. Morphology, Shape and Phylogeny (2002)
Edited by N. MacLeod and P.L. Forey
65. Developmental Genetics and Plant Evolution (2002)
Edited by Q.C.B. Cronk, R.M. Bateman and J.A. Hawkins
66. Telling the Evolutionary Time: Molecular Clocks and the Fossil Record (2003)
Edited by P.C.J. Donoghue and M.P. Smith
67. Milestones in Systematics (2004)§
Edited by D.M. Williams and P.L. Forey
68. Organelles, Genomes and Eukaryote Phylogeny
Edited by R.P. Hirt and D.S. Horner

*Published by Academic Press for the Systematics Association
†Published by the Palaeontological Association in conjunction with Systematics Association
‡Published by the Oxford University Press for the Systematics Association
**Published by Chapman & Hall for the Systematics Association
§Published by CRC Press for the Systematics Association

Milton Keynes UK
Ingram Content Group UK Ltd.
UKHW052021071024
449327UK00027B/2371

9 780367 394127